Cities and Mega Risks

Mohammad Aslam Khan

Cities and Mega Risks

COVID-19 and Climate Change

 Springer

Mohammad Aslam Khan
Department of Geography
University of Peshawar
Peshawar, Pakistan

ISBN 978-3-031-14090-7   ISBN 978-3-031-14088-4   (eBook)
https://doi.org/10.1007/978-3-031-14088-4

This Springer imprint is published by the registered company Springer Nature Switzerland AG
The registered company address is: Gewerbestrasse 11, 6330 Cham, Switzerland

Introduction

The book *Cities and Mega Risks* highlights the emergence of COVID-19 and climate change as twin Mega Risks in cities of the world. It discusses how the pandemic has transformed city functions—promoted remote working, affected socializing, education and learning patterns, recreation and exercise, shopping, and entertainment. A major part of the book focuses on the lessons learnt from the two Mega Risks, and the evolution of urban patterns and functions in their wake, and provides visionary thinking for the improvement of cities from the experiences gained.

The close relationship between the pandemic and climate change from ecological, economic, as well as social standpoint has been highlighted. On ecological grounds, both underscore the strong links between human, animal, and planetary health. For example, deforestation and a loss of wildlife habitat, due to economic pressures, is linked both to climate change and to disease outbreak. According to the US Center for Disease Control (CDC), "Scientists estimate that 3 out of every 4 new or emerging infectious diseases in people come from animals." Both risks are global in nature, have caused economic and employment problems, enhanced poverty and hunger, as well as created and sharpened inequalities and crisis of governance. Moreover, the best common response to both risks is action and appropriate interventions for a greener and better future.

Cities have been known to be the cradles of civilization. In terms of opportunities, they have been well recognized as hubs of commerce and birthplaces of innovations, designs, and ideas. They were the pivot of industrial revolution in the past, and today they are the focal points of technological revolution. Currently, cities accommodate 4.2 billion people or 55% of the world's population. Their economic importance can be gauged from the fact that they generate more than 80% of global GDP. However, serious concerns have also been expressed over the growing size of cities and their failure to develop adequate infrastructure. Social fragmentation, inequalities, racial and ethnic tensions, excessive consumption, and unsustainable production are other trepidations along with enhanced risks and disasters. Cities have always been epicenters of epidemics, pandemics, and risks since historic times. The recent coronavirus pandemic, however, has fully exposed the vulnerabilities of cities. Together with the enhancement of concurrent climate change–related hazards, it has minimized the myriad of challenges they were already facing in the form of housing, jobs, infrastructure, and service deficits. Hence, a survey conducted in fourteen advanced countries by PEW Research Center, a famous American Institution, infectious diseases, and climate change emerged as the top concerns or Mega Risks.

This makes it important for policy makers, planners, and practitioners not only to study the emerging urban Mega Risks but also to analyze experiences to cope with these in developed and developing countries alike. Such investigative case studies could provide guidance for investing in physical, social, and technological infrastructures that could enhance safety and security in cities, and help develop future visions, policies, and plans for resilience and sustainable development.

The word pandemic had its origin in the Greek word *pandēmos*—pan for (all), while demos for (people), together meaning "prevalent over a whole people or the world." Accordingly, climate change, which is prevalent globally, fits well in that definition just like the coronavirus and some other diseases like Spanish flue, plague, and cholera did. The main difference is that while COVID-19 has been a 2-year event, climate change spanned over several years or is more than a 100-year event—the incubation of which goes back to the Industrial Revolution. The twin pandemics or Mega Risks—coronavirus (infectious diseases) and climate change—are both posing serious threats to cities' future. Together, they demand changes in the ways cities' function and operate. This book has presented a case for a better understanding of the twin Mega Risks, the magnitude of their impacts, and the response of cities in terms of lessons learnt in combating these, and provides vision and planning strategies for preparing, mitigating, and adapting to these and future risks.

The work takes into consideration all aspects of sustainable development economic, social, and environmental as well as cross cutting issues like technology. Being thematic, it has no limitations in its geographical scope or focus. The Mega Risks present a serious threat to the entire world, but this threat takes varied shape in different cities and countries in terms of their development status (developing or developed), intensities of their impacts, and capacity to handle them. For example, cities that have less resources or inequalities and a larger proportion of urban poor and vulnerable population are potentially more vulnerable than those that are better resourced, more equal, and with less vulnerable population. The capacity of cities to respond and to endure the crises is also important. The book as far as possible has considered these differences in presenting case studies, best practices, and new urban models advocated for sustainability with incisive commentary.

The study is distinct in the sense that it is specifically targeting cities in relation to twin Mega Risks: coronavirus pandemic and climate change, which are both posing serious threats to their future. Cities have been targeted specifically, because they are (a) in the frontline of battle against the Mega Risks, (b) have a key role in implementing nation-wide measures, and (c) acting as test sites for groundbreaking and innovative strategies. The book while critically examining traditional urbanism also provides ideas on orchestrating the future of cities in response to contemporary situation. As part of the Sustainable Development Series, it addresses several Sustainable Development Goals (SDGs), simultaneously including goal 11—make cities inclusive, safe, resilient, and sustainable; goal 3—health; goal 13—climate change; goal 7—energy; goal 8—economic growth; goal 12—sustainable consumption and production; goal 4—education-; and goal 17—partnerships and several other goals.

Key Messages

Message 1. Twin Mega Risks, the coronavirus pandemic and climate crisis, along with their challenges, also offer opportunities in designing packages to "develop better" and make cities sustainable, climate friendly, and more humane rather than mechanical materialistic abodes. They call for investment and behavioral changes that, along with reducing the likelihood of future shocks, also promote sustainable development and help make a safer and better world.

Message 2. Recovery and long-term development packages must incorporate balanced economic development, social protection, and improvement of human capital together with environmental safeguard as a cornerstone of rebuilding city functions in a sustainable manner. *The best option is to follow the framework of Sustainable Development Goals of the 2030 Agenda for Sustainable Development.*

Message 3. Good governance, visioning, and enabling environment are crucial to overcome the negative impacts of twin Mega Risks and in implementing recovery as well as long-term development packages for the sustainability of cities. The pandemic has led to a new normal, and among others, lessons learnt from it should form the basis in designing the post-pandemic "next normal." It is critical to analyze how the economic and social order in which business and society have traditionally operated is being restructured. The experiences gained, should be effectively utilized in visioning, and planning new forms in the resolution of issues and promotion of sustainability and resilience in the "next normal."

Message 4. Successful emergence of cities from past pandemics, disasters, and crises including the recent crises of 9/11 and Global Recession 2008–11 depicts that cities have resilience to overcome shocks posed by twin Mega Risks, but it must be made stronger to face unforeseen future risks.

Message 5. The twin Mega Risks have alarmed both policy and decision makers as well as the public at large. However, being alarmed or "on the alert" is not the same as being prepared or equipped to prevent the risks or to mitigate their harmful effects and damages. The novel coronavirus and climate crisis have made it crucial *that "reactive approaches to emergencies" be changed to "proactive preparedness, planning, and decision-making."*

In sum, the main message for cities is that they need to accelerate efforts in coping with ongoing risks, whether they are pandemic or climate related. Simultaneously, actions need to be put in place to make cities more resilient, greener, compact, and smarter to face any new unanticipated crises.

Structure and Framework

There are several ways to structure an academic work. Traditionally a book on urban study starts with the basic description of urbanization and introduction of different type of cities and urban typologies. This book is different in the sense that it deals with cities and sustainable development in relation to Mega Risks, hence at the outset, it needs conceptual clarity. It is for the same reason that the book starts with the concept of Mega Risks first—why they are mega risks, how they have affected the whole world, and how one of them with the capability of spreading like wildfire paralyzed the functioning of both cities and nations. It also discusses why actions are necessary from local to global level and the ongoing agenda for the same. Logically, after the conceptual clarity, the next section of the book deals with the initial steps that are needed in a risk management cycle—that is, detecting the nature and intensity of problem. The idea is to put in place some form of monitoring and management mechanism. Both in case of coronavirus disease and climate change, this has been done through application of geospatial tools. Hence, the second section concentrates on the monitoring of Mega Risks using these tools. By the same token, the section also highlights that the cities' functioning would have been completely paralyzed in the wake of lockdown and other social distancing measures—if technology had not come to the rescue in the form of remote functioning. Operating digitally was the only way to stay in business through mandated shutdowns and restricted activity with rare exception. It amounted to stay digital to stay alive. A natural outcome of this was enhanced innovation, augmenting it to a higher level so that humanity and organizations can outmaneuver uncertainty.

The third section of the book deliberates on Mega Risks in relation to economic aspects of sustainable development—with two chapters on urban economy and power of its growth "energy," respectively. It also concentrates on how to make cities economically resilient in the wake of Mega Risks and the lessons learnt from the new adjustments that provide guidance for future development of linked city and national economies. The fourth section focuses on Mega Risks in relation to social development, concentrating on three interrelated issues health, education, and social protection—equally important for building socially resilient cities. The chapters critically assess cities' preparedness and response; containment and control measures adopted; pressures experienced by urban health, education, and social protection systems; and lessons learnt in resilience. A big lesson from the Mega Risks is to shift from "reactive approach to emergencies" to "proactive preparedness, planning, and decision-making." Cities and nations, to be truly resilient, need to be able to not just bounce back from a catastrophic event or mega risk but also move forward with resilience. This is the topic of the fifth section, which highlights the role of governance, long-term visioning, and appropriate planning in preventing and mitigating risks and adapting to them. It highlights the need for considering economic vs environmental and social trade-offs, as well as technology and urban design aspects that are important for creating resilience to risks. It advocates that improving sustainability of cities through selection of right trade-offs or making them smart by adopting appropriate technologies or making them resilient to mega risks by designing city, neighborhood, or building are not isolated ideas. They need to be implemented within one overarching framework that is available in the 17 Sustainable Development Goals of the 2030 Agenda for Sustainable Development. One that is based on inclusive, coordinated, and responsive governance across various jurisdictions—local, subnational, national, and international.

Acknowledgments

Adapting ideas into practice is not simple, especially when it comes to writing a book on global problems like coronavirus pandemic and climate change. Though challenging, the experience was no doubt rewarding. I wish to express my deep gratitude to the Senior Publishing Editor Zachary Romano, without whose encouragement and continuous support, I would not have been able to complete this work.

I would like to gratefully acknowledge the help of my mentors, colleagues, professional associates, family, and friends for their continuous encouragement and contribution in one way or the other while I was undertaking and finalizing the publication. Special thanks are due to my ex-professor deceased Dr. D. L. Linton for enhancing my skills in conducting research in connection with my doctoral thesis at the University of Birmingham, UK. Credit on my training for writing also goes to my former bosses and mentors Dr. Kazi F. Jalal and Mr. Ravi Sawhney, who guided and advised me on writing manuscripts and reports over the years while I was working for the United Nations.

I owe an extraordinary debt to my family for their constant support and assistance in writing this book. I would like to start with my beloved and caring wife Nasreen Aslam Khan. She helped me during all stages, beginning from reading the draft proposal for this book to advising me on the contents of the book and searching materials for it. She was as important in getting this work done as I was. My special thanks to my nephew Faheem Ahmed, as the original idea on the book emerged from my discussions with him. Many thanks to my son Umer Aslam Khan and his wife Asma Ahmad, my daughter Fatima Aslam Khan and her husband Mustafa Umar for encouraging me throughout the work and for helping me in collecting relevant information and/or reading the manuscript. Credit goes to my grandson Ibrahim Khan who assisted me in formatting the draft of the manuscript and its graphics as well as reading it and checking references.

Among colleagues and friends, I am grateful to Mr. Jawed Ali Khan, former director general in the Ministry of Environment and presently program manager of the United Nations Habitat Program, Pakistan, for guidance on the available materials on the work from UN Habitat. Prior to this, I had completed a book, *Environment and Climate Change Outlook of Pakistan*, for UNEP assigned to me by the Ministry, while he was its director general. This work prepared me for undertaking the present assignment. My former colleague and student Samiullah Khan helped me in preparing some figures and tables included in the manuscript. I would like to acknowledge his help with much appreciation. Several of my former colleagues in the United Nations, Dr. Toufiq Siddiqi, Mr. Adnan Aliani, and Dr. Amitava Mukharjee, and my professional friends including Dr. Sana Khan and Mr. Raziuddin also deserve my gratitude for their encouragement and contribution on undertaking this challenging job and completing it.

I would like to recognize the research conducted previously by individuals and organizations, from which I benefitted. Although due credit has been given to them through appropriate references, I owe them an additional note of thanks. I would be failing in my duty if I do not thank my publisher Springer, for stimulating and convincing me into writing this book.

Last but most important, I am truly grateful to God Almighty for providing me the knowledge and skills which enabled me to successfully undertake this project.

Contents

List of Figures

List of Tables

Part I

Overview: Risks, Cities, and Sustainable Development

This part of the book provides background information on mega risks. It also covers profile of cities as human habitat, analyzes their growth trend, and outlines the importance of strategies for risk management at the city level. It examines the risk trends, summarizes the approaches that have been used to reduce risks, and argues that sustainable development would be impossible if both cities and countries are not adequately prepared to face these risks. It also highlights risk reduction programs and international agendas for Disaster Risk Reduction and promotion of Sustainable Development highlighting the salient features of Sustainable Development Agenda 2030, Sendai Framework of Disaster Risk Reduction 2015–2030, New Urban Agenda and Paris Climate Agreement (2015), and WHO Health Emergency and Disaster Risk Management (EDRM) framework.

The pandemic caused by the coronavirus is not the first of its kind. Many momentous epidemics and pandemics as well as climate-related disasters have occurred over time and devastated cities and societies, altered the course of human history, and killed millions of people. Nevertheless, they are also witnesses to the fact that mankind has survived these epidemics, pandemics, and disasters. Although they have caused great human sufferings, they have also cleared the way for innovations and advances in science, economy, and governance systems and technology.

In conventional literature, only physical and climate-related risks are referred as risks, while epidemics and pandemics are excluded. The occurrence of coronavirus pandemic of the twenty-first century, however, has already gained a place as a mega risk event in modern history. The pandemic took its root in cities, and as the crisis unfolded, it exposed the fragility of present urban systems as never before. No doubt, the pandemic is distinct in contemporary history in bringing simultaneous global health and economic crises. In its harmful impacts, however, it shares marked similarities with the climate crisis which has been aggravated to climate emergency. A major indicator of this is doubling of climate-related extreme events over the last 20 years, compared to the previous 20 years—driving loss of life, property, and displacement in low- and middle-income countries and even in advanced countries like the USA to an all-time high level.

The chronology of events depicted in this part demonstrates that cities were not adequately prepared to face the mega risk events. This is very serious in a scenario where over half the world's population (55%) is living in urban areas now, which has been projected to become two-thirds of the global population in 2050. A more serious concern is that bulk of the new urban growth will take place in developing countries, where some 90% of fresh urban expansion is likely to be in or near hazard-prone areas and to be built in the form of informal and unplanned settlements. This is a grave concern in the wake of mega risks which have already affected disproportionately the vulnerable population living in slums and marginal settlements within developing countries cities.

1.1 Introduction

Risk is defined as the probability of future adverse effects, loss, or damage to humans, their livelihoods, infrastructure, and service support systems (IFRC 2010). The presence of risk depends upon three related components—existence of a hazardous episode or event (a storm or a flood), exposed things or elements (human beings, their assets, or their support systems or infrastructure), and conditions or state that may result in adverse effects on the exposed things (poor design or construction of a building or location of a residence in an active flood plain or a person already suffering from a disease, e.g., heart problem). The last component also referred to as vulnerability is a reverse of resilience. Resilience to a risk is a characteristic or characteristics which allow the absorbing of and recovery from the adverse effects of an external shock or disaster.

The term risk in this study articulates to disaster risk, which has periodic adverse effects resulting in loss and damage to humans, their assets, and their support systems. A risk event becomes a natural disaster when it causes physical or material damage. The United Nations Office for Disaster Risk Reduction and Center for Research on the Epidemiology of Disasters recognize the term "disaster" for natural risk or hazard-related disasters events (excluding epidemics or biological disasters) which kill ten or more people; affect 100 or more people; result in a declared state of emergency; call for international assistance (CRED and UNDRR 2020).

This chapter on risks, resilience, and sustainable development has been divided into six sections. After a brief introduction, the next section discusses the risks dimensions and trends including both climate change and infectious disease-related risks. The following section focuses on the emergence of COVID-19 and climate change as twin mega risks and brings forward synergies between them and the threats they have posed to sustainable development. Risk management and resilience to risks form the subject of the next section. The section after this includes the global agenda for risk management in relation to sustainable development. The findings of the chapter have been summed up in the concluding section.

1.2 Risk Trends and Dimensions

Risks result from a series of natural processes. However, they have often been triggered or aggravated by human activities—for example, increased consumption of fossil fuels in industrial processes and transportation leading to the increase in greenhouse gases (GHGs). This is enhancing global temperature triggering climate change and related disasters. The pattern of occurrence of climate-related natural disasters such as cyclones, floods, droughts, and wildfires is becoming more intense and more frequent, reportedly now averaging a disaster a week (Harvey 2019). Similarly, deforestation or destruction of wildlife habitat by humans is linked to both climate change and infectious disease outbreaks or epidemics. Experts believe that these diseases transfer from animal to human with increased contact between them as people encroach upon animal habitats (Boukerche and Roberts 2020). Once established in humans, these diseases quickly and increasingly spread across the world in the era of globalization, as depicted by COVID-19. An estimated 60% of known infectious diseases and up to 75% of new or emerging infectious diseases are attributed to animals or believed to have zoonotic origin (UNEP and ILRI 2020; Jones et al. 2008; Woolhouse and Gowtage-Sequeria 2005).

1.2.1 Risks in History

The risks and natural disasters have existed in the world since ages, probably ever since the origin of the earth and have affected humans since their existence. They can be traced in the destruction of volcanic disaster of the ancient Mediterranean island of Thera (now Santorini, Greece), which destroyed the entire Minoan civilization around 1600 BC (Kutschera 2020); the destruction of historic Indus Valley and Mesopotamian civilizations; Bubonic Plague of Europe of the fourteenth century; the Yellow River and Yangtze floods of 1887 and 1931,

respectively, in China; the Spanish Flu of 1918; tropical cyclone Bhola of Bangladesh (then East Pakistan) of 1970; Indian Ocean Tsunami of 2004; cyclone Nargis of 2008; and the fierce 2020 Atlantic hurricane season that generated an extraordinary 30 named storms (highest on record).

1.2.2 Risks and Disasters in the Twenty-First Century

A major rise in the disaster events and associated damages has occurred in the twenty-first century (Fig. 1.1). From the dawn of the century to 2019, over 7348 natural disaster events (excluding biological disasters/epidemics) were recorded worldwide, according to EM-DAT, the international database of such events (CRED and UNDRR 2020).

Overall, these disasters claimed approximately 1.23 million lives at an average of 60,000 annually and affected a total of over 4 billion people (many on more than one occasion). Economically, they costed US$ 2.97 trillion globally (CRED and UNDRR 2020). This is a big increase over the previous two decades 1980–1999 (Fig. 1.1), during which, 4212 disasters associated with natural hazards occurred world over, claiming some 1.19 million lives and affecting 3.25 billion people and resulting in economic losses worth US$1.63 trillion (CRED and UNDRR 2020).

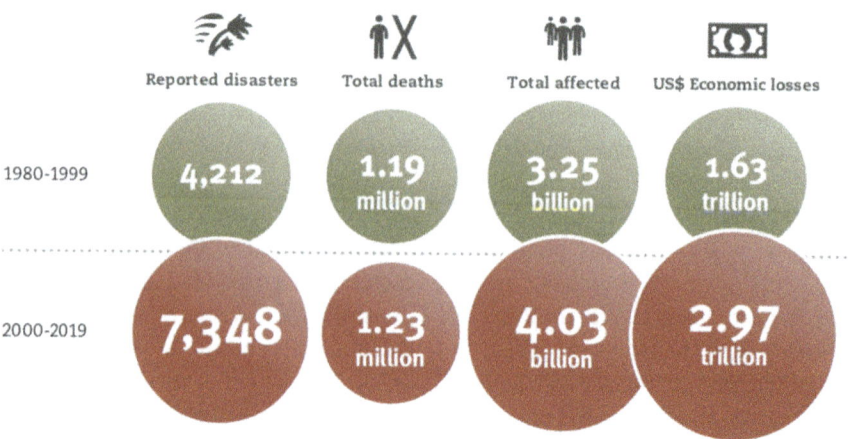

Fig. 1.1 Disasters impact 2000–2019 vs 1980–1999. (Source: CRED and UNDRR (2020))

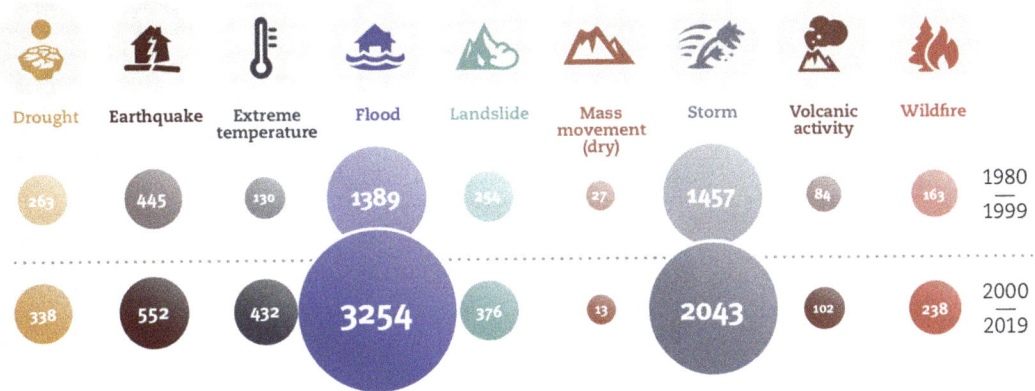

Fig. 1.2 World disaster events by type: 1980–1999 vs. 2000–2019. (Source: CRED and UNDRR (2020))

The key statistics on disaster events for 2000–2019 show that in terms of affected countries globally, China with 577 events was the world leader followed by the USA with 467 events. India was third with 321 events; the Philippines with 304 events and Indonesia with 278 events were fourth and fifth, respectively. All these countries have large and heterogeneous land-masses and relatively high population densities in risk-prone areas. Overall, eight of the top ten countries by disaster events were in Asia. It was the continent which recorded 3068 disaster events between 2000 and 2019—the highest number in the world—followed by 1756 events in the Americas and 1192 events in Africa (CRED and UNDRR, 2020).

In terms of type of disasters (Fig. 1.2), the extreme weather events or climate-related disasters topped the list. The last 20 years (2000–2019) saw the number of major floods more than doubled over the previous 20 years (1980–1999) from 1389 to 3254, while the incidence of storms increased from 1457 to 2034. Floods and storms were the most prevalent events. However, as Fig. 1.2 shows, there have also been major increases in other categories of climate-related events including droughts, wildfires, and extreme temperature events. Geophysical events such as earthquakes also increased, and tsunamis, particularly the Indian Ocean tsunami of 2004, killed more people than any of the other natural disasters in this period.

1.2.3   Risks and Disasters in 2020

In the present century, the year 2020 stands out as a very prominent year for disasters. Besides the COVID-19 pandemic, it stands out for climate-related disasters, which were responsible for the 389 recorded events that killed 15,080 people, affected 98.4 million, and caused minimum economic loss of US$ 171.3 billion (CRED and UNDRR 2021). It rivalled 2016 as the world's hottest recorded year even though there was no strong El Niño effect. The worldwide occurrence of disasters by type and number in 2020 compared to 2000–2019 annual average is presented in Fig. 1.3, while the economic losses caused by them are shown in Fig. 1.4.

When compared with the previous two decades, 2020 saw higher than the annual average in terms of disaster events that were 389 in number compared to the annual average of 368 in 2000–2019 (Fig. 1.3) and the annual average of economic losses, which was almost US$171 billion compared to annual average of about $152 billion between 2000 and 2019 (Fig. 1.4). The estimates of reinsurance company Munich Re worldwide in terms of economic losses were even higher amounting to $210 billion in 2020 (Munich Re 2020)—an increase of 26.5% compared to 2019's cost of $166 billion. About 40% or $82 billions of this damage was insured compared to $57 billion that was insured in 2019. Of the ten costliest global natural disasters in 2020,

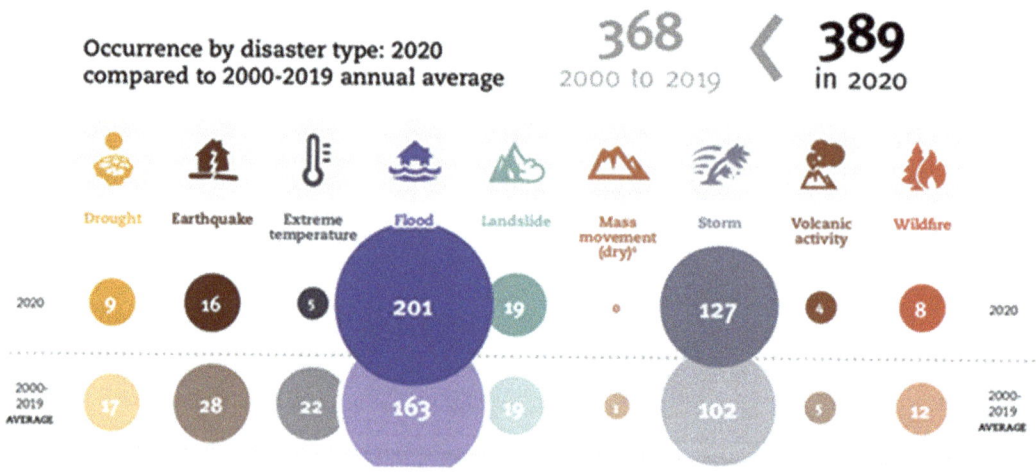

Fig. 1.3 Worldwide disasters occurrence by type in 2020 compared to 2000–2019 annual average. (Source: CRED and UNDRR (2021))

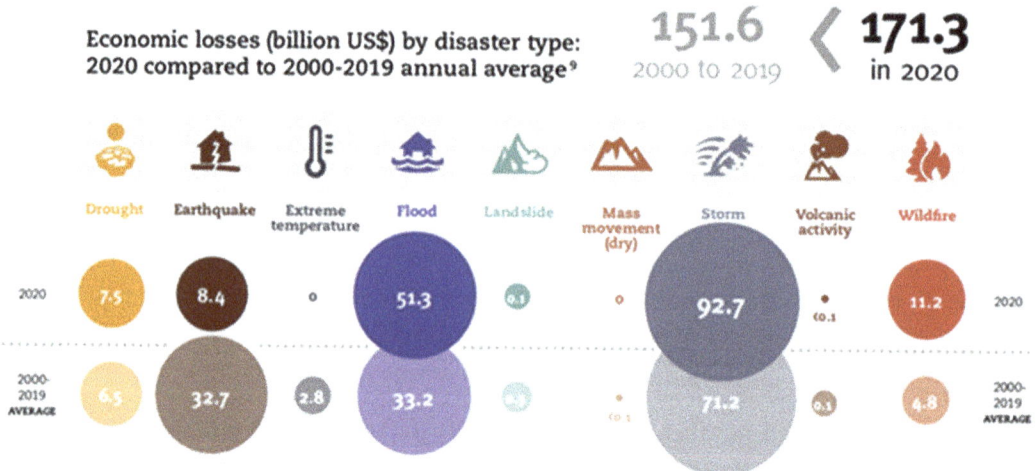

Fig. 1.4 Worldwide economic losses by disasters in 2020 compared to 2000–2019 annual average. (Source: CRED and UNDRR (2021))

six occurred in the USA. Hurricane Laura, which hit Louisiana in August, was the most destructive event in the country, with $13 billion in damage, of which $10 billion was insured. Disaster losses in Asia totaled $67 billion, of which only $3 billion were insured (Newburger 2021).

Among the leading climate-related disasters in 2020, floods were the most common with 201 events worldwide, while storms affected the highest number of people numbering 45.5 million and costed the most in terms of economic losses to the tune of US$92.7 billion (CRED and UNDRR 2021). The highest number of fatalities (excluding COVID-19) was caused by extreme temperatures that caused 42% of total casualties, followed closely by floods which accounted for 41% of all deaths (CRED and UNDRR 2021).

The year 2020 saw a total of 103 named storms around the globe (NOAA 2020). The North Atlantic played a major role in the global tropical cyclone activity with 30 named storms, which broke the previous record of 28 in 2005 (NOAA 2020). September 18 marked the second time on record with three named storms—

Number of disasters by continent and top 10 countries[4]

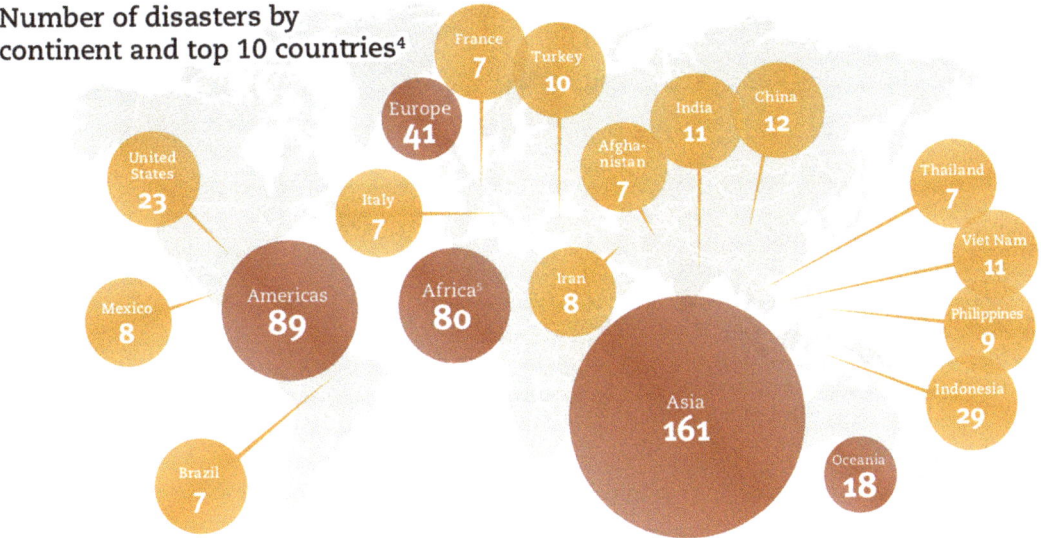

Fig. 1.5 Worldwide disaster occurrence by continent and top ten countries in 2020. (Source: CRED and UNDRR (2021))

Wilfred, Alpha, and Beta—all forming on the same day. The only other time this had happened was on August 15, 1893, before hurricanes received official alphabetical names (Thompson 2020). The year 2020 was also one of the worst for wildfires which affected the USA, Amazon, Russian Siberia, Australia, and Indonesia (Landscape News 2020). In the USA, the fires burned more than four million acres across California, about double the previous record of nearly two million acres set in 2018 (Thompson 2020).

The impacts of the disaster events also varied by the continent and countries (Fig. 1.5). Asia experienced 161 disaster events (with 41% share in number and 64% of total people affected). Heatwaves did the biggest damage in Europe with 42% of total reported casualties. The record-breaking storms and wildfires in the Americas caused 53% of total economic losses, largely in the USA which was hardest hit by the climate-related disasters. According to the US Billion-dollar disaster report (Smith 2021), it was a historic year of extremes. There were 22 separate billion-dollar weather and climate disasters across the USA, breaking the previous annual record of 16 events, which occurred in 2017 and 2011. The billion-dollar events of the year

included a record 7 disasters linked to tropical cyclones, 13 to severe storms, 1 to drought, and 1 to wildfire. Together, these 22 events costed $95 billion in damages (Smith 2021) (Fig. 1.6).

In summary, the most notable features of 2020 were significant flood events across East Africa, South Asia, and China; a record number of storms in the Americas; a series of storms that struck Southeast Asia; and the deadliest heat waves that struck Europe for the 2nd year in a row (CRED and UNDRR 2021).

1.2.4   Health Risks—Epidemics and Pandemics

Health risks have played an extremely important role in the disaster events throughout the history of mankind. Despite considerable improvements in human health over time and increase in the current average world life expectancy to 72 years, the humanity continues to face serious threats from infectious diseases particularly those causing epidemics and pandemics. Infectious diseases, whether bacterial or viral, are a never-ending challenge. They can (like cholera or TB or SARS) emerge or re-emerge/resurface in unpredictable places/regions and at unpredictable

Share of economic losses (%) by continent

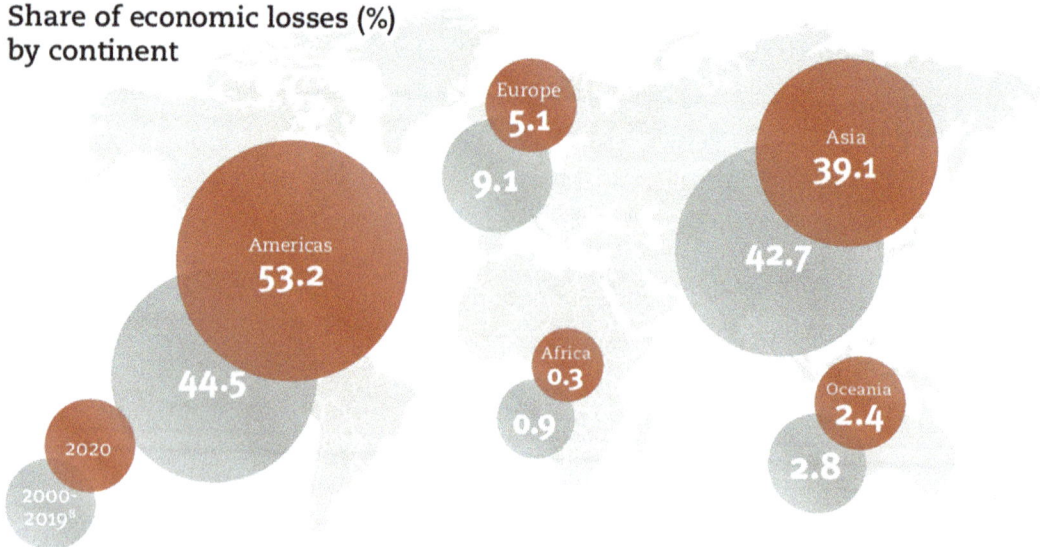

Fig. 1.6 Distribution of proportional share of economic losses by natural disasters in 2020 by continent. (Source: CRED and UNDRR (2021))

times (Jaijyan et al. 2018). An emerging viral disease can be newly evolved or newly recognized or not observed previously within a population or geographic location (Jaijyan et al. 2018; Morse 1995). Re-emergence of viral diseases results from resurgence or recurring outbreaks in a significant part of the population, after a previous decline in its incidence (Quaglio et al. 2012).

When an infectious disease affects many people in a community or population or a geographical region, it is termed as an epidemic, and when an epidemic spreads over multiple countries or continents, it is known as a pandemic (CDC 2012). In other words, a pandemic is a global outbreak of a disease. The earliest known pandemic was the Justinian Plague of 541 A.D., named after the Byzantine Emperor Justinian I. It was the bubonic plague that wiped out 25–50 million people in 1 year (Horgan 2014). Another bubonic plague pandemic or Black Death wiped about a third of Europe's population in the fourteenth century (Daily Sabah 2020). Its global outbreak originated in China in 1334, arrived in Europe in 1347, following the Silk Road (Huremović 2019).

Important historic pandemics/epidemics from the Antonine plague (165–180) to ongoing COVID-19 are shown in Fig. 1.7.

One of the most serious pandemics in modern history was the 1918 influenza or Spanish flu pandemic. It was caused by an H1N1 virus with genes of avian origin (CDC 2019). There is no consensus on the origin of this pandemic, but it was first publicly reported in Spain (Council on Foreign Relations 2020). Possible places of origin cited are the USA, China, Spain, France, or Austria (Huremović 2019). However, its outbreak quickly spread all over the world. Besides Europe, where massive military movements during the first world war and overcrowding contributed to its mammoth spread, the virus devastated the USA, Asia, Africa, and the Pacific Islands. The mortality rate of Spanish flu ranged between 10% and 20%. With over a quarter of the global population contracting that flu at some point, the death toll was immense well—over 50 million, possibly 100 million dead (Huremović 2019). It killed more individuals in a year than the Black Death had killed in a century (Flecknoe et al. 2018).

Fig. 1.7 History's deadly pandemics from the Antonine plague to COVID-19. (Source: LePan and Routley (2020))

1.2.5 Myth of the Twenty-First Century on Epidemics and Pandemics

When the "Health for All 2000" accord was signed in 1978 by the member states of the United Nations, it was believed that by the end of the century, infectious diseases would no longer pose a significant threat to human beings, even in the poorest countries. This myth was gone even before the coronavirus pandemic when outbreaks occurred due to the spread of drug-resistant microbes like *Mycobacterium tuberculosis* and the emergence of new infections with devastating effects. However, within the developed countries, most experts had a myth that epidemics and exotic pathogens particularly zoonotic diseases (transferable from animals to human) that cause so much misery in poverty ridden Africa and parts of Asia would never become a problem in cities of wealthy countries, with their high standards of living and well-developed health systems. Then came severe acute respiratory syndrome (SARS) of 2002–2003, a disease that took its heaviest toll on wealthy urban areas. SARS spread most efficiently in sophisticated hospital settings. That made this myth disappear.

The 2009 H1N1 influenza pandemic, the first of the twenty-first century, proved how very quickly a new virus can spread to every corner of the globe. But the biggest surprise that H1N1 virus brought was a lucky one—that "the pandemic was much milder than generally feared" (WHO 2015). The twenty-first century also brought a series of other viral epidemics to cities including Ebola, MERS (Middle East respiratory syndrome), West Nile, Swine flu, and Zika, which collectively infected millions of individuals. With the outbreak of these epidemics in the present century, it became clear that current demographic transitions—driven by population growth, rapid urbanization, deforestation, globalization of travel and trade, climate change, and political instability—also have fundamental effects on the dynamics of infectious diseases that are more difficult to predict (Bedford et al. 2019). Sensing this, even before the coronavirus pandemic struck, the Executive Director of the Health Emergencies Programme at the World Health Organization (WHO) stated, "We are entering a very new phase of high-impact epidemics... This is a new normal, I don't expect the frequency of these events to reduce" (BBC News 2019). He was very right, and it was not too far in the same year that the coronavirus attacked.

The contagion came as a surprise and then worsened because the authorities and the public were unaware of the risk and implications of its exponential spread. COVID-19 is a stark reminder to humanity that "infectious diseases haven't vanished. In fact, there are more new ones now than ever: the number of new infectious diseases like SARs, HIV and COVID-19 has increased by nearly fourfold over the past century. Since 1980 alone, the number of outbreaks per year has more than tripled" (Mair 2020). In terms of economic cost according to Smith et al. (2019), zoonotic diseases alone that did not evolve into pandemics amounted to a cost of 6.7 billion dollars between 1997 and 2009. The global damage of the ongoing coronavirus pandemic in terms of loss of lives had reached 5.7 million, by the beginning of February 2022, while the number of cases had climbed to 391 million according to global corona tracker. Moreover, economic damage has been enormous. The Economist (2021) magazine estimated the economic cost of the pandemic over 2020–2021 to more than 10 trillion US\$ in lost or foregone GDP (goods and services the world could have produced had it remained unafflicted). Only two largest economies in the world—America and China—have annual GDP greater than that, and 153 economies together produced less GDP than that in 2019. Commercially, this amount is enough to buy the ten biggest listed companies in the world, including Amazon, Apple, and Saudi Aramco (The Economist 2021). Estimates so far indicate that the virus reduced global economic growth to an annualized rate of −4.5% to −6.0% in 2020 (CRS 2021). The colossal damage caused certainly makes it a mega risk.

1.3 Coronavirus and Climate Change: The Twin Mega Risks

While the coronavirus mega risk fully exposed shortcomings in the current risk management, the huge increase in climate-related disasters entitles it as another mega risk. Together the twin mega risks pose a stark warning for sustained action to prevent catastrophic humanitarian impacts. The massive global humanitarian impact of COVID-19 can be gauged from the fact that it affected over 200 countries (CRS 2021) with 237 million confirmed cases until early October 2021 (over 390 million cases by early February 2022). Likewise, the climate change had a heavy toll affecting 1.7 billion people by its related disasters only during the past decade. Overall, there has been a big rise in climate-related disasters during the past 20 years—between 1980 and 1999, 3656 climate-related events occurred, as opposed to 6681 between 2000 and 2019 (CRED and UNDRR 2021). The differences are also reflected in the number of risk events like floods, which more than doubled in the same period—the incidence of storms increased from around 1457–2034. Even since the outbreak of the pandemic, a record-breaking climate-related disasters including hurricanes, flash floods, wildfires, and heat waves have plagued already reeling communities from India to the Russian Federation to the USA and the Caribbean, killing thousands of people.

Referring to climate change and its consequences for cities and regions globally, Banai (2020) calls it another pandemic, which appears to be true. The term "pandemic" had its origin in the Greek word pandēmos, pan for (all), and demos for (people) together, meaning "prevalent over a whole people or the world" (Banai 2020). Accordingly, climate change, which is prevalent globally, fits well in that definition, just like the coronavirus and some other diseases did. The main difference is that while COVID-19 has been a 2-year event, the climate change spanned over several years or is over a hundred-year event— the incubation of which goes back to the Industrial Revolution. However, while the vaccine is a solution for the coronavirus (Banai 2020), no such protection exists from the pandemic of climate change—which demands much wider action than merely a shot in the arm.

Both the coronavirus pandemic and climate change-related disasters have shown how vulnerable the world is to global risks and catastrophes. No doubt that it is for the same reason that the two risk perception surveys conducted in 2020 on risk horizon and high impact risks by the World Economic Forum (2021) and PEW Research Centre (Poushter and Huang 2020) of the USA both listed the infectious disease and climate change as the top two risks of the world.

1.3.1   Drivers of Mega Risks

Both mega risks are the result of reckless exploitation of environment and its resources; in the continuation of which, neither was unexpected. On ecological grounds, they underscore the strong links between human, animal, and planetary health. For example, deforestation and a loss of wildlife habitat, due to economic pressures, are linked both to climate change and to disease outbreak. According to the US Centers for Disease Control and Prevention (CDC), "Scientists estimate that 3 out of every 4 new or emerging infectious diseases in people come from animals" (CDC 2020). While removal of animal habitats enhances the likelihood of exchange of viruses and pathogens between animals and humans, the rising temperatures can also create suitable environments for the incubation of certain infectious diseases. Moreover, as the earth's temperature rises, animals, whether on land or in the sea, tend to move toward the polar areas to get out of the growing heat. This means more contacts between animals and between animals and humans creating chances for pathogens to attack new hosts.

Ecologically, curbing the drivers of climate change particularly deforestation for farming and increased encroachment into wildlife habitat will restrain the emergence and re-emergence of zoonotic diseases, by reducing the contact between human and animals and hence curbing zoonotic diseases (The Lancet 2021). It will also help in carbon sequestration, hence ameliorating climate change. Together, the two crises have raised the alarm that the threats of increasing catastrophes and mass extinction are real and true. Nevertheless, they have also depicted that it is still not too late and that there is room even now to greatly decrease the impact of or even diminish both crises. Responses to the pandemic by the governments, society, and individuals have shown that swift and decisive actions are possible for bringing in required changes and are also economically effective. Conversely, consequences of inaction can lead to grave consequences as was evident from the high number of cases in the USA, Brazil, and India. Early action is therefore necessary to avoid worst-case scenarios for both the mega risks.

1.3.2   Preparedness to Face Mega Risks

Notwithstanding numerous warnings from experts about the risks of a pandemic, the countries in the world were woefully unprepared for the coronavirus crisis. Unfortunately, the same is true for climate change. Like the chances of the occurrence of a pandemic, scientists have been continuously raising alarm about a climate emergency, which has already arrived. The warning signs are clear in the recent wildfire of Australia and California, melting of the Arctic ice, and the increase in the frequency, occurrence, and intensity of climate-related natural disasters. The IPCC (2022) Sixth Assessment Report and IPCC (2018) climate signal warn that the next few years are vital—the world has a decade left to achieve a low-carbon transition to bring its economy to a trajectory, limiting global warming to 1.5 °C above preindustrial levels, and save itself from catastrophic global warming. The economic shock caused by the pandemic reduced the global greenhouse gas emissions in the short term, but it started to rise again after a while. It is likely that there will be little direct effect in the long term, but indirect effect may be substantial—particularly in view of governments despite their predicament willing to promote green investments during economic recovery to meet or beat their Paris emission targets (Dwortzan 2021).

1.3.3 Similarities and Differences in Twin Mega Risks

There are many ecological, economic, and social similarities between the two mega risks—COVID-19 and climate crisis. The ecological similarities have already been explained above in the section on drivers of mega risks. Overall, the twin mega risks have two amazing commonalities—in their exponential spread and the colossal damages they incur. Further, they are both global in nature, have caused economic and employment problems, enhanced poverty and hunger, and created and sharpened inequalities and crisis of governance. The economic impacts of both are devastating and widespread across economies and markets, business, and industries as well as across varied sectors. Vulnerabilities to both have worsened in the wake of economic globalization and connectivity triggering multiplier effects.

The COVID-19 crisis for one has demonstrated how quickly a natural calamity can cause the collapse of national economies. Similarly, countries with climate vulnerability to their predicament have already faced disaster risks frequently for a long time. In the wake of inaction and lack of concerted efforts to significantly strengthen investment in mitigating and adapting to climate change, many more countries will fall into crisis mode. The few countries, spared for now, will also be unable to shield themselves in the future. Like the COVID-19 virus, which did not spare anyone and crossed borders, the impacts of climate change will transcend borders and move across the world—the least in the form of climate refugees and migrants. According to Global Report on Internal Displacement (IDMC 2020), extreme climate events already triggered around 25 million displacements in 140 countries. Without climate action and disaster risk reduction (DRR), related disasters could double the number of people requiring humanitarian assistance to over 200 million each year by 2050 (IFRC 2019).

Economically, prevention of mega risks whether COVID-19 or climate change makes sense because prevention or defenses against both cost a fraction of the loss resulting from the disaster events as demonstrated by several studies (World Bank 2014, 2019; Jonas 2014; Manzanedo and Manning 2020). The World Bank (2019) report on financial preparedness for pandemic estimates that every dollar invested in preparation will lead to saving 2 dollars in the future. Likewise, the Global Commission on Adaptation calculated that every dollar invested in building climate resilience now could score between $2 and $10 in net economic benefits later (Volz 2020). It has been estimated (Pueyo 2020) that delaying lockdown in the early stages of the pandemic by 1 day increased the number of cases by 40%. Likewise, Pei et al. (2020) estimated that starting social distancing 1 week earlier could have avoided 55% of deaths (36,000) in the USA between mid-March and early May. Further, often, a delayed lockdown also necessitated its maintenance for a longer time, increasing casualties as well as economic losses. Like COVID-19, delay in climate action is also expansive as it means locking economies in the long term into carbon-intensive infrastructure, reducing flexibility in future response options and enhancing differences in distributional impacts between countries (Klenert et al. 2020). In this connection one study showed that delaying climate action by 10 years could increase the cost of climate action by 37% (Furman et al. 2015). Another study estimated that delaying climate action compatible with 2° warming by the end of this century would mean an increase in mitigation costs by around 50% (Jakob et al. 2012). Besides mitigating cost, the impacts of extreme weather events like storms, wildfires, and sea level rise are only mild harbingers of projected future climate damages (IPCC 2022, 2014). The challenge is enormous, but the choice on how to build economies and societies around the globe is there—the best common response to both risks appears to be in timely actions and appropriate interventions leading to a greener and better future.

The pandemic and the climate crisis are also similar in their social impacts within and between cities and nations—most severely affecting vulnerable communities. Greater at risk in the twin mega risks are elderlies particularly with vulnerabilities like preexisting conditions or diseases, poor and slum dwellers, migrants, and refugee population, as well as ethnic minorities and marginalized communities. The poverty-ridden daily wage earners or those living paycheck-to-paycheck and marginalized groups suffered the most from lockdowns, rising unemployment and unexpected medical costs. Similarly, changes in climate regime and increase in extreme events are affecting most those socially marginalized groups who have little resources to meet their daily needs and to recoup from the emergencies.

Among countries, those low in economic resources, with social instability, and with lack of health infrastructure, have been at higher risks to both crises. Although initially some affluent nations bore the brunt of the pandemic because of their complacency, later, a clear drop occurred in the number of new COVID-19 infections in rich nations, while some relatively poor nations faced new waves of the crisis. Sadly, beside health, poor nations are likely to suffer more economically in the long term, as the damage caused to their economies by the pandemic is higher, and their recovery is likely to be slower—further sharpening their inequalities with wealthier nations. The climate crisis has worked in a similar fashion, whereby richer nations have been able to invest more in quick reconstruction after disaster as well as in climate adaptation, while poor nations have borne the worst of climate impacts due to higher vulnerabilities in their living conditions and lower capabilities for recovery, reconstruction, and adaptation (Althor et al. 2016).

Klenert et al. (2020) list some similarities between the two mega risks in terms of policy issues and response. They consider timely action, gaining public support for mitigation policy, addressing inequality, developing international cooperation, and clearly detailing the relationship between science and policy as the main policy challenges. They recognize that taking early action is difficult but argue that delay is costly. They emphasize that a major policy challenge on mitigation of mega risks whether COVID-19 or climate change is gaining public support. Building the trust between citizens and the government is an essential component of policies to tackle both COVID-19 and climate change (Klenert et al. 2020). Experience shows that during the pandemic, stringent measures for mitigation were possible to implement primarily because of public support. The same is true for climate change; the measures for its mitigation become easy to implement when people are more conscious and supportive as in most European countries. The challenge is to look for ways that can get citizens on board. The inequality between citizens also offers a great challenge to policymakers for both mega risks as the needs of have and have not vary. One case in point is formal and informal employment. It is rather easy to make policies to compensate those losing jobs from formal employment whether affected by COVID-19 or climate-related disasters. In contrast, it demands innovative efforts to rapidly compensate a huge number of citizens who are engaged in informal economies—which according to ILO (2020) engaged over 2 billion workers, 62% of all those working worldwide. The challenge is to design and implement equitable policy response that leaves no one behind and ensures respect for all human rights, equality of opportunity, and treatment without discrimination on all aspects of employment and occupation (ILO 2020).

There is also a commonality between COVID-19 and climate change in terms of interaction of science and policy. It is the outcome of scientific research which is often utilized in policy making. For example, in COVID-19, the policy on adoption of types of measures varied based on tracking of cases and their nature and severity in specific areas. Similarly, adoption of carbon pricing policy has often been based on the damages associated with emitting an additional ton of carbon. For example, the social cost of carbon (SCC) is used as a central economic measure of the damages associated with emitting one additional ton of carbon in the USA and Canada (Klenert et al. 2020). However, the scientific models used in forecasting or predicting the consequences of mega risks rely to a certain degree on assumptions and hence carry some uncertainty. Nevertheless, science has often been considered as a conclusive set of facts rather than a system or process of rigorous evidence-gathering to confirm or reject hypotheses in a setting of inherent uncertainty (Douglas 2017). This may cause disorientation of citizens due to changes in "facts" or by observing disagreements between experts. By the same token, a change in epidemiological or climatic projections can be easily interpreted as failure or incompetence of scientists. This can result in uncertainty or mistrust in science and provide fertile ground for "conspiracy theories" (Klenert et al. 2020).

Unfortunately, besides conspiracy theories, scientists, whether epidemiologists or climatologists, have faced the problems of misinformation (wrong information without ill will) and disinformation (deliberately spreading false information to misguide or mislead people), science skepticism, and science politicization. The world has seen a cascade of misinformation and conspiracy theories related to the pandemic. For example, one type of misinformation speculated about the origin of the virus or miracle cures of the disease,

while another spread doubts about the gravity of the situation and still other stir people to disregard government measures (WHO 2022; Tagliabue et al. 2020). Such misinformation has the potential to put people at risk (Bagherpour and Nouri 2020). The misinformation on climate change mainly in the USA is primarily based on political polarization along the party line. This is rather puzzling because climate change research is much more mature for two reasons—(a) due to multiple iterations of improving models with new evidence and (b) IPCC assessment reports that are issued not only after a rigorous and consolidated scientific process to evaluate the state of knowledge but also after the approval by officials from each national government (Klenert et al. 2020).

One difference between the two mega risks lies in the momentum with which they develop. The effects of the coronavirus, for instance, like other infectious diseases are instantaneous and more personal and frightening than climate change, one that brings rapid changes in people's lifestyle and poses an immediate threat to their families, neighborhoods, and communities. Climate change, on the other hand, resembles more an armageddon or disaster in slow motion, the effects of which appear more impersonal and make people think that it will not affect them directly (Harvard 2021). Further, just like the COVID-19 crisis can occur rapidly, it can also be controlled quickly through vaccination. In comparison, the more slowly looming climate crisis, the impacts of which may be even greater (Manzanedo and Manning 2020) requires more intense and consistent efforts in controlling GHG emissions. The pandemic is also more synchronous across large areas and nations than climate change, which varies massively in frequency, intensity, and timing across the world (Pachauri 2014).

There is also a variation in actions and their consequences in the two crises. For example, the link between personal actions (such as social distancing) and prevention of the spread of the virus is clear and compelling (Manzanedo and Manning 2020). In contrast, although personal actions, like use of energy, travel, and dietary decisions are important for climate change, the cause-effect relationship is less clear and more diffused compared to COVID-19. Also, as mentioned before, unlike virus crisis, there are no safe climate change vaccines that could "solve" the climate crisis, and any activities aiming for the reversal of climate change would likely take decades or more before they become effective. By the same token, climate change mitigation becomes a harder challenge because it demands consistent and major changes in the economic policy making. In contrast, the stringent measures adapted to control the pandemic may be lifted once the pandemic is over.

Finally, human perception to the two threats is different. The coronavirus threat as it became a pandemic received excessive media coverage that also made the fear of dying from the virus more horrific. People started conceiving it as an extremely bad death because of one's little control over catching the disease, its painful symptoms, and relatively agonizing process of death (Draulans 2020). It also generated willingness in the societies to forego certain rights to avoid the risk in terms of such restrictions as lockdowns, social distancing, and contact tracing. Since political judgment and decisions are often driven by moral convictions, not necessarily rational deliberation (Haidt 2001, 2017), therefore, the intuitive perceptions became important consideration in strong political responses to the pandemic compared to climate change.

1.4   Risk Management and Resilience

Although the risks may be inevitable, the damage they cause to human beings is not. The mega risks of the twenty-first century, both coronavirus and climate change, have shown how vulnerable communities, cities, and nations are to them in the absence of appropriate planning and management. They have also depicted the importance of creating resilience against them in both developing and developed countries alike. Moreover, they have underscored the need for tackling them at all levels—from cities and communities to the national and global scale.

1.4.1   Resilience Against Risks

Most crises and development losses from risks whether locally, nationally, or internationally occur due to lack of preparedness to face these or result from their mismanagement. It has therefore become extremely important to shift from unplanned and ad hoc responses in dealing with crises to proactive, systematic, and integrated approach to risk management and building resilience to them. Resilience is generally taken to mean resistance of an individual, community, city, country, or the world in terms of their respective ability to respond to external shocks by absorbing them and restoring their functions as soon as possible. According to a Japanese saying, "Resilience is like a bamboo, which bends under the weight of winter snow but stands tall again come springtime" (Mitchell 2013). In urban studies, according to Logan et al. (1987), resilience applies to the ability of an urban system to stay within its "basin of attraction," through its capability to cope with or adapt to impacts and benefit from opportunities. Meerow et al. (2016) conducted a review of "urban resilience" theory, concept, and definitions. Defining it comprehensively, they stated, "Urban resilience refers to the ability of an urban system-and all its constituent socio-ecological and socio-technical networks across temporal and spatial scales-to maintain or rapidly return to desired functions in the face of a disturbance, to adapt to change, and to quickly transform systems that limit current or future adaptive capacity." While the term resilience is still under debate (Romero-Lankao et al. 2016), the OECD (undated) defined a "resilient city" as the one that can absorb, recover, and prepare for future shocks (economic, environmental, social, and institutional). This practical definition has been adopted by the present study in discussing cities resilience to the two mega risks.

1.4.2   Risk Preparedness and Management

Bolstering the resilience of communities and cities through preparedness could help save lives during crisis or shocks, reduce potential economic losses, and help people undertake better decisions in facing future risks. No doubt, losses from mismanaged risks are expansive, but so are the measures needed for preparedness to face risks. Nevertheless, cost-benefit analyses suggest that risk preparation is often beneficial in averting costs (World Bank 2014). It has been estimated that just improving weather forecasting and public communication systems and providing earlier warning of climate-related or natural disasters in developing countries could yield benefits 4–36 times greater than the cost (Hallegatte 2012). A case in point for the same is Bangladesh, where improved preparedness for climate-related risks like cyclones drastically reduced losses in the past four decades. Three major cyclones of similar magnitude hit that country between 1970 and 2007—one in 1970 claimed more than 300,000 lives; but the second in 1991 took less than half of the first, 140,000; while the third in 2007 caused only 4000 deaths. The main reason for this reduction in fatalities was improved capacity for forecasting, an effective system for early warning to the population and a country-wide program to build shelters, the number of which increased from only 12 in 1970 to over 2500 in 2007 (Paul 2009).

Improved policies and capacity to face infectious diseases also pays off in saving life as was witnessed in Singapore, Hong Kong, and Taipei, all three of which had direct flights to and from Wuhan, the first outbreak center of the coronavirus in the world. However, their vigilant monitoring systems helped through which they started screenings of passengers from Wuhan in December 2019, much before the disease was declared as a pandemic by WHO. These monitoring systems were built as preparedness for infectious diseases over the years, after the failure of these cities to stop an earlier dangerous outbreak of SARS, 17 years ago (Barron 2020). In contrast, the USA disbanded its pandemic response unit in 2018 (Beech 2020) and suffered the worst outbreaks.

Risk management and creation of resilience to risks through policy and governance is increasingly being recognized as a key priority not only for saving lives but also for economic and social well-being. It is for the same reason that actors whether governments, civil society, entrepreneurs, or international organizations are switching their focus from ex post response to resilience and preparedness and prevention of risks. Risk management and governance at local and national levels has been discussed in detailed in Chap. 10. However, many unmanaged risks like infectious diseases or climate change transcend national boundaries, cause spill over problems through cause and effect in neighboring countries, and become mega risks. Such risks require actions at local as well as national and international level and require alignment between these.

1.5 Risk Management and Global Agenda on Risk and Sustainable Development

The international community has achieved considerable progress in providing an institutional mechanism and an action agenda for risk management. However, both COVID-19 and climate crises show that much more needs to be done on risks that transcend national and generational border.

1.5.1 Global Institutions

At the global level, the United Nations Office for Disaster Risk Reduction (UNDRR) coordinates efforts for risk preparedness, prevention, and management. It coordinates efforts of stakeholders including governments, communities, and other partners to reduce risks and disaster losses and to promote a better and sustainable future. In terms of relief, once a disaster occurs, the United Nations Office for Coordination of Humanitarian Affairs (UNOCHA) coordinates relief work of agencies concerned—including those of United Nations agencies, funds, and programs, the Red Cross movement, and NGOs. The coordination function of OCHA is primarily carried out through the interagency Standing Committee

(IASC), which is chaired by the Emergency Relief Coordinator. The IASC undertakes coordinated decision-making in response to emergencies. The response includes needs assessments, joint appeals, coordination in field, and the development of humanitarian policies.

Special international institutions also exist to deal with mega risks such as infectious diseases like COVID-19 and climate change crises. The one for infectious diseases is the World Health Organization (WHO), a specialized agency of the United Nations. The agency is primarily responsible of providing disease alerts and response assistance to countries and the global community. It also serves as the secretariat for the Global Outbreak Alert and Response Network (GOARN), established in 2000, comprising of technical institutions, research institutes, universities, international health organizations, and technical networks willing to contribute and participate in internationally coordinated responses to infectious disease outbreaks (Mackenzie et al. 2014). The WHO works actively with member states and donors to provide support whenever and wherever needed. Its role has become increasingly critical in the wake of newly emerging zoonotic diseases in the twenty-first century, not only for monitoring and surveillance but also in providing accurate information in the backdrop of an information revolution, which sometimes allows circulation of misinformation or information that is unverified or inaccurate (Mackenzie et al. 2014).

During the coronavirus crisis, WHO became the leading organization in the global coordination for mitigating the COVID-19 pandemic within the broader United Nations response to the pandemic led by the Secretary General. The WHO on January 5, 2020, alerted the world about "pneumonia of unknown cause" in China, conducted investigation, and on January 20 confirmed human-to-human transmission of this virus. It declared the outbreak a Public Health Emergency of International Concern on the 30th of January and warned all countries to prepare (Wikipedia, WHO's Response to the COVID-19 Pandemic). Subsequently on March 11, the WHO

declared the outbreak a pandemic. The Agency also spearheaded several initiatives like the COVID-19 Solidarity Response Fund to raise money for the pandemic response, the UN COVID-19 Supply Chain Task Force, and the solidarity trial for investigating potential treatment options for the disease. The WHO's COVAX vaccine-sharing program aimed to accelerate the development, production, and equitable access to COVID-19 tests, treatments, and vaccines.

United Nations Framework Convention on Climate Change (UNFCCC) is responsible for preventing "dangerous" human interference with the climate system (UNFCCC 2019a). The Convention entered into force on March 21, 1994, and has 197 countries as Parties to it. Among its important achievements are adoption of Kyoto Protocol, the first greenhouse gas emissions reduction treaty of the world, and two agreements—Cancun Agreements in 2010 and Paris in 2015 (Paris Agreement Summary). Most operational rules of the Paris Agreement are in place under the Paris rulebook. Among the achievements of UNFCCC Secretariat are creation of an international carbon market which enabled European Union Emissions Trading Scheme, the first and largest emissions trading scheme in the world. It has also facilitated stimulation of several national policies as well as establishment of institutional mechanisms to boost mitigation efforts. In terms of adaptation, the network of institutions created by UNFCCC during the last few years has led to considerable enhancement of initiatives related to climate change adaptation (UNFCCC 2019b). The Intergovernmental Panel on Climate Change (IPCC) is another United Nations body on climate change, which provides policymakers with regular scientific assessments on climate change, its implications, and potential future risks and presents adaptation and mitigation options. An important aspect of IPCC findings is that they form the scientific footing and basis of negotiation on climate change. The IPCC regular assessments provide the state of knowledge on climate change. The Sixth Assessment Report of IPCC was released recently (IPCC 2022).

1.5.2 International Agenda for Risk Management and DRR

Risk management and disaster risk reduction (DRR) have become key priorities for development, and many actors including governments and international organizations are working to shift the focus—from ex post or post disaster response primarily in the form of relief to disaster risk reduction through preparation and prevention. The scientific community has been using the term disaster risk reduction (DRR) since the 1970s. The concept has undergone transformation from investigating the causal factors to reduction of impacts of disasters. It is a systematic process of development of policies, plans, and strategies to reduce vulnerabilities and build resilience against the unforeseen events (UNISDR 2004). At the international level, the United Nations Office for Disaster Risk Reduction coordinates global efforts toward DRR (UNDRR 2021). The global agenda on disaster risk reduction has had several milestones including International Decade for Natural Disaster Reduction 1990–1999; International Strategy for Disaster Reduction; Hyogo Framework of Action (2008–2015); Sendai Framework for Disaster Rick Reduction 2015–2030; and Paris Agreement on Climate Change. Further, adoption of Sustainable Development Agenda 2015–2030 is a major effort for promoting all-encompassing sustainable development including risks especially mega risks. In addition, New Urban Agenda, Addis Ababa Action Agenda on Financing for Development, Agenda for Humanity, and WHO Health Emergency and Disaster Risk Management (EDRM) framework have also provided supplementary supportive frameworks of action in this regard.

1.5.2.1 International Decade for Natural Disaster Reduction (IDNDR) 1990–1999

The United Nations General Assembly adopted a resolution 42/169 on December 11, 1987, that declared the 1990s as the "International Decade for Natural Disaster Reduction" (IDNDR). The goal of the IDNDR was to promote disaster risk reduction through collaborative international actions. The idea was to improve capacity of countries to prevent or mitigate the adverse impacts of natural disasters and establish guidelines for the application of existing science and technology to reduce the impact of natural disasters. The focus behind International Decade for Natural Disaster Reduction (IDNDR) and its secretariat within United Nations was the creation of a global culture of prevention.

1.5.2.2 International Strategy for Disaster Reduction (ISDR)

After the completion of IDNDR, a successor arrangement was created by the United Nations in the form of International Strategy for Disaster Reduction (ISDR). An Inter-agency Task Force was established to implement it along with ISDR Secretariat in the United Nations. The World Summit on Sustainable Development held in Johannesburg in 2002 provided the ISDR with a concrete set of objectives within the sustainable development agenda to integrate and mainstream risk reduction into development policies and processes (UNGA 2006).

1.5.2.3 Hyogo Framework of Action (2005–2015)

Following the Indian Ocean Tsunami of 2004, the United Nations convened the World Conference on Disaster Reduction (UNWDR) in January 2005 at Kobe, Japan, the key outcome of which was Hyogo Framework of Action (2005–2015). The main objective of Hyogo Framework of Action (HFA) was to build the resilience of Nations and communities to disasters. In this connection, the HFA provided principles, guidelines, and priority actions for enhancing resilience. The goal of the HFA was to minimize the loss of life and assets of communities and nations. The HFA also established a Global Platform for Disaster Risk Reduction as the successor mechanism of the Inter-Agency Task Force for Disaster Reduction. The Global Platform was to enable member states of the United Nations and other stakeholders to assess progress made in the implementation of the Hyogo Framework of Action and enhance awareness of disaster risk reduction. It was also to serve as a forum to share experiences and learn from each other good practices, find gaps, and point out actions to accelerate national and local implementation. The HFA stimulated national and local DRR efforts and strengthened global cooperation as well as the development of regional strategies, plans, and policies. It led to global progress in developing institutions, policies, and legislation for disaster risk reduction.

1.5.2.4   Post HFA International Agenda on DRR

Despite some progress in setting up institutional mechanism and policy frameworks, managing underlying disaster risk drivers remained limited until the end of Hugo Framework of Action. In general, institutional, legislative, and policy frameworks in countries did not sufficiently facilitate the integration of disaster risk into development policies and decisions. Consequently, hazard exposure enhanced—with the development of new climate related risks and emerging and re-emerging bio risks in the form of zoonotic diseases (UNEP and ILRI 2020). In fact, these new risks generated faster than the reduction of existing risks.

Further, the efforts made during HFA neither reduced physical losses nor economic impacts, and it was concluded that the focus of national and international attention must shift from protecting social and economic development against external shocks to transforming growth and development to manage risks, in a holistic manner (Preventionweb 2019). The challenge implied to move from managing disasters themselves to managing risk and focusing on drivers of risks. Since climate change, poverty, and unsustainable development are the main drivers of disaster risks including biohazards that contribute to multiplication of zoonotic diseases around the world, they attracted major attention—leading to the development of three agreements to global agenda for risk management in 2015. These included Sendai Framework for Disaster Risk Reduction, Paris Agreement on Climate Change, and Sustainable Development Agenda with its 17 sustainable development goals. In addition to

these, there are three other agreements that have bearing on DRR including the New Urban Agenda (UNGA 2016a), the Addis Ababa Action Agenda (UNGA 2015), and the Agenda for Humanity (UNGA 2016b), all of which contain elements of DRR and resilience in their scope (Peters et al. 2016; Murray et al. 2017) and point to the connectivity between risks and sustainable development.

1.5.2.5   The Sendai Framework for Disaster Risk Reduction 2015–2030

The Framework was adopted in Sendai, Japan, at the Third UN World Conference on Disaster Risk in March 2015 and was endorsed by the United Nations General Assembly in June 2015. It outlines seven clear targets and four priorities for action to prevent new and reduce existing disaster risks (Table 1.1).

The Framework aims to achieve the substantial reduction of disaster risk and losses in lives, livelihoods, health, and assets of persons, businesses, communities, and countries over the next 15 years (UNISDR 2015a). The Sendai Framework puts the collective responsibility of people, governments, communities, the private sector, donors, and investors as well as media and civil society to effectively prevent and reduce disaster risks. Its principles demand accountability mechanisms to protect populations and ecosystems, simultaneously introducing risk-informed approaches to manage current and emerging risks better (UNDRR 2019).

1.5.2.6   Paris Agreement on Climate Change

The Paris Agreement, a legally binding international treaty on climate change, was adopted by 196 countries at the 21st session of the Conference of the Parties of UNFCCC in Paris, on December 12, 2015. Its goal is to limit global warming to well below 2 (preferably to 1.5) degrees Celsius, compared to pre-industrial levels (UNFCCC 2021). Disaster risk and resilience are engrained within this Agreement. Recognizing the fact that climate change-related disasters are seriously impeding the progress toward sustainable devel-

Table 1.1 Sendai Framework of Action on DRR: goals, targets, and priorities for action

Goal
Prevent new and reduce existing disaster risk through the implementation of integrated and inclusive economic, structural, legal, social, health, cultural, educational, environmental, technological, political, and institutional measures that prevent and reduce hazard exposure and vulnerability to disaster; increase preparedness for response and recovery and thus strengthening resilience

Targets						
Substantially reduce global disaster mortality by 2030; aiming to lower per 100,000 global mortality between 2020 and 2030 compared to 2005–2015	Substantially reduce the number of affected people globally by 2030; aiming to lower the average global figure per 100,000 between 2020 and 2030 compare d to 2005–2015	Reduce direct disaster economic loss in relation to global gross domestic product (GDP) by 2030	Substantially reduce disaster damage to critical infrastructure and disruption of basic services, among them health and educational facilities, such as by developing their resilience by 2030	Substantially increase the number of countries with national and local disaster reduction strategies by 2020	Substantially enhance international cooperation to developing countries through adequate and sustainable support to complement their national actions for implementation of this framework by 2030	Substantially increase the availability of and access to multi-hazard early warning systems and disaster risk information and assessments to people by 2030

Priorities for action
There is a need for focused action within and across sectors by states at local, national, regional, and global levels in the following four priority areas

Priority 1	Priority 2	Priority 3	Priority 4
Understanding disaster risk	Strengthening disaster risk governance to manage disaster risk	Investing in disaster risk reduction for resilience	Enhancing disaster preparedness for effective response and to "Build Back Better" in recovery, rehabilitation, and reconstruction

Source: UNISDR (2015b)

opment, the agreement has identified areas of cooperation central to DRR and called for investments to address the underlying risk drivers associated with rising greenhouse gas (GHG) emission levels and to inspire innovation and low-carbon growth (UNFCCC 2017). However, with nonlinear change in hazard intensity and frequency becoming a reality (IPCC 2018), much greater and ambitious efforts and accelerated action are required pre-2030, to converge with the goal, outcome, and targets of the Sendai Framework (UNDRR 2019).

Coherence between the Paris Agreement and the Sendai Framework exists primarily around commonalities of DRR and climate change adaptation (CCA). The two frameworks have the common objective of strengthening communities' resilience across the full range of hazards, so they

build back better. Support for the objective is demonstrated in coordinated action between the United Nations Office for Disaster Risk Reduction (UNDRR) and the Adaptation Committee of UNFCCC (UNFCCC 2017). Adaptation has multiple connections with risk reduction processes at the local and regional levels and will be most effectively pursued when integrated efforts reflect the important relationship between climate mitigation (and its associated risks), adaptation, hazard modification, and vulnerability reduction. Clear governance and accountability mechanisms will be the key to successful integration of the two frameworks along with collective action and joint monitoring processes, thus minimizing the reporting burden on countries while learning from previous successes (UNDRR 2019).

1.5.2.7 Sustainable Development Agenda 2030

The 2030 Agenda for Sustainable Development was adopted in the United Nations Sustainable Development Summit by 193 member countries in 2015. The new agenda recognizes the enormous challenges the world is facing from pervasive poverty, rising inequalities, and immense disparities of opportunity, wealth, and power to environmental destruction and the risks faced due to climate change and other natural events such as pandemics. The 17 Sustainable Development Goals (SDGs) of the agenda aim to end all forms of poverty and promote prosperity, peace, and partnerships while protecting the planets resources and environment (United Nations 2015a). The agenda recognizes and reaffirms the urgent need for DRR and realizes that risk reduction and resilience are extremely important and play a key role in sustainable development policy. It refers to the Sendai Framework of Action directly by adopting common indicators and by setting targets related to risk reduction in many SDGs (UNISDR 2015c). In addition, the agenda carries the specific opportunities for achieving SDGs through reducing disaster risk—for example, by decreasing poor's exposure and vulnerability to disasters or by building resilient infrastructure.

Both the Sustainable Development Agenda and the Sendai Framework recognize that their desired outcomes will be a product of complex and interconnected social and economic processes that overlap across the two agendas. Thus, there are synergies between the Sustainable Development Goals and Sendai Framework targets and indicators. For example, SDG 11 that aims to promote safe and resilient cities has similarity with the Sendai Framework which proposes to reduce the impact of disasters on cities and communities and strengthen their resilience. Likewise, the SDG on Health recognizes that health is not only a matter of biology but also an outcome of societal architecture. Hence, according to Aitsi-Selmi and Murray (2015), it is amenable to human intervention. Disasters, according to them, are not natural events—they are endogenous to society, and disaster risk often arises when hazards interact with the environmental, social, physical, and economic vulnerabilities and exposure of populations. Therefore, they argue that overall focus of disaster risk management should shift. Rather than focusing on protecting or shielding social and economic development against external events and shocks, it should concentrate more on transforming development to accept and manage risks. The focus should also be strengthening resilience and enabling development to be sustainable (Aitsi-Selmi and Murray 2015). Moreover, the adoption of common metrics for measuring the goals and targets of the two agreements could be mutually beneficial and greatly enrich data environment. According to UNDRR (2019), this will contribute to an improved understanding of the forensics of the multidimensional disruptions and a better anticipation of future opportunities, shocks, risks, and precursor signals (UNDRR 2019).

1.5.2.8 Addis Ababa Action Agenda

Among the other three agendas, Addis Ababa Action Agenda (AAAA), adopted at the Third International Conference on Financing for Development in July 2015 and endorsed by the United Nations General Assembly through resolution 69/313, provides a global framework for financing sustainable development including DRR. It commits to develop and implement holistic disaster risk management at all levels in line with the Sendai Framework in its paragraph 34. In its paragraph 62, the agenda mentions the importance of climate and disaster resilience in development financing. It also cites the need for innovative financing mechanisms to help countries prevent and better manage risks and improve the capacity of national and local entities and actors to manage and finance DRR (United Nations 2015b).

1.5.2.9 New Urban Agenda

New Urban Agenda, adopted at the Habitat III in Quito, Ecuador, in October 2016 and endorsed by the United Nations General Assembly, presents a shared vision for a better and more sustainable future. The agenda explicitly promotes SDG 11 and cites DRR and resilience in its vision, principles, and commitments. While promoting proactive risk-based, all-hazard, and all-of-society approaches, in its para 65, it calls for sustainable management of natural resources in cities to promote DRR by developing DRR strategies and assessing disaster risk periodically (UNDRR 2019). Moreover, in paragraph 67 and 77, it highlights commitment to foster cities' resilience to disasters by adopting approaches in line with the Sendai Framework (United Nations 2016).

1.5.2.10   Agenda for Humanity

Agenda for Humanity came out of the first ever World Humanitarian Summit held in Istanbul in May 2016 to reaffirm their commitment to humanity and address the suffering of the 130 million people affected by humanitarian crisis and the millions more living in constant risk and vulnerability. The main message of the agenda is to reduce risk and vulnerability on a global scale by anticipating for and preventing disaster and crises. It allocates five core responsibilities to achieve progress and reduce humanitarian need, risk, and vulnerability, namely, political leadership to prevent and end conflict, leaving no one behind, upholding the norms that safeguard humanity, changing people's lives from delivering aid to ending need, and investing in humanity (UNOCHA 2019).

In addition to the above six agreements, it is important to mention *International Health Regulations* and *WHO Health Emergency and Disaster Risk Management (EDRM)* framework in view of the ongoing COVID-19 pandemic and close relationship between health, DRR, and sustainable development. International Health Regulations (IHR) are the only international legal framework governing how WHO and its member States should respond to infectious disease outbreaks (Davies undated). In 2005, the IHR revisions were adopted by World Health Assembly Resolution 58.3. Its article 2 announced that the scope and purpose of the instrument was "to prevent, protect against, control and provide a public health response to the international spread of disease in ways that are commensurate with and restricted to public health risks" (WHO 2005). Since its entry into force in 2007, signatory States

have been working, individually and collectively, to meet their core capacity requirements under the new framework (Davies undated). The IHR provides basis for action to build resilience and health security in cities and nations.

Further, the growing recognition that health as a core dimension in disaster risk management has catalyzed the development of the WHO Health Emergency and Disaster Risk Management (EDRM) Framework, the WHO Thematic Platform for Health EDRM, and the WHO Health EDRM Research Network (Wright et al. 2020). The WHO Health Emergency and Disaster Risk Management (EDRM) Framework (WHO 2019) provides guidance to countries on putting in place the capacities and functions, within and across health and other sectors as well as, to reduce health emergency risks and impacts—a field that encompasses emergency and disaster medicine, DRM, humanitarian action, global health security, adaptation to climate change, and resilience of health systems, communities, and countries (Wright et al. 2020). The WHO Health EDRM Framework together with WHO work program is a critical tool for WHO member states to set and approve the priorities of the organization, define the targets to be delivered, and monitor their achievements.

1.5.3   Prospects in the Implementation of Global Agenda

The foregoing global agenda in its six-component agreements together with international health regulations and World Health and Emergency and Disaster Risk Management Framework carry the opportunity to address underlying risk drivers. A major step forward through these frameworks is the recognition of the link between the DRR and the sustainable development policy goals and the focus on resilience building and the forward-looking development domain on how to prevent the buildup of new risks. The relevance of DRR to the 2015 and later agreements and the links among them (Fig. 1.8) create opportunities to build international coherence and foster risk-

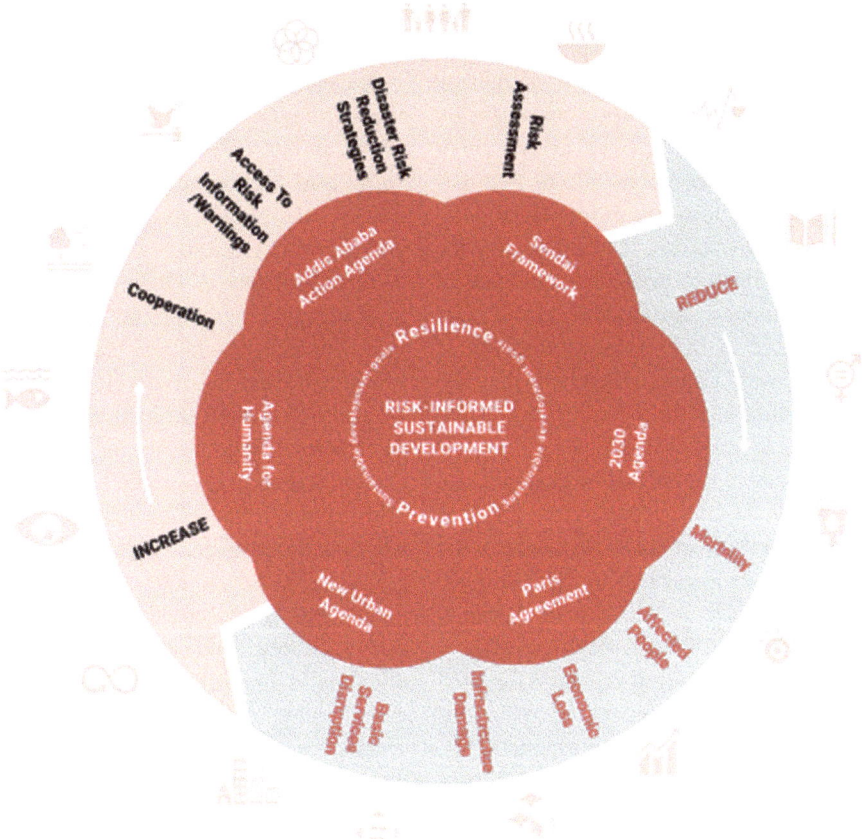

Fig. 1.8 Risk and sustainable development in light of the international agenda. (Source: UNDRR (2019))

informed policy and decision-making, promote multi-hazard and cross-sectoral approaches to assessing risk, and encourage a deeper understanding of socioeconomic and environmental vulnerability across different sectors and levels of government (Murray et al. 2017; United Nations 2018).

The urgent need now is to focus from rhetoric to action and foster risk-informed decision and investment focusing on issues such as poorly planned urbanization, climate change, environmental degradation, poverty, and health (UNISDR 2015a). In doing so, common actions will simultaneously support the achievement of the objectives or goals and targets of agreements as well as Sustainable Development Agenda 2030. Cities provide a great opportunity for experimenting and evaluating the joint implementations of these global agendas and to assess the convergence and

complementarities between them practically. Climate change, health, financing, and gender equality, among others, are a few urban related issues that need complimentary delivery and implementation in relation to the mentioned global agendas. For example, well-managed urbanization as articulated in the New Urban Agenda offers a good platform for implementing climate change adaptation and mitigation as outlined in the Paris agreement. It also provides a chance to advance SDG5 gender equality, through provision of facilities that are more responsive to the needs of women and the caring economy in cities. Likewise, cities can also provide good venues for developing partnerships for innovative mechanisms for climate financing.

One major weakness in controlling risks is weak global risk governance mechanism particularly when it comes to mega risks like climate change and infectious diseases like COVID-19, which not only transcend national boundaries but also generational border. Today's globalized world demands forging not only a consensus on dealing with such mega risks but also creation or designation of an international body that has appropriate accountability and enforcement powers over sovereign nations. Unfortunately, the international architecture needed to take care of the global public goods and address global risks has not been able to follow the momentum of growing globalization and connectivity that ties the world together and the complexities that it creates (Goldin 2013; Hale 2011).

Climate change represents a key example of a global public good that needs some form of global enforcement mechanism to achieve the goals and targets set to combat it. "The absence of a global authority to enforce cooperation across nations undermines collective efforts, combined with the free-riding problems, as each country hopes that others will bear the cost of climate change mitigation" (World Bank 2014). There are some other problems related to climate change mitigation too—for example, despite increased trust in climate models, some scientific uncertainty exists on the critical warming thresholds (so-called tipping points). Moreover, climate change effects vary across countries, creating varied incentives for action. Further, short-term outlook and different valuations of natural goods breed inaction and result in passing the risk to future generations. Despite consensus that climate change is a serious threat, and after decades of debate and negotiations, climate change-related risks are growing and will continue to grow until the problem is effectively addressed.

For infectious diseases that are likely to become epidemic or pandemic, the World Bank and the WHO in response to recommendations by the UN Secretary General's Global Health Crises Task Force in 2017 created the Global Preparedness Monitoring Board (GPMB) as an independent monitoring and accountability body. Some progress, according to the World Bank (2021), was made in this regard since the West Africa Ebola crisis in 2014–2015, but the GPMB (2020) annual report, *A World in Disorder*, noted

that the COVID-19 pandemic has revealed a collective failure to take pandemic prevention, preparedness, and response seriously and prioritize it accordingly. GPMB warned that diseases like Ebola, influenza, and SARS prone to epidemics were also becoming increasingly difficult to manage due to prolonged conflict, fragile states, and forced migration.

One problem in dealing more aggressively with mega risks at city and national level is the cost. For many countries, taking already scarce resources away from development efforts has slowed or reversed difficult development gains. Even before the pandemic, lack of resources constituted a major constraint to meet the cost of mitigation and undertake adaptation measures. It was also an obstacle to reducing greenhouse gas emissions in developing countries and to reach agreements in global negotiations (World Bank 2014; Barrett and Dannenberg 2012). The dilemma is that while climate change crisis is global, vulnerability as well as efforts to mitigate or adapt to it is local. Besides financial resources, national and local authorities' capacity to implement necessary corrective actions is another constraint. In addition, lack of technology and available knowledge also hampers appropriate risk management action at local and national level. Further, lack of proper incentives, accountability, and effective enforcement mechanisms undermine international cooperation. Imprecise and delayed information in early warning systems for natural hazards related to climate change in some developing countries is also a serious constraint. Control of infectious diseases, as in the case of the coronavirus, gets undermined by weak communication between public health authorities and citizens. Thus, delays in testing and lack of tracking and surveillance in the Pandemic led to late and more costly control measures. Failure to translate scientific knowledge for use as public information and limited systematic dissemination of the key messages also hampered effective action.

Perception and political economy are other factors that have played important role in translating disaster risk information into practice or action. For example, despite extensive information on the evidence of drivers of climate change and their related risks and disasters, individuals, communities, and governments continue to discount the future and overlook their potential exposure to what they view as rare or distant events, underestimate their potential cost, and fail to insure or otherwise protect themselves and others. Similarly, small-probability, high-impact risks are often ignored in the face of short-term challenges, resulting in under-investment in preventive steps (World Bank 2014).

1.6 Conclusion

Risks and disasters have affected humans since their existence. They can be traced in the destruction of volcanic disaster of the ancient Mediterranean island of Thera (now Santorini, Greece), which destroyed the entire Minoan civilization around 1600 BC to fierce 2020 Atlantic hurricane season that generated extraordinary 30 named storms (highest on record). They are caused by natural processes which often get triggered or aggravated by human activities, for example, increased consumption of fossil fuels leading to the increase in greenhouse gases (GHGs). Likewise, deforestation or destruction of wildlife habitat is linked to both climate change and to infectious disease epidemics or outbreaks.

The twenty-first century has seen considerable rise in the disaster events—from the dawn of the century to 2019, 7348 natural disaster events (excluding biological disasters/epidemics) were recorded worldwide claiming 1.23 million lives, affecting 4.2 billion people (many on more than one occasion) resulting in approximately US$2.97 trillion in global economic losses. The new century also saw a series of viral epidemics like Ebola, MERS (Middle East respiratory syndrome), West Nile, Swine flu, Zika, and finally coronavirus pandemic that became a once-in-a-century-giant event like the Spanish flu of the last century. The year 2020 stands out very prominent for natural disasters in the present century. The pandemic alone claimed millions of lives and resulted in estimated economic damages of over 10 trillion US dollars. The year also stands out for climate-related disasters, which were responsible for the 389 recorded events that killed 15,080 people, affected 98.4 million, and caused minimum economic loss of US$ 171.3 billion (CRED and UNDRR 2021). No wonder that the two risk perception surveys conducted in 2020 on risk horizon and high impact risks by the World Economic Forum (2021) and Poushter and Huang (2020) of the USA both listed the infectious disease and climate change as the top or mega risks of the world.

The twin mega risks—coronavirus crisis, a zoonotic disease, and climate crisis—have many synergies and commonalities from ecological, economic, as well as social standpoint. For one, they are both potentially devastating global problems and require prompt remediating. Both are the result of reckless exploitation of environment and its resources; in the continuation of which neither was unexpected. On ecological grounds, both mega risks underscore the strong links between human, animal, and planetary health. For example, deforestation and a loss of wildlife habitat, due to economic pressures, are linked both to climate change and to disease outbreak. Both mega risks are causing economic and employment insecurity, enhancing hunger and poverty, and creating crisis of governance. Together, they have led to the preventable loss of lives through actions that are delayed, insufficient, or mistaken. Their prevention makes sense because defenses against both cost a fraction of the loss resulting from the disaster events. A World Bank report (2019) on financial preparedness for a pandemic estimates that every dollar invested in preparation will lead to saving two dollars in the future. Likewise, the Global Commission on Adaptation calculated that every dollar invested in building climate resilience now could score between $2 and $10 in net economic benefits later. The pandemic and the climate crisis are also similar in their social impacts within and between cities and nations—most severely affecting poor countries and vulnerable communities such as elderlies particularly with vulnerabilities like preexisting conditions or diseases, poor and slum dwellers, migrants, and refugee population, as well as ethnic minorities and marginalized communities.

The outbreaks of infectious diseases and climate-related disasters to some extent may be inevitable, but the damage they cause to human is not. They also depict the importance of risk management and promoting resilience against risk in the world. Therefore, risk management, as a policy, is increasingly being recognized as a key priority not only for saving lives but also for economic and social development. The cost of the pandemic has been estimated at over 10 trillion dollars, and likewise, under a high-emissions scenario by 2050 that "sees global warming reach 2 °C, global GDP losses could range between 2.5% to 7.5%, according to World Economic Forum" (WEF 2019). Since both mega risks have transboundary and transgenerational impacts, therefore, they need both national and international institutions and global agenda for actions on risk management and disaster risk reduction (DRR). At the global level, the United Nations Office for Disaster Risk Reduction (UNDRR) coordinates efforts for risk preparedness, prevention, and management. Special international institutions also exist to deal with mega risks such as infectious diseases like COVID-19 and climate change crises. The one for infectious diseases is the World Health Organization (WHO), a specialized agency of the United Nations, while the United Nations Framework Convention on Climate Change Secretariat is responsible for dealing with the Climate Crisis.

The global agenda on disaster risk reduction has evolved over time and has resulted in development of six global agreements recently. Three of these agreements concluded in 2015 are Sendai Framework for Disaster Rick Reduction 2015–2030, Paris Agreement on Climate Change, and adoption of Sustainable Development Agenda 2015–2030. In addition to these three, there are three other agendas linked to DRR including Addis Ababa Action Agenda (financing for sustainable development) 2015, New Urban Agenda 2016, and Agenda for Humanity 2016. The commonalties between Sendai Framework of Action for disaster risk reduction and other agreements lie around managing risk drivers such as climate change, urbanization, health, and other social and economic risk drivers. Further, the Sendai Framework and growing recognition of health as a core dimension in disaster risk management catalyzed the development of the WHO Health Emergency and Disaster Risk Management (EDRM) Framework which together with International Health Regulations provides guidance to countries on putting in place the capacities and functions, within and across health and other sectors, to reduce health emergency risks and impacts. Cities provide a great opportunity for experimenting and evaluating the joint implementations of these global agendas—and to assess the convergence and complementarities between them practically. Climate change, health, financing, and gender equality, among others, are a few urban-related issues that need complimentary delivery and implementation in relation to the mentioned global agendas.

Finally, a major step forward through these frameworks is the recognition for integration of DRR in the sustainable development policy goals, putting focus on resilience building, and the forward-looking development domain in preventing the buildup of new risk. The urgent need now is to focus from rhetoric to action and foster risk-informed decision and investment targeting issues such as poorly planned or unplanned urbanization, climate change, environmental degradation, poverty, and health (UNISDR 2015c). A major weakness toward implementation of DRR frameworks is the lack of an effective global risk governance mechanism on mega risks. Today's globalized world demands forging not only a consensus on dealing with climate change and infectious disease risks but also creation or designation of an international body that has appropriate accountability and enforcement powers. Unfortunately, the international architecture needed to take care of the global public goods and address global risks has not been able to address it. Finally, although the main responsibility for tackling risks belongs to the national governments, combating them effectively demands cooperation from a wide variety of actors both public and private and national as well as international including private sector, businesses, civil society, as well as the public at large.

References

Aitsi-Selmi and Murray (2015) Disaster risk reduction; a cross-cutting necessity in the SDGs. https://sustainabledevelopment.un.org/content/documents/6724139-Aitsi-Selmi-DRR_A%20cross-cutting%20necessity%20in%20the%20SDGs.pdf

Althor G, Watson JE, Fuller RA (2016) Global mismatch between greenhouse gas emissions and the burden of climate change. Sci Rep 6:20281. https://doi.org/10.1038/srep20281

Bagherpour A, Nouri A (2020) COVID misinformation is killing people. Scientific American. https://www.scientificamerican.com/article/covid-misinformation-is-killing-people1/

Banai R (2020) Pandemic and the planning of resilient cities and regions. Cities 106:102929. https://doi.org/10.1016/j.cities.2020.102929

Barrett S, Dannenberg A (2012) Climate negotiations under scientific uncertainty. PNAS 109(43):17372–17376

Barron L (2020) What we can learn from Singapore, Taiwan and Hong Kong about handling coronavirus. Time Magazine, March

BBC News (2019) Large Ebola outbreaks new normal, says WHO. https://www.bbc.com/news/health-48547983

Bedford J, Farrar J, Ihekweazu C et al (2019) A new twenty-first century science for effective epidemic response. Nature 575:130–136. https://doi.org/10.1038/s41586-019-1717-y

Beech H (2020) Tracking the coronavirus, how crowded Asian cities tackled an endemic. The New York Times, 17 March

Boukerche S, Roberts RM (2020) Fighting infectious disease: the connection to climate change. World Bank Blog. https://blogs.worldbank.org/climatechange/fighting-infectious-diseases-connection-climate-change

CDC - Centers for Disease Control and Prevention (2012) Lesson 1: Introduction to epidemiology. Section 11: Epidemic disease occurrence. Level of disease. https://www.cdc.gov/csels/dsepd/ss1978/lesson1/section11.html

CDC Center for Disease Control and Prevention (2019) Remembering the 1918 influenza pandemic. https://www.cdc.gov/features/1918-flu-pandemic/index.html

CDC Center for Disease Control and Prevention (2020) Zoonotic diseases, one health. https://www.cdc.gov/onehealth/basics/zoonotic-diseases.html

Council on Foreign Relations (2020) Major epidemics of the modern era 1899–2020. https://www.cfr.org/timeline/major-epidemics-modern-era

CRED & UNDRR (2021) The non-COVID year in disasters. CRED, Brussels. Available at https://www.undrr.org/publication/2020-non-covid-year-disasters. Accessed on 25 Feb 2022

CRED Center for Research on the Epidemiology of Disasters, and UNDRR, United Nations Center for Disaster Risk Reduction 2020, Human Cost of Disasters, Geneva. https://reliefweb.int/sites/reliefweb.int/files/resources/Human%20Cost%20of%20Disasters%202000-2019%20Report%20-%20UN%20Office%20for%20Disaster%20Risk%20Reduction.pdf

CRS Congressional Research Service (2021) Global economic effect of COVID-19. https://fas.org/sgp/crs/row/R46270.pdf

Daily Sabah (2020) 10 deadliest infectious disease outbreaks in history that predate COVID-19. https://www.dailysabah.com/life/health/10-deadliest-infectious-disease-outbreaks-in-history-that-predate-covid-19

Davies SE (undated) National security and pandemics. United Nations. https://www.un.org/en/chronicle/article/national-security-and-pandemics

Douglas H (2017) Science, values, and citizens. In: Eppur si muove: doing history and philosophy of science with Peter Machamer. Springer, Cham, pp 83–96. http://joelvelasco.net/teaching/2330/douglas17-science_values_citizens.pdf

Draulans D (2020) 'Finally, a virus got me.' Scientist who fought Ebola and HIV reflects on facing death from COVID-19. Sciencemag. https://doi.org/10.1126/science.abc7042

Dwortzan M (2021) How will COVID-19 ultimately impact climate change. MIT Office of Sustainability. https://sustainability.mit.edu/article/how-will-covid-19-ultimately-impact-climate-change

Flecknoe D, Charles Wakefield B, Simmons A (2018) Plagues & wars: the 'Spanish flu' pandemic as a lesson from history. Med Confl Surviv 34(2):61–68. https://doi.org/10.1080/13623699.2018.1472892

Furman J, Shadbegian R, Stock J (2015) The cost of delaying action to stem climate change: a meta-analysis. VoxEU. https://voxeu.org/article/cost-delaying-action-stem-climate-change-meta-analysis

Goldin I (2013) Divided nations: *why global governance is failing and what we can do about it.* Oxford University Press, New York

GPMB (2020) Global Preparation Monitoring Board. A world in disorder. World Health Organization, Geneva. License: CC BY-NC-SA 3.0

Haidt J (2001) The emotional dog and its rational tail: a social intuitionist approach to moral judgment. Psychol Rev 108(4):814

Haidt J (2017) The new synthesis in moral psychology. Science 316(5827):998–1002

Hale T (2011) A climate coalition of the willing. Wash Q 34(1):89–101. Winter

Hallegatte S (2012) A cost-effective solution to reduce disaster losses in developing countries: hydrometeorological services, early warning, and evacuation. Policy Research Working Paper 6058. World Bank, Washington, DC

Harvard (2021) Coronavirus, climate change, and the environment, school of public health. https://www.hsph.harvard.edu/c-change/subtopics/coronavirus-and-climate-change/

Harvey F (2019) One climate crisis disaster happening every week, UN Warns. The Guardian, 07 July 2019. https://www.theguardian.com/environment/2019/jul/07/one-climate-crisis-disaster-happening-every-week-un-warns

Horgan J (2014) Justinian's plague 541–542 CE. World History Encyclopedia. https://www.ancient.eu/article/782/justinians-plague-541-542-ce/

Huremović D (2019) Brief history of pandemics (pandemics throughout history). In: Psychiatry of pandemics, pp 7–35. Published 2019 May 16. https://doi.org/10.1007/978-3-030-15346-5_2

IDMC – Internal Displacement Monitoring Center (2020) Global report on internal displacement. Norwegian Refugee Council.

IFRC (2019) Cost of doing nothing, the humanitarian price of climate change, how it can be avoided. Geneva

IFRC - International Federation of Red Cross and Red Crescent Societies (2010) World disaster report. Geneva

ILO – International Labour Organization (2020) COVID-19 crisis and the informal economy: immediate responses and policy challenges. https://www.ilo.org/wcmsp5/groups/public/%2D%2D-ed_protect/%2D%2D-protrav/%2D%2D-travail/documents/briefingnote/wcms_743623.pdf

IPCC (2014) Climate change 2014: impacts, adaptation, and vulnerability. Part A: global and sectoral aspects. In: Field CB, Barros VR, Dokken DJ, Mach KJ, Mastrandrea MD, Bilir TE, Chatterjee M, Ebi KL, Estrada YO, Genova RC, Girma B, Kissel ES, Levy AN, MacCracken S, Mastrandrea PR, White LL (eds) Contribution of working group II to the fifth assessment report of the intergovernmental panel on climate change. Cambridge University Press, Cambridge, UK/New York

IPCC (2018) Global Warming of 1.5°C. An IPCC Special Report on the impacts of global warming of 1.5°C above pre-industrial levels and related global greenhouse gas emission pathways, in the context of strengthening the global response to the threat of climate change, sustainable development, and efforts to eradicate poverty

IPCC (2022) Climate Change 2022, impacts, adaptation and vulnerability. Working Group II contribution to Sixth Assessment Report (AR6). https://www.ipcc.ch/report/ar6/wg2/

Jaijyan DK, Liu J, Hai R, Zhu H (2018) Emerging and re-emerging human viral diseases. Ann Microbiol Res 2:31–44

Jakob M, Luderer G, Steckel J, Tavoni M, Monjon S (2012) Time to act now? Assessing the costs of delaying climate measures and benefits of early action. Clim Chang 114(1):79–99

Jonas O (2014) Global Health threats of the 21st century. Finance and Development 51(4)., International Monetary Fund

Jones KE, Patel NG, Levy MA, Storeygard A, Balk D, Gittleman JL et al (2008) Global trends in emerging infectious diseases. Nature 451:990–993. https://doi.org/10.1038/nature06536

Klenert D, Funke F, Mattauch L, O'Callaghan B (2020) Five lessons from COVID-19 for advancing climate change mitigation [published online ahead of print, 2020 Aug 3]. Environ Resource Econ (Dordr):1–28. https://doi.org/10.1007/s10640-020-00453-w

Kutschera W (2020) Proc Natl Acad Sci 117(16):8677–8679; first published April 14, 2020. https://doi.org/10.1073/pnas.2004243117

Landscape News 2020, World on Fire 2020. https://news.globallandscapesforum.org/47794/fires-2020-experts-explain-the-global-wildfire-crisis/

LePan N, Routley M (2020) Visualizing the history of pandemics. Visual Capitalist, COVID-19. https://www.visualcapitalist.com/history-of-pandemics-deadliest/

Logan JR, Molotch HL, Fainstein S, Campbell S (1987) The city as a growth machine. In: The gentrification debates: a reader. Routledge, Abingdon-on-Thames

Mackenzie JS et al (2014) The global outbreak alert and response network. Glob Public Health 9(9):1023–1039. https://doi.org/10.1080/17441692.2014.951870

Mair S (2020) How will coronavirus change the world. BBC Future. https://www.bbc.com/future/article/20200331-covid-19-how-will-the-coronavirus-change-the-world

Manzanedo RD, Manning P (2020) COVID-19: lessons for the climate change emergency. Sci Total Environ 742:140563. https://doi.org/10.1016/j.scitotenv.2020.140563

Meerow S, Newell JP, Stults M (2016) Defining urban resilience: a review. Landsc Urban Plan 147:38–49

Mitchell (2013) Risk and resilience: from good idea to good practice. OECD, Working Paper 13. https://doi.org/10.1787/5k3ttg4cxcbp-en

Morse SS (1995) Factors in the emergence of infectious diseases. Emerg Infect Dis 1:7–15

Munich Re (2020) Record hurricane season and major wildfires – The natural disaster figures for 2020. https://www.munichre.com/en/company/media-relations/media-information-and-corporate-news/media-information/2021/2020-natural-disasters-balance.html

Murray V, Maini R, Clarke L, Eltinay N (2017) Coherence between the Sendai Framework, the SDGs, the Climate Agreement, New Urban Agenda and World Humanitarian Summit, and the Role of Science in Their Implementation. https://www.preventionweb.net/publications/view/53049

Newburger E (2021) Disasters caused $210 billion in damage in 2020, showing growing cost of climate change. CNBC. https://www.cnbc.com/2021/01/07/climate-change-disasters-cause-210-billion-in-damage-in-2020.html

NOAA (2020) Tropical cyclones – annual 2020. https://www.ncdc.noaa.gov/sotc/tropical-cyclones/202013

OECD (undated) Resilient cities. https://www.oecd.org/cfe/regionaldevelopment/resilient-cities.htm

Pachauri RK, Allen MR, Barros VR, Broome J, Cramer W, Christ R, Church JA, Clarke L, Dahe Q, Dasgupta P, Dubash NK, IPCC; 2014. Climate change 2014: synthesis report. Contribution of working groups I, II and III to the fifth assessment report of the Intergovernmental Panel on Climate Change. [Google Scholar]

Paul BK (2009) Why relatively fewer people died? The case of Bangladesh's Cyclone Sidr. Nat Hazards 50(2):289–304

Pei S, Kandula S, Shaman J (2020) Differential effects of intervention timing on COVID-19 spread in the United States. MedRxiv. https://doi.org/10.1101/2020.05.15.20103655

Peters K, Langston L, Tanner T, Bahadur A (2016) Resilience across the post-2015 frameworks: towards coherence? Overseas Development Institute. https://www.odi.org/sites/odi.org.uk/files/resource-documents/11006.pdf

Poushter J, Huang C (2020) Despite pandemic many Europeans still see climate change as greatest threat to their countries. PEW Research Center. https://www.pewresearch.org/global/2020/09/09/despite-pandemic-many-europeans-still-see-climate-change-as-greatest-threat-to-their-countries/

Preventionweb (2019) Hyogo framework of action. https://www.Preventionweb.net/sendai-framework/hyogo/

Pueyo T (2020) Coronavirus: why you must act now. Medium. Retrieved on 21 April 2020 from https://medium.com/@tomaspueyo/coronavirus-act-today-or-people-will-die-f4d3d9cd99ca

Quaglio G, Demotes-Mainard J, Loddenkemper R (2012) Emerging and re-emerging infectious diseases: a continuous challenge for Europe. Eur Respir J 40:1312–1314

Romero-Lankao P, Gnatz DM, Wilhelmi O, Hayden M (2016) Urban sustainability and resilience: from theory to practice. Sustainability 8(12):1224. https://doi.org/10.3390/su8121224

Smith BA (2021) 2020 U.S. billion-dollar weather and climate disasters in historical context. NOAA Climate.gov. https://www.climate.gov/news-features/blogs/beyond-data/2020-us-billion-dollar-weather-and-climate-disasters-historical

Smith KM, Machalaba CC, Seifman R, Feferholtz Y, Karesh WB (2019) Infectious disease and economics: the case for considering multi-sectoral impacts. One Health (Amsterdam, Netherlands) 7:100080. https://doi.org/10.1016/j.onehlt.2018.100080

Tagliabue F et al (2020) The "pandemic" of disinformation in COVID-19. SN Compr Clin Med 1–3. 1 Aug 2020. https://doi.org/10.1007/s42399-020-00439-1

The Economist (2021) What is the economic cost of COVID-19. Finance and Economics, January 7th, 2021 Edition. https://www.economist.com/finance-and-economics/2021/01/09/what-is-the-economic-cost-of-covid-19

The Lancet (2021) Climate and COVID-19: converging crises. Lancet 397(10269):71. https://doi.org/10.1016/S0140-6736(20)32579-4. Epub 2020 Dec 2. PMID: 33278352

The World Bank (2014) World development report: risk and opportunity. Washington DC

The World Bank (2019) Pandemic preparedness financing status update. Prepared for Pandemic Preparedness Monitoring Board, Washington DC

The World Bank (2021) Pandemic preparedness and COVID-19. https://www.worldbank.org/en/topic/pandemics

The World Economic Forum 2021. The global risks report 2021, 16th Edition. http://wef.ch/risks2021

Thompson A (2020) A running list of record-breaking natural disasters in 2020. Scientific American. https://www.scientificamerican.com/article/a-running-list-of-record-breaking-natural-disasters-in-2020/

UNDRR (2019) Global assessment report on disaster risk reduction. Geneva

UNDRR United Nations Office for Disaster Risk Reduction (2021) History. https://www.undrr.org/about-undrr/history. Accessed 23 May 2021

UNEP - United Nations Environment Program and ILRI International Livestock Research Institute (2020) Preventing the next pandemic: zoonotic diseases and how to break the chain of transmission. Nairobi

UNFCCC (2017) Opportunities and options for integrating climate change adaptation with the sustainable development goals and the Sendai framework for disaster risk reduction 2015–2030. FCCC/TP/2017/3. https://www.preventionweb.net/publications/view/55605

UNFCCC (2019a) 25 years of efforts and achievement. Key milestones in the evolution of international climate policy. https://unfccc.int/timeline/

UNFCCC (2019b) 25 years of adaptation under the UNFCCC. Report of the Adaptation Committee ISBN no. 978-92-9219-188-7. Bonn

UNFCCC (2021) The Paris agreement. https://unfccc.int/process-and-meetings/the-paris-agreement/the-paris-agreement

UNGA – United Nations General Assembly (2006) Resolution 60/195, International Strategy for Disaster Reduction, A/RES/60/195, New York

UNISDR (2015a) Sendai framework for disaster risk reduction 2015–2030. Geneva

UNISDR (2015b) Chart of Sendai framework for disaster risk reduction 2015–2030. https://www.preventionweb.net/files/44983_sendaiframeworksimplifiedchart.pdf

UNISDR (2015c) Global Assessment Report 2015. Making development sustainable: the future of disaster risk management

UNISDR - United Nations International Strategy for Disaster Risks Reduction (2004) Terminology: basic terms of disaster risk reduction. Geneva

United Nations (2015a) Transforming our world: the 2030 agenda for sustainable development.

https://sustainabledevelopment.un.org/post2015/transformingourworld

United Nations (2015b) Addis Ababa action agenda. https://sustainabledevelopment.un.org/index.php?page=view&type=400&nr=2051&menu=35

United Nations (2016) The new urban agenda. http://habitat3.org/the-new-urban-agenda/

United Nations (2018) Implementation of the Sendai framework for disaster risk reduction 2015–2030. A/73/268

United Nations General Assembly (2015) A/RES/69/313, Addis Ababa action agenda of the third international conference on financing for development. New York

United Nations General Assembly (2016a) A/RES/71/256. New Urban Agenda, New York

United Nations General Assembly (2016b) A/71/353, Outcome of the World Humanitarian Summit. Report of the Secretary-General

UNOCHA - United Nations Office for the Coordination of Humanitarian Affairs (2019) Agenda for humanity. https://agendaforhumanity.org/resources.1.html

Volz U (2020) Investing in a green recovery. Finance and Development. Fall 2020 Issue. IMF

WEF - World Economic Forum (2019) Outbreak readiness and business impacts. White Paper. https://www.weforum.org/agenda/2020/03/coronavirus-global-epidemics-health-pandemic-covid-19/

WHO (2015) How the 4 biggest outbreaks since the start of this century shattered some long-standing myths. https://www.who.int/news/item/01-09-2015-how-the-4-biggest-outbreaks-since-the-start-of-this-century--shattered-some-long-standing-myths

WHO (2019) Disease outbreaks by year. Geneva. http://www.who.int/csr/don/archive/year/en/. Accessed 31 Mar

WHO (2022) Infodemic. https://www.who.int/health-topics/infodemic#tab=tab_1

WHO World Health Organization (2005) Revision of the international health regulations, WHA58.3. Fifty-eighth World Health Assembly 2005

Woolhouse ME, Gowtage-Sequeria S (2005) Host range and emerging and reemerging pathogens. Emerg Infect Dis 11:1842–1847. https://doi.org/10.3201/eid1112.050997

Wright N, Fagan L, Lapitan JM et al (2020) Health emergency and disaster risk management: five years into implementation of the Sendai framework. Int J Disaster Risk Sci 11:206–217. https://doi.org/10.1007/s13753-020-00274-x

Cities' Sustainability in a Risk-Prone World

2.1 Introduction

Cities have been known to be the cradle of civilizations such as those of Indus Valley, Mesopotamian, and Nile Valley civilizations. They have also been well recognized as hubs of commerce and birthplace of innovations, designs, and ideas. They were the pivot of industrial revolution three centuries ago, and today they are the focal points of technological revolution. Currently they accommodate 4.2 billion people or 55% of the world's population (UNDESA 2018a). Besides, being abode to human beings, their economic importance can be gauged from the fact that they generate more than 80% of global gross domestic product (GDP) (World Bank 2020). Further, the economies of scale allow provision of infrastructure and services comparatively much cheaper in cities because of their high density of population. Ironically, along with their dynamism in economic production, cities have also been big consumers of resources and emitters of pollutants. They devour two thirds of global energy consumption and release more than 70% of greenhouse gas emissions that are contributing to change in the global climate (World Bank 2020). On the positive side, this clearly demonstrates the opportunity cities have in climate change mitigation.

In their negative connotation, cities have been epicenters of epidemics, pandemics, and risks. Historically risks had a strong urban component, and one finds many striking examples of "urban disasters" (IFRC 2011). They can be traced in the destruction of historic cities in Indus Valley Civilization—like Mohenjo-Daro and Harappa in 1900 BCE due to climate change, the Athenian plague of 430 BCE, volcanic disasters of Pompeii in 79 CE, the Bubonic Plague of European cities of the fourteenth century; the floods that played havoc with the Chinese cities in 1887 and 1931, Spanish Flu of 1918, Hurricane Katrina's disaster in New Orleans of 2005, Cyclone Nargis destruction in Yangon, Hurricane Maria's catastrophe of Houston in 2017, and 20 separate billion dollar weather and climate-related disasters in the USA in 2021 (Smith 2022). The recent coronavirus pandemic has fully exposed the vulnerabilities of cities. Together with the concurrent climate change-related hazard, it has minimized the myriad of challenges the cities were already facing in the form of housing, jobs, infrastructure, and service deficits.

This chapter has been divided into six sections. After this short introduction, the next section highlights the growing importance of cities as human habitat and as focal points of economic and political power in the present world. The following section on cities and risks mainly focuses on the emergence of COVID-19 and climate change as twin mega risks. It gives an overview of the magnitude of their impacts and how they have transformed city functions. The section after it deals with urban risk management which focuses on capacities of cities in both developed and developing countries to respond to and endure the crises related to mega risks. The fifth section highlights the preparedness aspects for handling mega risks, while the concluding section summarizes the findings of the chapter.

2.2 Perspectives on Cities

Only one third of the world's population lived in cities or urban areas in 1950. One hundred years later, the ratio is set to reverse, meaning two thirds of the world population have been forecasted to live in cities in 2050. The turning point in this ratio was the year 2007, which brought it to equilibrium—half and half or 50% on each side (UNDESA 2018a). Since then, the reversal continues in favor of urbanization. Cities are growing rapidly both in terms of inhabitants and the space they occupy. Every minute, 10,000 sq. meters of city space is added. Every 5 days, a new Paris is created, and each year, a Japan is built (Fausing 2020). The land consumption in terms of spatial expansion of cities outpaces population growth by about 50%, which means that 1.2 million km^2 of new urban built-up area would be added to the world in the next three decades (World Bank 2020).

2.2.1 Cities as Human Habitat

Currently, there are some 10,000 cities in the world today, half of which became cities in the

last 40 years according to research released at the 10th World Urban Forum held in February 2020 (Scruggs 2020). According to UN-Habitat (2020a), the biggest concentration of urban population is in large cities (more than 300,000 people) or metropolises. Currently there are 1934 metropolises in the world representing approximately 60% of the world's urban population. At least 2.59 billion people lived in metropolises in 2020 (Fig. 2.1), equivalent to one third of the global population. Among these, 34 surpassed ten million inhabitants (gaining megacity status), while 51 had a population of 5–10 million, 494 had 1–5 million, and 1355 had 300,000–1 million (UN Habitat 2020a).

It is projected by the United Nations that the number of people living in metropolises in 2035 will increase to 3.47 billion, representing 39% of the global population and 62.5% of the world's urban population. Almost 1 billion people will become metropolitan inhabitants in the next 15 years (UN Habitat 2020a). Half of metropolitan areas emerged in developing or less developed countries, due to city growth since 1975. This is in sharp contrast to high-income countries of OECD and European Union, where most metropolitan areas existed even prior to 1975 (OECD and European Commission 2020). More than half of metropolitan areas in low-income countries, primarily in sub-Saharan Africa, were towns with a population below 50,000 inhabitants in 1990. This trend is likely to continue in the future with

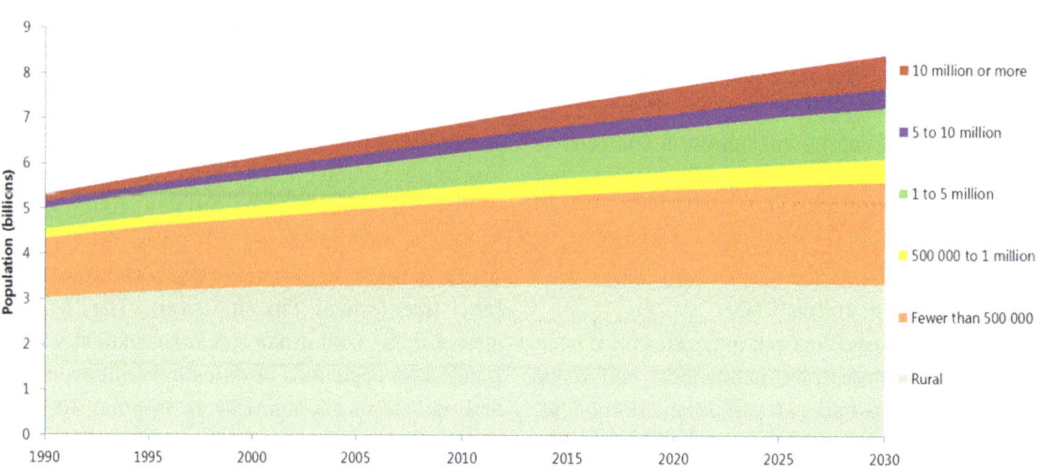

Fig. 2.1 World's population by size class of human settlements, 1990–2030. (Source: UNDESA 2016)

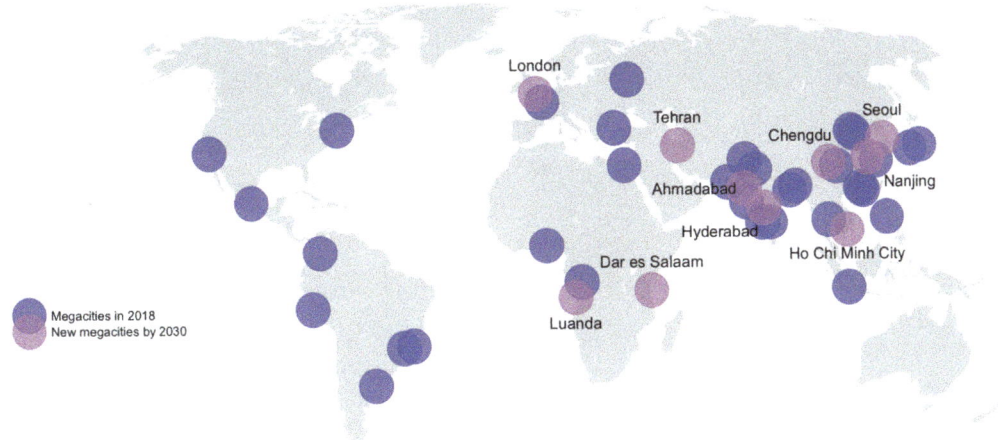

Fig. 2.2 Megacities of the world 2018 and 2030. (Source: UNDESA (2018b))

most metropolises, and almost all megacities according to projections will grow in developing nations.

Globally, the number of megacities is projected to rise to 43 by 2030, compared to 33 in 2018 (UNDESA 2018b). Most of the present ones are already located in global south or developing countries, and they will continue to emerge there in the near future (Fig. 2.2). As the map indicates, all the metropolises projected to become megacities in 2030 are in developing nations. They include Lahore, Pakistan; Hyderabad, India; Bogotá, Colombia; Johannesburg, South Africa; Bangkok, Thailand; Dar es Salaam, Tanzania; Ahmadabad, India; Luanda, Angola; Ho Chi Minh City, Vietnam; and Chengdu, China. Despite a projected decline of nearly one million inhabitants, Tokyo is expected to remain the world's largest city in 2030, followed by Delhi, which is projected to add nearly ten million people by 2030 (UNDESA 2016).

An important concern in this regard is that according to the World Bank (2020), some 90% of the urban expansion in developing countries is likely to be near hazard-prone areas and built in the form of slums and unplanned settlements. Therefore, as the world continues to concentrate in cities, sustainable development in developing countries will increasingly hinge upon the effective management of urban growth. Along with

disproportionate growth of metropolitan areas, urban growth has also continued in other towns/cities (Fig. 2.1). However, it is important to note that while population of most metropolitan areas is increasing globally, some metropolitan areas have been shrinking since 2000 (Fig. 2.3), and by 2050, almost a third of metropolitan areas have been projected to shrink (OECD and European Commission 2020). Metropolitan areas with less than a million inhabitants are the most susceptible to population loss. In some cities, population decline occurred in response to a natural disaster. This has been the case, for example, in New Orleans, USA, which lost population after Hurricane Katrina in 2005, and in Sendai, Japan, following the 2011 earthquake and tsunami. Economic contraction has also contributed to population decline in some places. For example, Buffalo and Detroit, both located in the USA, experienced population decline associated with the loss of industry and jobs. In most cases, however, declining or stagnating populations have been associated with persistent low fertility rates, particularly in Europe. The 52 cities with declining populations were home to 59 million people in 2018, down from more than 62 million in 2000 (UNDESA 2018b). It is yet to be seen what effects the ongoing pandemic and climate-related disasters will have in the future.

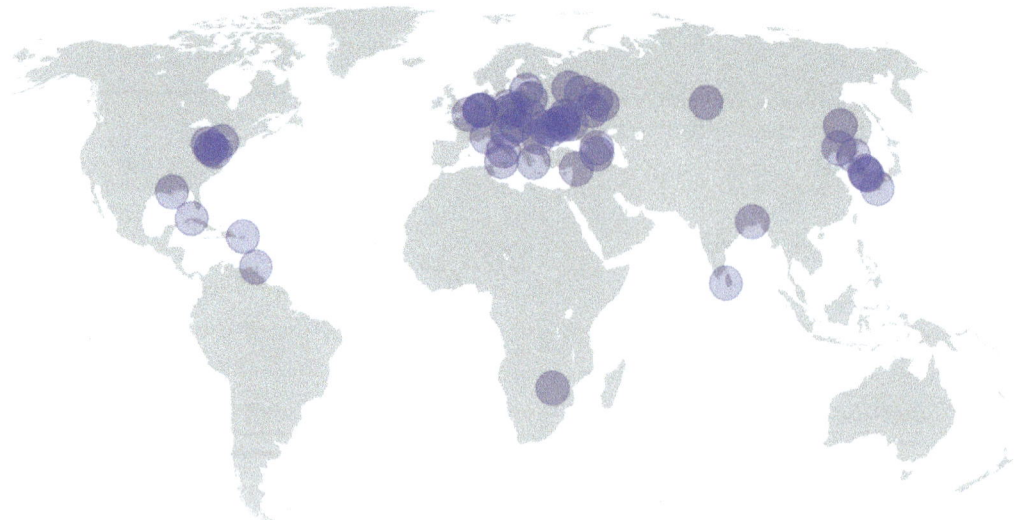

Fig. 2.3 Cities recording a decline of population between 2000 and 2018. (Source: UNDESA (2018b))

2.2.2   Cities as Centers of Economic and Political Power

Presently generating 80% of global GDP (World Bank 2020), cities are at the heart of economic globalization today. Major cities of developed world no doubt are economic giants. The 380 developed region cities accounted for 50% of global GDP in 2007, with more than 20% of global GDP coming from 190 North American cities alone (Dobbs et al. 2011). The 220 largest cities in developing regions contributed another 10%—China's cities generated 4% and Latin America's largest cities another 4%. Across all regions, 23 megacities— with ten million or more inhabitants—generated 14% of global GDP in 2007 (Dobbs et al. 2011). Currently, the top 780 global cities produce almost 60% of all world economic activities, and they are set to grow in importance as urbanization continues (Oxford Economics 2018).

In terms of projection, Global Cities report of Oxford Economics (2018) depicts significant growth of disparities between cities of the world to 2035. It predicts the balance of urban economic power shifting eastward with Chinese cities emerging as the leaders, doubling their economies to outstrip the European and North American cities by 2035 (Fig. 2.4). However,

New York, Tokyo, London, and Los Angeles are projected to stay as world's urban superpowers. Paris is projected to drop out of the top 5 cities and be displaced by Shanghai and Beijing (Oxford Economics 2018).

The top 10 cities by gross domestic product (GDP) in 2035 will be widespread. Three of these are expected to be in the USA—New York, Los Angeles, and Chicago. Four cities will be in China, while London, Paris, and Tokyo are set to be the remaining three. However, several variables could impact these 2035 projections, from mega risks like COVID-19 to financial recessions and political uncertainty, rapid urbanization, and technological advances. But one thing is certain in the coming decades, cities are the places where many of the economic factors will converge and play out (Ghosh 2019).

Cities are also important politically and socially as meeting points for goods, people, and ideas. They are pivotal as instigators and connectors in a growingly networked globe. The rise of digital technology may allow many of the functions traditionally provided only by cities to happen anywhere, but there is little sign of the reduction in growth or power of cities (Beall and Adam 2017). In fact, the role of cities on political and social issues has enhanced particularly in controlling risks, be they pandemic or climate

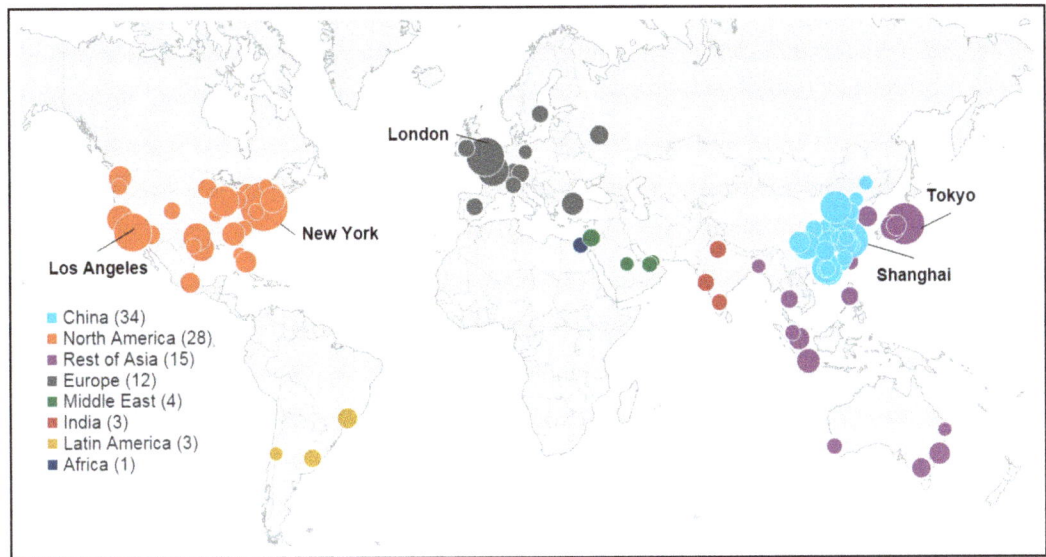

Fig. 2.4 World leading cities by GDP 2035. (Source: Oxford Economics (2018))

change. This has become particularly important in the wake of nation-states' struggle to retain control within and across their sovereign boundaries over borderless sociopolitical trends in terms of migration, travel, and trade.

2.3   Mega Risks and Cities Transformation

As stated earlier, the cities have faced risks and disasters throughout history, but the way the twin mega risks of coronavirus and climate change has transformed cities in the twenty-first century was unpredictable and unimaginable. They not only inflicted colossal damages to their vital systems and infrastructures but paralyzed the very functioning of cities and brought major lifestyle changes. Today newspaper headlines and articles and magazine stories are replete with the growing incidence of COVID-19 pandemic and climate-related risks such as rising sea levels and storm surges, wildfires, heat stress, extreme precipitation, inland and coastal flooding, landslides, increased water scarcity, and air pollution. No doubt, the occurrence of these harmful events has alarmed both policy- and decision-makers and the public at large. However, being alarmed or

"on the alert" is not the same as being equipped to prevent or reduce those risks or to mitigate their harmful effects and damages. The recent chronology of events has amply demonstrated that neither cities nor countries have adequately prepared to face these risk events.

Mitigating climate change is extremely important for cities' own sake because their exposure to climate related risks like storms, floods, and sea level rise has increased over time. According to the World Meteorological Organization (WMO) and UN Office for Disaster Risk Reduction (UNDRR) climate/weather-related disasters have seen unprecedented rise—surging fivefold over the last 50 years (United Nations 2021). With 136 biggest coastal cities accommodating 100 million people and containing $4.7 trillion in assets exposed to coastal floods (World Bank 2020), city-based disaster risk management with a central focus on climate-related risk reduction needs to occupy a central place in city management.

The frontline challenge to cities, however, that emerged with full swing in 2020 and beyond was COVID-19. It was a massive challenge to cities in both rich and poor countries alike. The spread of coronavirus, lockdown, and other measures taken to control the spread have had massive implications on cities and their economies, public health and service delivery systems and social welfare. It also exposed the extent to which urban population's health and livelihoods are vulnerable. Urbanites, across the world, changed their lifestyle during the pandemic. They made efforts to adapt themselves to new realities of living and working, in some ways completely different from traditional regimes. In some ways, as discussed in the previous chapter, cities' response to the coronavirus pandemic was inextricably linked to their response to climate change for a greener and better future for both.

2.3.1 COVID-19

With an estimated 90% of all reported COVID-19 cases in 2020 (UN Habitat 2020b), urban areas had become the hotspots of the pandemic during the first wave. Although rural areas were also affected later, cities remained the epicenters even in the subsequent waves of the pandemic.

2.3.1.1 Triggering Factors

The individual cities in each country were hit differently by COVID-19, depending on varied factors and circumstances. Different theories have been put forward attributing different factors to link COVID-19 and cities (Nathan 2021). One of these attributes international connectivity or fast mobility between cities was a cause of rapid spread. After the first case of COVID-19 was identified in Wuhan City of China in December 2019, the disease spread rapidly across the cities of China and the world through connectivity. Human mobility placed doctors and scientists at a serious disadvantage in slowing potential spread or stopping a chain effect. Consequently, in the current globalization era, a traveler picking up a virus in Wuhan shared it with another person or persons in a city or cities at remote location within hours. The last trains and scheduled domestic and international flights before lockdown left Wuhan in the morning of January 23, 2020, putting an end to a surge of outbound Chinese Lunar New Year travel that had started 3 days earlier. However, by that time, superspreaders were already on the move globally,

infecting people across cities of the world. The strength of the impact can be gauged from the fact that an estimated five million people who were potentially exposed to the virus had already left Wuhan before lockdown (Kinetz 2020). The international connectivity together with other operating factors including weak institution and leaderships, structural inequalities, crowding of population particularly in slums, sustained interconnectivity or close contacts in indoor spaces and public transport, etc. have been pointed out to be other factors that contributed to the spread. Discussed elsewhere (Nathan 2021; Florida et al. 2021; Alsan et al. 2021; Almagro et al. 2020; Almagro and Orane-Hutchinson 2020; Gaskell et al. 2020; Harris 2020) and in Chap. 7 of this book in more detail, these factors in one way or another resulted in the failure of the disease containment or stopping both global and local outbreaks. Thus, one place after another started falling victim to the virus, and by early February 2022, over 390 million cases, including 5.7 million deaths, had been reported worldwide. Among countries, the six hardest hits up to 2021 were the USA, India, Brazil, Mexico, Peru, and the UK (BBC News 2021).

2.3.1.2 City Size and COVID-19

The coronavirus pandemic has no doubt triggered a wave of investigation and research. According to one estimate, well over 100,000 articles about the coronavirus pandemic were published in 2020, while another count shows them exceeding 200,000 (Else 2020). However, little is known about the effects of city size on the propagation of this disease. One study (Stier et al. 2020) showed that COVID-19 spread faster on average in larger cities with the additional implication that, in an uncontrolled outbreak, larger fractions of the population were expected to become infected in more populous urban areas. Another study of Brazilian cities (Ribeiro et al. 2020) indicated that small towns were proportionally more affected by COVID-19 during the initial spread of the disease, and the cumulative numbers of cases and deaths per capita initially decreased with population size. However, in the long term, this advantage vanished, and large cities started to exhibit higher incidence of cases and deaths, such that every 1% rise in population got associated with a 0.14% increase in the number of fatalities per capita after about 4 months since the first two daily deaths (Ribeiro et al. 2020). However, in the absence of more studies, it is difficult to draw firm conclusions. Nevertheless, large cities like New York, London, and Madrid were worst hit by the pandemic up to the end of May 2020. London recorded 21% of the total number of COVID-19 deaths in England and Wales until May 1, 2020, while having only 15% of the population (Morris and Barnes 2020). Madrid recorded almost a third of all deaths in Spain and New York at its peak with 2% of national population recorded almost 15% of all deaths in the USA due to the pandemic (Serhan 2020).

In the USA, large cities like New York, Los Angeles, San Francisco, Washington DC, Chicago, New Orleans, etc. were hard hit. According to the New York Times (2020), all the nation's most populous places suffered a lot. In Cook County, Illinois, which includes Chicago, more than 5200 people died. In Los Angeles County, California, at least 278,000 people had the virus attack, more than in most states in 2020. While in New York City, about 1 of every 354 residents died. But unlike in the early days of the pandemic, it is not so simple to say that big cities have been hit hardest. On a per capita basis, many places with the most cases have been small and mid-sized metros in the southwest of the USA with large Native American or Hispanic populations (New York Times 2020). In Yuma County, Arizona, along the country's border with Mexico, for example, about 1 of every 17 residents had the virus attack. In McKinley County, New Mexico, which includes part of the hard-hit Navajo Nation, one of every 277 residents died (New York Times 2020).

In developing countries, Indian cities were hit hard. According to the newspaper *Hindustan Times* dated September 20, 2020, the top 10 cities severely affected by COVID-19 were Delhi, Mumbai, Chennai, Bengaluru, Bhubaneswar, Jaipur, Hyderabad, Mohali, Faridabad, and Indore. However, by June 2021, Delhi retained the top position, while Pune and Bengaluru became the second and third highest hit cities (Sethi 2021). In Brazil, again, all major cities including Sao Paulo, Rio de Janeiro, Fortaleza, and the capital Brasilia suffered most.

2.3.1.3 Economic and Social Effects

Besides affecting health, COVID-19 caused a major global economic contraction. According to International Monetary Fund (Gopinath 2020), it has been worst than the Great Depression and far exceeded those of the Great Recession that ended in 2009. The economic impacts of COVID-19 on cities were enormous and extended far beyond their boundaries (United Nations 2020a). It has also occurred at a much faster rate, hitting all sectors and many of the world's largest employers (Sand 2020). Capitals and large cities often lost or had a larger share of jobs potentially at risk due to severely reduced activity in local services. Tourism (including transport and civil aviation), retail business (other than food and pharmacies), hospitality (hotels and restaurants), and personal services (hairdressers, beauticians, gyms, etc.) were worst affected. According to OECD (2020a), in eight countries, the Czech Republic, Denmark, Finland, France, Lithuania, Norway, Romania, and Sweden, the capital regions faced the highest job losses in the short term. Greece and Spain followed the same pattern due to the decline in tourism. In most cases, the higher risk observed in capitals, or other large cities, was due to their specialization in retail and wholesale trade, as was the case for Athens, Bucharest, Prague, Helsinki, Oslo, Stockholm, and Vilnius. On the contrary, large cities with a diverse economy and a more skilled labor force than average fared better as they had the ability to adapt to shock and had a good future potential for the economic bounce back because they also hosted a larger share of high-skilled workers whose jobs were more compatible with teleworking (OECD 2020a).

Globally, the informal sector in the cities was especially affected severely. This sector represents 90% and 67% of total employment in low- and middle-income countries, respectively (ILO 2020). In the first month of the crisis, on average, informal workers worldwide lost as much as 60% of their earnings. In Africa and Latin America, this figure was nearly 80% (United Nations 2020b). This had devastating impacts on women, as they had dominant share in the informal economy and were predominantly engaged in hardest hit sectors like tourism, hospitality, and services.

Beyond the immediate impacts on health, jobs, and incomes of urbanites, the pandemic affected people's mindset by increasing anxiety and worry. It also adversely impacted on their social relations with each other and resulted in undermining their trust in institutions, their personal security, and sense of belonging (OECD 2020b). The short- and medium-term impacts on people varied depending upon where a person lived in a city, for example, it had a disproportionate impact on groups living in slums and squatter settlements in cities that already were subject to greater vulnerability. It also varied due to other factors such as age, gender, social situation, etc. According to the United Nations (2020b), tackling COVID-19 became more challenging in urban areas with high levels of crime and violence, poor infrastructure and housing, and/or weak local governance and with ill-equipped or under-resourced frontline workers. Limited access to healthcare and basic services and inadequate housing and/or public space further undermined COVID-19 responses (United Nations 2020b).

2.3.2 Climate Change

Climate change mega risk is also threatening the welfare, livelihood, and assets of city dwellers, and many key and emerging global climate risks are concentrated in urban areas. Droughts are causing the water taps in Mexico City to run dry, forcing millions of people to pay many times more for tanker truck water deliveries. Heat waves have struck cities from India and Pakistan to Europe and the USA. Torrential rains have triggered killer landslides in La Paz and Durban, while fierce storms and floods have destroyed factories in the Chinese cities of Guangzhou and

Dongguan and brought New York City subways to a halt (Global Commission on Adaptation and WRI 2019). The pattern of occurrence of climate-related hazards like cyclones, floods, droughts, and wildfires is becoming more intense and more frequent, reportedly now averaging a disaster a week (Harvey 2019). Polar ice is melting more quickly than anticipated (The Economist 2019) with serious implications for sea level rise and coastal cities.

2.3.2.1 Range of Impacts
The impacts of climate change on cities range from loss of life and property to infrastructure failures as well as water and food shortages. A full picture of physical impacts of climate change is available in the report by the Urban Climate Change Research Network (UNCCRN 2018), which is an outcome of a project being implemented to enhance understanding and communicating key challenges cities are facing, and will continue to face, due to climate change (Table 2.1).

2.3.2.2 Economic Losses
The economic losses due to climate change are also colossal. If determined efforts are not made to adapt to climate impacts, the economic toll and human suffering in cities will inevitably mount enormously. The double threat of rising seas and greater storm surges alone could force hundreds of millions of people out of their homes in coastal cities, with a total cost to coastal urban areas of more than $1 trillion each year by midcentury (UN Habitat 2016). According to a UNCCRN (2018) report, the total human cost to urban areas from sea level rise, if emissions are not reduced, could be enormous—putting over 800 million people at risk in 570 cities by 2050. Under high emission scenarios, even wealthy megacities like New York and Shanghai will face grave risks from sea level rise (Fig. 2.5). Overall cost of climate-related extreme events will be devastating for cities. Worldwide economic stress and damage from climate-related disasters in 2020 alone, as presented in the previous chapter, was over US$170 billion. According to a report by CDP (an environmental disclosure platform) and University College London (CDP and UCL

Table 2.1 Physical impacts of climate change risks by their magnitude and time

Vulnerability	Time Period	Population Estimate	City Estimate
EXTREME HEAT	Present Day	Over 200 million people	Over 350 cities
	2050s	Over 1.6 billion people	Over 970 cities
EXTREME HEAT AND POVERTY	Present Day	Over 26 million people	Over 230 cities
	2050s	Nearly 215 million people	Over 490 cities
WATER AVAILABILITY	2050s	Over 650 million people	Over 500 cities
FOOD SECURITY	2050s	Over 2.5 billion people	Over 1,600 cities
SEA LEVEL RISE	2050s	Over 800 million people	Over 570 cities
SEA LEVEL RISE AND POWER PLANTS	2050s	Over 450 million people	Over 230 cities

Source: UNCCRN (2018)

Fig. 2.5 Impact of 0.5 meter sea level rise on coastal cities by 2050. (Source: C40 Cities (2022))

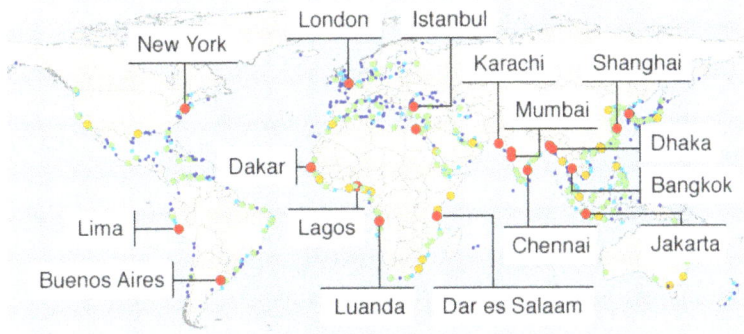

Urban populations at risk by the 2050s
● 10million +   ● 5m - 10m   ● 1m - 5m   · 0.5 - 1m   ·0.1m - 0.5m

2020), estimated mean damage costs due to climate change in "business as usual" scenario will be $5.4 trillion a year up to 2070 and $31 trillion a year by 2200. It amounts to a 10% reduction in GDP growth rate by 2050 and 25% by 2100, respectively. This cost appears to be too high when compared with a 2 °C scenario of the Paris Agreement—where policy action and investing in mitigation activities will result in a fraction (one third) of the costs of damage at US$1.8 trillion by 2070 plateauing beyond that year.

In a survey of corporate establishments at city level (CDP 2019a), over 215 of the world's largest firms estimated that climate change would cost them a combined total of nearly US$1 trillion in the business as usual or non-action scenario. Among the companies reporting, more than 86% identified exposure to substantive risks from climate change (Fig. 2.6). Conversely, in the same survey, 225 of the world's 500 biggest companies reported that climate-related opportunities also have great financial potential totaling over US$2.1 trillion dollars (CDP 2019a).

Like other corporates, even central banks increasingly see climate change as a systemic risk to the global capital market and recognize that non-action is not an option (World Economic Forum 2020). As the extreme weather events become more common, the problems in insurance may also increase by making it unaffordable or even unavailable for individuals and businesses. Already globally, the "catastrophe protection gap"—what should be insured but is not—reached US$280 billion in 2018 (World Economic Forum 2020). The mortgage market may also face problems particularly in cities vulnerable to climate risk. Thus, in risk-prone cities located in areas like Florida, 30-year mortgages could default enmasse if homes become uninsurable over time (Dellink et al. 2017).

Looking at above facts, it has become critical for cities to build resilience against climate change and protect their inhabitants. The first vital step in this direction is to understand and assess their full vulnerabilities. The assessments should not only consider current risks but also those which are expected in the near and distant future. This will allow to plan in the right direction for both the present and the future and ensure proper investment in the infrastructure of the future, an investment that will not compound the climate crisis and stand the test of time in the new normal resulting from climate crunch (CDP 2019b).

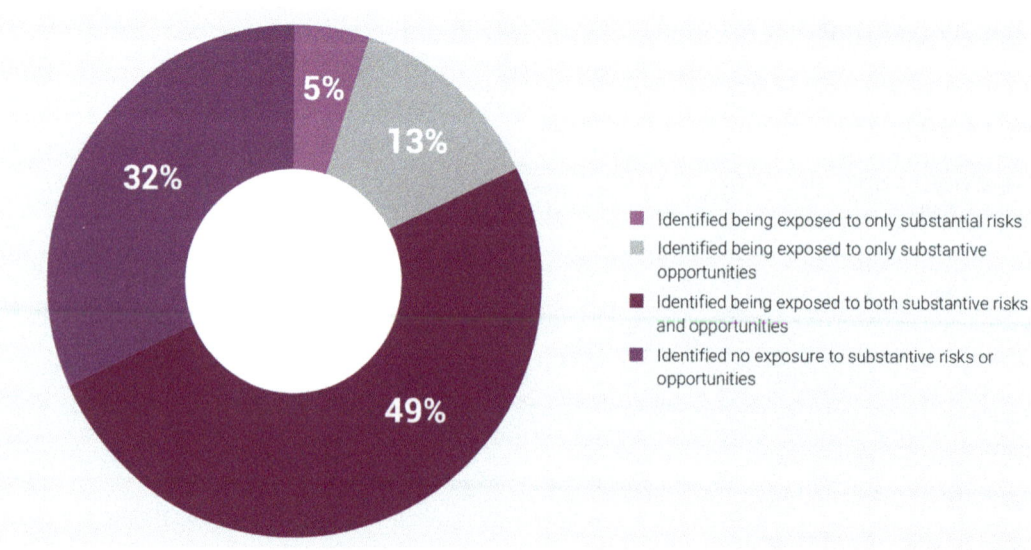

Fig. 2.6 Companies reporting exposure to climate-related risks and opportunities. (Source: CDP (2019a))

2.4 Cities' Response to Mega Risks

In response to mega risks, whether coronavirus or climate change, cities have been working on solutions both individually and through networks.

2.4.1 Coronavirus Pandemic

The cities were the first to be affected by the coronavirus, and they were the first to respond. With their creative potential, combined with vibrant economic and cultural life, they devised numerous means and mechanisms to solve at least some of the problems.

2.4.1.1 Immediate Response

Their immediate response in tackling the COVID-19 pandemic was to overcome the health emergency. Regulatory measures they took included making face masks and social distancing mandatory, enhancing testing, closing of schools and educational institutions, closing of public transport, initiating quarantine and stay-at-home requirements, unleashing public information campaigns, stopping or limiting public gatherings, closing some public services, and enforcing partial or full lockdown and even curfews. Some cities took leadership in developing or advocating emergency health measures. For example, Manchester, London, Birmingham, and Liverpool in the UK advocated successfully to the national government to make masks mandatory on public transport (Lemonde 2020). Seoul (Korea) took a leadership role in the immediate response, through the installation of crisis centers, social distancing, and mandatory masks in public transport. Such measures were later adopted nationally in Korea (Lemonde 2020).

The seriousness of cities' situations depended on how they responded to the pandemic. Those cities that responded in a timely, effective, and firm manner fared better than those which delayed action or lacked coordination in response. The case in point is the difference between New York City and San Francisco; the former suffered a lot because of a lack of firm response. In contrast, the Bay Area City responded to the pandemic earlier and more decisively imposing social distancing measures much before major cities on the East Coast (Wiener and Iton 2020). Apart from San Francisco, some other cities in the world—Singapore, Taipei, Hong Kong, and Seoul—have been much more successful than other less dense, more sprawling cities in containing the virus. Although some of these cities struggled later to contain the virus, their firm, science-based steps to stop it from cascading into disaster initially yielded positive results (Wiener and Iton 2020).

2.4.1.2 Constraints/Predicaments

Putting cities under lockdown and mandatory regulatory measures to deal with COVID-19 health emergency triggered the greatest economic, financial, and social shock of the twenty-first century. This systemic shock prompted a halt in city production, hit their supply chains, and brought a steep drop in consumption together with a collapse in confidence and a sharp decline in services. The pandemic plunged the cities into a severe economic crisis affecting not only local authorities but also businesses, industries, professionals, and laborers. The most serious affect was on individual citizens particularly low-paid workers, who had fewer savings and were less likely to be able to telework. The pandemic's socioeconomic consequences in cities have been especially devastating in developing countries. It deepened poverty and inequalities across cities in the world and potentially derailed efforts to achieve the United Nation's Agenda 2030 on Sustainable Development Goals.

Even in one of the richest countries of the world, the USA, the pandemic reduced local government revenues enormously. Therefore, in response to COVID-19, initially, cities/municipalities made cuts, had to freeze spending and hiring, laid off workers, and drew down on rainy day funds, where available (Siripurapu and Masters 2021). The pandemic drained state and local coffers due to lost sales tax revenues, deferred tax payments, and reduced income and property taxes (Green and Loualiche 2021), and the federal government had to step in to provide substantial aid. With COVID-19 requiring residents to stay at home and stores to shutter, the bulk of reduction in local revenues of cities came from a slump in local sales taxes. Declines through US cities, according to some estimates, were to continue into 2021 (Davidson and Ward 2020). With States facing similar problems, cities needed to rely more and more on Federal Government's help. Despite the Federal assistance, however, every city's budget was not fixed. Even cities in relatively good financial health—including those with rainy day funds to help them through an emergency—faced significant changes to staffing and services (Davidson and Ward 2020).

The local governments while enduring revenue losses also faced higher demands on spending because of the pandemic. A huge fall in employment also occurred due to the amount of fiscal stress (Sheiner 2021). According to Brooking Institution (Sheiner 2021), there is also evidence that tight fiscal conditions led to employment declines in local education, which accounted for roughly 5% of the overall US workforce. Furthermore, scarred by the Great Recession (2007–2009), the local governments were cautious about spending given the tremendous uncertainty on the economic outlook, and therefore states and cities, particularly those relying on tourism, for example, were clearly facing very difficult budget conditions.

2.4.1.3 Rescue and Recovery

Globally, cities resorted to work with several organizations and actors, including the national and regional governments as well as international organizations, urban stakeholders, and citizens, to design and implement short-term economic rescue packages as well as long-term multiple dimensional measures in response to the crisis. Such response and recovery processes also brought out the important role of domestic and international city networks as an important factor in peer learning, exchanging knowledge and experience, and taking leadership in policy making. The OECD Champion Mayors for Inclusive Growth Initiative, Global Resilient Cities Network, C40 Cities, Eurocities network, UNESCO Cities Platform, and UNDP City2City Platform specially provided important curated information and brought together cities and experts to design solutions for challenges raised by the pandemic. They have also been playing a key role in their dialogue at the regional and national levels and on the international stage, to call for immediate assistance, for coordinated actions during and after lockdowns, and for a holistic and integrated approach to long-term urban recovery and resilience.

The grave economic crises have led cities to start defining recovery strategies. The most immediate and pressing among these was the implementation of economic rescue packages to help households and ongoing businesses. This together with the recovery and rehabilitation packages forms the subject of Chap. 5 of this book. One important question the coronavirus crisis has raised is, "Will this pandemic, like those before it, inspire a new blueprint for urban planning and make urbanites invest in increased resilience, sustainability, and equity, or will it drive people away from cities for good?" Regarding this, researchers at the Harris Poll—a firm of social scientists and strategists—conducted an online survey of 2050 US adults nationwide from April 25 to 27, 2020. Nearly 40% of those adults living in urban areas indicated that they would consider moving "out of populated areas and toward rural areas" (Newman 2020). However, Axios reports that people are less likely to pack up during an economic recession (Hart 2020). Historic experience also makes it less likely because better job opportunities, higher wages, and attraction to urban amenities have always attracted residents to cities.

2.4.2 Climate Change

Cities have become important leaders in the struggle against climate change because of their role as laboratories, incubators, and implementers of climate solutions (C2ES Undated). Further, city governments are closer to their citizens, and they can tap better into public sentiment compared to national governments (Halais 2020). Moreover, city networks are playing a prominent role at the international stage in stimulating and galvanizing climate action through showcasing case studies.

2.4.2.1 Cities and Climate Emergency

Cities have been the prime mover in associating the word "climate" with "emergency" in 2019 (Zhou 2019), and over 1200 local governments around the world had already signed the Climate Emergency Declaration by October 2020. Many of the world's most influential mayors have already announced to give it a practical shape by a Global Green New Deal (C40 Cities 2019a). These mayors are members of C40, a network of 94 large cities—Paris, Copenhagen, Los Angeles, Shanghai, Tokyo, Sydney, Lagos, and Rio de Janeiro, to name a few—committed to meeting the goals of the Paris Agreement of limiting global warming to 1.5 °C over preindustrial levels and reducing global greenhouse gas emissions by 50% up to 2030 (C40 Cities 2019a).

2.4.2.2 Climate Action and Mitigation

Under the Global Green New Deal, cities have reaffirmed their commitment to protecting the environment, strengthening the economy, and promoting a sustainable future by cutting emissions from the sectors most responsible for the climate crisis—transportation, buildings, industry, and waste. This means placing climate action at the center of all urban decision-making to secure a just transition for those working in high-carbon industries and correcting long-running environmental injustices for those disproportionately impacted by the climate crisis—people living in the Global South generally and the poorest communities everywhere (C40 Cities 2019a).

Similar commitments made by cities, regions, and businesses can help countries meet or surpass pledged emission cuts. According to the Climate Action Report 2019 (New Climate Institute and others 2019; United Nations Climate Change 2019), one example of such commitments is that made individually by more than 6000 cities and regions and 1500 companies in 10 of the world's biggest-emitting economies. This alone could reduce greenhouse gas emissions by 1.2–2 gigatons of carbon dioxide equivalent ($GtCO_2e$) per year by 2030. Even in the USA, these commitments could put the country in range of meeting its climate pledge for 2025, despite efforts by the previous Trump administration rollback policies.

Coalition for Urban Transition (2019) advocates that it is possible to cut 90% of emissions from cities by adopting technologies and practices already in use—including carbon savings from buildings, transport, materials efficiency, and waste—while also delivering a significant economic return. It would require an investment of US$1.8 trillion (approximately 2% of global GDP) per year, which would generate annual returns worth US$2.8 trillion in 2030 and US$7.0 trillion in 2050 based on cost savings alone. Many of these low-carbon measures would pay for themselves in less than 5 years, including more efficient lighting, electric vehicles, improved freight logistics, and solid waste management.

The CDP and ICLEI (2021) data on climate action also shows that cities are building resilience against climate threats and over the previous 10 years undertook thousands of actions, ranging from tree planting and greening (20% of cities) to developing hazard-resistant infrastructure (10% of cities). The amount of emission reduced by just five cities—Denver, London, Madrid, Durban, and Taipei—between 2009 and 2014 was 13.1 million tons of CO_2 equivalent (CDP, AECOM, and C40 Cities 2014). The capabilities of cities for carbon action, however, vary between developed and developing countries. Moreover, there has been a variation in performance of various cities even within and between developed countries.

The difference is depicted by a report of CDP (2019c), which released the first ever ranking of urban areas on climate action—596 cities for 2018. Each city was awarded an "A" to "D" score based on how effectively they were managing, measuring, and tackling greenhouse gas emissions and adapting to climate risks. About 7% of reporting cities—43 including 24 in North America—received an A for climate leadership and action (CDP 2019c). An "A" score, according to CDP "meant a city demonstrated strong climate adaptation and mitigation strategies and consistently tracked its emissions." they had also specified ambitious emissions reduction targets—80% emissions reduction by 2050 at the latest.

One year later in 2019 ranking for climate action (CDP 2020), 105 cities were included in the A list. North America continued as the leader with 41 cities (with 34 cities in the USA and 7 in Canada). Cities on A list were found to be taking three times more action than others. Some examples of city climate leadership were the following cities:

- Fayetteville, Arkansas, USA: committed to convert all facilities to 100% renewable energy by 2030 and reduce emissions 40% by 2030 and 80% by 2050.
- Greater Manchester, UK: set the target to become carbon neutral by 2038—12 years ahead of the UK government's nationwide target—in a move that required it to cut emissions by 15% per year.
- Thekwini, South Africa: the municipality that includes the city of Durban was aiming for 40% renewable electricity by 2030 and for 70% of private electricity demand to be supplied by self-generated renewable energy by 2050.
- Petaling Jaya, Malaysia: was the only local council in Asia that had launched a tax rebate to residents who implemented low-carbon building retrofit measures and made more sustainable lifestyle choices such as ownership of hybrid/electric vehicles. Since its inception, the project has cut reductions by around 200t CO_2e per annum.
- Windsor, Canada: was fitting playgrounds with features like misting stations, drinking fountains, and tree planting to increase canopy cover and sequester carbon, thereby lowering city emissions.

As the COVID-19 pandemic shook the world in 2020, 88 cities qualified for the Climate Action A status (CDP 2021). These cities made major progress in climate action since the signing of the Paris Agreement by setting ambitious targets and

adapting to the impacts of climate change (Table 2.2)

In the "A" List of 2020, which came 5 years after the signing of the Paris Agreement of 2015, the USA topped with the highest number of cities (25). Overall, 26 cities on the 2020 list were targeting to be powered by 100% renewable energy up to 2050 or earlier. A total of eight cities, including Copenhagen, Stockholm, and San Francisco, achieved 50% or more of their targets. All 88 of the A List city reported building resilience to climate change through adaptation plans. In 2015, only 30% had reported such plans. The number of A List cities dropped to 88 in 2020 compared to 105 in 2019 because of the coronavirus pandemic and due to CDP tightening the criteria on the 2020 scoring methodology to better align with the climate emergency. Despite these factors, however, 34% of cities were new to the 2020 A List. These included Newcastle (UK), Louisville (USA), Firenze, (Italy), Municipalidad de Peñalolén (Chile), Pingtung (Taiwan), and Quezon City (Philippines). European cities had made greater strides, and as a result they together surpassed the number of North American cities on the A List.

European cities on the CDP A List for 2018 were only nine. These went up to over 35 by 2020, including most capital cities like Athens, Copenhagen, Helsinki, London, Moscow, Paris, and Stockholm. The example of action for London included declaration of ultralow emission zones, under which drivers of older, more polluting cars are to pay more to drive in the center of the city. The Hague, in the Netherlands, has built, in a seaside resort within its boundaries, a kilometer-long dike. This is built underneath a new waterfront boulevard to add another layer of protection against the increased flooding that stronger, more frequent storms and sea level rise

Table 2.2 Number of CDP cities with set targets and adaptation plans based on the Paris Agreement 2015–2020

Type of climate action	2015	2020
With set targets for GHG reduction	44	88
With adaptation plans	26	88

Source: CDP (2021)

may bring. In terms of meeting their climate-related target, over 60% of the more than 1000 European cities that have monitored their performance are on track (Hsu et al. 2020). Cities in this dataset have, on average, committed to reduce emissions by 24% compared to their baselines. Cities that monitored their progress and reported inventory data have already achieved about a 15% reduction in emissions, totaling 51 m tons of CO_2 between 2008 and 2019.

Asia lags in the number of cities on the A List. Most cities from Asia falling on the A List are from East Asia including Hong Kong, Seoul, and cities of Taiwan. In Taipei, the Taiwanese capital, city authorities have focused on fixing water leaks to tackle droughts. The initiative has saved more than 600,000 tons of water per year, since 2015. Only Hong Kong among Chinese cities was included in the A List. However, several Chinese cities are committing now to deliver on their share of Paris Agreement—Nanjing, Chengdu, and Qingdao, for example, have publicly confirmed their commitment on this. They became the first Chinese cities to join a coalition of over 70 C40 cities from across the globe to meet the objectives of the Paris Agreement (C40 Cities 2019b). C40 cities network is also helping Jakarta, Hanoi, Ho Chi Minh, and Kuala Lumpur on climate action.

2.4.2.3 Cities and Climate Adaptation

Besides some of the cities mentioned above under the A List, an increasing number of other cities are also taking lead on adaptation plans to face climate change. Some 207 cities across the world reported to CDP (CDP, AECOM, and C40 Cities 2014) that they had already undertaken action on climate adaptation to protect 394 million people from the effects of climate change and to build resiliency for doing business. The total adaptation activities reported were 757, and 102 cities among reporting ones had their adaptation plans.

In their adaptation activities, cities are taking different approaches to reduce the impacts of climate risks depending primarily on the kind of risk involved. For example, Wuhan, which is experiencing increasingly frequent flooding

events, has taken on this threat by constructing sponge infrastructure to absorb and redirect excessive water. It has also created riverfront parks to avoid waterlogging and promote better rainwater drainage (Badia 2018). The sponge initiative (LaMonica 2017) was launched in 2015 in 16 cities within China to reduce the intensity of rainwater runoff by enhancing and distributing absorption capacities more evenly across targeted areas. The resulting groundwater replenishment increased availability of water for various uses. The approach while reducing flooding also augments water supply security. Cape Town's historically most severe drought between 2015 and 2018 led to the use of measures like free plumbing, repairing of water leaks, and removal of invasive non-native plants that consumed almost 10% of the city's water supply (Badia 2018). Buenos Aires, which along with Mexico City qualified for CDP 2020 A List, confronted the problem of higher vulnerability to climate change of its low-income communities residing around lake Soldati by converting it into a natural reservoir with proper stormwater drainage. The scheme was reinforced by an outreach campaign to help some 700,000 inhabitants of the area (Badia 2018).

New York City's plan to adapt to climate change launched in 2013 is unique. The plan was developed in response to Superstorm Sandy, which pummeled 1000 miles of the Atlantic coastline in October 2012 and costed $19 billion in damage and economic losses to the city. It involved protection of 520 miles of coastline, through a multilayered approach focusing on the most vulnerable areas first. The actions under the plan ranged from providing incentives for residents' voluntary relocation at their own cost to funding and coordinating the relocation of entire communities. The plan also restricted development in high flood-prone areas, designated as Special Coastal Risk Districts, which included parts of Queens and Staten Island. The city's Climate Resiliency Design Guidelines, based on future climate projections by the New York City Panel on Climate Change, direct the design, engineering, and construction of capital projects through such efforts as the Waterfront

Fig. 2.7 Staten Island seawall. (Source: CNN (2019))

Revitalization Program. Among others, it included building a protective sea wall, 20 feet high above the sea, spanning for about 5 miles (Fig. 2.7). It is likely to save the island 30 million dollars yearly in flood-related damages.

Lessons from the New York City experience provide a good practice case to a variety of other coastal cities facing similar risks within the USA (Houston, New Orleans, Miami) as well as in developing countries for cities like Karachi and Dares Salam.

Several other cities have climate change adaptation plans and strategies. Rotterdam in Holland and London and Manchester in England also provide good practice examples for other cities in adapting to changing climate. In Rotterdam, the target is to create a city that is climate proof for the people, both present and future, aesthetically attractive and economically prosperous. Rotterdam's trio of floating pavilions, the bubble-shaped domes, is the most notable project of city's adaptation. Anchored off the city's waterfront, measuring 12,000 square feet, it serves as a pilot project for future floating urban districts that will be able to rise with the changing sea levels (Gallucci 2013). In London, based on national climate change projections, the adaptation plan identifies the residents and infrastructure that are most vulnerable and proposes initiatives to protect and strengthen the city, through three key actions: managing the risk of surface water flooding, increasing the number of parks and vegetation in the city, and retrofitting more than one million homes to improve water and energy efficiency (Gallucci 2013). Some other adaptation action cities are undertaking (Scott 2020) include sophisticated flood mapping systems as in Fort Worth; tree plantation initiatives as in Louisville, Kentucky, and Milan (targeted to erect three million trees by 2030); heat mapping as in Richmond and Washington in the USA; planting mangroves as in Guangzhou; promoting green buildings or overhauling existing buildings to ensure a more sustainable future through investment in green roofs and siding and improved insulation; and undertaking energy efficiency measures such as smart heating and cooling systems (Scott 2020).

Most municipal or local authorities in cities of developing countries are grappling with large deficits in infrastructure and services and do not see climate change adaptation as a priority or as their responsibility. Nevertheless, there are some examples of good practices in adaptation from cities even in developing countries like Hanoi, Jakarta, Durban, etc. The C40 network of cities has developed Good Practice Guides from the experience of cities around the world to offer mayors and urban policy makers roadmap for tackling climate change (C40 Cities, Good Practice Guides). It identifies nearly 70 categories of good practice for climate change actions in adaptation and finance, energy, transport, waste, and urban planning.

Climate change adaptation plans and strategies prepared and implemented carefully, simultaneously considering persistent problems of poverty, inequality, and infrastructure deficits, can steer cities on a resilient, safer, and economically stronger path. Additionally, adaptation action in cities can even mitigate climate change, for example, better public transit infrastructure can both improve resilience and cut carbon emissions (Global Commission on Adaptation and WRI 2019). Further, early adaptation actions are better for development because they reduce risks and costs associated with asset losses from climate-related disasters or reducing infrastructure repair costs. They can also create new opportunities—investing in mangrove replanting, for instance, may protect a local community against sea level rise and storm surges while also creating new opportunities for ecotourism and fisheries. Early and proactive resilience-building actions are cost-effective, compared to waiting to address worst impacts when they occur. It has been estimated that the annual cost of global adaptation is one-tenth of the total cost of no action (Leon-Guerrero 2018) that causes recurrent damages.

2.5   Preparedness for Handling Urban Mega Risks

Considering the wide-ranging and colossal humanitarian and socioeconomic impacts of mega risks, preparedness is a crucial investment because its cost is small compared with the unmitigated impact of a health or climate emergency. The financing gap for preparedness, estimated at US\$4.5 billion per year for health emergency, is miniscule (Lee et al. 2020) compared with estimated pandemic cost of over \$10 trillion. The same is true for climate emergencies.

Ensuring better preparedness requires a fresh emphasis on strengthening efforts and capacities in cities to deal with risks and emergencies. Many of these efforts are applicable across both mega risks, such as having a good understanding of the local socioeconomic and cultural milieu and an active involvement of stakeholders, communities, and local leaders in both planning and implementation for combating mega risks. Availing existing set of opportunities is crucial in this regard. Some of these opportunities include securing the commitment of local leaders and strengthening networks particularly public health networks to prevent, detect, and respond to disease threats early and business networks to assist in climate change mitigation and adaptation. It is also vital to establish early warning systems, strengthening means of risk communication, and learning from good practices of both mega risks. The actions range from containment measures to surveillance system and control and relief and rehabilitation in disease outbreaks, while in climate emergency, they span from establishing early warning system to developing mitigation and adaptation measures and action plans.

Whether preparedness or adaptation to mega risks, both require not only identifying options and assessing their costs and benefits but also exploiting available mechanisms for expanding the capacity of human and technological systems in cities. A growing body of literature particularly from the WHO (on disease outbreak) and IPCC (on climate change) and city networks provides guidance on how enabling conditions for preparedness or adaptation can be developed and how constraints can be reduced. Continued development of this knowledge through research and practice could accelerate more widespread and successful preparedness and adaptation outcomes. However, seizing opportunities, overcoming constraints, and avoiding limits can involve complex challenges including both resources and technology, which cities in developing countries lack; hence, their dependence on international cooperation will continue for the purpose.

According to an IFRC (2020) report, the resources needed to adapt to current and imminent climate-related disaster risks are not beyond reach. It has been estimated at 50 billion US dollars annually to meet the adaptation requirements set out by 50 developing countries for the coming decade. This amount, according to the report, is dwarfed by the global response to the economic impact of COVID-19 which has already passed 10 trillion US dollars.

2.6   Conclusion

The two mega risks—the coronavirus pandemic and climate crisis—have revealed the vulnerabilities of cities to risks and disasters. The pandemic has shown that the risk of viral infectious diseases can spark in one city and quickly flash into an international disaster in an increasingly globalized world. The climate crisis has brought in a bigger and longer-lasting catastrophe, but the perspective on its gravity is slow to get its due attention. The virus appeared deadly as it played havoc in claiming millions of lives and affecting billions in the short span of a few months and triggered immediate actions. The climate crisis has not been able to get similar attention even though it has kept killing many more people over time, devastating lives, and destroying livelihoods year after year. Regretfully, it will worsen unless determined actions are taken in a timely manner.

The cities have faced risks and disasters throughout history, but the way the twin mega risks have transformed cities in the twenty-first century was unimaginable. They are not only inflicting colossal damages to their vital systems and infrastructures but paralyzing the very functioning of cities. The growing impacts of COVID-19 and climate-related risks such as rising sea levels and storm surges, wildfires, heat stress, extreme precipitation, inland and coastal flooding, landslides, increased water scarcity, and air pollution are the topics of the day in both electronic and print media. The occurrence of these harmful events has alarmed both policy- and decision-makers as well as the public at large. However, being alarmed or "on the alert" is not the same as being equipped to prevent or reduce those risks or to mitigate their harmful effects and damages. The chronology of events depicted in this chapter has amply demonstrated that cities are not adequately prepared to face these risk events. This is very serious in a scenario where over half the world's population (55%) living in urban areas now have been projected to become two thirds of the world population in 2050. According to the World Bank (2020), some 90% of new urban expansion in developing countries is likely to be near hazard-prone areas and to be built in the form of informal and unplanned settlements. This is very serious in terms of mega risks which have already disproportionately affected the vulnerable population living in slums and marginal settlements within developing countries cities.

With the continuous world urbanization, sustainable development will increasingly hinge upon the successful management of urban growth, especially in developing countries, where the pace of urbanization is projected to be the fastest. The challenge is to make the ballooning urban population part of the solution rather than the problem. Besides being abode to human beings, cities also generate more than 80% of the global GDP and are at the heart of economic globalization today; therefore, urbanization of mega risks carry major threats for the global economy. This further raises the tackling of urban vulnerability to these risks an urgent necessity.

In relation to the pandemic, cities were the first to be affected by the coronavirus, and they were the first to respond. With their creative potential, they devised numerous means and mechanisms to solve at least some of the problems. Their immediate response in tackling the pandemic was to overcome the health emergency through regulatory measures like making face masks and social distancing mandatory, enhancing testing, closing educational institutions and public services, initiating quarantine and stay-at-home requirements, unleashing information campaigns, and enforcing partial or full lockdown and even curfews. Despite budgetary constraints, they also took rescue and recovery measures to control economic fallout. It is commendable that some cities like Manchester, London, Birmingham, and Liverpool in the UK and Seoul in Korea took leadership in developing or advocating health emergency measures like face masks at the national level. The seriousness of cities' own situation depended on how they responded to the pandemic; those responding in a timely, effective, and firm manner fared better than those which delayed action or lacked coordination in response. The case in point was the difference between New York City and San Francisco; the former suffered a lot because of a lack of firm response. Other cities of the world—such as Singapore, Taipei, Hong Kong, and Seoul—have been much more successful due to firm and early action in containing the virus.

Meanwhile, the impacts of the climate crisis, the other mega risk, continue to have substantial effects across urban areas in the world. Droughts are causing water taps in Mexico City to run dry, forcing millions of people to pay many times more for tanker truck water deliveries. Heat waves have struck cities from India and Pakistan to Europe. Torrential rains have triggered killer landslides in La Paz and Durban, while fierce storms and floods have ravaged cities like New Orleans and Houston in the USA, destroyed factories in the Chinese cities of Guangzhou and Dongguan, and brought New York City subways to a halt. If not tackled in a timely manner, it is likely to have worse consequences in:

- The rising sea level that will affect over 800 million urban residents in 570 cities by 2050
- Rising temperatures that will put 1.6 billion people in 970 cities to face extreme heat by 2050
- Decreasing water availability that will affect 650 million people in 500 cities by 2050
- Decreasing crop yields by at least 10% that will impact 2.5 billion people in 1600 cities by 2050

Cities have also been in the forefront in terms of climate action individually as well as through their networks. They played and are playing a prominent role at the international stage in stimulating climate action through showcasing case studies. They also spearheaded the move on declaration of Climate Emergency in 2019— over 1200 local governments around the world have already signed the Climate Emergency Declaration. One of the world's most influential mayors' network, C40 Cities, has already given it a practical shape through a Global Green New Deal. Under this deal, they are committed to meeting the goals of the Paris Agreement of limiting global warming to 1.5 °C over preindustrial levels and reducing global greenhouse gas emissions by 50% up to 2030. Cities' resolve to climate action is also evident in commitment made individually by more than 6000 cities and regions and 1500 companies in 10 of the world's biggest-emitting economies. This could reduce greenhouse gas emissions by 1.2–2 gigatons of carbon dioxide equivalent per year by 2030. Even in the USA, these commitments are likely to put the country in range of meeting its climate pledge for 2025, despite efforts that were made by the previous Trump administration to roll back policies. Many cities have also embarked on ambitious mitigation and adaptation plans.

Mega risks whether in the form of infectious disease outbreak or climate crisis carry wide-ranging colossal humanitarian and socioeconomic impacts. Therefore, preparedness and adaptation plan for them is a crucial investment because its cost is small compared with the unmitigated impact of a health or climate emergency. The financing gap for preparedness, estimated at US$4·5 billion per year for health emergency, is miniscule compared with estimated pandemic cost of over $10 trillion. The same is true for climate emergency. Early and proactive resilience-building actions are cost-effective, compared to waiting to address worst impacts when they occur. Ensuring better preparedness requires a fresh emphasis on strengthening efforts and capacities in cities to deal with risks and emergencies. Many of these efforts are applicable across both mega risks; having a good understanding of the local socio-economic and cultural milieu and an active involvement of stakeholders, communities, and local leaders in planning and implementation are applicable toward combating both mega risks.

References

Almagro M, Orane-Hutchinson A (2020) JUE insight: the determinants of the differential exposure to COVID-19 in New York City and their evolution over time. J Urban Econ. https://doi.org/10.1016/j.jue.2020.103293

Almagro, M, Coveny, J, Guptaz, A, et al (2020) Racial disparities in frontline workers and housing crowding during COVID-19: evidence from geolocation data. Mimeo. Institute Working Paper 37. Minneapolis: Federal Reserve Bank of Minneapolis. DOI: 10.21034/iwp.37

Alsan M, Chandra A, Simon KI (2021) The great unequalizer: initial health effects of COVID-19 in the United States, National Bureau of Economic Research Working Paper Series, No. 28958. NBER, Cambridge, MA

Badia M (2018) Cities leading the Fight Against Climate Change. https://www.citiestobe.com/cities-leading-the-fight-against-climate-change/

BBC News (2021) COVID map: coronavirus cases, deaths, vaccination by country. https://www.bbc.com/news/world-51235105

Beall J, Adam D (2017) Cities, prosperity and influence. British Council. https://www.britishcouncil.org/sites/default/files/g229_cities_paper.pdf

C2ES-Center for Climate and Energy Solutions (undated) City climate policy. https://www.c2es.org/content/city-climate-policy/

C40 Cities (2019a) Mayors announce support for global green new deal; Recognize global climate emergency. https://www.c40.org/press_releases/global-gnd

C40 Cities (2019b) Chinese cities commit to the Paris agreement. https://www.c40.org/news/chinese-cities-commit-to-the-paris-agreement/

C40 Cities (2022) Sea level rise and coastal flooding. https://www.c40.org/what-we-do/scaling-up-climate-action/adaptation-water/the-future-we-dont-want/sea-level-rise/

CDP Carbon Disclosure Project (2019a) Major risk or rosy opportunity. London

CDP (2019b) Cities at risk: dealing with the pressures of climate change. https://www.cdp.net/en/research/global-reports/cities-at-risk

CDP (2019c) 24 cities in North America score an A grade in new cities climate change ranking. https://www.cdp.net/en/articles/media/24-cities-in-north-america-score-an-a-grade-in-new-cities-climate-change-ranking

CDP (2020) Top 105 global cities for climate action revealed by CDP. https://www.cdp.net/en/articles/media/top-105-global-cities-for-climate-action-revealed-by-cdp

CDP (2021) Cities scores - a list 2020. https://www.cdp.net/en/cities/cities-scores

CDP, AECOM, and C40 Cities (2014) Protecting our capital. https://www.c40.org/researches/protecting-our-capital

CDP and ICLEI (2021) Cities on the Route to 2030. London. https://cdn.cdp.net/cdp-production/cms/reports/documents/000/005/759/original/CDP_Cities_on_the_Route_to_2030.pdf?1621329680

CDP and UCL – University College London (2020) Costing the earth. Climate damage costs and GDP. https://www.cdp.net/en/research/global-reports/costing-the-earth

CNN (2019) Staten Island Seawall: Designing for Climate Change. https://edition.cnn.com/style/article/staten-island-seawall-climate-crisis-design/index.html

Coalition for urban transition (2019) Climate emergency, urban opportunity. September. https://urbantransitions.global/en/news/climate-emergency-urban-opportunity-new-flagship-global-report-launched/

Davidson M, Ward K (2020) Next COVID casualty: cities hit hard by the pandemic face bankruptcy. https://www.usnews.com/news/cities/articles/2020-07-30/next-covid-casualty-cities-hit-hard-by-the-pandemic-face-bankruptcy

Dellink R, Hwang H, Lanzi E, Chateau J (2017) International trade consequences of climate change, OECD trade and environment working papers 2017/01. OECD Publishing, Paris. http://www.fao.org/3/a-bu414e.pdf

Dobbs R, Smit S, Remes J, Manikya J, Roxburgh C, Restrepo A (2011) Urban World: Mapping the economic power of cities, Mckinsey global institute. https://www.mckinsey.com/featured-insights/urbanization/urban-world-mapping-the-economic-power-of-cities

Else H (2020) How a torrent of COVID science changed research publishing - in seven charts. NEWS, Nature. https://www.nature.com/articles/d41586-020-03564-y

Fausing K (2020) Climate emergency: how our cities can inspire change. World Economic Forum. https://www.weforum.org/agenda/2020/01/smart-and-the-city-working-title/

Florida R, Rodríguez-Pose A, Storper M (2021) Cities in a post-COVID world. Urban Studies. June. https://doi.org/10.1177/00420980211018072

Gallucci M (2013) 6 of the world's most extensive adaptation plans. Inside Climate News. https://insideclimatenews.org/news/20130620/6-worlds-most-extensive-climate-adaptation-plans

Gaskell J, Stoker G, Jennings W et al (2020) Covid-19 and the blunders of our governments: long-run system failings aggravated by political choices. Polit Q 91(3):523–533

Ghosh I (2019) Where will the top ten cities be in 2035? The Word Economic Forum. https://www.weforum.org/agenda/2019/10/cities-in-2035/

Global Commission on Adaptation and WRI (2019) Adapt now, a global call for leadership on climate resilience. https://www.wri.org/initiatives/global-commission-adaptation/adapt-now-report

Gopinath G (2020) The great lockdown: the worst economic downturn since the great depression. https://blogs.imf.org/2020/04/14/the-great-lockdown-worst-economic-downturn-since-the-great-depression/

Green D, Loualiche E (2021) State and local government employment in the COVID-19 crisis. J Public Econ 193(C)

Halais F (2020) Cities race to slow climate change and improve life for all. https://www.wired.com/story/cities-climate-change-equity-paris-agreement/

Harris JE (2020) The subways seeded the massive coronavirus epidemic in New York City, National Bureau of Economic Research Working Paper Series, No. 27021. NBER, Cambridge, MA

Hart K (2020) Coronavirus may prompt migration out of American cities. https://www.axios.com/coronavirus-migration-american-cities-survey-aba181ba-a4ce-45b2-931c-6c479889ad37.html

Harvey F (2019) One climate crisis disaster happening every week UN warns. The Guardian. 07 July 2019. https://www.theguardian.com/environment/2019/jul/07/one-climate-crisis-disaster-happening-every-week-un-warns

Hsu A, Goyal N, Weinfurter A (2020) Are European cities delivering on their climate commitments. Carbon Brief. https://www.carbonbrief.org/guest-post-are-european-cities-delivering-on-their-climate-commitments

IFRC - International Federation of Red Cross and Red Crescent Societies (2011) No time for doubt: tackling urban risks. A glance at urban interventions by Red Cross Societies in Latin America and the Caribbean. Geneva

IFRC - International Federation of Red Cross and Red Crescent Societies (2020) World Disaster Report 2020. https://www.ifrc.org/sites/default/files/2021-09/IFRC_WDR_ExecutiveSummary_EN_Web.pdf. Geneva

International Labor Organization (ILO) (2020) ILO monitor: COVID-19 and the world of work, fifth edition. 30 June 2020

Kinetz E (2020) Where did they go? Millions left city before quarantine. AP news. https://apnews.com/article/c42eabe1b1e1ba9fcb2ce201cd3abb72

LaMonica M (2017) China's 'sponge cities' aim to re-use 70% of rainwater – here's how. The Conversation. https://theconversation.com/chinas-sponge-cities-aim-to-re-use-70-of-rainwater-heres-how-83327

Lee VJ, Ho M, Kai CW et al (2020) Epidemic preparedness in urban settings: new challenges and opportunities. Lancet 20(5):527–529. https://doi.org/10.1016/S1473-3099(20)30249-8

Lemonde (2020) June 16. https://www.lemonde.fr/smart-cities/article/2020/06/16/andy-burnham-maire-de-manchester-la-crise-due-au-covid-19-a-souligne-nos-hypocrisies_6042975_4811534.html

Leon-Guerrero A (2018) Social problems: community, policy, and social action. Sage, Thousand Oaks

Morris C, Barnes O (2020) Coronavirus, which regions have been worst hit. BBC Reality check. https://www.bbc.com/news/52282844

Nathan M (2021) The city and the virus. Urban Stud. https://doi.org/10.1177/00420980211058383

New Climate Institute, Data-Driven Lab, PBL, German Development Institute/Deutsches Institut für Entwicklungspolitik (DIE), Blavatnik School of Government, University of Oxford (2019). https://newclimate.org/wp-content/uploads/2019/09/Report-Global-Climate-Action-from-Cities-Regions-and-Businesses_2019.pdf. Accessed 16 Apr 2022

New York Times (2020) Covid in the US: the latest map and case count. 14 October 2020. https://www.nytimes.com/interactive/2020/us/coronavirus-us-cases.html

Newman K (2020) Survey: amid the COVID-19 pandemic, urbanites are eyeing the suburbs. https://theharrispoll.com/survey-amid-the-covid-19-pandemic-urbanites-are-eyeing-the-suburbs/

OECD (2020a) Coronavirus, COVID 19, from pandemic to recovery, local employment and economic development. https://read.oecd-ilibrary.org/view/?ref=130_130810-m60ml0s4wf&title=From-pandemic-to-recovery-Local-employment-and-economic-development

OECD (2020b) Protecting people and societies. https://www.oecd.org/inclusive-growth/resources/COVID-19-Protecting-people-and-societies.pdf

OECD & European Commission (2020) Highlights: Cities in the world: a new perspective on urbanization

Oxford Economics (2018) Global Cities: the future of the world's leading economies to 2035. https://www.oxfordeconomics.com/recent-releases/7fa5c39e-6603-433c-9e59-9dd928fa2415

Ribeiro HV, Sunahara AS, Sutton J, Perc M, Hanley QS (2020) City size and the spreading of COVID-19 in Brazil. PLoS One 15(9):e0239699. https://doi.org/10.1371/journal.pone.0239699. PMID: 32966344

Sand D (2020) COVID-19 and the great reset: Briefing note #26, October 7, Mackinsey. http://www.chamber.lk/images/COVID19/pdf/COVID19Briefingnote26October7.pdf

Scott M (2020) Cities step up emissions cuts as climate change starts to bite. Forbes. https://www.forbes.com/sites/mikescott/2020/02/21/cities-step-up-emissions-cuts-as-climate-change-starts-to-bite/

Scruggs. G (2020) Urban planet. https://nextcity.org/daily/entry/there-are-10000-cities-on-planet-earth-half-didnt-exist-40-years-ago

Serhan Y (2020) Vilnius Shows How the Pandemic Is Already Remaking Cities, The Atlantic, 9 June. https://www.theatlantic.com/international/archive/2020/06/coronavirus-pandemic-urban-suburbs-cities/612760/

Sethi V (2021) COVID-19: these are the most affected cities in India with the highest number of cases. Business Insider. https://www.businessinsider.in/india/news/these-are-the-most-affected-cities-in-india-with-the-highest-number-of-cases-after-covid-19-second-wave/slidelist/82082660.cms

Sheiner L (2021) Why is state and local employment falling faster then revenues. Brookings Institution. https://www.brookings.edu/blog/up-front/2020/12/23/why-is-state-and-local-employment-falling-faster-than-revenues/

Siripurapu A, Masters J (2021) How COVID 19 is harming state and city budgets. Council of Foreign Relations. https://www.cfr.org/backgrounder/how-covid-19-harming-state-and-city-budgets

Smith AB (2022) 2021, US billion dollar weather and climate disasters in historical context. NOAA. https://www.climate.gov/news-features/blogs/beyond-data/2021-us-billion-dollar-weather-and-climate-disasters-historical

Stier AJ, Berman MG, Bettencourt L (2020) COVID-19 attack rate increases with city size. https://arxiv.org/abs/2003.10376

The Economist (2019) The Greenland ice sheet is melting unusually fast. 17 July 2019. https://www.economist.com/graphic-detail/2019/06/17/the-greenland-ice-sheet-is-melting-unusually-fast

UNCCRN- Urban Climate Change Research Network (2018) The future we don't want: how climate change could impact the world's greatest cities. Technical Report, February, in collaboration with C40 Cities. Global Covenant of Mayors for Climate and Energy and Acclimatize

UN Habitat (2020a) Global state of metropolis 2020 population data booklet. HS/013/20E Nairobi

UN Habitat (2020b) Opinion: COVID-19 demonstrates urgent need for cities to prepare for pandemics, 15 June 2020. https://unhabitat.org/opinion-covid-19-demonstrates-urgent-need-for-cities-to-prepare-for-pandemics

UN-Habitat (2016) Urbanization and development: emerging futures. World Cities Report 2016. https://unhabitat.org/world-cities-report

UN-DESA, United Nations, Department of Economic and Social Affairs (2016) The world's cities in 2016 – Data Booklet (ST/ESA/ SER.A/392)

UN-DESA, United Nations, Department of Economic and Social Affairs (2018a) The World Urbanization Prospect 2018

United Nations, Department of Economic and Social Affairs (2018b) *The World's Cities in 2018—Data Booklet (ST/ESA/ SER.A/417)*

United Nations (2020a) COVID 19 in an urban world, July, New York. https://www.un.org/en/coronavirus/covid-19-urban-world

United Nations (2020b) Policy brief: the world of work and COVID-19. Available at https://www.un.org/sites/un2.un.org/files/the_world_of_work_and_covid-19.pdf

United Nations (2021) Climate and weather-related disasters surge five-fold over 50 years, but early warnings save lives - WMO report. UN news. https://news.un.org/en/story/2021/09/1098662

United Nations Climate Change (2019) Cities, region and businesses can help countries surpass pledged emission cuts https://unfccc.int/news/cities-regions-and-businesses-can-help-countries-surpass-pledged-emissions-cuts

Wiener S, Iton A (2020) A backlash against cities will be dangerous, Atlantic, May 17. https://www.theatlantic.com/ideas/archive/2020/05/urban-density-not-problem/611752/

World Bank (2020) Urban development overview, April. https://www.worldbank.org/en/topic/urbandevelopment/overview

World Economic Forum (2020) The global risk report. Geneva. https://www.weforum.org/reports/the-global-risks-report-2020

Zhou N (2019) Oxford Dictionaries declares 'climate emergency' the word of 2019. https://www.theguardian.com/environment/2019/nov/21/oxford-dictionaries-declares-climate-emergency-the-word-of-2019

Part II

Overview: Mega Risks Monitoring, Management, and Sustainable Development

Monitoring has a massive role in preparing for and managing risks. This part has two chapters. Chapter 3 covers tools for monitoring and assessment such as GIS and remote sensing, which are needed and/or have been applied for monitoring and managing both COVID-19 and climate change crises. With case studies, it exhibits how these geospatial tools have become prominent at every stage of disaster risk management cycle, from preparedness and planning to response, rehabilitation, and recovery as well as development and implementation of policies on disaster prevention and management.

Chapter 4 highlights the massive growth in the use of digital technologies in the wake of coronavirus crisis. These technologies besides rescuing city functions during the pandemic brought forward unparalleled and multiple benefits—ranging from urban health surveillance to improving efficiency in public administration; allowing continuity of business and education through remote operation; and maintaining events in social spheres to integrating global supply chains. Moreover, they played crucial role in delivery of health services and relief as well as communication with public at large. As mobility restrictions were enforced to contain the spread of the novel coronavirus (COVID-19), more and more of the estimated 4.2 billion city dwellers started working, studying, and interacting from home availing virtual opportunities for engagement, entertainment, and conferencing. Even critical international policy coordination was being conducted online in fora such as the G7 or G20. Operating digitally was the only way to stay in business through mandated shutdowns and restricted activity with rare exception. The proven application of these technologies (particularly ICT) in climate change monitoring, early warning, disaster management, mitigation, and adaptation has also been presented in the part. Among the changes brought about by the mega risks, the most important is the fast-tracking of digital revolution with significant implications for the future. The prospects for post-pandemic likely continuation and enhancement of the use of digital technology have been discussed in the light of mechanisms and means that would alleviate its demonstrated deficiencies and shortcomings.

Mega Risks: Monitoring for Management and Sustainability

3.1 Introduction

Monitoring and assessment play a fundamental role in management and control of urban risks. Hence, geospatial tools for purposes – including GIS, remote sensing, navigation, and earth observation satellites have been extensively used and relied upon in dealing with both mega risks COVID-19 pandemic and climate change crisis. Their use became prominent at every stage of risk management cycle, from development and implementation of policies on risk prevention and preparedness to planning the response in the form of rehabilitation and recovery. The main objective behind their use is that they minimize damage through informed decision-making while allowing the involvement of stakeholders through interactive communication.

Methodology and application of technological tools for monitoring, assessment, and management of risks have undergone many breakthroughs and improvements over the last few years. They have progressed from digitizing and analyzing information and presenting it through maps, graphics, and satellite images to interactive dashboards in multi-layers of data or geographical information system. This has certainly paid off a great deal during the COVID-19 pandemic and in monitoring, mitigating, and adapting to climate change crisis.

This chapter after this introduction in the next two sections discusses the use of monitoring, assessment, and management tools during the pandemic and for climate change, respectively,

while the findings of the chapter have been summed up in the concluding section.

3.2 Monitoring and Managing COVID-19

Understanding the outbreak of COVID-19 and its spatiotemporal dynamics forms a key component of disease control and prevention. Mapping and monitoring tools such as geographical information system (GIS) enable disease tracking, surveillance, and identifying hotspots as well as vulnerable population. Moreover, by facilitating to relate outbreaks with public health facilities, medical resources, and equipment, it allows better response planning. It also became a very important means for outreach and communication with relevant stakeholders and citizens at large during the lockdown through utilization of intelligent web maps and dashboards. In addition, it served as an extremely important mechanism for analyzing the impacts on the local economy and society particularly those relating to unemployment, bankruptcies, lifestyle changes, and providing social protection.

3.2.1 GIS Use in Monitoring/Managing Diseases

3.2.1.1 Origin and History

It is interesting to note that it was a health issue which led to the use of GIS in its initial form. It was a British doctor, John Snow, who used GIS in its simplest form while tracking the cause for the spread of the London cholera epidemic of 1854. Using two layers of map information, one on the incidence of cholera cases and the other on the distribution of water pumps, he discovered a mutual relationship. It led him to conclude that water was the main cause of the spread of the disease. The spatial connection between the water pumps and the disease made a big difference in preventing further spread of the disease. Since that early work, GIS mapping approach has been used not only to study other disease outbreaks but also to plan the response or treatment (Waller and Gotway 2004) as "the science of its day" (Koch and Kenneth 2005). It was, for example, used to delineate the geographical distribution of rickets (weakening of bones due to deficiency of vitamin D, calcium, or phosphate) in 1890 and later for studying the relationship between sunlight and cancer (Waller and Gotway 2004) as well as other cancer studies (National Cancer Institute 2006). In recent times, it has been used in identifying areas with high incidence of mosquito-borne diseases like malaria and dengue, relating asthma to air pollution, targeting anti-smoking campaigns by mapping high smoking areas, mapping and analyzing influenza and tracking distance to healthcare center; and vaccination reach in vulnerable areas (GISgeography 2020). It was also utilized in the study of Ebola virus (Gleason et al. 2014) and management of HIV/AIDS (Kandwal et al. 2009).

Progress in computer technology brought a major boost to GIS. It opened new possibilities to analyze, visualize, and detect patterns of infectious diseases and plan for their mitigation. The intensity of its enhanced usage can be gauged from the fact that out of 865 papers published in health GIS literature, 248 (28.7%) focused on mapping infectious diseases (Lyseen et al. 2014). During its growing application through web-based tools, it also saw many innovations in animation and diversification of themes (Boulos et al. 2010; Gong et al. 2015; Tanser 2002). Based on its successful application in past health studies, it is no wonder that professionals increasingly turned to GIS for monitoring and managing of the coronavirus pandemic (Tribolet 2020).

Chinese cities which were hit early by the pandemic have made various applications of geospatial technology in the fight against COVID-19 (Kamel Boulos and Geraghty 2020). Notable use of the technology was for monitoring and managing critical medical supplies, detecting outbreak sources (Yu et al. 2020), disinfection, and selecting appropriate sites for construction of health facilities. Today, GIS-driven spatial health information, together with remote sensing, satellite communication, and global navigation satellite systems, are playing a major role in dealing with the COVID-19 crisis in many places, supporting a wide range of activities such as monitoring, diagnosis, screening, quarantining, disinfecting, and transport operations (Eurisy 2020).

3.2.1.2 Multidimensional Use in the Pandemic

GIS communities ranging from local governments and academics to media and businesses have used it extensively for monitoring and health-related planning during the COVID-19 pandemic as highlighted in the following examples with associated case studies in parenthesis:

- Identifying outbreaks and showing spatial distribution of cases (Kamel Boulos and Geraghty 2020; Coccia 2020; Yu et al. 2020)
- Multi-scale dynamic mapping for tracking the disease, relating it to environmental factors and making projections (Rossman et al. 2020; Su et al. 2020)
- Implementing lockdown and other social distancing measures and evaluating control and prevention measures and reopening after lockdown (Esri 2020a, b; Samuelsson et al. 2020; Gibson and Rush 2020)
- Vulnerability assessment—spatial segmentation of disease risk and prevention (Lakhani 2020; Leger 2020; Gibson and Rush 2020; Geraghty and Lanclos 2020)
- Spatiotemporal mapping to orient the data in terms of time, place, and person and modeling (Gross et al. 2020; Burton 2020)
- Planning to balance supply and demand of medical resources (Kent 2020; Geraghty and Lanclos 2020)

- Assessment of the supply of materials, transportation, and mobility (Warren and Skillman 2020)
- For communication with city residents through dashboards and maps to impart information on local circumstances, location of food assistance programs, existing emergency medical facilities, and location of city resources, through interactive maps (Leger 2020; Esri 2020c)
- For evaluating socioeconomic and environmental impacts—recovery and rehabilitation planning and implementation (Diffenbaugh et al. 2020; Leadbeater 2020; Cooke 2020)

The multifaced use of geospatial technology (GIS and remote sensing) in tackling the coronavirus pandemic has depended on how a country, city, or community wanted to respond. It also depended on the need which varied from community to community and the purpose for which it was used—communicating, planning, or management—and by field or theme whether used for health, social welfare, business and economics, transport, mobility, etc. Whatever the case may be, it has led to the development of some extremely good case studies whereby cities/local governments and professional institutions developed interactive maps to:

- Help public assess magnitude of risk
- Share critical information with residents
- Assist people in locating essential resources and services
- Evaluate economic impact and facilitate business continuity
- Assist in social distancing capabilities after easing lockdown
- Focus on and deal with service disparities and inequities
- Assess preparedness related to the pandemic

3.2.2 Observational, Analytical, and Display Tools

The data display has also been made more user-friendly with the passage of time. It started in the form of maps, which were made more and more interactive with time for display on dashboards and platforms using apps which also increased in geometric progression.

3.2.2.1 Maps

Historically, individual or series of maps have been an important source of information on disease outbreaks and their growth and for identifying hotspots and tracking vulnerable population as well as to study the pattern of diffusion or spread of virus through space and time. Recently they have also been used to advise people on crowding status/social distancing or as locater of essential resources and services. The methodology of analysis has involved varied techniques such as static map series, animations, and linked interactive micro maps. Many times, advanced tools are used to create animated and visualized curve of an epidemiology for all the spatial units within a city or in a case study area. Curve magnitude and direction are shown in maps to help in visualizing the outbreak stage, magnitude, and geographic distribution.

The following series of self-explanatory maps show different facets of coronavirus-related information at the city level. The first map (Fig. 3.1) constructed by Lall and Wahba (2020) using their hotspots methodology predicted that 5.2 million people in Mumbai were at risk of infection, even with lockdown measures in place. The map shows Mumbai's containment zones as of May 9, 2020, overlaid with predicted hotspots.

The second map (Fig. 3.2) of Tokyo shows underlying risk intensity to Tokyo population from diseases using an urban health index. Similar urban health vulnerability index maps for US cities have been prepared by RS 21 (2021)—examples of eight cities including Alhambra, California, are available on their website (https://covid.rs21.io/map/AlbuquerqueNM). The methodology combined health and demographic characteristics/indicators such as age, chronic illness, etc., to generate an aggregate score for the vulnerability index by census tract for each city. The next map (Fig. 3.3) shows the distribution of coronavirus cases in Dublin, Ireland. This map clearly shows the difference between the cases in the north,

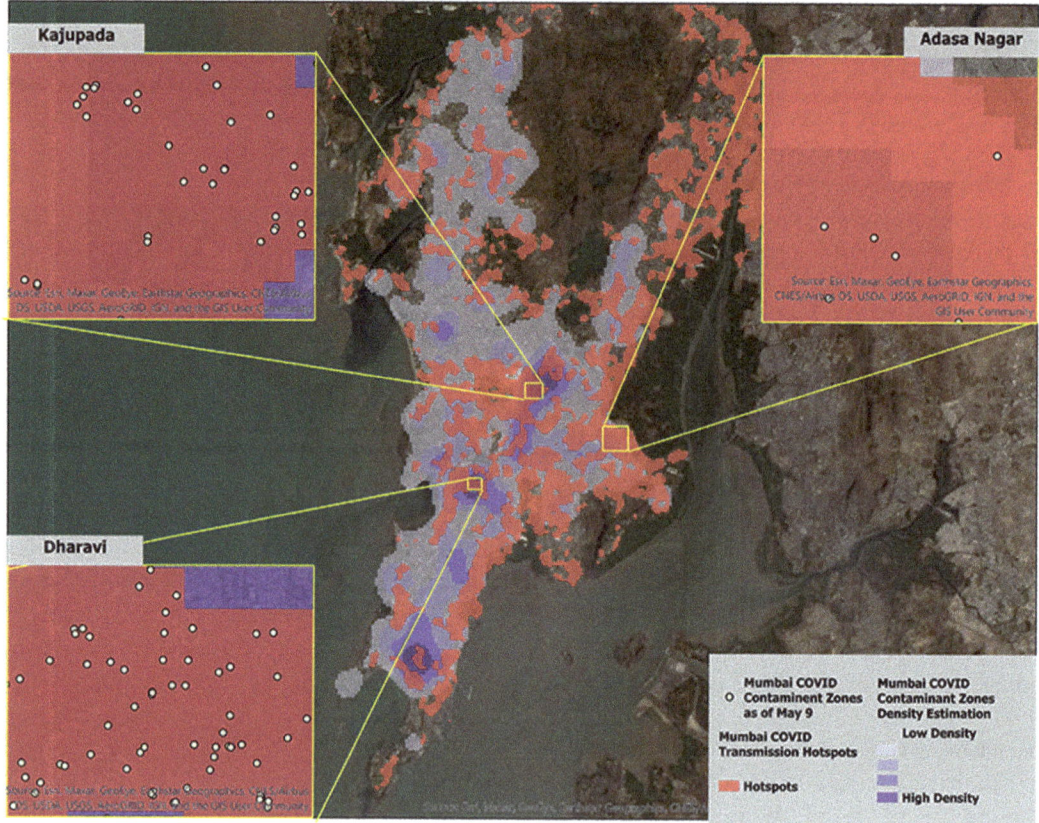

Fig. 3.1 Mumbai's COVID-19 infections in crowded and underserviced neighborhoods. (Source: Lall and Wahba (2020))

south, and city center. Cases are mostly absent in the northern and southern parts of the city inhabited by white well-off people (Bralic 2020).

The fourth map (Fig. 3.4) of Singapore appeared on the dashboard of the city. The purpose of this map was to manage park visitors by showing the intensity of crowdedness of a specific park at a given time. A web application was used that allowed the park managers to track park visitor data, conduct basic spatial analysis, and present it in real time. The fifth map is of Tioga County of New York which shows Grab-and-Go meal pick-up sites for people needing food during the lockdown (Figs. 3.4 and 3.5).

3.2.2.2 Web-Based Dashboards and Interactive Maps

Web-based, coronavirus GIS data dashboards organized, stored, and displayed important infor-

mation on the pandemic from multiple data sources into one, easy-to-access place.

International Level At the international level, the world-famous Johns Hopkins University Coronavirus Dashboard was the first of its kind that started giving most updated information and picture of the global spread of the disease. The dashboard, first shared publicly on January 22, 2020, illustrated the location and number of confirmed COVID-19 cases, deaths, and recoveries for all affected countries. It was developed to provide researchers, public health authorities, and the public a user-friendly tool to track the outbreak as it unfolded. All data collected and displayed were made freely available, initially through Google Sheets and later through a GitHub repository, along with the feature layers of the dashboard, which afterward were included

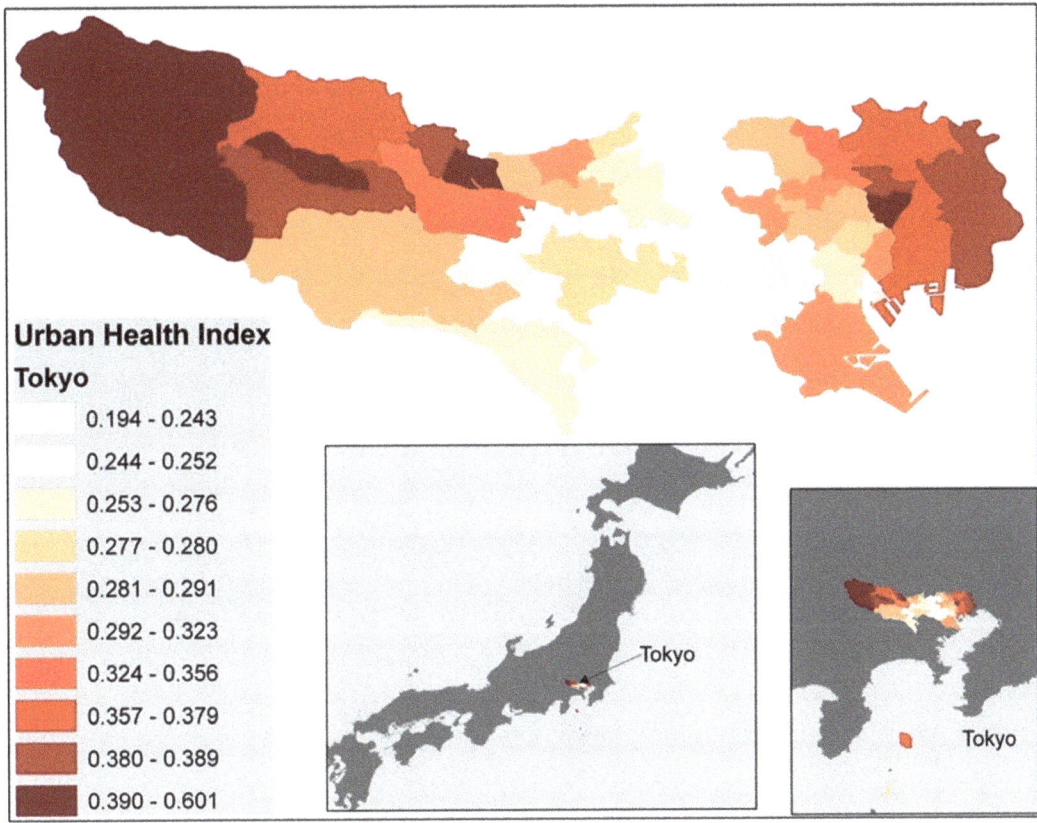

Fig. 3.2 Vulnerability of Tokyo Prefecture population to diseases indicated by Urban Health Index. Note: The darker shades represent areas that are more vulnerable compared to lighter ones. (Source: Rothenberg et al. (2014))

in the Esri Living Atlas. The dashboard, besides providing information at the national level, also reported cases at the provincial level in China, as well as at the city level for the USA, Australia, and Canada (Dong et al. 2020). In addition to Johns Hopkins, there were other dashboards which displayed interactive maps and continuously updated them. Among these important ones included the World Health Organization's novel COVID-19 Situation Dashboard; United Nations Dashboard that focused on Italy but also displayed the outbreak at a global level; and European Centre for Disease Control and Prevention COVID-19 Situation Dashboard that displayed COVID-19 cases in Europe and worldwide. There were other dashboards that provided information at the international level, details of which can be seen in Kamel Boulos and Geraghty (2020) and Kaser (2020).

Since cities were in the forefront in facing the pandemic, therefore monitoring the local impacts and patterns of the virus was key to national-level responses. Hence, UN-Habitat and CitiIQ (https://unhabitat.citiiq.com/) were motivated to develop a COVID-19 tracker platform (Fig. 3.6) for 1200 cities initially for daily monitoring. The coverage increased to include 1700 cities.

The platform presented the city data trends through color coding on a world map simplifying the process of understanding those dynamics that were changing daily at the city level. For example, upward or downward trends resulting from different waves of COVID-19 were visible on this tracker by city depending on data availability. The platform was also enabled to demonstrate the results of the positive local actions such as changed or lowered rate of COVID-19 infections due to

Fig. 3.3 COVID-19
cases in Dublin.
(Source: Bralic (2020))

lockdown implementation or other measures. In other words, the platform showed cities' progress over time, tor example, the European cities progress was shown through colors—red for cities moving from increasing cases to a stable level and green color for subsequent declines in cases.

City/Local Level Data for individual city and its parts at neighborhood and community level becomes more important for application and practical use. It was often displayed on dashboards. An example of a local dashboard is given in Fig. 3.7. The concerns at this level were different and sometimes unique depending on the vulnerability of the communities and their resources. For example, some communities were more vulnerable to COVID-19 infection due to poverty or disproportionate share of old-age population or people with preponderance of adverse health conditions.

These dashboards proved very important for local authorities to communicate with city resi-

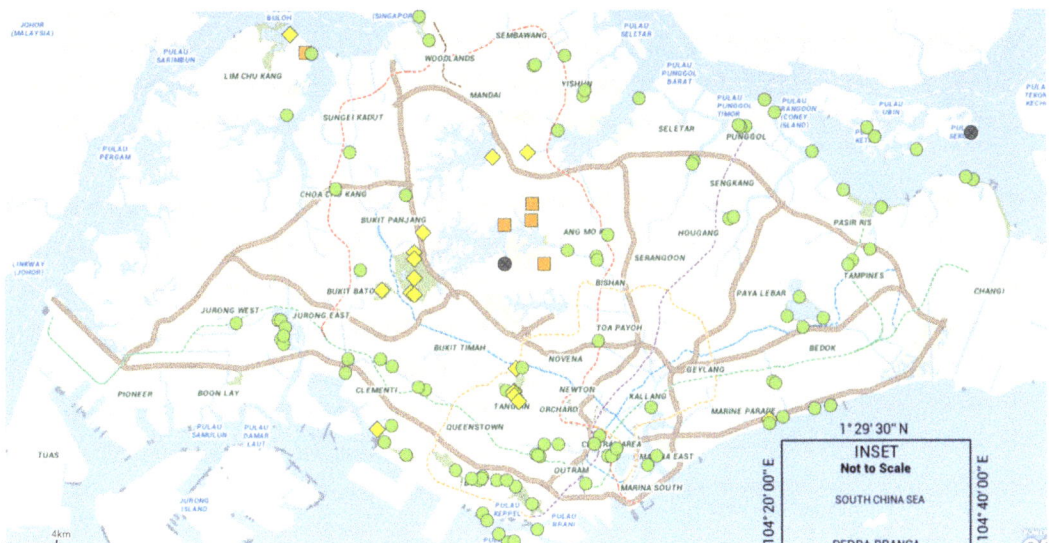

Visitorship Status

High – This area is currently experiencing high visitorship. Please refrain from visiting this area now.

Moderate – This area is currently experiencing moderate visitorship. If you are visiting, please practise safe distancing.

Low – If you intend to visit, please ensure a safe distance of at least a metre apart from others.

Closed – This area is now temporarily closed. Please check our website for the latest updates.

Fig. 3.4 Helping people assess risk: Singapore map with park visit vulnerability (Zheng 2020)

dents on aspects that demanded collective action and/or helped them locate essential services or facilitated their business operations safely and/or assisted in their economic recovery. This is where maps sharing information at the local level through this mechanism played an extremely important role.

Disease Tracking Initially the local and city dashboards tracked city-wide COVID-19 cases, recoveries, and death data and kept the citizens informed about ongoing trends, disseminated precautionary steps, and communicated mitigation measures. Soon many cities also started employing GIS technologies and web-based tools, improved data sharing, presented spatiotemporal patterns, and shared real-time information with communities. Many American counties (Fig. 3.7) and European cities developed extremely useful and interesting dashboards as did some Asian cities.

The city of New Orleans, for example, with some experience in using GIS for flood control in the USA, initiated a dashboard to track the trends of the disease for enforcing and easing control measures. One dashboard of the city contained data on daily and cumulative death rates, test numbers, and positive diagnosis figures. This was overlaid with geographical data to highlight hotspots for attention. A second dashboard mapped the data against baseline milestones set by the city, including the 5-day rolling average of new COVID-19 cases and testing, as well as the availability of ventilators, hospital beds, and intensive care spaces (Wray 2020). Many other city dashboards in the world had information on disease situation in their own areas (Fig. 3.8).

Community Vulnerability Mapping Appropriate response to the outbreak of an infectious disease ahead of reaching its peak demands (a) understanding of what preexisting conditions make a certain

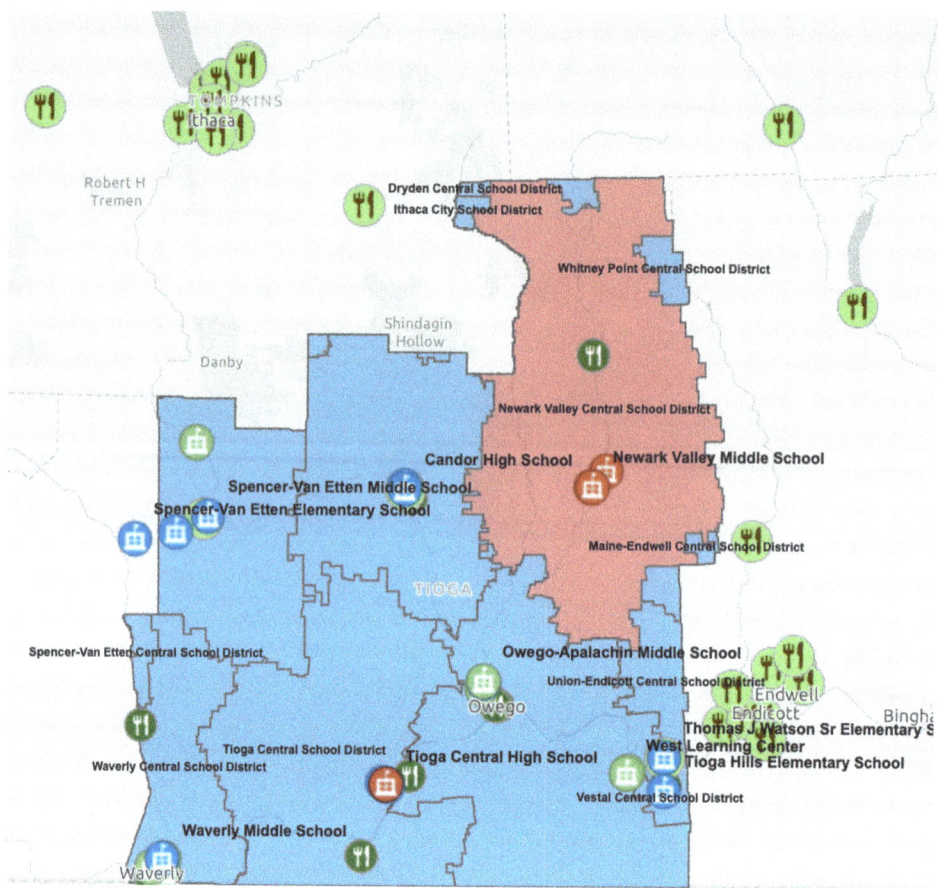

Fig. 3.5 Tioga County, New York, Grab-and-Go meal sites. (Source: https://www.arcgis.com/apps/webappviewer/index.html?id=89a9dcfc8f8e4a1eb63d42ee8474e4ba)

community more vulnerable to the pandemic and (b) the health facilities for the communities present in the area to face the disease. This was accomplished in Canada using GIS software and was presented on a map of Canada displayed on a dashboard. The results for Toronto are displayed in Fig. 3.9. Using data from a community health survey, the map classified resident population at different degrees of risk (vulnerability) in small areal units within the city.

The indicators for population at risk chosen were preponderance of aged population (70 plus) and high proportion of population with preexisting conditions such as chronic obstructive pulmonary disease (COPD), cancer, or hypertension. The interactive map of Canada allowed users to zoom in to any area/province or community

across the country. It also highlighted nearby medical facilities and the services provided. Among developing countries, Indonesia's National Institute of Aeronautics and Space (LAPAN) calculated vulnerability levels in three major cities of the country, Jakarta, Bandung, and Surabaya, based on zones of population density, road access conditions, and strategic locations against COVID-19 spread (UNESCAP 2020).

Resources and Facilities Mapping The interactive maps on resources and facilities were compiled for city residents to inform them on facilities available during the lockdown. For example, the city of Baltimore displayed an online user-friendly interactive map of the city (Fig. 3.10) in March 2020 with information on resources for children, seniors, people with compromised

Fig. 3.6 International platform for city COVID-19 data. (Source: https://unhabitat.citiiq.com/)

immune systems, and people who were uninsured at that time. The map allowed the residents to locate resources, including youth and senior food sites, healthcare clinics, and shopping options for seniors by City Council districts.

Likewise, LA county had several dashboards to display facilities and resources. One of these on education displayed the status of school opening, whether open or closed, as well as their mode of operation, whether it was operating on-site or by distance education or online learning mode (Fig. 3.11).

Social Distancing GIS was also utilized in managing social distancing after relaxation in lockdown to determine and promote social distancing. Two good examples for the management of social distancing in UK included (a) control of overcrowding on beaches and (b) analyses of pavement widths for social distancing (SmartCitiesWorld 2020). Esri UK carried out a spatial analysis of UK's ten most popular beaches including Bournemouth and Brighton by geographic information system (GIS) software to find how many people could fit on the sand while adhering to social distancing guidelines. The study revealed that only about 78,000 people could be accommodated following the social distancing rules on beaches in Bournemouth, whereas 500,000 people had descended there in the month of June 2020. The method used was based on placing a person inside a 2 m diameter circle following the 2 m social distancing rule. Another 2 m was added between circles to allow people to move around on the beach (The Geospatial 2020). In another study, a pavement width dataset was developed for social distancing. The GIS analysis revealed that in the UK, only 30% of pavements were suitable to follow social distancing rules. Around 30 local authorities started using the pavement maps as a part of the COVID-19 management. A web survey app was also developed for estimating capacity of parks (The Geospatial 2020). An interactive map highlighting real-time crowding status of a park in Singapore has already been displayed above in Fig. 3.5. Visitors to this park could pull up the map before heading there to see the crowding status.

Fig. 3.7 County of Los Angeles COVID-19: GIS dashboards. (Source: https://covid19.lacounty.gov/dashboards/)

Cities in developing countries generally lagged in utilizing GIS during the pandemic, but even they utilized it based on their technological capacity. The mayor's office in Bogotá, Colombia, for example, monitored the spread of the virus in each neighborhood of the city based on movements, travel history, and contact tracing (Fig. 3.12). In Indonesia, staff at the geospatial agency Badan Informasi Geospasial (BIG) conducted spatial analysis with daily reports on the BIG website. Pakistan National Coronavirus Dashboard showed virus situation in the city of Lahore (Fig. 3.13).

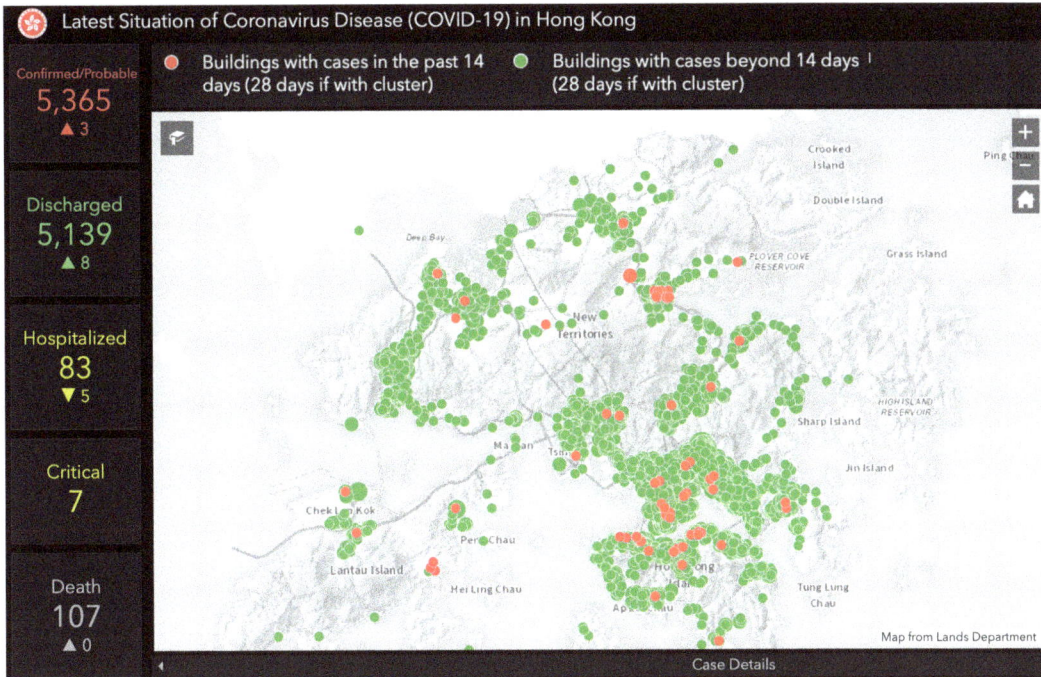

Fig. 3.8 Coronavirus dashboard in Hong Kong displaying disease situation. (Source: Normile (2020))

Fig. 3.9 Canadian dashboard displaying high-risk communities in Toronto. (Source: https://covid19canada.maps.arc-gis.com/apps/opsdashboard/index.html#/f781ff7d6f6d4a1d925aacef97330d93)

3.2.2.3 Satellite Navigation and Communication-Based Devices

Mobile telephone and other devices embedding satellite navigation also opened a new avenue to deal with the pandemic by contact tracing. This method was used in Taipei (Fig. 3.14) for tracing those locals who had interface or contact with passengers from the Diamond Princess Cruise Ship affected by the coronavirus (Chen et al.

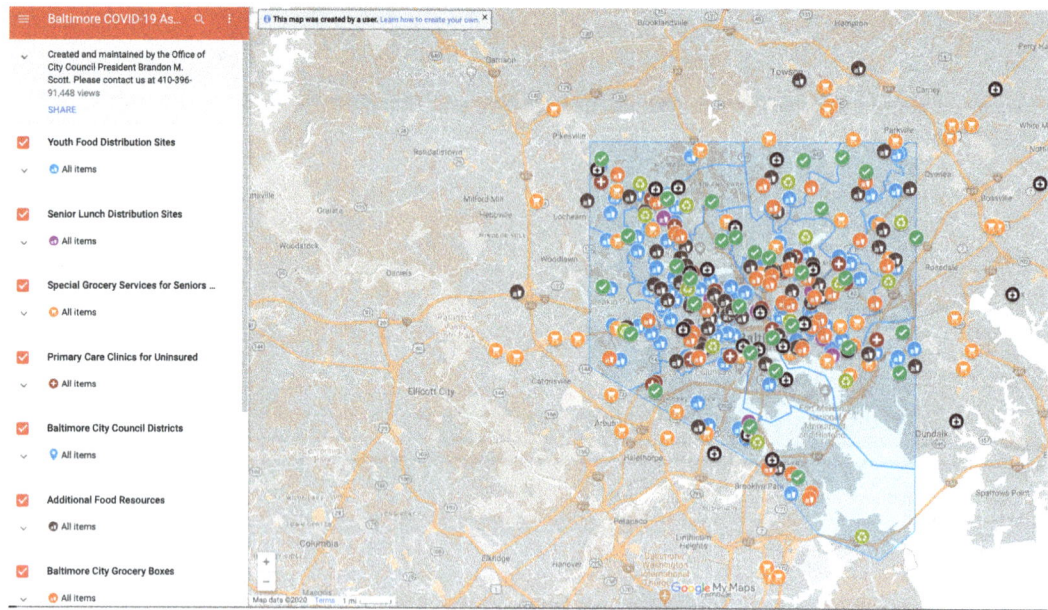

Fig. 3.10 Community resources in Baltimore. (Source: https://www.google.com/maps/d/u/0/viewer?mid=1LSfov2kKi_ipyJpj4D2L4zm4mtS2bRbu&ll=39.304348193682976%2C-76.64160660546527&z=12)

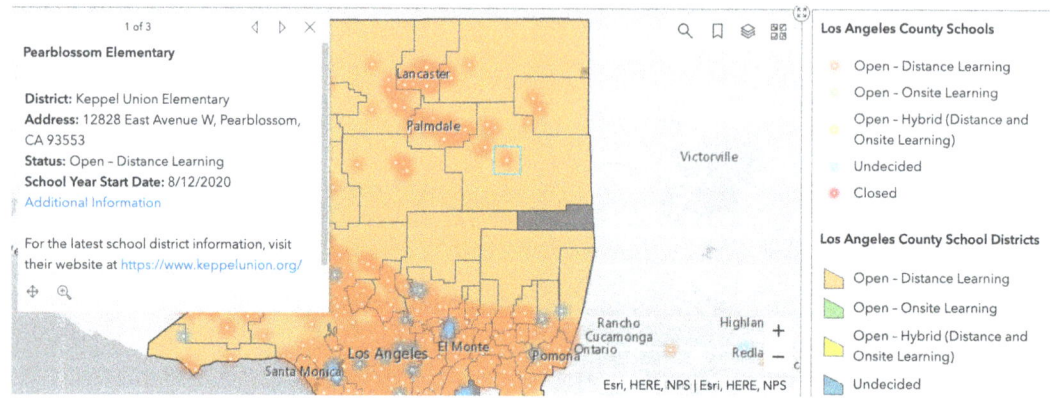

Fig. 3.11 Los Angeles County schools opening and teaching mode status. (Source: LA County, https://experience. arcgis.com/experience/b52442649c6546fb8798eb3e871ea854)

2020). The method was cross verified with bus routes followed by the passengers and by their credit card use. China, the USA, Singapore, Poland, Israel, and South Korea are some other countries that used smartphone tracking.

While there is no denying that the seriousness of the pandemic had some justification for the contact tracing measures, it is however essential not to completely override privacy. The ramification of adoption of such steps by countries and corporations could be ominous for citizen liberty and make surveillance a new normal, even in the most democratic establishments (Chaturvedi 2020a).

The progress in space technology and its application has also facilitated considerably monitoring mega risks. There were more than 4500 satellites in orbit, registered in more than 80 countries in 2020, and public and private investments are increasing in the technology

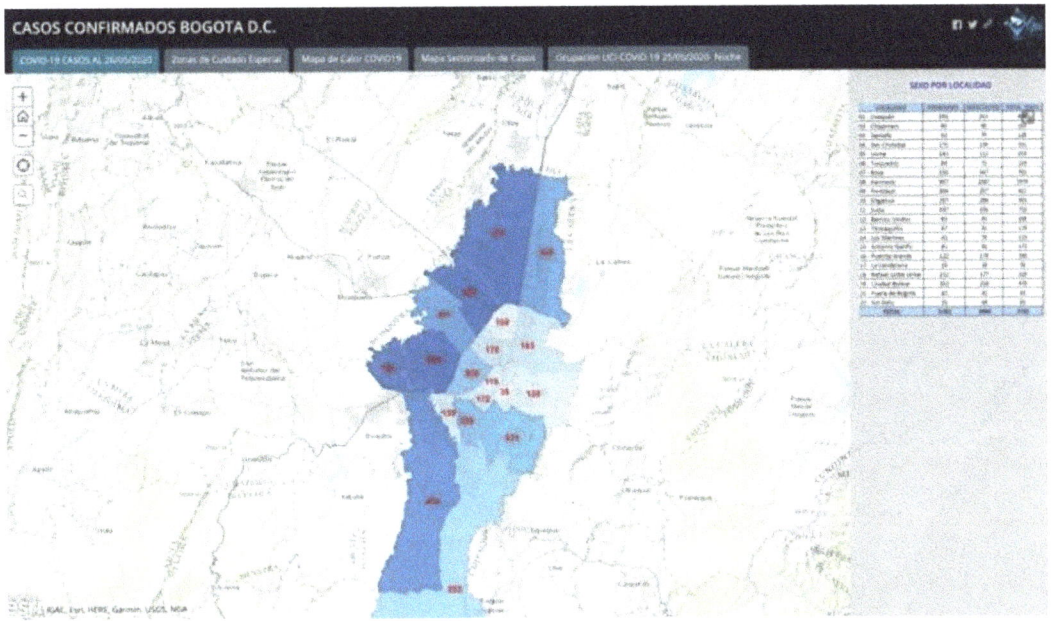

Fig. 3.12 Bogota (Colombia): Number of COVID-19 cases, deaths, and recovery by age and sex. (Source: Rabiee (2020))

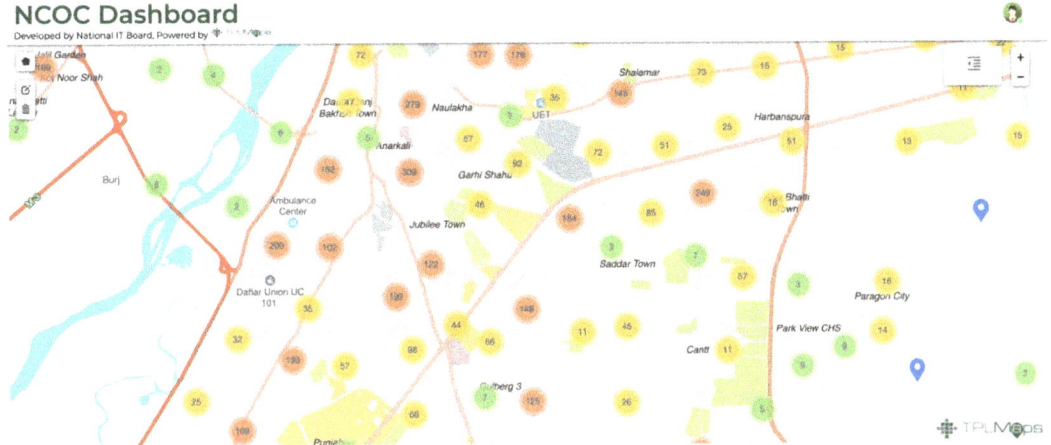

Fig. 3.13 National Corona Dashboard in Pakistan displaying virus situation in Lahore City. (Source: Warriach (2020))

(UNESCAP 2020). The data obtained from the earth observation satellites as well as navigation satellite system has been used effectively in the pandemic at least in some countries. Navigation technology, for example, proved extremely useful and was used extensively in China—from tracking patients' position to the monitoring of cargo transport. Global Navigation Satellite System (GNSS) was utilized for guiding drones during large-scale disinfection missions (Eurisy 2020). Further, satellite positioning and other modern technologies proved to be a major asset for Chinese cities in containing the highly contagious coronavirus. Excluding Wuhan and its surrounding cities in Hubei Province, which together accounted for 83% of all Chinese cases of COVID-19 by March 2020, the Chinese authorities using technology and all other means did a

Fig. 3.14 Routes of
passenger from the
Diamond Princess
Cruise Ship and contact
tracing. (Source: Chen
et al. (2020))

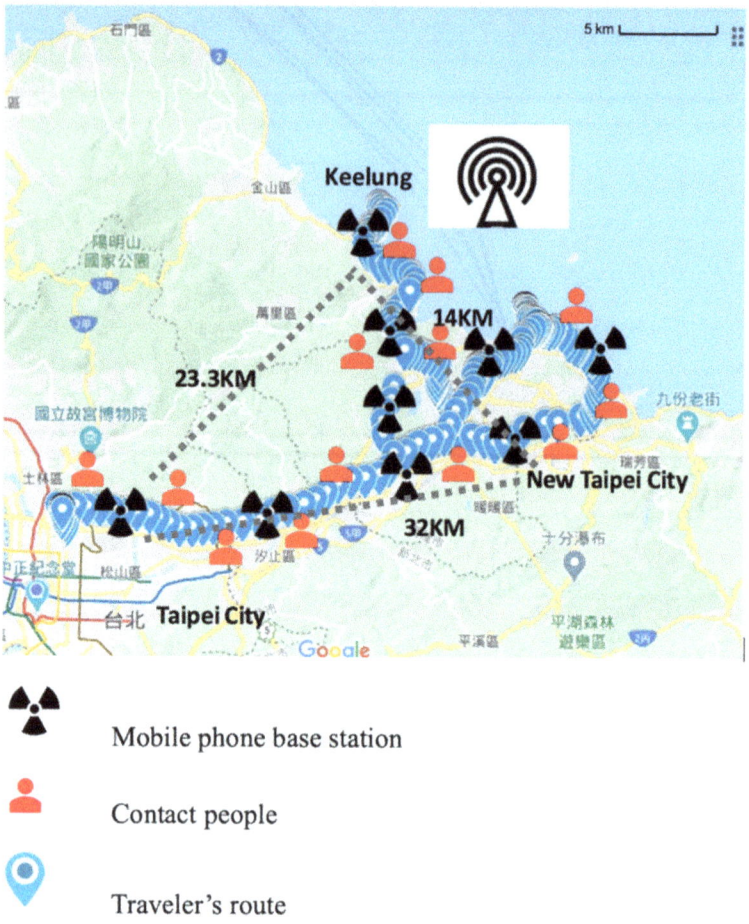

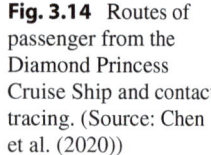 Mobile phone base station

Contact people

Traveler's route

remarkable job in containing the disease, which did not spread exponentially to other cities outside Hubei province (Rice Kinder Research Institute 2020).

Eurisy (2020) and Chaturvedi (2020a) have given details on how satellite or space-based technologies helped control the virus in Chinese cities. They especially mention the use of BeiDou Navigation Satellite System (Chinese own GNSS), which proved crucial in tracking patients' position to the monitoring of cargo transport. As mentioned earlier, it was also used to guide drones in undertaking large-scale disinfection missions. China was able to dispatch more than 10,000 disinfection drones to perform nationwide aerial disinfectant spraying with precise accuracy. Moreover, the HaiGe Smart Epidemic Prevention Management Platform, a BeiDou-based system for health monitoring, had the

capability of not only displaying real-time maps of infested areas but also guiding drivers to navigate out of these localities. The system's capability allowed the auto drivers to avoid contagious areas and follow safe routes. In fact, China's transport management department equipped all vehicles going to Wuhan with BeiDou mobile satellite positioning devices for the purpose. Since satellite connection works in remote locations, it was also used for real-time communication with isolated drivers. In addition, contactless delivery missions, operated by indoor autonomous robot, relied extensively on satellite navigation for the delivery of medical supplies and groceries in the designated hospitals and to local communities during quarantine (Eurisy 2020).

Satellite-based remote sensing from earth observation satellites in the form of images also provided big time help to cities in locating and tracking the construction of testing and health facilities, new hospitals, and other public health infrastructure. For example, Chinese cities used images from Gaofen, a constellation of high-resolution satellites in nonstop monitoring of construction of makeshift hospitals, which were constructed at swift pace to deal with the spread of the outbreak. Wuhan University actively collected and analyzed multiple data sources and identified sites best suitable for hospital locations (Chaturvedi 2020a). Satellite images were also used to observe air pollution during the pandemic. The images on NASA's earth observatory show that after the outbreak and subsequent lockdown, air quality in Chinese and Italian cities prompted pollution to plummet. Among others, nitrogen dioxide emissions declined considerably following quarantine measures. A reduction in air pollution to this extent has never happened in such a short period in cities across the world (Eurisy 2020). Like the air quality, remote sensing was also used in tracking the movements of goods and vehicles on terminals, ports, and elsewhere, particularly for measuring the economic impacts of the outbreak and to help in monitoring the pace of eventual recovery and accelerating recovery efforts.

3.2.2.4 Robotics and Drones

Robots and drones have also benefitted from growth in information technology especially rapid innovations and advancements in smartphones, which have brought cheap cameras and sensors, fast wireless communications, as well as powerful, smaller computer chips. Likewise, advancement in machine learning has added software that are making robots better informed on their surroundings and enabled them to make better decisions (The Economist 2022). Armed with these innovations, robots were deployed extensively in Chinese cities during the pandemic. They assisted in many activities ranging from preparing meals at hospitals to working as waiters in restaurants, spraying disinfectants to vending rice, and dispensing hand sanitizers. In many hospitals, robots were also performing diagnosis and conducting thermal imaging. Shenzhen-based company Multicopter used robots to trans-

port medical samples. One hospital in Wuhan, the epicenter of the outbreak, was being staffed entirely by robots (Chaturvedi 2020a). Regarding drones, they were used as rescue measure in severely affected areas, where humans were at a risk of catching the virus. In such areas, they were transporting both medical equipment and patient samples, saving time and enhancing the speed of deliveries while preventing contamination of medical samples (Chaturvedi 2020a). Extensive use of robotics in Chinese cities during the pandemic has shown their great potential for future use in epidemics and pandemics.

3.3 Monitoring and Management of Climate Change

Geospatial technologies such as remote sensing and GIS have also been of great help in dealing with climate change crises in cities from monitoring to mapping, modeling, mitigation, and adaptation.

3.3.1 Multidimensional Use of GIS

Some of the areas in which geospatial technologies have been deployed in climate change-related work are given below with case studies in parenthesis:

- Measuring land surface temperatures and heat island effects and their mitigation (Bennet 2019)
- Exploring the causes and impacts of climate change in urban areas (Ali-Toudert and Ji 2019; Tomlinson et al. 2011)
- Simulation, scenario building, and forecasting (Szymanowski and Kryza 2009)
- Vulnerability and risk mapping (Feyissa et al. 2018)
- Carbon sequestration and renewable energy development to achieve carbon neutral goal (Esri 2010)
- Greening buildings, roofs, and infrastructure for climate change mitigation (Berger 2013)

Mapping Portal

Explore the Interactive
Decision-Support Tool

Story Map

Green Infrastructure Planning
for a Low-Carbon, Resilient
Boston

Downloadable PDF Maps

View key project maps

User Guide

Learn how to use the
Decision-Support Mapping
Tool

ABOUT

The City of Boston is tackling the challenges associated with climate change head on. In a city where a densely populated highly urbanized meets the sea, the city is doing the hard work to prepare for sea level rise, increasing urban temperatures, greater precipitation, and other risks through their robust and comprehensive climate analysis and action plan, Climate Ready Boston. The Trust for Public Land's national Climate-Smart Cities program is providing key planning and decision-making support to help the city achieve the goals laid out in Climate Ready Boston. Together with city and community partners, we're bringing cutting-edge science, Geographic Information Systems (GIS) planning, and innovative design to prepare the entire region for a climate-resilient future with a particular emphasis on underserved populations and critical at-risk infrastructure.

Our Climate-Smart Cities process bring together a team of experts to research, design, and build the infrastructure and tools that help increase our resilience to climate change. Our strategy is to:

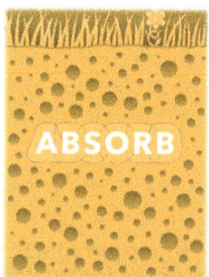

Fig. 3.15 City of Boston: climate dashboard. (Source: https://web.tplgis.org/boston_climate/)

- Sea level rise (Malik and Abdalla 2016)
- Adaptation to climate change (Becker 2018; City of Boston 2019b)

The geospatial technology's variety of uses in climate crisis has depended on the need of a city or community—whether it was used for mitigation, adaptation, communication, planning, or management. It has also varied by field or theme—to assess vulnerability to hazard or for health, social welfare, business, and economics, etc. Whatever the case, it has led to the development of some extremely good case studies and important observational, analytical, and display tools.

Fig. 3.16 Boston: land surface temperature in open spaces overlaid with parks in green. (Source: City of Boston (2019a))

Fig. 3.17 Boston: 2070 coastal storm flooding conditions overlaid on top of areas with a higher concentration of older adults in green color. (Source: City of Boston (2019a))

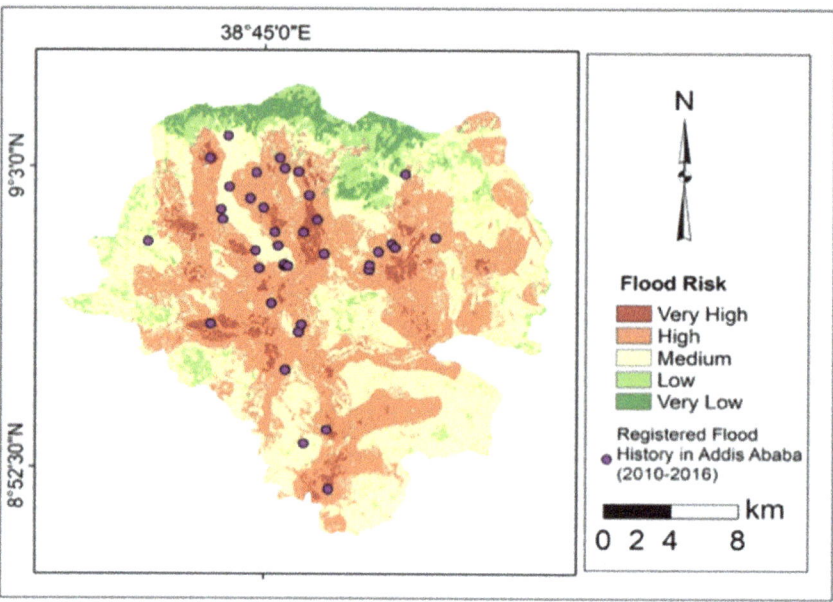

Fig. 3.18 Flood-risk map of Addis Ababa overlaid by frequency of flood hazard records. (Source: Feyissa et al. (2018))

3.3.2 Observational, Analytical, and Display Tools

The cities/local governments and professional institutions have used these tools including computer cartography and apps to develop interactive maps and illustrated information and placed these on dashboards to:

- Highlight the impact of climate change
- Prepare the community for climate change
- Share critical information with residents and to consult them in planning/implementation
- Emphasize the magnitude of risk to guide people for appropriate actions
- Highlight climate action plans or mitigation and adaptation strategies
- Establish mechanisms for early warning and provision of shelter in emergency
- Enabling businesses to adapt effectively
- Investing in climate resilience

Since climate apps have been used over several years, they were considerably improved over time in terms of information and data display and made more user-friendly. The concerns at local and city level vary depending upon the type of climate risks or hazards and vulnerability of the communities and their resources—for example, some communities are more vulnerable to one risk like rising temperatures and others to sea level rise, still others to cyclones and so on. Likewise, poor or dwellers of slums are more vulnerable compared to well-off communities. In addition, cities also vary in their resources and technology to deal with the crisis. Local and city government efforts have therefore varied—from carefully worked-out adaptation plans in some developed countries at one end of the scale to meet emergency needs when faced with climate-related disasters in cities of poor developing countries (on firefighting basis) at the other end of the scale. Over time however they have led to the development of many best practices.

Mapping Climate Change Maps have traditionally been used to monitor weather conditions. Many cities of developed countries have developed interactive climate-related maps and displayed these on dashboards. Often these are used as an avenue to communicate with stakeholders and residents of the city on climate-related issues. An excellent example of this is the dashboard of the city of Boston (City of Boston 2019a) shown in Fig. 3.15. The story map on the dashboard highlights how the city government is building defense against climate change by using flooding, heat, and social vulnerability information in resilience planning. Another dashboard, Map Explorer, features spatial data from Climate Ready Boston (City of Boston 2019b). It shows areas in Boston that are projected to be at risk of flooding (due to sea level rise and increased precipitation) and extreme heat (due to rising temperatures and the urban heat island effect). It allows combining layers of risks with demographic features and social factors to better understand vulnerability of the city as well as its parts through interactive maps (Fig. 3.16 and 3.17).

The first map (Fig. 3.16) compares land surface temperatures with open spaces. At this detailed resolution, one can notice that areas around parks tend to be cooler. To compare specific data sets, one or more data layers can be turned on or off in the "Layers" tab. The second map shows 2070 coastal storm flooding conditions overlaid on top of areas with a higher concentration of older adults (+65) shown in green color. These aged people have excessive physical vulnerabilities in a climate event as they often suffer from higher rates of chronic medical illness than the rest of the population and may also have some functional limitations in an evacuation scenario.

Mapping Climate Vulnerability A climate vulnerability GIS study conducted in Addis Ababa provides a good example of the case. The study (Feyissa et al. 2018) quantified, mapped, and categorized climate change vulnerability in terms of exposure, sensitivity, and adaptive capacity. It used 15 layers of vulnerability indicators with biophysical, social, and economic layers at sub-city level. The study followed the IPCC's (2014) climate change vulnerability analyses and Sullivan and Meigh (2005) model for selection of indicators developed to prepare climate change vulnerability index. Out of all the maps produced by the study, one showed flood-risk vulnerability (Fig. 3.18) and the other adaptive capacity (Fig. 3.19).

Vulnerabilities have also been displayed on websites and dashboards often along with warning

systems. For example, Digital Earth Australia Hotspots (GeoScience, Australia 2019) is a monitoring system for bushfires in the country that provides timely information on hot spots or risk zones to Emergency Service Managers across Australia (Fig. 3.20). The mapping system uses satellite sensors to detect areas that are producing high levels of infrared radiation (called Hotspots), allowing Emergency Service Managers to identify potential fire locations.

Mapping and Tracking Climate-Related Extreme Events Mapping of climate-related extreme events and tracking them in space and time have also become quite common. Based on images from airborne sources like satellite images, changes in intensity and extent of climate-related extreme events have been monitored through dynamic mapping. For example, using the method, NASA (2021). showed the way climate change is affecting forest fires by supercharging heat waves and droughts in California (Fig. 3.21). Since 1970, the total area burnt in the State increased from 3% to 11%, - the shift toward larger fires is clear in the decadal map. Two recent incidents of 'gigafires' (Gabbert 2020) - the Dixie fire in 2021 and the fire of August 2020 stand out for their size (NASA 2021) - Each of these burned nearly 1 million acres - an area larger than Rhode Island.

Aerial photographs and satellite data/images obtained by sensors aboard earth observing satellites have also allowed measurements of land surface temperature (Fig. 3.22) and greenhouse gas emissions. Currently, there are some 162 satellites in-orbits that measure various indicators related to climate change. New generation of satellites while considerably enhancing optical and temporal resolutions have also led to great improvements in weather forecasting, climate modeling, and obtaining real-time data (Chaturvedi 2020b). In addition to predicting future of cities in relation to climate change, this data and information has helped considerably in tackling extreme climate events such as floods, forest and bush fires, and heat waves. The flood risk maps drawn from such data and information

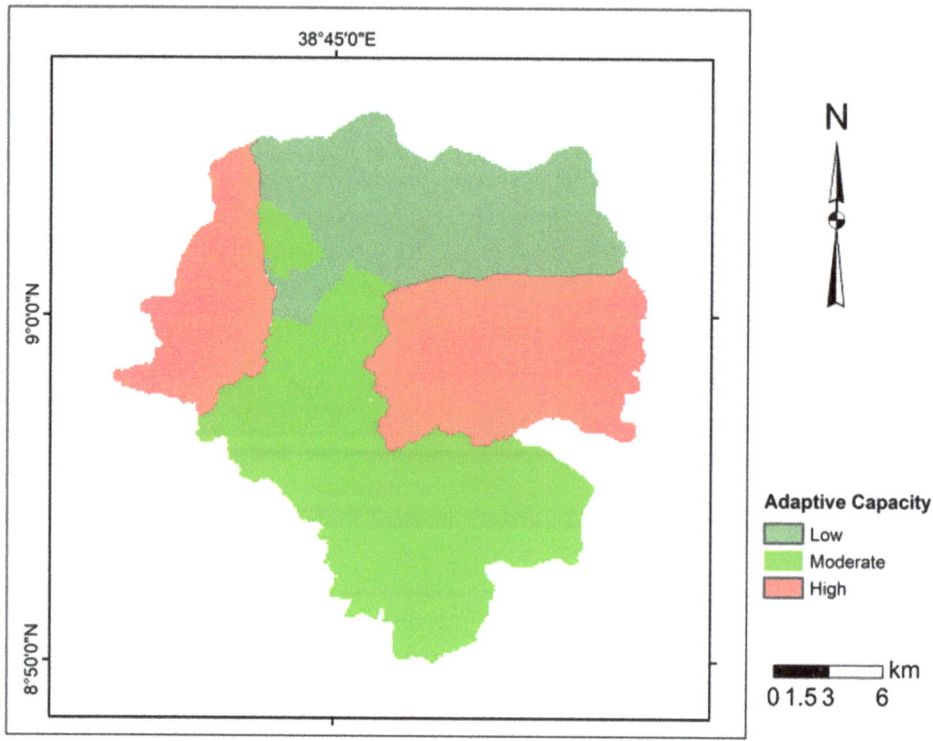

Fig. 3.19 Adaptive capacity map of Addis Ababa. (Source: Feyissa et al. (2018))

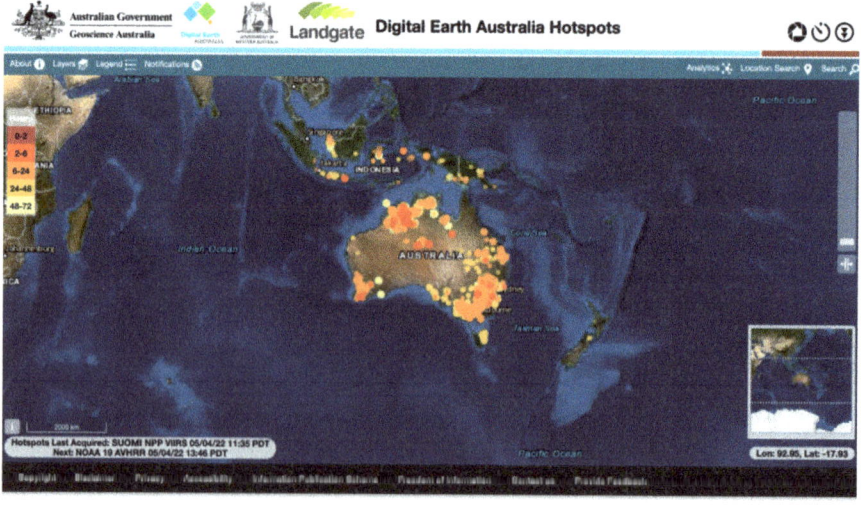

Fig. 3.20 Australia: hotspots portal. (Source: GeoScience Australia, https://hotspots.dea.ga.gov.au)

for four cities of the USA including Columbia (South Carolina), Houston, New York, and Tampa (Florida) along with Silicon Valley are available on the Temblor website (Temblor 2016). Remotely sensed data and images are also very useful in an emergency response to climate-related extreme events—for assessing evacuation and other civilian needs in anticipation of an expected disaster. Thus, overall, combining satellite images and data from other sources, a rich

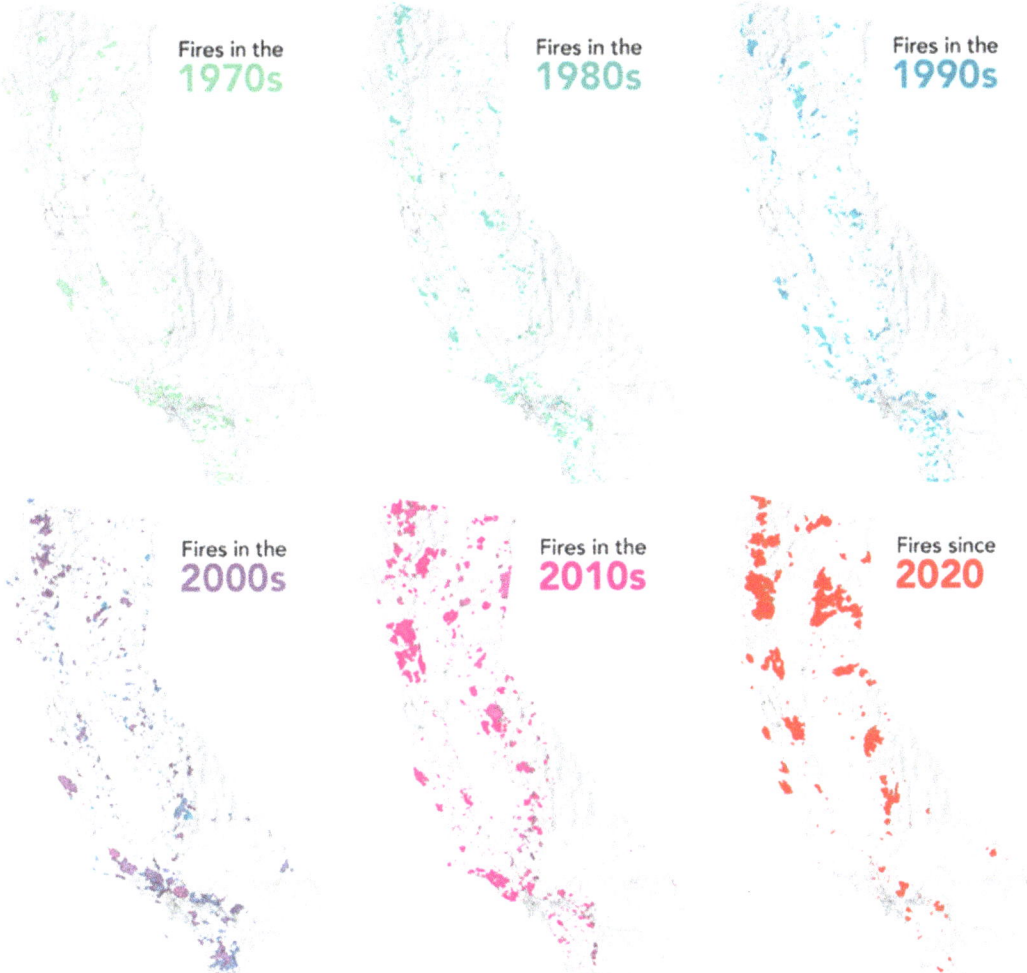

Fig. 3.21 Growing Intensity of Forest Fires in California (USA). Note: Only 3% of the state's land burned between 1970 and 1980 compared to 11% from 2010 to 2020. (Source: NASA (2021))

geospatial dataset, and multilayer maps can be generated for monitoring and managing risks—assessing vulnerabilities, enforcing restrictive measures in hazard-prone zones, and better designing critical infrastructure to face climate-related hazards.

3.3.3 Monitoring and Preparedness to Face Risks

Monitoring and assessment tools that keep track of risks like pandemic and climate change can also help build safer urban centers, minimize the threat of global and community spread of diseases, and make appropriate preparation to face any possible disaster. This information becomes more useful when combined with indices used to measure the difference in vulnerability between various part of cities. Global Health Security (GHS) Index (JHU and NTI and EIU 2019), developed in 2019, is the first comprehensive index to assess and benchmark health security and related capabilities at the country level to prevent and mitigate epidemics and pandemics. Likewise, the Notre Dame Global Adaptation Index (ND-GAIN) brought together over 74 variables to form 45 core indicators to measure vulnerabilities

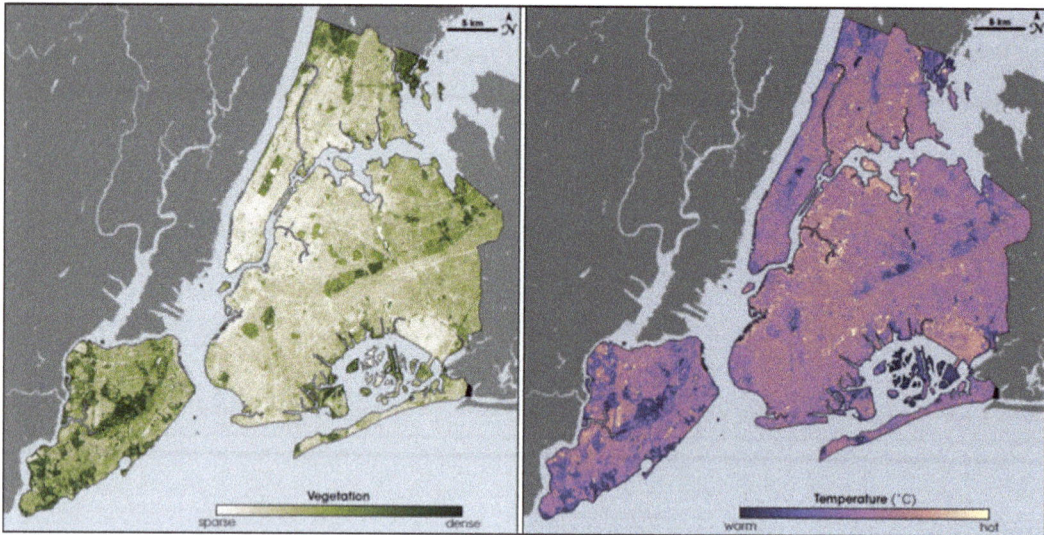

Fig. 3.22 New York City satellite images depicting temperatures and cooling effect of plants. (Source: NASA (undated))

Note: While displaying land surface temperature, the images also show the cooling effect of plants. Dark-green patches in the left image

and readiness of 184 countries to climate disruptions (Chen et al. 2015). Both these studies provide stimulus for developing health security and climate-related security indices for cities.

3.4   Conclusion

Monitoring, assessment, and management of risks have seen many breakthroughs and improvements over the last few years with the progress in the application of geospatial technology-related tools. They have made headway from digitizing and analyzing information and presenting it through maps, graphics, and satellite images to interactive dashboards in multilayers of data or geographical information system. This has paid off very well during the COVID-19 pandemic and in monitoring, mitigating, and adapting to climate change crisis. Their utility has particularly remained extremely valuable throughout the risk management cycle of climate-related extreme events from preparedness and pre-disaster capacity building to post-disaster relief and response.

As the pandemic spread within and across cities, communities from local governments to pro-

fessional and academics and to media and businesses used the geospatial technology extensively for various purposes from planning and management to communicating important information, assisting vulnerable population, seeking public cooperation, and running businesses. The governments and local authorities relied on these for identifying and displaying disease outbreaks spots and the patterns of disease spread. Additionally, they were used for surveillance, quarantining, and social distancing. In cities, multi-scale dynamic spatiotemporal maps were displayed on dashboards to display the rapidly changing situation and highlighting the need for the enforcement of administrative measures and to assist citizens in locating facilities. Their benefits to local government were immense in implementing lockdowns and other social distancing measures and evaluating control and preventive measures as well as in reopening after lockdowns.

In terms of management of medical resources, they were used to balance supply and demand by assessing supply chains, as well as transportation and mobility mechanisms. In addition, the geospatial tools enabled communication with city residents through interactive dashboards and maps—to impart information on local circumstances; on location of food assistance programs; and for emergency medical facilities and educational and other resources, etc. Last but not the least, they are being utilized in recovery and rehabilitation planning and implementation and for evaluating the socioeconomic and environmental impacts.

Geospatial tools, in their application, utilize data from multiple sources including temporal data from earth observation satellites as well as navigation satellite systems. Remote sensing from earth observation satellites in the form of images provided big time help to cities during the pandemic for locating and tracking the construction of testing and health facilities like makeshift hospitals and other public health infrastructure. Likewise, navigation technology proved extremely beneficial in contact tracing, tracking patients' position, monitoring, movement of medical supplies, and cargo transport. It was also applied in guiding drones assigned for large-scale disinfection missions using Global Navigation Satellite System (GNSS). Additionally, satellite positioning and other modern technologies helped in containing the highly contagious virus particularly in China.

Geospatial technologies such as remote sensing and GIS have also proved extremely helpful in dealing with climate crisis from monitoring to mapping, modeling, mitigation, and adaptation. Some of the areas in which geospatial technologies have been deployed in climate change-related work include measuring land surface temperatures and heat island effects in cities and their mitigation; exploring the causes and impacts of climate change in urban areas; simulation, scenario building, and forecasting; greenhouse gas monitoring; vulnerability and risk mapping; carbon sequestration and renewable energy development to achieve carbon neutrality goals; greening buildings and climate proofing roofs and infrastructure; sea level rise monitoring; etc. New generations of satellites have considerably enhanced optical and temporal resolutions leading to improvements in weather forecasting, climate modeling, and obtaining real-time data.

The use of geospatial technologies in tackling climate crisis has largely depended on the need of a city or community—whether it was being used for monitoring and assessment, preparedness, mitigation, adaptation, emergency relief, communication, and dissemination of information, planning, and implementation for recovery and rehabilitation or developing capacity for resilience. Remotely sensed data has proved particularly useful in emergency response, for example, in assessing evacuation and other civilian needs in anticipation of an expected disaster. Moreover, combining satellite image data with those from other sources can generate a rich geospatial dataset and multilayer maps for city or cities' preparedness. This becomes an asset in monitoring and managing climate-related risks, assessing vulnerabilities of communities to the risks, enforcing restrictive measures in hazard-prone zones, and better designing critical infrastructure to face these.

The analysis in this chapter shows that the availability and accessibility of quality geospatial data is of fundamental importance in managing both mega risks. Geospatial technologies together with digital innovation including smartphones and frontier technologies such as big data, Internet of Things (IoT), crowdsourcing, and cloud computing are increasingly being used in integrated and innovative ways to address disaster risk management and resilience (see Chap. 4). However, while creating opportunities, they are also generating new challenges particularly in the form of availability and accessibility due to digital divide, which will be discussed in greater detail in the next chapter. Nonetheless, it is important to mention here that the existing challenges and inequalities have sharpened with the introduction of novel technologies, especially in terms of access to digital hardware and software. What to say of cities of least developed and Small Island Developing States, many cities in developing countries lack even basic Internet connectivity. With the new wave of innovations in geospatial and digital technologies, it is vital that these less connected countries are provided access, support, and guidance through international cooperation so that they can benefit from these innovative technologies.

References

Ali-Toudert F, Ji L (2019) A multi-scale and GIS-based investigation of climate change effects on urban climate and building energy demand for the city of Stuttgart. J Phys Conf Ser 1343(1):012008. https://doi.org/10.1088/1742-6596/1343/1/012008

Becker S (2018) How New Orleans is using map for climate change. https://www.govtech.com/fs/How-New-Orleans-is-Using-Maps-to-Plan-for-Climate-Change.html

Bennet N (2019) Urban heat islands – a natural solution. https://storymaps.arcgis.com/stories/5f8ccbcbc1cd44f4b9e4e035eff90979

Berger D (2013) A GIS based suitability analysis of the potential for rooftop agriculture in New York city. Ph D thesis, Columbia University

Boulos M, Warren J, Gong J, Yue P (2010) Web GIS in practice VIII: HTML5 and the canvas element for interactive online mapping. Int J Health Geogr 9(1):14. https://doi.org/10.1186/1476-072X-9-14

Bralic M (2020) The hidden patterns behind the COVID 19 map of Dublin. http://newslab.ie/ddjucd/hidden-patterns-behind-the-covid-19-map-of-dublin/

Burton K (2020) Mapping the curve: how GIS is helping the Covid-19 response. Geographical. https://geographical.co.uk/people/development/item/3676-how-gis-is-helping-the-covid-19-response

Chaturvedi A (2020a) The China way; use of technology to combat covid 19. Geospatial World. https://www.geospatialworld.net/article/the-sino-approach-use-of-technology-to-combat-covid-19/

Chaturvedi A (2020b) How satellite imagery is crucial for monitoring climate change. Geospatial World. https://www.geospatialworld.net/blogs/satellites-for-monitoring-climate-change/

Chen C, Noble I, Hellmann J, Coffee J, Murillo M, Chawla N (2015) University of Notre dame global adaptation index, country index technical report. https://gain.nd.edu/assets/254377/nd_gain_technical_document_2015.pdf

Chen CM, Jyan HW, Chien SC, Jen HH, Hsu CY, Lee PC, Lee CF, Yang YT, Chen MY, Chen LS, Chen HH, Chan CC (2020) Containing COVID-19 among 627,386 persons in contact with the diamond princess cruise ship passengers who disembarked in Taiwan: Big data analytics. J Med Internet Res 22(5):e19540. https://doi.org/10.2196/19540

City of Boston (2019a) Climate smart cities Boston. https://web.tplgis.org/boston_climate/

City of Boston (2019b) Climate ready Boston map explorer. https://www.boston.gov/departments/environment/climate-ready-boston-map-explorer

Coccia M (2020) Factors determining the diffusion of COVID-19 and suggested strategy to prevent future accelerated viral infectivity similar to COVID Sci. Total Environ 729:138474, https://www.ncbi.nlm.nih.gov/pmc/articles/PMC7169901/

Cooke K (2020) Economic restoration in the wake of COVID 19. https://www.naco.org/blog/economic-restoration-wake-covid-19

Diffenbaugh NS, Field CB, Wong-Parodi G (2020) The covid 19 lockdowns: a window into the earth system, earth review. https://www.nature.com/articles/s43017-020-0079-1

Dong D, Du H, Gardner l. (2020) An interactive web-based dashboard to track coronavirus in real time. Lancet Infect Dis 20(5):533–534. https://doi.org/10.1016/S1473-3099(20)30120-1

Esri (2010) GIS best practices: GIS for renewable energy. https://www.gisday.com/content/dam/esrisites/en-us/about/events/gis-day/renewable-energy.pdf?rmedium=gisday-www

Esri (2020a) UK analysis reveals beach capacity ahead of holiday season. https://www.geospatialpr.com/2020/07/02/esri-uk-analysis-reveals-beach-capacity-ahead-of-holiday-season/

Esri (2020b) Map of pavement widths helps local authorities with social distancing plans. https://www.directionsmag.com/pressrelease/986

Esri (2020c) Geographic information systems for coronavirus planning and response. Esri White Paper. https://www.esri.com/content/dam/esrisites/en-us/media/pdf/geographic-information-systems-for-coronavirus-planning-response-white-paper.pdf

Eurisy (European International Space Year Association) (2020) What we can learn from the coronavirus crises with satellite data. https://www.eurisy.eu/what-we-can-learn-from-the-corona-crisis-with-satellite-data_46/?__ART_URL__=what-we-can-learn-from-the-corona-crisis-with-satellite-data_46

Feyissa G, Zeleke G, Gebremariam E et al (2018) GIS based quantification and mapping of climate change vulnerability hotspots in Addis Ababa. Geoenviron Disasters 5:14. https://doi.org/10.1186/s40677-018-0106-4

Gabbert (2020) Where did the term "gigafire" originate. https://wildfiretoday.com/2020/10/07/where-did-the-term-gigafire-originate/

GeoScience Australia (2019) Digital health australia hotspots wms. https://researchdata.edu.au/digital-earth-australia-hotspots-wms/1239334

Geraghty E, Lanclos R (2020) Smart maps guide COVID-19 investigations and actions, and monitor effectiveness. https://www.esri.com/about/newsroom/blog/gis-helps-guide-covid-19-response/

Gibson L, Rush D (2020) Novel coronavirus in Cape Town Informal Settlements: feasibility of using informal dwelling outlines to identify high risk areas for COVID-19 transmission from a social distancing perspective. JMIR Public Health Surveill. 6(2):e18844. https://doi.org/10.2196/18844

GISgeography (2020) 1000 GIS applications and uses – how GIS is changing the world. https://gisgeography.com/gis-applications-uses/

Gleason BL, Foster S, Wilt GE, Miles B, Lewis B, Cauthen K, King M et al (2014) Geospatial analysis of household spread of ebola virus in a quarantined village – Sierra Leone, Abstract. U.S. National Library of Medicine. https://www.ncbi.nlm.nih.gov/pubmed/28826426

Gong J, Geng J, Chen Z (2015) Real-time GIS data model and sensor web service platform for environmental data management. Int J Health Geogr 14(1):2. https://doi.org/10.1186/1476-072X-14-2

Gross B, Zheng Z, Liu S, Chen X, Sela A, Li J, Li D, Havlin S (2020) Spatio-temporal propagation of COVID-19 pandemics. EPL (Europhys Lett) 131(5):58003

RS21-Revolutionary Socialism in 21st Century (2021). https://www.rs21.org.uk/about/

John Hopkins Center for Health Security (JHU), Nuclear Threat Initiative (NTI), The Economic Intelligence Unit (EIU) (2019) Global health index. https://www.ghsindex.org/wp-content/uploads/2019/10/2019-Global-Health-Security-Index.pdf

Kamel Boulos MN, Geraghty EM (2020) Geographical tracking and mapping of coronavirus disease COVID-19/severe acute respiratory syndrome coronavirus 2 (SARS-CoV-2) epidemic and associated events around the world: how 21st century GIS technologies are supporting the global fight against outbreaks and epidemics. Int J Health Geogr 19:8. https://doi.org/10.1186/s12942-020-00202-8

Kandwal R, Garg PK, Garg RD (2009) Health GIS and HIV/AIDS Studies: Perspective and Retrospective. Journal of Biomedical Informatics 42(4):748–755. https://doi.org/10.1016/j.jbi.2009.04.008

Kaser R (2020) 7 best coronavirus dashboards to map the spread of COVID 19. https://thenextweb.com/corona/2020/03/10/best-coronavirus-dashboards-map-spread-covid-19/

Kent J (2020) Data platform tracks hospital bed capacity during COVID 19 outbreak. https://healthitanalytics.com/news/data-platform-tracks-hospital-bed-capacity-during-covid-19-outbreak

Koch T, Kenneth D (2005) Cartographies of disease: maps, mapping, and medicine. ESRI Press, Redlands, CA. https://esripress.esri.com/storage/esripress/images/321/cartofdisease_two_chap.pdf

Lakhani A (2020) Which Melbourne metropolitan areas are vulnerable to COVID-19 based on age, disability and access to health services? Using spatial analysis to identify service gaps and inform delivery

Lall S, Wahba S (2020) A journey to economic development. https://www.worldbank.org/en/news/immersive-story/2020/06/18/no-urban-myth-building-inclusive-and-sustainable-cities-in-the-pandemic-recovery

Leadbeater R (2020) COVID 19: Pennsylvania counties assess economic fallout with shared maps. Esri.

https://www.esri.com/about/newsroom/blog/covid-19-pennsylvania-counties-assess-economic-fallout/

Leger M (2020) COVID 19 response: GIS best practices in local government. Harvard Kennedy School. https://datasmart.ash.harvard.edu/news/article/covid-19-response-gis-best-practices-local-government

Lyseen AK, Nøhr C, Sørensen EM, Gudes O, Geraghty EM, Shaw NT, Bivona-Tellez C (2014) A review and framework for categorizing current research and development in health-related geographical information systems (GIS) studies. Yearb Med Inform 23(01):110–124. https://doi.org/10.15265/IY-2014-0008

Malik A, Abdalla R (2016) Geospatial modeling of the impact of sea level rise on coastal communities: application of Richmond, British Columbia, Canada. Model Earth Syst Environ 2:146. https://doi.org/10.1007/s40808-016-0199-2

NASA (2021) What's behind California's surge of fires, earth observatory. https://earthobservatory.nasa.gov/images/148908/whats-behind-californias-surge-of-large-fires

NASA (undated) Can you explain the urban heat island effect. https://climate.nasa.gov/faq/44/can-you-explain-the-urban-heat-island-effect/

National Cancer Institute (2006) Geographic information system (GIS) and cancer research. US Department of Health and Human Services

Normile D (2020) 'Suppress and lift': Hong Kong and Singapore say they have a coronavirus strategy that works. Science. https://www.sciencemag.org/news/2020/04/suppress-and-lift-hong-kong-and-singapore-say-they-have-coronavirus-strategy-works

Rabiee M (2020) COVID-19: a look at global geospatial challenges and achievements. Esri Blog. https://www.esri.com/about/newsroom/blog/covid-19-global-geospatial-challenges-achievements/

Rice Kinder Research Institute (2020) Cities and coronavirus: some thoughts. https://kinder.rice.edu/urbanedge/2020/03/19/cities-and-coronavirus-some-thoughts

Rossman H, Keshet A, Shilo S, Gavrieli A, Bauman T, Cohen O, Shelly E, Balicer R, Geiger B, Dor Y, Segal E (2020) A framework for identifying regional outbreak and spread of COVID-19 from one-minute population-wide surveys. Nat Med 26(5):634–638

Rothenberg R, Weaver SR, Dai D, Stauber C, Prasad A, Kano M (2014) A flexible urban health index for small areas. J Urban Health. https://doi.org/10.1007/s11524-014-9867-6

Samuelsson K, Barthel S, Colding J, Vetenskapsakademien K, Giusti M (2020) Urban nature as a source of resilience during social distancing amidst the coronavirus pandemic. Research Gate Pre-Print. https://doi.org/10.31219/osf.io/3wx5a

SmartCitiesWorld (2020) Analysis examines UK's beach capacity. https://www.smartcitiesworld.net/news/news/analysis-examines-uks-beach-capacity-5437

Su L, Hong N, Zhou X, He J, Ma Y, Jiang H, Han L, Chang F, Shan G, Zhu W, Long Y (2020) Evaluation of the secondary transmission pattern and epidemic prediction of COVID-19 in the four metropolitan areas of China. medRxiv. https://doi.org/10.1101/2020.03.06.20032177

Sullivan C, Meigh J (2005) Targeting attention on local vulnerabilities using an integrated index approach: the example of the climate vulnerability index. Water Sci Technol 51(5):69–78. https://www.ncbi.nlm.nih.gov/pubmed/15918360

Szymanowski M, Kryza M (2009) GIS-based techniques for urban heat island spatialization. Clim Res 38:171–187. https://doi.org/10.3354/cr00780

Tanser FC (2002) The application of geographical information systems to infectious diseases and health systems in Africa. Int J Health Geogr 9:1–9. https://www.ij-healthgeographics.com/content/1/1/4

Temblor (2016) US flood maps: do you live in a flood zone? https://temblor.net/flood-insights/us-flood-maps-flood-zone-1146/

The Economist (2022) The world should welcome the rise of the robots. https://www.economist.com/leaders/2022/02/26/the-world-should-welcome-the-rise-of-the-robots

The Geospatial (2020) Managing social distancing post lockdown using GIS. https://medium.com/@the-geospatialnews/managing-social-distancing-post-lockdown-using-gis-f1ad23e17616

Tomlinson CJ, Chapman L, Thornes JE, Baker CJ (2011) Including the urban heat island in spatial heat health risk assessment strategies: a case study for Birmingham, UK. Int J Health Geogr 10. https://doi.org/10.1186/1476-072x-10-42

Tribolet A (2020) The role of GIS during a pandemic: why the "where" is so important for the coronavirus. https://gis.utah.gov/the-role-of-gis-during-a-pandemic-why-the-where-is-so-important-for-the-coronavirus/

UNESCAP, United Nations Economic and Social Commission for Asia and the Pacific (2020) Geo spatial practices for sustainable development in Asia and the Pacific, Bangkok

Waller LA, Gotway CA (2004) Applied spatial statistics for public health data. Wiley, Hoboken, NJ

Warren MS, Skillman SW (2020) Mobility changes in response to COVID-19. Descartes Labs. https://arxiv.org/pdf/2003.14228.pdf

Warriach F (2020) Current virus hotspots status in Lahore leads to partial lockdown in city. https://pk.mashable.com/coronavirus/3653/current-virus-hotspots-status-in-lahore-leads-to-partial-lockdown-in-the-city

Wray S (2020) How cities are using data to plan for COVID 19. Cities for Global Health. https://cities-today.com/cities-look-to-data-to-inform-next-steps-on-covid-19/

Yu W-B, Tang G-D, Zhang L, Corlett RT (2020) Decoding the evolution and transmissions of the novel pneumonia coronavirus (SARS-CoV-2/HCoV-19) using whole genomic data. Zool Res 41(3):247–257. https://doi.org/10.24272/j.issn.2095-8137.2020.022

Zheng Z (2020) NParks develops nifty real-time map showing crowd levels at most green spaces in S'pore. https://mothership.sg/2020/04/nparks-map-parks-crowd-level/

Mega Risks: Digital Transformation and Sustainability

4

4.1 Introduction

The digital technology has provided the tools that permitted cities to keep functioning during lockdowns and social distancing norms in the wake of the coronavirus pandemic. It rescued cities from becoming totally paralyzed by allowing people and organizations to remain active in all major cities of the world by introducing a new norm or lifestyle of remote functioning. As mobility restrictions were enforced to contain the spread of the novel coronavirus (COVID-19), more and more of the estimated 4.2 billion city dwellers started working, studying, and interacting from their homes, availing virtual opportunities for engagement, entertainment, and conferencing. Even critical international policy coordination was conducted online in fora such as the G7 or G20 (OECD 2021a). Operating digitally was the only way to stay in business through mandated shutdowns and restricted activity with rare exception. It amounted to stay digital to stay alive.

This chapter on cities, risks, and digitalization has been divided into five parts. After a brief introduction, the next part discusses the digitalization with respect to the coronavirus pandemic. The first part has been further divided into two sections—the first one highlights practical examples on how digitalization enabled cities to perform various functions during the pandemic through remote working, electronic public administration, online business and shopping, distance education, and health solutions and for social engagements. The second section focuses on how the massive jump in teleworking, teleservicing, and online conferencing amplified the network pressure and the ways human efforts have been able to deal with it. The third part of the chapter on digitalization and climate change has been divided into three sections. The first of which discusses the role of digital technologies in climate monitoring and early warning—how it is helping in monitoring the variables related to climate change and assisting in obtaining vital information pertaining to climate, weather, and related risks and disasters. The second and third sections describe the role of digital technologies in mitigation and adaptation to climate change, respectively. The fourth part of the chapter focuses on the emerging social and ethical issues in the use of digital technology and the efforts needed to resolve these. The findings of the chapter have been summed up in the conclusion.

4.2 Cities, Digitalization, and COVID-19

The pandemic has brought an enormous growth in the use of digitalization—a term that refers to digital tools including information, computing, communication, and connection technologies (Guo et al. 2020; Vial 2019). A natural outcome of which is enhanced innovation, augmenting it to a higher level so that humanity and organizations can outmaneuver uncertainty. This has also led to a massive change in the pattern of usage of technical devices, information technology systems, and networks. The rapid diffusion of digital devices and their usage had already started in the twenty-first century even before the pandemic. By 2019, 67% of the global population had subscribed to mobile devices, of which 65% were smartphones—with the fastest growth in sub-Saharan Africa (GSMA 2020). However, the pandemic has added fuel to the fire as indicated by the following pointers:

- The world digital population reached 4.8 billion by the third quarter of 2020. Their numbers shot up considerably during the pandemic—between 300 and 350 million people came online for the first time in 2020 over the previous 12 months (Kemp 2020).
- Presently over 4.5 billion people actively use social media, and more than 180 million people started to use social media between July and September 2020, amounting to two million new users every day (Kemp 2020).
- The number of global Internet users in 2010 was at 1.9 billion, per Statista and Internet World Stats data. It increased to 3.2 billion by 2015, a growth of 66% in 5 years and 4.66 billion by 2020 (Statista 2020a). By October 2021, the number has increased to 4.88 billion (Kemp 2021) and by 2022 to 4.95 billion (Kemp 2022).
- Currently there are 5.29 billion mobile phone users in the world; their number increased enormously, and their current growth rate is about 2% per year (Kemp 2021).
- In terms of (all) devices, there are more connected mobile devices in the world (7.94 billion) than there are people.

4.2.1   Bridging the Functional Gaps

Digitalization has helped cities to bridge the functional gaps created by mandated shutdowns and social distancing measures. Without digital tools and technologies, cities would be grossly handicapped to operate activities in public administration, business, and finance and in social and health spheres. No doubt, it has brought enor-

mous benefits such as enhanced connectivity, financial inclusion, and access to trade as well as public services. It has broadened outreach to education by making possible virtual learning environments and distance learning—opening opportunities for working people as part-time students who would otherwise be excluded. It has also made city public services more accessible, accountable, and less bureaucratically burdensome.

During the pandemic, a major factor contributing to cities' resilience was the increased adoption of digital or smart technologies—gadgets or devices, apps, data, and systems such as the Internet of things (IoT), big data, blockchain, and artificial intelligence (AI) as well as robotics. From a technological standpoint, a technologically smart city is equipped with IoT (Internet of Things) and devices (including sensors, processors, wearables, electronics, software, actuators, vehicles, cell phones, and computers). These devices collect data and information from different areas in the city and analyze it using AI or artificial intelligence—which is a machine that derives patterns and trends from the data within seconds. The intelligent machines continuously self-learn and improve their performance based on big data. Big data refers to the large amount of data generated from IoTs (Kummitha 2020). Blockchain is also a kind of database which stores data in blocks—fresh data enters a new block, and upon getting filled, it is chained onto the previous block, which makes the data chained together in chronological order (Hays, 2022).

Emerging technologies such as artificial intelligence (AI) have helped in fast development of vaccine against the disease, designing lab tests, analyzing CAT scans, etc. (Chun 2020). These also helped in predicting efficacy of public health measures and keeping the public updated with scientific information (UNESCO 2020a). Above all, digital technologies allowed people to move much of their activities online and maintained world's operating systems in a situation where most people while staying home remained connected to each other.

Obviously, cities, organizations, companies, and businesses which had developed digital strategies and started executing them prior to the pandemic were able to face the COVID-19-related challenges better because they had significantly more tools at their disposal to face the storm and to come out of it stronger. Cities across developed world were therefore in a better position compared to those in the developing world. Even within developing countries, cities in fast-growing economies performed better. Cities like Hong Kong, Singapore, Taipei, and those of Korea and mainland China moved quickly in response to COVID-19 to transition their employees to remote work, move services online, use conferencing technology, provide virtual opportunities for urbanites engagement, and utilize social media and web platforms better to deliver critical information to their residents.

4.2.2   Digital Technologies in the New Normal

The novel coronavirus has brought forward a "new normal" driven by digital technologies—digital government, digital economy, digital finance, digital health, digital education, digital conferences, and digital recreation. Many governments and businesses moved quickly to digital solutions and digital platforms. Urbanites made efforts to adjust to evolving patterns of functioning in a "new normal"—with activities shifting online in most spheres of life: work, public administration, business and shopping, health, education, recreation, etc.

4.2.2.1   Remote Working

Remote or teleworking had already started in the twenty-first century, and the trend was rising even before the COVID-19 pandemic (Krantz-Kent 2019). ILO (2020) estimates that globally among employees, 2.9% were working exclusively or mainly from their home even before the pandemic. However, the restrictions imposed during the pandemic brought a major boost. Results from various surveys (Kamouri and Lister 2020; Day et al. 2020) conducted during the shutdown provided a snapshot of the new reality and its challenges. A March 23, 2020, survey of 250 large firms in Argentina, for example, indicated that 93% of companies had adopted teleworking as a policy in response to the COVID-19 crisis (Berg et al. 2020). In the case of the USA, according to Stanford Research (Wong 2020), 42% of labor force worked from home full-time, while some 33% did not work, a testimony to the serious impact of the lockdown-triggered recession. The remaining quarter of workers that mostly belonged to essential services were working on their business premises. Thus, in a way, the USA became a working-from-home economy—with roughly twice as many employees working from home as those who presented themselves at the workplace (Wong 2020).

China presented an interesting case for growth in remote working. The cities there used a variety of online platforms for work from home such as DingTalk by Alibaba, Feishu/Lark by ByteDance, and WeChat Work by Tencent. All these platforms offered the same services as Google Drive, Dropbox, Zoom, and Doodle Survey, apart from supporting daily check-ins and payroll systems (Ling 2020). Companies' eagerness to continue their activities online generated a huge market for remote working tools in Chinese cities. According to estimates, the size of China's remote working industry grew by 104% in 2020 (Ling 2020). The massive jump in remote working in the country is also reflected in the huge surge in download of work-from-home apps—with Lark recording a growth of 6085%, Dingtalk 1446%, and WeChat Work 572% (Statista 2020b).

In Europe, according to Eurofound, an agency of European Union, only about 15% of all workforces had ever worked remotely before the pandemic, but early estimates indicated that up to 40% of all EU-based workforce switched into the remote mode (Calamari 2020). A Eurofound (2020) electronic survey brought some useful insights for remote work. It reported that the experience of working from home during the COVID crisis for many employees was positive, even though over half (53%) of them pointed out that their employers had not provided the equipment needed to work from home—as the transition to working from home was ad hoc and unplanned. Over three-quarters of employees indicated a preference to work from home at least occasionally if there were no COVID-19 restrictions. The preferred type of remote or telework cited was several times a week—very few respondents indicated that they would like to telework daily. The teleworking arrangement, preferred by many respondents, was hybrid or a mix of teleworking and presence at the workplace. In terms of satisfaction, employees who worked exclusively from home scored worse than those working from other locations. This observation appears to be important because working from home has its limitations among others due to family distractions.

Remote work carried several benefits including reduced office cost as well as commuting time and costs, flexible working hours, better work-life balance, less air pollution, and reduced traffic as well as flexibility of working even from abroad. However, it had limitations as many jobs like food services, cleaning, maintenance, etc. do not allow to work from home. Moreover, it has challenges in terms of family distractions particularly for female workers with younger children, shortcomings related to technical equipment and the Internet, problems of security and compliance, as well as deficiency related to professional connection. It has also been reported to cause management problems (Parker et al. 2020) as it calls for different management skills than face-to-face management. Moreover, the adjustment to teleworking is not always straightforward. Therefore, even though many urban institutions and businesses recognized the benefits of teleworking, some had difficulty in transitioning due to activity restrictions or digital limitations.

Regarding activity restrictions, a Stanford Survey (Wong 2020) revealed that only 51% of the respondents who could carry out their jobs on computers were able to work from home at an efficiency rate of 80% or more. The remaining half could not work remotely because their jobs required their presence at work as in retail, healthcare, transport, and business services (Wong 2020). The adoption of remote work by organizations, whether in business, society, or government, depended largely on their readiness. The change was swift, and it barely gave time to organizations and people to plan, prepare, and implement the new normal. The only option they had was to adjust, try, experiment, and adapt to new ways.

Remote Work Potential It is important to note that the potential of cities and countries for remote work depended on several factors including location, occupation, capacity to work remotely, as well as availability of and accessibility to technology. Potential for remote work is higher in cities, but it is not the same across all cities even within the same country. For example, capital cities, in most cases, have the highest share of employment in occupations that can potentially be performed remotely. This share according to Özgüzel et al. (2020) is, on average, 9 percentage points higher in capitals than in their respective country taken as a whole. Another study in the USA by Dingel and Neimann (2020) also showed that remote working potential differed across locations in the country.

Remote work potential also depended a lot on type of occupation. Those jobs which demand heavy manual or physical labor input at a spot or site for creating a product are not suitable for remote work. Thus, accommodation (hotel) and food services demanding largely onsite services are not so suitable and according to some sources have potential of only employing less than 10% of total employees online (Consultancy.eu 2020). Similarly, manufacturing, transport, warehousing, and construction relying heavily on high levels of physical input from in-person labor cannot use bulk of their employees for remote work. In contrast, finance and insurance companies at the other end of the scale can spare more than three-quarters of their workers to work from home (Consultancy.eu 2020). Likewise, management, business services, and information technology also carry heavy potential for remote work. The details on suitability of jobs for work from home are available in a White Paper published by Dingel and Neimann (2020). Figure 4.1 shows teleworking or remote work rates by industry/sector for selected countries—at the time when each country's total teleworking rate reached its

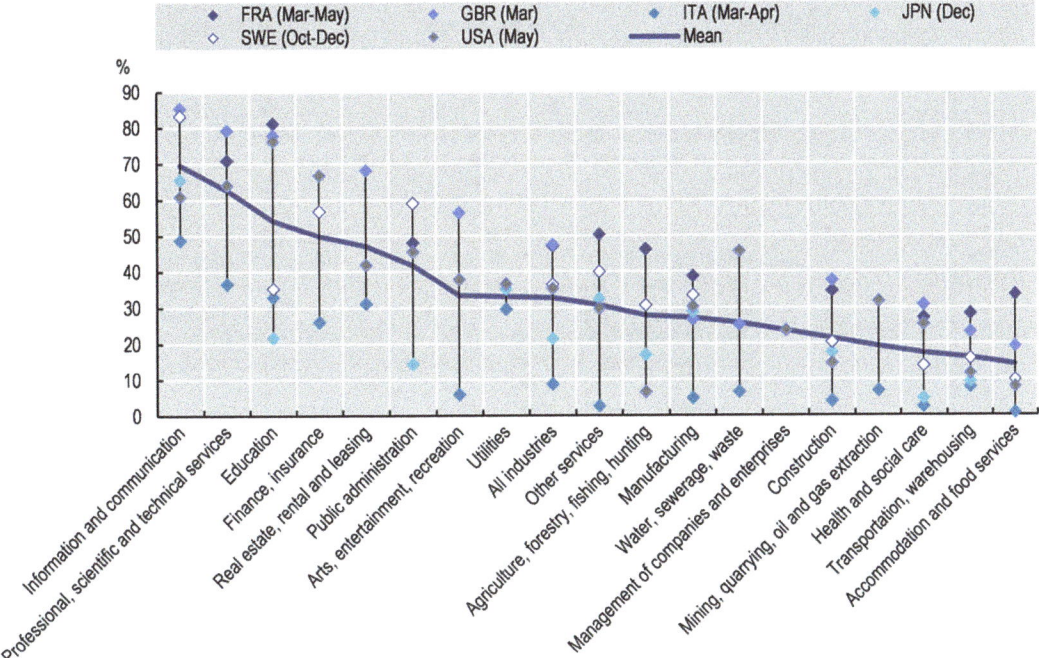

Fig. 4.1 Teleworking peaks during the pandemic by industry in selected countries (% employed). (Source OECD 2021b)

peak. Although teleworking rates by industry varied across countries, the trends were generally similar (OECD 2021b).

Technological and digital facilities are also extremely important in remote working. In this regard, both individual constraints—such as a lack of necessary equipment or working environment—and place-based constraints, such as the availability of a high-speed Internet connection, constitute important factors in determining potential of cities and the capacity for people and firms to adopt remote work (Özgüzel et al. 2020). Hence, in identifying potential and preparedness of a city or a country, it is important, among others, to use both place-based and individual constraints in ascertaining the potential for remote work. It is important to note that even in this era of rapid globalization, there exists a significant disparity in the use of digitalization not only between developed and developing countries/cities but even between many economies belonging to the developed world.

A comparative study of remote work potential in 27 European Union member countries, Switzerland, Turkey, and the United States was carried out by OECD (2020a). The potential of remote work varied greatly between and within the selected countries. For example, 50% of jobs could potentially be done from home in Luxembourg, but only 21% in Turkey (Fig. 4.2). The study also provided an assessment of the capacity of places to adapt to remote working during lockdowns. It revealed that capitals have, in most cases, the highest rate of potential remote working capacities than their country's average (Fig. 4.2).

Remote Work Preparedness In terms of preparedness for remote work, no studies targeted cities specifically, but one study on preparedness of countries to remote work yielded some interesting results (Chakravorti and Chaturvedi 2020). The authors based this study on digital limitations of remote work factored into three main issues: robustness of key telecommunication platforms and digital foundations, resilience of Internet infrastructure, and resilience of digital payment transactions. Using these variables, the

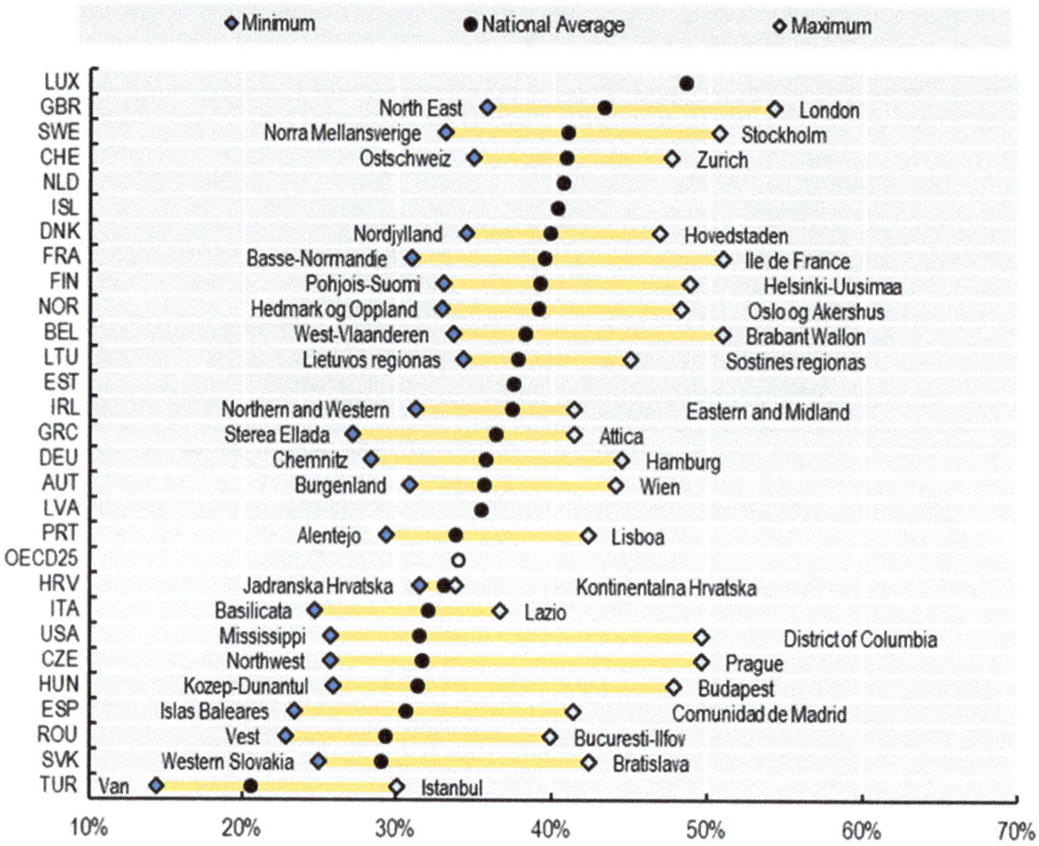

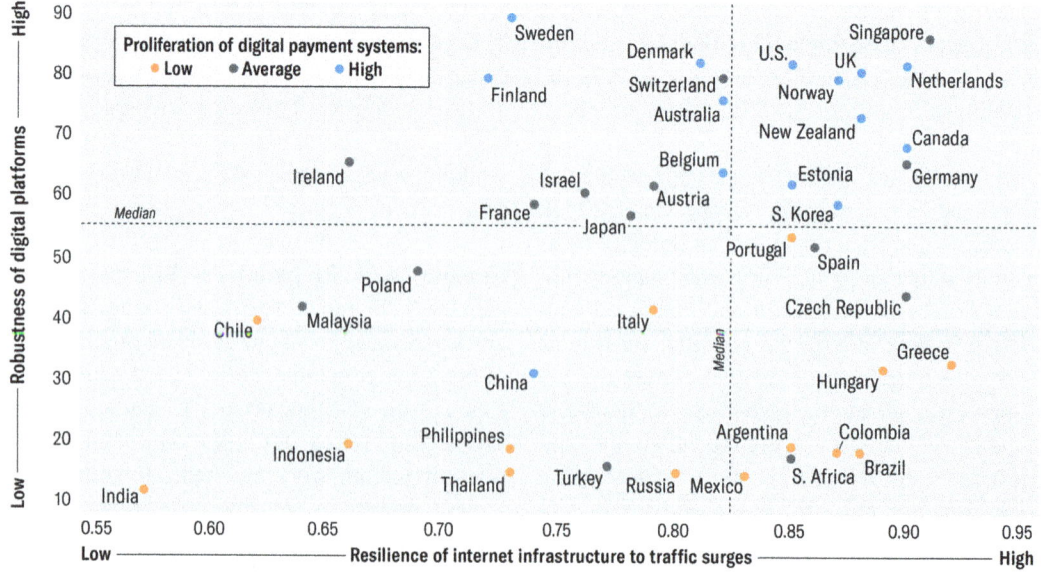

Fig. 4.2 Variation in remote work potential between and within selected countries. The number of jobs in each country or city/region that can be carried out remotely as the percentage of total jobs. (Source: OECD 2020a)

Fig. 4.3 Preparedness of selected countries to remote working. (Source: Chakravorti and Chaturvedi (2020))

authors scored the remote work preparedness for a cross section of 42 countries (Fig. 4.3) that were significant in the global economy and had also enacted social distancing measures.

Countries at the top of the graph had more robust digital platforms, and therefore their cities were better prepared for the pivot to online work compared to developing economies in the bottom. They also had stronger digital payments infrastructure. However, Internet resilience varied widely, meaning the infrastructure struggled to support the increased demand. Countries on the left of the graph like India, with low resilience infrastructure, were expected to experience jittery connections and slowdowns, which are hurtful for productivity. This also indicated a demand for long-term investments in needed infrastructure (Chakravorti and Chaturvedi 2020).

Among Asian countries, Singapore was on top. Other countries like Korea and China presented interesting cases. Korea, with one of the best Internet resilience in the world, also had strong digital payment system along with robust platforms and appeared ready for the change. China's preparedness for remote work was much better than most developing countries, which were ill-prepared (Chakravorti and Chaturvedi 2020). This is also evident from the use of digital work application from WeChat, Tencent, and Ding (Fig. 4.4) which took off sharply at the end of January when lockdown measures started to take effect.

Countries in Southern Europe had low robustness of the platforms compared to countries in Northern Europe. The preparedness of the USA is clear from the graph, where millions of white-collar workers in technology, finance, and media shifted to work from home early on.

Remote Work Impacts on Economy The combination of virus and remote work had many implications for economy and land use in cities. One study in the USA (Thompson 2020), for example, has shown that virus resulted in cutting down business travel and affected related hotel and restaurant business enormously. Spending on leisure and hospitality, which were booming before the pandemic, was also cut drastically hitting the downtowns of supercities. With dwindling businesses and corporate giants welcoming remote work, real-estate markets in the superstar cities also tumbled (Thompson 2020). Barely, 3 months after its work-from-home (WFH) announcement, Facebook leased a massive 730,000-square-foot office in mid-town Manhattan (Young 2020). It is speculated that the WFH may also enhance ongoing demographic shift from expansive large metros to cities in the sunbelt.

Remote Work Prospects The various surveys conducted on the remote work globally (Chavez-Dreyfuss 2020; Murray and Meyer 2020; Dahik et al. 2020), in regions (Eurofound 2020), or in individual countries, show that its prospects are good. A global survey of US-based Enterprise

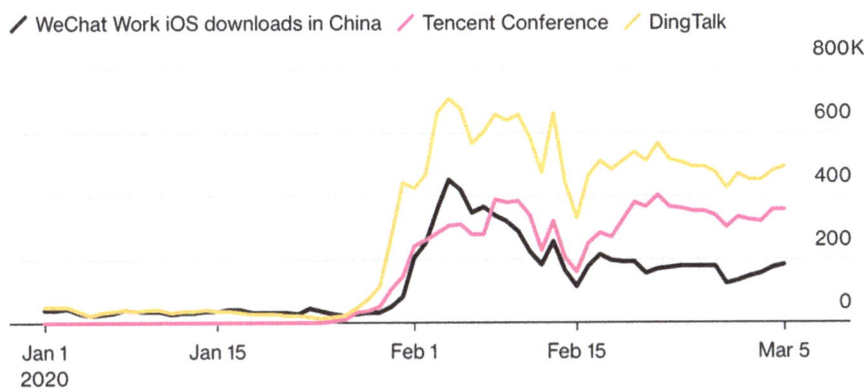

Fig. 4.4 China's lockdown triggered a spike for remote work apps. (Source: Sensor Tower)

Technology Research (Chavez-Dreyfuss 2020) conducted in September 2020 of about 1200 chief information officers (CIOs) from the world across different sectors/industries showed that the percentage of employees working from home was likely to double in 2021. Among CIOs, nearly half (48.6%) reported that productivity had improved since the start of remote work, while about a quarter (28.7%) indicated a decline in productivity. IT divisions in telecommunication, financial, and the insurance sectors reported large productivity boosts, whereas energy, utilities, and education sectors did not see much benefit. Another global survey (Murray and Meyer 2020) was conducted of 4700 remote workers in the USA, the UK, Germany, Japan, and Australia by Slack's Future Forum. The research found that most remote workers did not want to return to the old way of working. About three-quarters of them (72%) wished to follow a hybrid remote office model moving forward, while only 12% wanted to return to full-time office work.

With the likelihood of remote work continuing in the future, there are many challenges for managers and managing organizations. Some are more demanding than others. Dahik et al. (2020) conducted a survey of more than 12,000 professionals to assess these during May and June 2020, in the USA, Germany, and India. Their findings suggest that the future of work will be increasingly hybrid, which entails developing new hybrid working models to enable employees to shift between onsite and remote work. It also demands new thinking on the allocation of appropriate physical space in terms of both its size and shape for the hybrid work (Dahik et al. 2020). The survey confirmed the ongoing trend of heavy investment in the tools needed to work remotely and pointed out that 87% of employers surveyed indicated that they anticipate prioritizing technical and digital infrastructure investments that support sustained remote work. Some three-quarter of employees indicated that during the first few months of the pandemic, they were able to maintain or improve productivity on their individual tasks. The survey found that a powerful driver of productivity was satisfaction with tools

of remote work such as videoconferencing, computer monitors, team communication platforms, security of remote connections, virtual whiteboards, and project management software. Social connectivity and mental well-being were also pointed out to be important factors in maintaining productivity. This brought forward the fact that while investing in infrastructure and digital technologies by organizations, it is also important to focus on employees' physical and mental well-being and virtual social connections (Dahik et al. 2020).

4.2.2.2 Business and Industries

Digital technology has also helped in promoting e-commerce. It has helped businesses and industries such as aviation, shopping, and catering. Even in sectors which traditionally relied on "brick-and-mortar" models to deliver services, many businesses shifted to online channels. They used innovative ways to engage with customers and provide products and services. For example, information technology giants like Google and Alibaba took big advantage from the opportunity. Amazon and Alibaba became bigger and stronger during the pandemic, while brick-and-mortar stores failed to compete during lockdowns (Saliola and Islam 2020). Even in China, over 30 financial institutions worked with Alipay to roll out "contactless wealth management products" for customers. Property companies also changed their approach by using artificial intelligence (AI), virtual reality (VR), live streaming, and Internet celebrities to sell apartments online (Li 2020).

The digital transformation was already underway in business and industries, but the pandemic accelerated the paradigm shift. A global online survey of executives of varied industries and companies by McKinsey and Company (2020) confirmed the rapid shift toward interacting with customers through digital channels. It showed that rates of digital adoption were years ahead of where they were when previous surveys were conducted—and even more so in developed Asia than in other regions. Respondents indicated that their customer interactions were more digital in nature than before. Nearly all respondents had the opinion that their companies had worked out at least temporary solutions. In addition, they were able to meet many of the new demands much more quickly than what they had thought possible before the crisis.

The rates for developing digital products across companies differed, according to the survey. Respondents on consumer-packaged goods (CPG) and automotive and assembly plants, for example, reported relatively low levels of change in their digital-product portfolios. While healthcare, pharma, financial services, and professional services saw much more significant increases. They saw a quantum jump nearly twice as large as those in CPG companies was reported. The customer-facing elements of organizational operating models were not the only ones affected. Respondents reported similar accelerations in the digitization of their own internal operations (such as office work, production, and R&D processes) as well as for interactions in their supply chains.

Digital technology also enabled companies to take advantage of new market demand through quickly shifting their production lines to new products. For example, many of them switched to manufacturing medical supplies and PPE including masks, gowns, gloves, etc., as the demand for these skyrocketed during the pandemic. Chinese companies including BYD, Gree, and Foxconn took lead in this respect. Such transition would not be possible without digitalized management of inventories, manufacturing materials, as well as a high level of industrial automation based on digital design, modelling, and 3D printing (Li 2020). While use of digital capabilities contributed to manufacturers' ability to respond to the pandemic, it also gained further importance in the recovery phase to better manage supply chains, build resilience, maintain sustainable growth, and meet changing customer demands.

In retail ecommerce, demand for contactless delivery services during the pandemic increased all over the world. Some popular companies included DoorDash for food delivery, Instacart for groceries, and Amazon for all kinds of shopping goods in the USA; HelloFresh, Deliveroo, and Glovo in Europe; and Uber Eats and Uber Delivery throughout the globe. Choosing groceries or food, paying online, and leaving feedback all on one app made life easier. Businesses also started blending the physical and the digital mechanisms through such delivery methods as curbside pick-up and contactless delivery.

The retail e-commerce trend greatly accelerated by the pandemic is likely to continue. According to eMarketer (Lebow 2021), a global marketing company, the global retail e-commerce sale increased from $3.4 trillion in 2019 to $4.2 trillion in 2020, 17.8% of the total retail sale. It is projected to grow rapidly by double digit rate and likely to reach 7.4 trillion by 2025, almost a quarter of the total retail sale from 13.8% in 2019 (Fig. 4.5). Although with the vaccine rollout, the retail in-person market environment may improve, but online future of retailing still appears bright due to improvements in the ways of conducting business such as extended options for selection and quick delivery. Moreover, factors such as convenience particularly to elderly, paying online, and contactless delivery all in one package; enhanced online competition between companies to woe their newly acquired customers through loyalty programs, subscription models, promotions, and expansion of the product range; and improved logistic and allowing same-day delivery are likely to encourage not only current but new customers to shop online.

Shutting down of many stores has also encouraged customers to turn more and more to online shopping. Many big-time American retail stores including discount retailer Stein Mart having 279 branches, JCPenney, J.Crew Group, Inc., Neiman Marcus Group Inc., and Ascena Retail Group Inc. filed for bankruptcy protection in 2020. According to S&P Global Market Intelligence, corporate bankruptcies in 2020 hit the highest levels in the USA in more than a decade (Irum and Hudgins 2021). In addition, MarketWatch quoting Coresight Research reported 8401 store closures by December 4, 2020 (Garcia 2020).

A "COVID-19 and E-commerce" survey conducted by the United Nations Conference on Trade and Development (UNCTAD) and NetComm Suisse eCommerce Association of some 3700 consumers in 9 emerging and developed economies (UN News 2020) provides

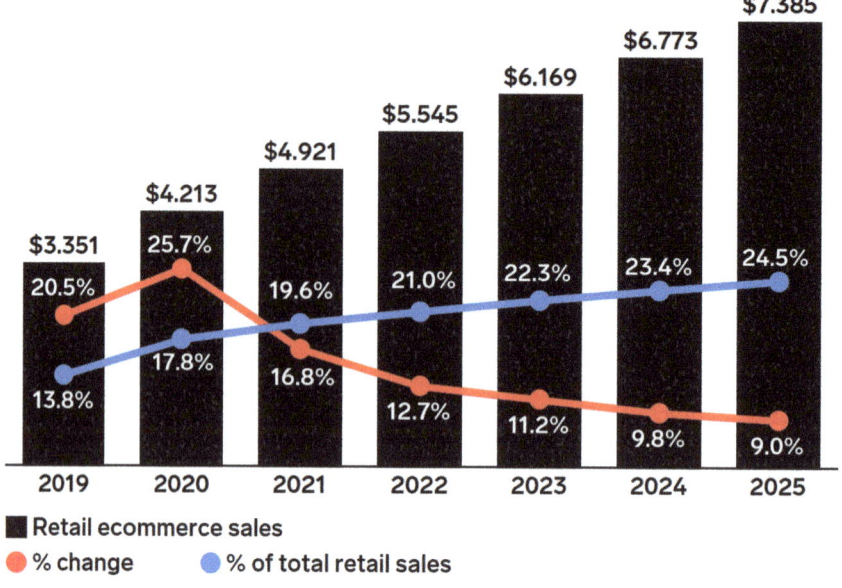

Fig. 4.5 Worldwide retail e-commerce sale 2019–2025. (Source: Lebow (2021))

some insights into consumers' behavior change toward e-commerce during the pandemic. The survey showed that online average monthly spending per shopper had dropped markedly, but online purchases had increased by 6–10 percentage points across most product categories. More than half of the survey's respondents shopped online more frequently. However, online shopping varied between countries (UNCTAD 2020), with the strongest rise occurring in China and Turkey while the weakest in Switzerland and Germany, where more people were already engaged in e-commerce even before the pandemic. Small merchants in China were better equipped to sell their products online, while those in South Africa were least prepared.

4.2.2.3 Education
Education is another domain in which there was a dramatic shift to the online mode due to widespread school closures in cities of 185 countries, where learners were forced to stay home (UNESCO 2020a). In response to the disruption, governments preferred technology as the primary—and in many cases the only—channel to maintain the continuity of formal learning.

Teaching remotely, from primary schools to universities, became the new normal shifting from lectures in classrooms to online teaching using web-based courses (Gewin 2020). However, provision of digital tools for online education became a major constraint in cities of developing countries. Countries deficient in digital networks and educational platforms used TV and radio broadcast channels (Tawil 2020) or mainstream social media platforms or standard email rather than shifting to unfamiliar technologies, platforms, and software. Even at the university level in developing countries like Egypt, teachers and students used social media platforms such as Facebook or WhatsApp for communication (Sobaih et al. 2020). Basically, the types of technologies deployed to maintain learning continuity during the pandemic largely reflected a country's development status. According to UNESCO-UNICEF-World Bank survey, two thirds of low-income countries used radio for primary education, compared to less than half of upper- and middle-income countries (UNESCO 2020b). The mechanisms of distance learning in cities of both developed and developed countries have been discussed in more details in Chap. 8 on Education in this book.

In terms of technology-enabled distance learning responses, governments' focus in cities was on connected digital technologies including videoconferencing platforms like Skype, Zoom, and Google Meet—the ones people were more familiar with, understood, and could access seamlessly. Additionally, other education-focused platforms like Google Classroom and Moodle enabled learning continuity while offering greater functionality to support learning (Tawil 2020). Among developing countries, the most interesting case was China where at school level (displayed on the website of the Australian Government), the Ministry of Education, together with the Ministry of Industry and Information Technology, launched an online portal in February 2020 for primary- and secondary-level students. The platform provided digital materials for schools to conduct teaching online. It had the capacity to support learning of 50 million students simultaneously. The platform had been visited over 2 billion times by citizens up to May 2020 from all 31 mainland provinces. China also issued guidelines for use by higher educational institutions on remote study, which recommended 22 online platforms in the country capable of providing 24,000 higher education courses for free (Australian Government 2020).

Along with synchronous modes of teaching, asynchronous platforms of EdTech (a combination of IT tools and educational practices) also saw enhanced surge during the pandemic. For example, edX and Coursera witnessed an increase in enrolments during the pandemic (Shah 2020). According to the reports by KPMG and Google, EdTech will boom and is likely to boom in the future (Dhawan 2020). Some of the famous EdTech start-ups, asynchronous platforms, listed by Dhawan (2020) were Byju's, Adda247, Alolearning, AptusLearn, Asmakam, Board Infinity, ClassPlus, CyberVie, Egnify, Embibe, ExtraaEdge, iStar, Jungroo Learning, GlobalGyan, Lido Learning, Pesto, Vedantu, EduBrisk, ZOOM Classroom, ZOOM Business, Toppr, Unacademy, Kahoot, Seesaw, Khan Academy, e-pathshala, and GuruQ.

Digital divide posed a major problem in digital education primarily in developing countries, but it also posed problems even in developed countries. Technical barriers such as insufficient access to electricity and shortage of human resources in terms of digital literacy (Vegas 2020) have also been blamed for the shortfall in digital education in developing countries. In terms of future of digital education, a survey of about 1200 school administrators and teachers was conducted (Castelo 2020). Eighty-two percent of the respondents said that combining technology use with traditional resources and teaching methods will be the most likely trend in the next 10 years. They expected to see the biggest growth in remote learning (63%), virtual learning (54%), and online content and resources (50%) in schools over the next 3 years. In the wake of growing tele-education, the respondents called for effective technology training for teachers as a crucial move forward (Castelo 2020).

4.2.2.4 Health

Technology was also deployed by healthcare organizations, hospitals, and even city administrations and city residents for implementation of health-related measures and needs during COVID-19 in cities worldwide—from disease tracking to surveillance and clinical management (Table 4.1). Moreover, technology played a vital role in promoting telehealth and provided digital solutions to meet the shortages of medical equipment and their parts through 3D printing. Robotics also enabled performing several functions—from contactless delivery to patient services and in conducting medical and disinfectant missions. The research data (Jiang and Ryan 2020) indicates that even Internet use showed the difference in resilience to health crisis—countries with wider Internet access and safer Internet servers were more resilient to the COVID-19 pandemic compared to others (Fig. 4.6). Overall, the pandemic has become a game-changer for healthcare technology as indicated by massive use and promotion of telehealth alone. Statista (2020c) forecasted that with rapid growth of telehealth, its revenue globally will rise to around $332.7 billion by 2025.

Table 4.1 Digital technology initiatives used for health in pandemic preparedness and response

	Functions	Digital technology	Countries	Advantages	Disadvantages
Tracking	Tracks disease activity in real time	Data dashboards; migration maps; machine learning; real-time data from smart phones and wearable technology	China, Singapore, Sweden, Taiwan, USA	Allows visual depiction of spread; directs border restrictions; guides resource allocation; informs forecasts	Could breach privacy; involves high costs; requires management and regulation
Screening for infection	Screens individuals and population for disease	Artificial intelligence; digital thermometers; mobile phone applications; thermal cameras; web-based toolkits	China, Iceland, Singapore, Taiwan	Provides information on disease prevalence and pathology; identifies individuals for testing, contact tracing, and isolation	Could breach privacy; fails to detect asymptomatic individuals if based on self- reported symptoms or monitoring of vital signs; involves high costs; requires management and regulation; requires validation of screening tools
Contact tracing	Identifies and tracks individuals who might have come into contact with an infected person	Global positioning systems; mobile phone applications; real-time monitoring of mobile devices; wearable technology	Germany, Singapore, South Korea	Identifies exposed individuals for testing and quarantine; tracks viral spread	Could breach privacy; might detect individuals who have not been exposed but could have had contact; could fail to detect individuals who are exposed if application is deactivated, the mobile device is absent, or Wi-Fi or cell connectivity is inadequate
Quarantine and self-isolation	Identifies and tracks infected individuals and implements quarantine	Artificial intelligence; cameras and digital recorders; global positioning system; mobile phone applications; quick response codes	Australia, China, Iceland, South Korea, Taiwan	Isolates infections; restricts travel	Violates civil liberties; could restrict access to food and essential services; fails to detect individuals who leave quarantine without devices
Clinical management	Diagnoses infected individuals; monitors clinical status; predicts clinical outcomes; provides capacity for telemedicine services and virtual care	Artificial intelligence for diagnostics; machine learning; virtual care or telemedicine platforms	Australia, Canada, China, Iceland, USA	Assists with decision-making, diagnostics, and risk predictions; enables efficient service delivery; facilitates patient-centered, remote care; facilitates infection control	Could breach privacy; fails to accurately diagnose patients; involves high costs; equipment may malfunction

Source: Whitelaw et al. (2020)

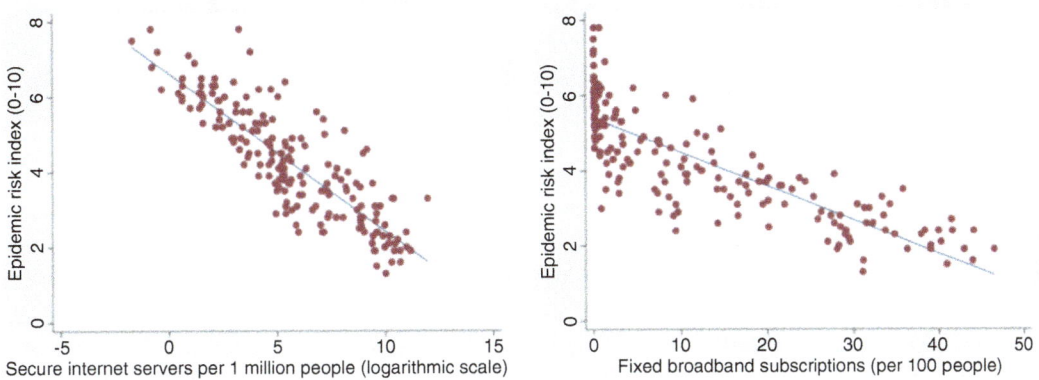

Fig. 4.6 Safer Internet servers and wider Internet access are strongly correlated with lower risk for epidemic. (Source: Jiang and Ryan (2020))

Diagnostic, Contact Tracing, Testing, and Surveillance Digital tools were used for screening, contact tracing, tracking, surveillance, quarantining, and evaluation of interventions based on mobility data and communication with the public (Budd et al. 2020). Together, these constituted basic public health measures to keep the outbreak within a manageable scale. Data-driven digital technologies were extensively used to implement these basic health measures by China, the Republic of Korea, Singapore, and Taiwan during the pandemic (Huang et al. 2020). The solutions leveraged billions of mobile phones, large online datasets, connected devices, relatively low-cost computing resources, and advances in machine learning (Budd et al. 2020). Among various means, mobile phone apps, mobile payment apps, credit card records, social media applications, AI-powered surveillance cameras, drone-borne cameras, and GPS data were used to collect real-time data in cities on the location of individuals and to track and restrict the movement of individuals within a monitored or lockdown area.

A widely used app for this purpose in China allowed the public to trace if they were ever on the same vehicle, i.e., a train or a plane or in proximity with any confirmed cases in the previous 2 weeks. The app, first developed by an independent software developer, initially used coarse data from social media and websites. It was later refined and made more reliable by retrieving data from all-level public surveillance systems and the national transportation authorities, including the Ministry of Transport, Railway, and Civil Aviation Administration after approval by the State Council (WHO 2020). Both China and the Republic of Korea also used several tools for contact tracing including security camera footage, facial recognition technology, bank card records, and global positioning system (GPS) data from vehicles and mobile phones. These provided real-time data and detailed timelines of people's travel. Cities in European countries such as Germany, Norway, Italy, Iceland, England, and Poland among others also used smartphone-enabled apps to detect COVID-19's chain of contagion (Gupta et al. 2021). Within the USA, in May 2020, Apple and Google launched their jointly developed, interoperable API, which allowed government and health agencies to track the spread of COVID-19 using Bluetooth technology, and Salesforce within New York City to build a city-wide contact tracing program (Drees et al. 2020).

Diagnostic tools application was also used for improving accuracy. Hospitals in Wuhan and nationwide in China deployed AI-powered CT imaging interpretation tools, which helped radiologists reduce CT scan reading time from hours to seconds. Some other tools allowed patients at community clinics to have their CT scan read by medical experts' miles away, which expanded capacity for diagnostics and helped avoid overwhelming the healthcare workforce (WHO 2020). Blockchain technology provided further help in diagnostics, while artificial intelligence (AI)-coupled self-testing and tracking systems for COVID-19 also proved beneficial (Mashamba-Thompson and Crayton 2020). Sensors, including thermal imaging cameras and infrared sensors, were deployed (Budd et al. 2020) particularly at the airports to identify potential cases based on febrile symptoms. Wearable technologies were also used for monitoring the virus in populations (Armitage 2020).

An interesting digital tool for the diagnosis and triage of patients was chatbot which besides diagnostic was used for a broad range of healthcare issues (e.g., patient triage, clinical decision support for providers, directing patients and staff to appropriate resources, and mental health applications (Golinelli et al. 2020). In the USA, CDC launched its chatbot "Clara," a "Coronavirus Self-Checker." A series of questions on demographic and medical aspects allowed it to assess users' risk of having contracted COVID-19. Microsoft also launched a chatbot to identify patients who had recovered from COVID-19 and were good candidates to donate plasma for treating the disease.

Tracking and surveillance facilitated preparedness, clinical management, and control of the disease. China, for example, used daily aggregated origin-destination data from Baidu (China Data Lab 2020), a Chinese multinational company, to evaluate the effect of travel restrictions (Chinazzi et al. 2020) and quarantine measures (Kraemer et al. 2020) on COVID-19 transmission in the country. Further, tools such as migration maps, which use mobile phones, mobile payment applications, and social media to collect real-time data on the location of people, allowed Chinese authorities to track the movement of people who had visited the Wuhan market, the pandemic's epicenter. With these data, machine learning models were also developed to forecast the regional transmission dynamics of COVID-19 and guide border checks and surveillance (Whitelaw et al. 2020). In Italy, smart phone data

was used to monitor mobility, and its analysis showed an estimated reduction of 50% in the total trips between Italian provinces in the week after the announcement of lockdown in March 2020 (Pepe et al. 2020).

Interventional Planning and Clinical Management Interventional planning and clinical management also used technological measures to mitigate transmission and decrease fatality rates. One effort in that direction prompted innovation in remote monitoring platforms for improving management in hospitals and health systems across the world. In Sweden, a platform was developed as an information resource for healthcare workers to report real-time data on volumes of patients with COVID-19, assess status of personal protective equipment and staffing, ventilator usage, and other services. This information assisted healthcare authorities in tracking the status of facilities, allocating healthcare resources, and increasing hospital bed capacity. A smartwatch application was used in Germany, which collected pulse, temperature, and sleep pattern data to screen for signs of viral illness. Data thus obtained were displayed on an online, interactive map that enabled authorities to assess the likelihood of COVID-19 incidence across the nation. With such digital health interventions and extensive testing, Germany successfully maintained a low per capita mortality rate, relative to other countries, despite a high prevalence of cases (Whitelaw et al. 2020).

Telehealth Due to either total closure of many outpatient services or because of capacity issues or safety concerns, health authorities in many countries promoted policies to foster telehealth, which became a lifeline for patient care (Drees et al. 2020) during the lockdowns. In terms of telemedicine and healthcare consultation and delivery apart from "video visits," other options used include email and mobile phone apps as well as wearable devices, chatbots, artificial intelligence (AI)–powered diagnostic tools, voice-interface systems, and mobile sensors such as smart watches, oxygen monitors, or thermometers (Golinelli et al. 2020). Online consultation services were boosted either by revamping exist-ing privately owned online telemedicine plat-forms or equipping public hospitals for such functions. Health and human services even allowed providers to use such platforms as Facetime and Zoom for virtual visits covering a wide range of conditions, from urgent care, pri-mary care check-ups, medication follow-up, to COVID-19 screenings (Drees et al. 2020).

Robotics and 3D Printing Health systems also deployed robotic technology to treat coronavirus patients. Notable examples of this included New Brunswick-based Robert Wood Johnson University Hospital's "video robots" that trans-mitted videos to physicians inside the emergency room. Nuro robots were used in California to deliver food, fresh linens, and protective gear. Hospitals also used robotics to directly fight the novel coronavirus—thus Charlottesville-based University of Virginia decontamination robot used 3D imaging and ultraviolet light to kill COVID-19 pathogens. Robots also played a role in patient care at some hospitals (Drees et al. 2020). Regarding 3D printing, it was introduced in hospitals to help address equipment shortages and immediate PPE needs and shortages (Drees 2020). The potential of the 3D printing industry covered a wide range from helping in machine conversion to making oxygen tubes and nasal swabs. It was also used to make device compo-nents including parts for ventilators (Drees et al. 2020).

4.2.2.5 Governance and Public Administration

The governments at city, national, and other lev-els also switched to digital technologies to deal with the immediate problems generated by the crisis in the short term, recover from socioeco-nomic effects in the mid-term, and formulate policies and plans for future development in the long term. To start with, the digital communica-tion channels provided to cities and local govern-ments essential tools to communicate with citizens and give them reliable information on the disease and available essential and basic services in the cities and their neighborhoods. It also

allowed communication with competent authori-ties (UCLG, CGLU, Metropolis, and UN-Habitat 2020). In short term, it enabled healthcare man-agement and in providing immediate relief assis-tance to citizens through cash transfer and food supply. In addition, governments also leveraged digital technologies for its own business—remote work for its employees and for other public administration measures. They also used it to facilitate citizens and private organizations or businesses in filing taxes online as well as for conducting business registration, etc. and for pro-viding at least some public services online. In the long term, city governments as leaders will need to continue many aspects of public administra-tion by digital means at least for two reasons. Firstly, the officials within the government enti-ties are likely to or at least would wish to main-tain the advances made in digital transformation during the pandemic. Secondly, the citizens and private sectors who have acquired the habit of working online are likely to become more demanding because of easiness of access to ser-vices and the usability of digital apps.

4.2.2.6 Social Media

Digital technology–operated social media became a very important source of news, infor-mation, and communication during the pandemic and made people a little less lonely while practic-ing social distancing. Further, an increasing num-ber of people started relying on the Internet-supported social media for digital enter-tainment. According to Statista, 51% of global Internet users started watching more shows on streaming services due to the coronavirus (BDO 2020). Netflix alone saw 16 million new sign-ups for its service in the first 3 months of 2020 (BBC News 2020). That is almost double the new sign-ups it recorded in the final months of 2019. According to a WARC (2020) study, "daily online content use or consumption around the world soared from the start of the COVID-19 crisis, doubling on average from 3 hours 17 minutes to 6 hours 59 minutes." Some 48% of those sur-veyed indicated enhanced use of social media platforms. Overall, YouTube saw the biggest growth in audience interest, with 43% of consumers spending more time on the platform. Meanwhile, TikTok saw the highest growth among those aged 18–24. Businesses and organizations of different types also turned to digital platforms in response to the pandemic—for example, the fitness industry switched to holding virtual classes on streaming services, both live and pre-recorded. Big conferences and events also were held virtually (BDO 2020), and the New York Stock Exchange moved entirely initially to online trading to cope with the pandemic.

The surge on the use of social media was unparalleled; a glimpse of the same taken from Digital 2020 (Kemp 2020) is given below:

- Globally, 4.14 billion people were using social media in October 2020, more than half the world population.
- In terms of growth, social media users surged by more than 12% over the 12-month period (October 2019 to October 2020), amounting to an average of 14 new users every second.
- Among the world Internet users, about nine out of ten already use social media each month.
- The following six social media platforms now claim to have more than 1 billion monthly active users:
  1. Facebook has 2.701 billion monthly active users.
  2. YouTube has 2 billion monthly active users.
  3. WhatsApp has 2 billion monthly active users.
  4. Facebook Messenger has 1.3 billion monthly active users.
  5. WeChat (inc. Weixin 微信) has 1.206 billion users.
  6. Instagram's potential advertising reach is roughly 1.16 billion.
- More than half of the combined global audience of Facebook, Instagram, and Facebook Messenger live in cities with at least half a million inhabitants.

As people around the world got used to new normal during the pandemic, their digital social life will probably never be the same again. The habits they acquired of accessing social media for contacting friends and family, entertainment, making purchasing decisions without leaving their homes, and virtual reality may continue at least for a while. It is therefore likely that the use of social media will endure and the trends toward improving online streaming and other services will continue.

4.2.3 Pressure on Networks and Resilience

The leap in Internet users worldwide together with teleworking, teleservicing, online conferencing, and increasing need to access social media during the pandemic developed tremendous pressure on the networks and the Internet. It also amplified the demand for apps and online conferencing software such as Microsoft Teams, Skype, CISCO's Webex, and Zoom. Statista estimates that the total download of mobile apps in 2020 was 218 billion or 597 million per day (Website Builder 2021).

4.2.3.1 Intensity of Pressure

A logical outcome of the above was a massive increase in the use of the Internet and Internet-based services from 40% to 100%, compared to pre-lockdown levels (De' et al. 2020). Videoconferencing services like Zoom's increased ten times, and content delivery services like Akamai increased by 30% (Branscombe 2020). The number of users of Microsoft's new business messaging and collaboration tool, Microsoft Teams, grew 37% only in 1 week in March 2020 to more than 44 million daily users—at least 900 million meeting and call minutes were recorded on Teams every day (Wakabayashi et al. 2020). Stay-at-home orders also enhanced traffic on video-streaming sites and social media platforms. Downloads of Netflix's app—a proxy for traffic from the streaming site—jumped 66% in Italy, according to data from Sensor Tower, an app data company. In the USA, where Netflix was already popular, there was a 9% surge (Wakabayashi et al. 2020). Traffic also soared on

Facebook and YouTube, while cloud computing became essential to home workers. All this brought a huge pressure on the urban Internet—Bangalore in India, for example, recorded a 100% increase in Internet traffic. So did other cities across Asia, Europe, North America, and elsewhere.

According to OECD (2020b), many countries reported increase in traffic. In Korea, operators reported traffic increases by 13%; in Japan, NTT Communications reported an increase in data usage around 30–40%. Similarly, in the UK, British Telecom reported a 35–60% increase in daytime weekday fixed broadband usage. Telecom Italia in Italy saw a traffic increase of 63% and 36% in the fixed and mobile network, respectively. In the USA, both Verizon and AT&T reported increases. Verizon reported a 47% increase in use of collaboration tools and a 52% increase of virtual private network traffic, while AT&T reported that its core network traffic was up 23%.

While overall telecommunication networks exhibited resilience despite the changes in traffic (Sinibaldi 2020; Cooper 2020), accessible ultrabroadband technologies such as fiber to the home (FTTH) appeared better prepared to respond to spikes in broadband traffic. Countries with the largest deployments of accessible ultrabroadband exhibited less slowdown in latency and download speed (Katz et al. 2020). Wi-Fi capacity was also stressed by an 80% increase in PC uploads to cloud computing platforms (Fig. 4.7) with additional peaks from videoconference calls (Gil et al. 2020), requiring additional spectrum to be assigned for unlicensed use. Adaptive Spectrum and Signal Alignment, Incorporated (ASSIA 2020), vendor to over 35 service providers worldwide with more that 125 million broadband and Wi-Fi lines under contract, in 17 countries, reported that in March 2020, they were contacted by multiple carriers and service providers to help troubleshoot connectivity issues.

One critical element of network infrastructure is Internet exchange points (IXPs), which are bulk traffic exchange crossroads where multiple networks connect (to exchange traffic). IXPs report recorded net increases of up to 60% in total bandwidth handled by selected country from December to March 2020 (Fig. 4.8). Individual IXPs have also reached new records of peak traffic. DE-CIX Frankfurt, one of the largest IXPs in the world, at one time was regularly peaking over 9.1 terabits per second (Tbps) data, which equals a simultaneous transmission of up to two million high-definition videos.

To cope with the significant traffic increases, network operators and governments across the globe worked continuously to ensure that connectivity and communication services operated in a reliable, stable, and secure manner. Fixed and mobile broadband operators, as well as content providers, were able to successfully manage their networks to accommodate changes in utilization patterns, responded to overall increased demand, and avoided congestion that would have impacted working and studying from home and operation of critical services such as telemedicine and emergency response.

4.2.3.2 Resilience to Stress

The above discussion shows that the coronavirus crisis thrust the cities around the globe suddenly into physically isolated, online existence. All facets of online technology, including the Internet backbone, broadband and wireless infrastructure, cloud resources, and online services, underwent a stress test that no one had predicted, and very few providers were fully prepared for it. With several months of hindsight, however, it appears that the Internet, in the collective sense of the term, passed the test amicably (Kurt 2020). Regarding the digital technology itself, the world no doubt increasingly relied upon digital technology to help weather the COVID-19 storm and several waves of the contagion, and its need continues in recovery and overall future development.

4.3   Climate Change, ICT, and Digital Technologies

Technologies including satellites, mobile devices, and the Internet are already being used to address challenges related to climate change. They have shown great potential in monitoring climate

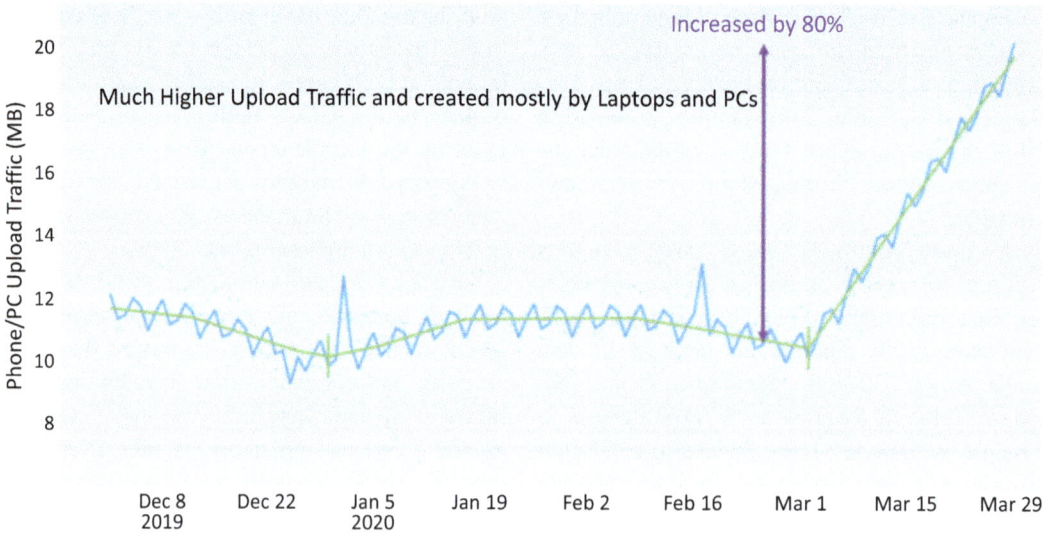

Fig. 4.7 Increase in PC/phone upload traffic during early period of COVID-19. (Source: ASSIA (2020))

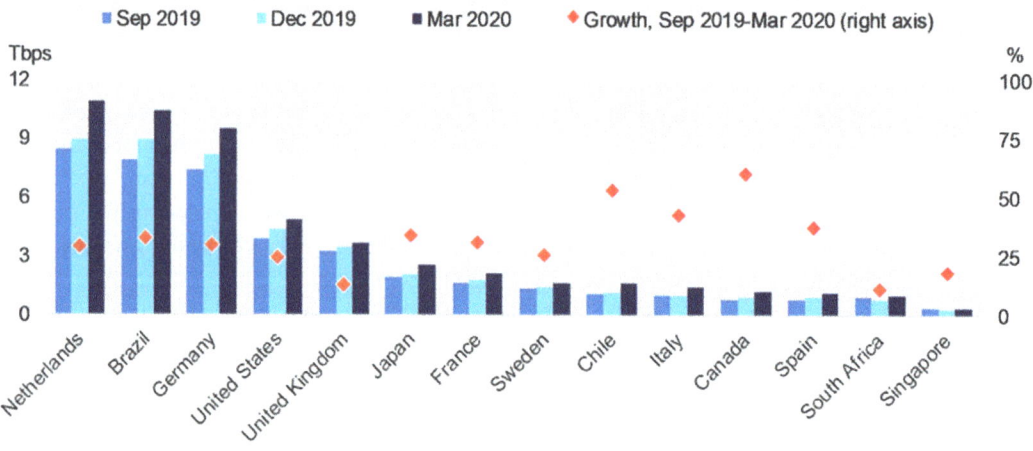

Fig. 4.8 Internet bandwidth use growth at Internet exchange points in selected countries. (Source: OECD (2020b))

change, its mitigation, and managing its impacts. In addition, they carry great potential to build climate-resilient cities and have a crucial role in adaptation. In addition, the role of digital technology in developing networks of cities and societies for exchange of information and experience and collective action is invaluable. Case studies presented in Chap. 3 of this book have already shown their significance in monitoring and managing climate change using GIS and remote sensing.

Besides bolstering connectivity in and between cities because of exponential increase in devices and tools, digital technology is also producing societal impacts in exchange of ideas on a scale that was never experienced before. It is contributing to greater awareness and education, with positive effect on fostering an environmentally responsible mindset among all stakeholders—including policy makers at city and national levels, business and industries, NGOs, citizens, and academia. This is extremely important as everyone needs to come together to take responsibility for the very complex problem of climate change and to take concerted and timely action (ITU 2019) in handling it. The role of digital tools is important in dealing with all aspects of climate change-related issues including mitigation and adaptation. Their role is particularly vital even in handling a climate-related disaster, whereby improved communications help in handling these more effectively—from carrying out evacuation and emergency relief to conducting efforts for recovery. In summary, the key areas in which digital technology helps address climate actions are monitoring and warning, mitigation, adaptation, and handling extreme climate events.

4.3.1 Monitoring and Early Warning

Digital and ICT technologies have assisted in obtaining vital information pertaining to climate, weather, and related risks and disasters through World Weather Watch (WWW).

4.3.1.1 World Weather Watch

The World Weather Watch has three component systems: the Global Observing System, the Global Telecommunication System, and the Global Data-processing and Forecasting System. These three individual components deliver their achievements in combination for monitoring and relaying weather-related information and to prepare for climate-related risks. The Global Observation System (GOS) of the World Meteorological Organization (WMO) is the foundation of the WWW. The GOS is critically important in understanding both short-term climate events and long-term climate change and its impacts on cities and human habitat. The tools utilized for the purpose (Fig. 4.9) include satellite and surface-based remote sensors, equipment, and accessories that include the following: weather satellites that track the progress of hurricanes and typhoons; weather radars that track the progress of tornadoes, storms, and the effluent from forest fires; radio-based meteorological aid systems that collect and process weather data;

and Earth Observation Satellite System, which obtains environmental information such as atmospheric composition (e.g., carbon dioxide, vapor, ozone concentration), ocean parameters (like temperature, surface-level change), soil moisture, vegetation including status of forests, and other data related to it (WMO and ITU 2017; ITU 2017).

The sensors, accessories, and their associated processes spot and forecast as well as issue warning for extreme weather events. Collected data by the Global Observing System (GOS) enables studying both short-term weather conditions and long-term global climate change. The GOS has also become part of the Global Climate Observing System (GCOS).

4.3.1.2 Global Climate Observing System

Global Climate Observing System is together co-sponsored by WMO, the Intergovernmental Oceanographic Commission of the United Nations Educational, Scientific and Cultural Organization (IOC-UNESCO), the United Nations Environment Program (UNEP), and the International Science Council (ISC). Established in 1991, the GCOS supports all the components of the World Climate Program and associated activities.

Following the Paris Agreement of 2015, GCOS is in the process of working on three aspects—(a) observational requirements to monitor emissions and emission reductions, (b) information needs for assessing adaptation to climate change, and (c) climate resilience and data needs for public awareness and capacity development (WMO, IOC-UNESCO, UNEP, and ISC 2016). It also regularly assesses the status of global climate observations of the atmosphere, land, and ocean and produces guidance for its improvement. The status reports and implementation plan are provided to the United Nations Framework Convention on Climate Change (UNFCCC), the focal point of climate action in the United Nations. The last status report (GCOS 2021) was prepared in 2021, and the implementation plan in response to this report will be released in 2022. The outline of the implementation plan

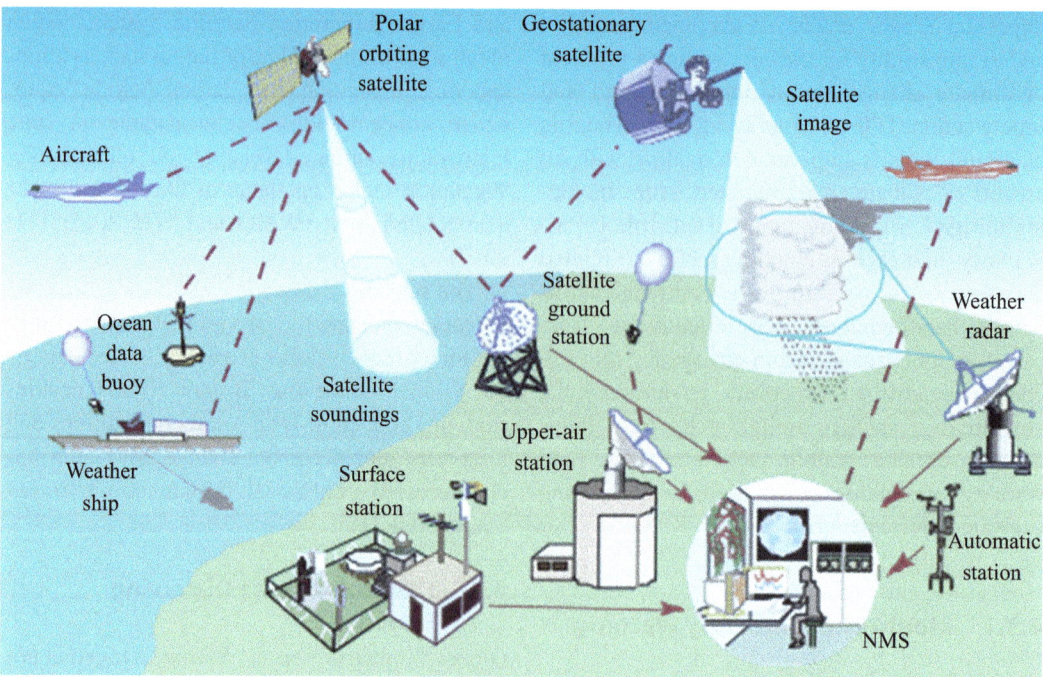

Fig. 4.9 Global observing system of the World Meteorological Organization. (NMS National Meteorological Service/ Station). (Source: WMO (2022))

has already been drafted (WMO, IOC-UNESCO, UNEP and ISC 2021).

The Global Climate Observation System (GCOS) while developing implementation plan gives due consideration to the climate monitoring needs of the Climate Convention in the context of the Paris Agreement, other related multilateral agreements, as well as the Sustainable Development Agenda 2030 particularly its Goal 13 related to climate change. It also sets out the framework for the science community to provide the data and information to implement the global climate observing system, advance scientific research knowledge, and support climate services and the development of climate indicators (UNFCCC undated). The GCOS plans and progress have been documented in a publication (GCOS 2021). One of the most important achievements of the GCOS is the development of WMO's Global Observing System and its potential funding mechanism—the Systematic Observations Financing Facility (SOFF), which supports countries to generate and exchange

basic surface-based observational data critical for improved weather forecasts and climate services. If fully implemented, it would lead to significant improvements in climate modeling.

4.3.2 Mitigation

IPCC (2014a) defines climate change mitigation as limiting and preventing the emission of greenhouse gases by enhancing activities that remove these from the atmosphere. In cities, the main sources of GHG are energy, transportation, buildings, industry, and waste management activities. Digital technologies have been used for mitigation actions in all these sectors—for improving energy efficiency in buildings; facilitating data center management; improving traffic congestion and public transportation in metropolitan areas; and operating smart grids (ITU 2019) as well as for smart waste disposal. Further, smart sensors and Internet of things (IoT) applications have allowed to enhance the capability to collect, transfer, and analyze a vast amount of data on these in near real time. In addition, when used in building and urban construction, they have enabled to measure energy use, optimize energy distribution, and automate building operations, thus, contributing to improve overall energy performance of buildings. They have also been used to manage traffic flows in cities such as Moscow (Smiciklas and Imran 2018) and in the Rapid Transit System such as Belfast (ITU 2019).

The United for Smart Sustainable Cities (U4SSC), a global platform initiated by the United Nations and coordinated by the International Telecommunication Union (ITU), United Nations Economic Commission for Europe (UNECE), and UN-Habitat, has also advocated for the use of digital technology in facilitating the transition to smart, energy-efficient, climate-resilient, and sustainable cities (ITU 2019). With inputs and expertise gathered from telecom providers, policy makers, civil society, and city stakeholders, the U4SSC has developed technical reports and guidelines that support cities in connecting digitally, making them resilient to climate change and promoting sustainable development goals. To encourage cities to identify and explore the potential of digital technologies in addressing the most pressing urban issues, the U4SSC has developed the key performance indicators (KPIs) for Smart Sustainable Cities (ITU 2019) and published case studies that document the experiences of the participating cities particularly those of Dubai, Singapore, and Moscow in implementing the U4SSC key performance indicators for smart sustainable cities (the concept of Smart Cities has been discussed in detail in Chap. 11 of this book).

While discussing mitigation, it is important to note that the digital sector by itself has also been contributing to global GHG emissions. The carbon footprint of digital technology comes from manufacturing devices (phones, tablets, laptops, etc.), replacement rather than repair of devices, and consumption of electricity for digital technology uses especially their high demand in new or frontier technologies (artificial intelligence, machine learning, blockchain, etc.). With more and more people coming online, more data being generated, and more devices connecting to the network, the proliferation of frontier technologies GHG emission from the digital ecosystems has increased exponentially. Moreover, information and communication component of digital technologies is powered by hundreds of thousands of data centers. These data centers are already consuming more than 2% of the world's electricity and are emitting roughly as much emissions as the aviation industry (IPCC 2014a). As the world is moving faster in a digital age, this percentage continues to increase and is expected to enhance further. Hence, it has become imperative to monitor this growth closely to assess its benefits and costs and mitigate its adverse impacts. As part of this effort, ITU has recently established a focus group that will provide a global platform to raise awareness of the environmental impacts of frontier technologies, as well as disseminate information on the ability of these technologies for the achievement of the Sustainable Development Goals and the objectives of the Paris Climate Agreement (ITU 2019). Along with this credible effort however, it is also important that some of the largest digital or ICT companies take steps toward adapting renewable energy to power data centers and further lower their carbon emission by improving overall energy efficiency (Aréstegui 2018). Microsoft, Deutsche Telecom, and Lyft are some of the leading ICT companies which are striving for this and have joined RE100, a global initiative that brings together businesses that are committed to using 100% renewable energy (2030 Vision 2020). Along with adoption of renewable energy, adherence to international standards for energy-efficient ICTs can help further in this effort.

4.3.3   Adaptation

Digital technologies can help cities adapt to climate in many ways. Firstly, they play a vital role in establishing early warning systems. Digital networks and infrastructure are critical in disseminating and circulating weather and climate-related information. It is with this information that cities and their disaster response units can prepare for appropriate response. Many countries and cities vulnerable to climate-related hazards have already put in place early warning systems and implementing strategies to limit the impacts of climate change. Some notable examples of countries are Peru (Aréstegui 2018; IRC 2018), the Philippines (Malano 2015; PAGASA 2019), Malawi (IRC 2018; Savvides 2018), and Mozambique (Rae 2019).

Digital and ICT technologies, through their role in information and communication, have helped migrants and refugees—those forced to flee their homes due to recurring climate-related extreme events. For example, they are enabling these vulnerable communities to communicate with each other, stay informed on current affairs, keep in touch with family and friends, and acquire information to access the basic services during disasters. They have also helped humanitarian organizations as well as local and national governments in gathering critical information, for improving their internal services to better support those in need. In addition, mobile phone data are being used to assess climate change and migration patterns in Bangladesh (Lu et al. 2016). Most importantly, in some climate emergencies, they are the only provider of lifeline or basic connectivity (Granryd 2017).

The role of remote sensing and geographic information systems in developing adaptation plans has already been explained in Chap. 3. These technologies have made possible risk mapping, scenario development, modelling, and development of contingency plans. Risk analysis involving risk and hazard mapping as well as scenario development is a key component in developing a disaster risk reduction strategy by establishing the links between exposure to hazards, level of vulnerabilities, and the ability to cope (Bueti and Faulkner, undated). It is for the same reason that the Nairobi Framework, adopted in 2006, aims to assist all UNFCCC reporting Parties, in particular developing countries, including least developed countries (LDCs) and Small Island Developing States (SIDS), to improve their understanding of risks, vulnerabil-

ity, and adaptation. This would help them make informed decisions on practical adaptation actions and measures to respond to climate change on a sound scientific, technical, and socioeconomic basis, considering current climate change realities.

4.4 Digital Technology, Ethical Considerations, and Other Issues

This chapter has so far discussed how the use of digital technology has increased during the prevalence of mega risks. With mass vaccination on the way and relaxation and lifting of stay-at-home, social distancing, and lockdown measures, what will be people's future digital activity pattern is difficult to predict. However, the path to digitalization had already started even before the pandemic, and the pandemic only accelerated it. Most studies and surveys, as mentioned before, have predicted that the foot that has been put on the accelerator is likely to remain there, though there may be some variation in expectations regarding the speed. The continuation and most likely acceleration of their use, however, makes it important to appropriately examine their limitations, as well as their negative issues that cropped up in terms of infringement of privacy and personal autonomy, inequality, insecurity resulting from associated criminal activities or cybercrimes, spreading of misinformation, etc. These necessitate putting in place mechanisms that could mitigate or control these.

4.4.1 Infringement of Privacy and Personal Autonomy

Digital tools have the potential to undermine not only privacy but also personal autonomy particularly if their applications during the pandemic is transformed into standard surveillance protocols. The most obvious form of violation of personal autonomy is the mandatory use of digital public health technologies. For example, India's home ministry has required that all local workers, public or private, use a government-backed COVID-19 tracking application called Aarogya Setu (Fingas 2020). Less explicit threats to autonomy are raised by smart phone applications that include permissions to collect data beyond the stated purpose of the application (Gasser et al. 2020). In some cases, the effectiveness of quarantine might be undermined if it remains voluntary rather than mandatory. It may however overreach autonomy if a government overdo it—for example, the action of Polish government in creating shadow profiles for returning citizens (Hamilton 2020). Likewise, tools using aggregate mobile phone tower data are, on average, less privacy-invasive compared with tools based on Global Positioning System data and sensor tracking for individual users (Amnesty International 2020a). Concerns have been particularly raised about centralized systems and GPS tracking. Norway halted data collection, and its use from Smittestopp app after the country's data protection watchdog objected to the app's collection of location data as "disproportionate to the task," and they recommended a Bluetooth-only approach. According to Amnesty International (2020b), this is a major win for privacy. The good news is that several international frameworks with varying levels of privacy preservation are also coming out like Decentralized Privacy-Preserving Proximity Tracing (DP-3T/ documents 2020), the Pan-European Privacy-Preserving Proximity Tracing initiative, and the joint Google-Apple framework (Goldberg 2020).

A major problem arises when private data is acquired for one objective or reason, say, health, and then later it is shared with other public agencies—like police or agencies responsible for bio-surveillance (Kim and Tak 2019) for a different reason, say counter-terrorism. Likewise, data on demographics acquired through digital technologies on race, ethnic group, gender, political affiliation, and socioeconomic status, which stratify populations, is also ethically sensitive as it may lead to stigmatization of one or more ethnic or socioeconomic group for discrimination against them (Quinn 2018).

4.4.2 Digital Divide

One unfortunate aspect of digitalization is the existing digital divide, which continues to accentuate inequalities between cities and social groups. This is because not all nations, places, and social groups are able to harness the potential of digital technologies alike to combat the risks. According to the State of Broadband Report (Broadband Commission 2019), although Internet user penetration rate is 51.2%, it is only 45% in developing countries and 20% in least developed countries. Although the digital and knowledge divides have always existed, in a situation where many people must stay at home, it transforms from a disadvantage to a debilitating disability. It is therefore important that access to digital technologies is increased at a faster pace in the long term. Meanwhile, in the short term, it should be made sure that the digital divide does not translate into an inability to continue daily life (UNESCO 2020a). The citizens affected most by digital divide or who remain cut off from the benefits of the digital era are women, the elderly, persons with disabilities or from ethnic or linguistic minorities, indigenous groups, and residents of poor section of cities or remote towns. The pace of connectivity is slowing, and even reversing, among some of these communities. For example, globally, the proportion of women using the Internet is 12% lower than that of men (ITU 2017). While this gap narrowed in most regions between 2013 and 2017, it widened in the least developed countries from 30% to 33% (United Nations 2020a).

Digital divide or inequality in the context of harnessing the potential of digital technologies for education, telehealth, remote work, social networking, etc. is also a serious concern. It could result from lack of access to technology due to unavailability of a device or the Internet. For example, smart phones—one most used digital device—are very unevenly distributed in cities and countries. In 2019, two thirds of the world's population did not own a mobile phone technology, and one third did not own any mobile phone (Gasser et al. 2020). Smart phone ownership disparities are particularly noticeable in developing countries. For instance, in India, the world's second most populous country accounting for more than 17% of the global population, only 24% of adults owned a smart phone (Silver 2019). Even in developed world with high smart phone ownership rates, not all age cohorts possess them. Thus, in Japan, Italy, and Canada, most citizens older than 50 years did not own a smart phone in 2018 (Bradshaw 2020). Hence, they, like other marginalized groups, are automatically excluded from any digital public technology solution for health, education, and other services that rely on smart phones or devices. In the USA, a survey by the Urban Institute of chief technology officers, chief information officers, chief innovation officers, and digital inclusion leaders in cities across the country indicated that although Internet access is expanding, the digital divide is more consequential than ever (Brown et al. 2020).

4.4.3 Cybersecurity

Cybersecurity is another serious issue for all, while "cyber incidents are consistently ranked at the top of business concern" (Brown et al. 2020). The global cost of cybercrimes is predicted to inflict damages totaling six trillion dollars in 2021 (Morgan 2020). In the future, these costs are expected to grow by 15% annually in the next few years, peaking to $10.5 trillion by 2025 (Morgan 2020). The number of cyberattacks during COVID-19 increased several-fold (Williams et al. 2020). Two major victims were Marriott and Honda customers. The former experienced a data breach that affected its 5.2 million customers, and a ransomware attack forced the later (Honda) to shut down its global operation (Dooley and Ueno 2020). Hackers also attacked collaboration platforms—a data breach affected more than 500,000 Zoom users in April 2020. Due to another attack in the same month, a zoom videoconference meeting of the Milwaukee Election Commission had to be shut down (Richardson and Mahle 2020). According to Federal Trade Commission, there had been more

than 172,000 fraud reports related to the pandemic itself, at the cost of about $114.4 million by mid-August 2020 (Brown 2020). The pandemic provided new opportunities to hackers, crooks, and criminals. The massive increase in the use of the internet resulted in a hike of malicious attacks by phishing emails and messages, SMS text, malware inserted into COVID-19-related resources, the sale of fake products and phony cures for COVID-19, and fraudulent links posing to give vital information and messages impersonating high-profile organizations like WHO, CDC, and websites embedded with hackers' codes (Saeed et al. 2020; Brown et al. 2020; Davies 2020). The combination of virus, remote work, and online business changed the adversary's opportunities and shifted their focus on some of the tools that the security personnel use (Brown et al. 2020).

Changing cybercrime tactics and a shift to employees working from home and more businesses expected to come online in the future raise the importance of security strategies, according to experts from MIT Sloan School and cybersecurity executives from companies such as Mastercard, Booz Allen, Liberty Mutual, and the Mars Co. Experts talked about the expected security threats and how their strategies have shifted during the pandemic in two recent webinars—offering some best practices for cyber resilience (Brown et al. 2020). With the likelihood of enhanced use of digital technology in the future, it is imperative to reinforce cybersecurity through technical means, appropriate policies, and practices as well as laws and rules to regulate cybersecurity. In terms of technical means, the businesses and organizations and individuals can safeguard against cyberattacks by following the Trusted Cyber-Infrastructure (TCI) considerations. They also need to make sure that laptops, desktops, cell phones, and apps are updated and installed with every patch that is required, regularly updating the browsers. Further, they should develop secure collaboration tools for the transfer of data, shifting away from a perimeter-based security model to protecting access to information with emphasis on identity by adopting a zero-trust architecture and using multifactor authentication for accessing particularly the organizational data.

In terms of policies and practices, it is important to review and update data security policies to ensure that they are compatible with the remote work setup and that all employees are aware of it. It is also important to limit access to protected and confidential information; encourage employees to use virtual private network or other secure form of remote access; and remind and keep employees alert on scams, phishing emails, fraudulent links and websites, etc. Further, video-conferencing and meetings should be accessed by invitation and made secure by using videoconferencing software with security features, locking it once initiated, disabling screen sharing when practical, and using a virtual waiting room before the host joins (Richardson and Mahle 2020). "Bait-phishing" exercises can also be useful in which companies send a phishing-type email to their own employees to make sure that they remain alert to potential scams. Moreover, companies can benefit from working together and sharing best practices.

In terms of laws and regulation for cyberspace, the role of governments becomes very important. They would need to adjust national legal and regulatory frameworks to the need of the time while working together to increase international cooperation. Their agility in updating or developing national cybersecurity legal and regulatory framework in collaboration with the technical community and the private sector is extremely important. International cooperation is equally important for cybersecurity, which demands increased trust, at all levels, between countries and industries. Collaboration at the policy, technical, and law enforcement levels is vital to protect common interest at regional and international levels (Contreras 2020). Here, the United Nations should play an important role. It is already providing a platform for cooperation and making an objective assessment of the impact that emerging technologies have on cybersecurity as well as overall sustainable development outcomes. Moreover, its mandate allows it to bring together people, businesses, and organizations from across the world to build strong consensus-led agreements.

4.4.4 Spreading of Misinformation

Another major ethical issue in the use of digital technology is spreading of misinformation (Hart et al. 2020). While both the pandemic and climate crisis pose significant risks, national responses and public perceptions have sometimes been politicized. This raises questions about the role of politicians as well as media in amplifying politicization and polarization of COVID-19 and climate change. This is extremely important in terms of message conveyed on digital platforms like Twitter, Facebook, Google, etc. as well as news coverage in digital newspapers. These have been particularly influencing public attitudes particularly toward COVID-19 in countries like the USA and Brazil. It is encouraging to note that the digital platforms have started labeling misleading contents and removing posts that violate community welfare.

4.4.5 Environmental and Social Impacts

The use of digital technology is not without environmental and social impacts, which are both positive and negative. Among positive environmental impacts are dematerialization such as digitization of books and literature. Moreover, growth in online activities has also made positive changes in land use and resulted in fewer offices and malls, more warehouses occupying less spaces, reduced traffic, and increased environmental awareness and hence adoption of environmentally friendly lifestyle. The negative footprint comes from growth in use and manufacturing of devices (phones, tablets, laptops, network infrastructure, etc.), replacement rather than repair of devices, higher consumption of electricity by data centers, and emerging frontier technologies (artificial intelligence, machine learning, blockchain, etc.). Among positive social impact is enhanced connection of people and removal of their social isolation. However, overdependence has had severe negative impacts during Internet shutdowns which does happen in many developing cities—as documented in the case of Kashmir Shutdowns (De' et al. 2020). There are also additional issues related to technostress—particularly work overload and excess supervision.

4.5   Conclusion

The information presented in this chapter shows the importance of digital technologies in managing mega Risks—both the coronavirus pandemic and climate crisis. The technology acted as lifeline for cities during the coronavirus pandemic when physical contacts became minimal in the wake of lockdown and social distancing orders. Companies, businesses and retailers, educational institutions, and health centers which embraced digital weathered the storm better and are prepared for the future. The pandemic brought forward the true potential of digital technologies for individuals, governments, and businesses in cities around the world. Even early in the pandemic, the World Bank's Digital Development Global Practice identified more than 300 government and private sector initiatives in different parts of the world in which COVID-19 response encompassed actions on digital ICT infrastructure and digital services for health, education, and payments. The database did not even encompass the whole world but covered responses in 30 high-income countries and 53 low- and middle-income countries, including more than 20 countries in Africa (Begazo 2020).

Along with the enhancement of their use, the pattern of utilization of digital technologies also got extended, diversified, and transformed by the events of 2020. Their innovative applications led to the utilization of frontier technologies such as blockchains, artificial intelligence (AI), machine learning (ML), cloud computing, and smart technologies, commonly referred to as the Internet of Things (IoT). They are paving the way for automation, enabling surveillance, reducing transaction time, helping in making the system robust in increased traffic, and offering a mean to reimag-

ine and transform physical space in homes, offices, factories, healthcare facilities, and public and civic buildings and centers. A logical outcome of the above was a massive increase in Internet and Internet-based services. Videoconferencing services like Zoom, Teams, and Slack flourished. Zoom's use increased ten times, and content delivery services like Akamai increased by 30%. The number of users of Microsoft's new business messaging and collaboration tool, Microsoft Teams, grew 37% only in 1 week in March 2020 to more than 44 million daily users—at least 900 million meeting and call minutes were recorded on Teams every day. This immense growth put all components of online technology—the Internet, broadband and wireless infrastructure, cloud resources, as well as online services under a stress test no one had predicted and few if any were fully prepared for. However, with several months of hindsight, apart from a few glitches, the resilience shown in the collective sense was admirable—the degree of performance was good, and reliability degradation was far less than that feared in the early days.

Digital technologies, besides rescuing city functions during the pandemic, brought forward unparalleled and multiple benefits—ranging from urban health surveillance to improving efficiency in public administration; allowing continuity of business and education through remote operation; maintaining events in social spheres; delivering relief; and maintaining communication with public at large. Their proven application (particularly ICT) in climate change monitoring, early warning, disaster management, mitigation, and adaptation is also well known—a witness to that is the Global Observation System (GOS), an efficient mechanism for, among others, releasing early warnings on climate-related hazards. It is important to mention however that the application of digital technologies while proving their benefits has also brought forward many shortcomings and gaps in their usage, management, and governance. Among these are inequity due to digital divide; infringement of privacy and personal autonomy through misuse of personal data collected; hazards of cyber insecurity; and the threat of misinformation. In addition, like any other industry or business, the digital technologies by themselves carry beneficial as well as harmful social and environmental impacts, which demand their diligent mitigation.

The likely continuation and enhancement of the use of digital technology in the future or new normal demand implementation of methods and mechanisms that would alleviate its demonstrated deficiencies and shortcomings. Adoption of appropriate policies and practices to at least reduce if not alleviate digital divide is needed. In addition, regulatory laws and rules and technical means as well as enhanced collaboration between cities and countries are imperative for the enhancement of cybersecurity. Realizing the importance of digital technologies in the promotion of future sustainable development at all levels, the Secretary General of the United Nations had established a High-level Panel on Digital Cooperation which submitted its report in 2019 (UN High Level Panel on Digital Cooperation 2019), following which round-table discussion groups of subject-matter experts were constituted to address the Panel's recommendations. Subsequently, the Secretary General's report was published (United Nations 2020b) proposing further line of action on digital cooperation in the following areas: global connectivity; digital public goods; digital inclusion; digital capacity building; digital human rights; artificial intelligence; digital trust and security; and global digital cooperation. Hopefully, it would lead to meaningful action at the ground level.

References

2030 Vision (2020) Climate crisis: how digital technologies can help. https://www.2030vision.com/news/climate-crisis-how-digital-technology-can-help

Amnesty International (2020a) COVID-19, surveillance and the threat to your rights, April 3. https://www.amnesty.org/en/latest/news/2020/04/covid-19-surveillance-threat-to-your-rights/

Amnesty International (2020b) Norway: halt to COVID-19 contact tracing app a major win for privacy. https://www.amnesty.org/en/latest/news/2020/06/norway-covid19-contact-tracing-app-privacy-win/

Aréstegui M (2018) Intermediate climate information systems for early warning systems. Practical Action, June

Armitage H (2020) Stanford Medicine scientists hope to use data from wearable devices to predict illness, including COVID-19. Stanford Medicine News Center. http://med.stanford.edu/news/all-news/2020/04/wearable-devices-for-predicting-illness-.html

ASSIA (2020) The new normal, holiday level upload. https://www.assia-inc.com/the-new-normal-holiday-level-wi-fi-upload/

Australian Government (2020) China's education arrangements during COVID-19 pandemic period. https://internationaleducation.gov.au/International-network/china/PolicyUpdates-China/Pages/China%27s-education-arrangements-during-COVID-19-pandemic-period.aspx

BBC News (2020) Netflix gets 16 million new sign-ups thanks to lockdown. https://www.bbc.com/news/business-52376022

BDO Binder Dijker Otte (2020) COVID-19 is accelerating the rise of the digital economy. https://www.bdo.com/insights/business-financial-advisory/strategy,-technology-transformation/covid-19-is-accelerating-the-rise-of-the-digital-economy/

Begazo T (2020) COVID-19: we're tracking digital responses worldwide. Here's what we see. https://blogs.worldbank.org/digital-development/covid-19-were-tracking-digital-responses-worldwide-heres-what-we-see

Berg J, Bonnet F, Soares S (2020) Working from home: estimating the worldwide potential. Vox. https://voxeu.org/article/working-home-estimating-worldwide-potential

Bradshaw T (2020) 2 billion phones cannot use Google and Apple contact-tracing tech. https://arstechnica.com/tech-policy/2020/04/2-billion-phones-cannot-use-google-and-apple-contract-tracing-tech/

Branscombe M. The new stack; 2020. The network impact of the global COVID-19 pandemic, https://thenewstack.io/the-network-impact-of-the-global-covid-19-pandemic/. Accessed on 6 June 2020

Broadband Commission (2019) State of Broadband, ITU and UNESCO, Geneva

Brown S (2020) How to think about cybersecurity in the era of COVID- 19, MIT Management Sloan School. https://mitsloan.mit.edu/ideas-made-to-matter/how-to-think-about-cybersecurity-era-covid-19

Brown M, Ezike R, Stern A (2020) How cities are leveraging technology to meet residents' needs during a pandemic, June. https://www.urban.org/sites/default/files/publication/102355/how-cities-are-leveraging-technology-to-meet-residents-needs-during-a-pandemic_1.pdf

Budd J, Miller BS, Manning EM et al (2020) Digital technologies in the public-health response to COVID-19. Nat Med 26:1183–1192. https://doi.org/10.1038/s41591-020-1011-4

Bueti C, Faulkner D (undated) ICTs as a key technology to help countries adapt to the effects of climate change. World Resources Institute. https://www.wri.org/our-work/project/world-resources-report/icts-key-technology-help-countries-adapt-effects-climate

Calamari (2020) Remote work in Europe before COVID 19 pandemic. https://calamari.io/blog/remote-work-in-europe-before-covid

Castelo M (2020) The state of educational technology in a post-pandemic world. https://edtechmagazine.com/k12/article/2020/11/state-educational-technology-post-pandemic-world

Chakravorti B, Chaturvedi RS (2020) Which countries were (and weren't) ready for remote work? Harvard Business Review, https://hbr.org/2020/04/which-countries-were-and-werent-ready-for-remote-work

Chavez-Dreyfuss G (2020) Permanently remote workers seen doubling in 2021 due to pandemic productivity: survey. Reuters. https://www.reuters.com/article/us-health-coronavirus-technology/permanently-remote-workers-seen-doubling-in-2021-due-to-pandemic-productivity-survey-idUSKBN2772P0

China Data Lab (2020) Baidu mobility data. Harvard Dataverse. https://doi.org/10.7910/DVN/FAEZIO

Chinazzi M et al (2020) The effect of travel restrictions on the spread of the 2019 novel coronavirus (COVID-19) outbreak. Science 368:395–400

Chun A (2020) In a time of coronavirus, China's investment in AI is paying off in a big way. South China morning Post, 18 March

Consultancy.eu (2020) One-third of work in Europe could be remote post-pandemic. https://www.consultancy.eu/news/5436/one-third-of-work-in-europe-could-be-remote-post-pandemic

Contreras B (2020) 3 Ways Governments can address cybersecurity in the post-pandemic world. https://www.weforum.org/agenda/2020/06/3-ways-governments-can-address-cyber-threats-cyberattacks-cybersecurity-crime-post-pandemic-covid-19-world/

Cooper T (2020) Internet speed analysis, top 200 cities, March 15th – March 21st. Broadbandnow. https://broadbandnow.com/report/internet-speed-analysis-march-15th-21st/

Dahik A et al (2020) What 12,000 employees have to say about the future of remote work. BCG. https://www.bcg.com/publications/2020/valuable-productivity-gains-covid-19

Davies N (2020) Are we ready for a post-COVID-19 cybersecurity landscape? http://www.circleid.com/posts/20200501-are-we-ready-for-a-post-covid-19-cybersecurity-landscape/

Day M, Frazis H, Loewenstein MA, Sun H (2020) Ability to work from home: evidence from two surveys and implications for the labor market in the COVID-19 pandemic. US Bureau of Labor Statistics. https://www.bls.gov/opub/mlr/2020/article/ability-to-work-from-home.htm

De' R, Pandey N, Pal A (2020) Impact of digital surge during Covid-19 pandemic: a viewpoint on research and practice. Int J Inf Manag 55:102171. https://doi.org/10.1016/j.ijinfomgt.2020.102171

Dhawan S (2020) Online learning: a panacea in the time of COVID-19 crisis. J Educ Technol Syst, SAGE 49(1):5–22. https://doi.org/10.1177/0047239520934018

Dingel JI, Neimann B (2020) How many jobs can be done at home? white paper. Chicago: Becker Friedman Institute for Economics at the University of Chicago, April. https://bfi.uchicago.edu/wp-content/uploads/BFI_White-Paper_Dingel_Neiman_3.2020.pdf

Dooley B, Ueno H (2020) Honda hackers may have used tools favoured by countries. The New York Times, 12 June. https://www.nytimes.com/2020/06/12/business/ransomware-honda-hacking-factories.html

DP-3T/documents (2020) Decentralized privacy-preserving proximity tracing–documents. The DP-3T Project. https://github.com/DP-3T/documents

Drees J (2020) How hospitals are tapping into 'endless potential' of 3D printing to combat COVID-19, 13 April. https://www.beckershospitalreview.com/healthcare-information-technology/how-hospitals-are-tapping-into-endless-potential-of-3d-printing-to-combat-covid-19.html

Drees J, Dyrda L, Adams K (2020) 10 big advancements in healthcare tech during the pandemic. Becker's Health IT. https://www.beckershospitalreview.com/digital-transformation/10-big-advancements-in-healthcare-tech-during-the-pandemic.html

Eurofound (2020) Living, working and COVID-19, COVID-19 series. Publications Office of the European Union, Luxembourg

Fingas J (2020) India requires all workers to use its COVID-19 tracking app. https://www.engadget.com/india-requires-workers-to-use-covid-19-app-042811484.html

Garcia T (2020) Retail bankruptcies in 2020 hit the highest levels in more than a decade, and experts say there are more to come. https://www.marketwatch.com/story/retail-bankruptcies-in-2020-hit-the-highest-levels-in-more-than-a-decade-and-experts-say-there-are-more-to-come-11608151350

Gasser U, Ienca M, Scheibner J, Sleigh J, Vayena E (2020) Digital tools against COVID-19: taxonomy, ethical challenges, and navigation aid. The Lancet Digital Health 2(8):e425–e434, ISSN 2589-7500. https://doi.org/10.1016/S2589-7500(20)30137-0

GCOS, The Global Climate Observing System (2021) The GCOS status report, GCOS-240. https://ane4bf-datap1.s3.eu-west-1.amazonaws.com/wmod8_gcos/s3fs-public/gcos-status_report_full_text-240_lr_compressed.pdf?FDdn12yqICpIxugb2V7hTQ9lTIcMRQFd=

Gewin V (2020) Five tips for moving teaching online as COVID-19 takes hold. Nature 580(7802):295–296

Gil T et al (2020) The new normal: holiday level Wi-Fi upload

Goldberg J (2020) Apple and Googles contact tracing is privacy preserving. https://blog.1password.com/contact-tracing/

Golinelli D, Boetto E, Carullo G, Nuzzolese AG, Landini MP, Fantini MP (2020) Adoption of digital technologies in health care during the COVID-19 pandemic: systematic review of early scientific literature. J Med Internet Res 22(11):e22280. https://doi.org/10.2196/22280. PMID: 33079693

Granryd M (2017) Five ways mobile technology can help in humanitarian emergencies. World Economic Forum. https://www.weforum.org/agenda/2017/08/mobile-technology-humanitarian-crisis/

GSMA Global System for Mobile Association/Communication (2020) The mobile economy. https://www.gsma.com/mobileeconomy/

Guo H, Yang Z, Huang R, Guo A (2020) The digitalization and public crisis responses of small and medium enterprises: implications from a COVID-19 survey. Front Bus Res China 14(1):19. https://doi.org/10.1186/s11782-020-00087-1

Gupta M, Shoja A, Mikalef P (2021) Toward the understanding of national culture in the success of non-pharmaceutical technological interventions in mitigating COVID-19 pandemic. Ann Oper Res. https://doi.org/10.1007/s10479-021-03962-z

Hamilton IA (2020) Poland made an app that forces coronavirus patients to take regular selfies to prove they're indoors or face a police visit. https://www.businessinsider.com/poland-app-coronavirus-patients-mandaotory-selfie-2020-3

Hart PS, Chinn S, Soroka S (2020) Politicization and polarization in COVID-19 news coverage. Sci Commun 1075547020950735. Published 2020 Aug 25. https://doi.org/10.1177/1075547020950735

Hays A (2022) Blockchain Facts: What is It, How It Works and How It Can Be Used, Investopedia. https://www.investopedia.com/terms/b/blockchain.asp

Huang Y, Meicen Sun, and Yuze Sui (2020) How digital contact tracing slowed Covid-19 in East Asia. Harvard Business Review. https://hbr.org/2020/04/how-digital-contact-tracing-slowed-covid-19-in-east-asia

ILO International Labour Organization (2020) Working from home: estimating the worldwide potential. ILO policy brief

IPCC Intergovernmental Panel on Climate Change (2014a) AR5 Climate Change 2014: mitigation of climate change. www.ipcc.ch/report/ar5/wg3/

IPCC Intergovernmental Panel on Climate Change (2014b) AR5 synthesis report: climate change. www.ipcc.ch/report/ar5/syr/

IRC, The International Red Cross and Red Crescent Movement (2018) Early warning can transform lives, 17 October. media.ifrc.org/ifrc/2018/10/17/early-warning-can-transform-lives/

Irum T, Hudgins C (2021) US corporate bankruptcies end 2020 hit at 10-year high amid COVID – 19 pandemic. S&P Global Market Intelligence. https://www.spglobal.com/marketintelligence/en/news-insights/latest-news-headlines/us-corporate-bankruptcies-end-2020-at-10-year-high-amid-covid-19-pandemic-61973656

ITU (2017) ICT facts and figures. https://www.itu.int/en/ITU-D/Statistics/Documents/facts/ICTFactsFigures2017.pdf

ITU (2019) Turning digital technology innovation into climate action. https://www.uncclearn.org/wp-content/uploads/library/19-00405e-turning-digital-technology-innovation.pdf

Jiang N, Ryan J (2020) How does digital technology help in the fight against COVID-19? World Bank Blog. https://blogs.worldbank.org/developmenttalk/how-does-digital-technology-help-fight-against-covid-19

Kamouri A, Lister K (2020) Global work-from-home experience survey. https://globalworkplaceanalytics.com/global-work-from-home-experience-survey#AboutSurvey

Katz R, Jung J, Callorda F (2020) Can digitization mitigate COVID-19 damages? Evidence from developing countries. SSRN

Kemp S. Digital 2020: October Global Statshot digital overview. DataReportal. https://datareportal.com/reports/digital-2020-october-global-statshot

Kemp S. Digital 2021: October Global Statshot Report. DataReportal. https://datareportal.com/reports/digital-2021-october-global-statshot

Kemp S. Digital 2022, Global Overview Report. DataReportal. https://datareportal.com/reports/digital-2022-global-overview-report

Kim AJ, Tak S (2019) Implementation system of a biosurveillance system in the republic of Korea and its legal ramifications. Health Secur 17:462–446. https://www.ncbi.nlm.nih.gov/pmc/articles/PMC6964808/

Kraemer MUG et al (2020) The effect of human mobility and control measures on the COVID-19 epidemic in China. Science 368:493–497

Krantz-Kent (2019) Where did workers perform their jobs in the early 21st century? Monthly Labor Review, July. https://doi.org/10.21916/mlr.2019.16

Kummitha R (2020) Smart technologies for fighting pandemics: the techno- and human- driven approaches in controlling the virus transmission. Gov Inf Q 37(3):101481. https://doi.org/10.1016/j.giq.2020.101481

Kurt M (2020) How the internet bent, but didn't break under COVID-19 pressure. Diginomica. https://diginomica.com/how-internet-bent-didnt-break-under-covid-19-pressure

Lebow S (2021) Worldwide ecommerce continues double - digit growth following pandemic push to online. https://www.emarketer.com/content/worldwide-ecommerce-continues-double-digit-growth-following-pandemic-push-online

Li B (2020) Going digital during Covid-19 and beyond, EURObizonline, 13 June. https://www.eurobiz.com.cn/going-digital-during-covid-19-and-beyond/

Ling Z (2020) How China's nationwide remote working is changing communication, biz analysis, 22 March. https://news.cgtn.com/news/2020-03-21/How--China-s-nationwide-remote-working-is-changing-communication-P2UeVp3LMc/index.html

Lu X, Wrathall DJ, Sundsøy PR et al (2016) Detecting climate adaptation with mobile network data in Bangladesh: anomalies in communication, mobility and consumption patterns during cyclone Mahasen. Clim Chang 138:505–519. https://doi.org/10.1007/s10584-016-1753-7

Malano V (2015) Philippine early warning system. Philippine Atmospheric, Geophysical and Astronomical Services Administration (PAGASA), 16 March. www.wmo.int/pages/prog/drr/events/WCDRR-MHEWS/documents/VicenteMalano.pdf

Mashamba-Thompson TP, Crayton ED (2020) Blockchain and artificial intelligence technology for novel coronavirus disease-19 self-testing. Diagnostics (Basel) 10(4):198. https://doi.org/10.3390/diagnostics10040198. [FREE Full text] [Medline: 32244841]

McKinsey & Company (2020) How COVID-19 has pushed companies over the technology tipping point and transformed business forever. https://www.mckinsey.com/business-functiocns/strategy-and-corporate-finance/our-insightshow-covid-19-has-pushed-companies-over-the-technology-tipping-point-and-transformed-business-forever/

Morgan S (2020) Cybercrime to cost the world $10.5 trillion annually by 2025. Cybercrime Magazine, November. https://cybersecurityventures.com/cybercrime-damages-6-trillion-by-2021/

Murray A, Meyer D (2020) Slacks remote-work research yielded interesting results. Fortune. https://fortune.com/2020/10/07/slack-future-forum-research-results-ceo-daily

OECD (2020a) Capacity for remote working can affect lockdown costs differently across places. https://www.oecd.org/coronavirus/policy-responses/capacity-for-remote-working-can-affect-lockdown-costs-differently-across-places-0e85740e/

OECD (2020b) Keeping the Internet up and running in times of crisis, updated May 4. https://www.oecd.org/coronavirus/policy-responses/keeping-the-internet-up-and-running-in-times-of-crisis-4017c4c9/

OECD (2021a) Fostering Economic Resilience in a World of Open and Integrated Markets. https://www.oecd.org/newsroom/OECD-G7-Report-Fostering-Economic-Resilience-in-a-World-of-Open-and-Integrated-Markets.pdf

OECD (2021b) Teleworking in the COVID-19 pandemic: trends and prospects. https://www.oecd.org/coronavirus/policy-responses/teleworking-in-the-covid-19-pandemic-trends-and-prospects-72a416b6/

Özgüzel C, Veneri P, Ahrend R (2020) Potential for remote working across different places. VOX EU. https://voxeu.org/article/potential-remote-working-across-different-places

PAGASA. 'Philippine Atmospheric, Geophysical and Astronomical Services Administration' (2019) Department of Science and Technology. https://pubfiles.pagasa.dost.gov.ph/pagasaweb/files/climate/elninolanina/PAGASA_MEMO_-_Adoption_of_the_PAGASA_ENSO_ALERT_AND_WARNING_SYSTEM.pdf

Parker SK, Knight C, Keller A (2020) Remote managers are having trust issues. Harvard Business Review. https://hbr.org/2020/07/remote-managers-are-having-trust-issues

Pepe E et al (2020) COVID-19 outbreak response: a first assessment of mobility changes in Italy following national lockdown. Preprint at medRxiv. https://doi.org/10.1101/2020.03.22.20039933

Quinn P (2018) Crisis communication in public health emergencies: the limits of 'legal control' and the risks for harmful outcomes in a digital age. Life Sci Soc Policy 2018(14):4

Rae T (2019) Drones to the rescue as cyclone desmond storms Mozambique. Medium, WorldFood Programme Insight, 24 Jan. insight.wfp.org/drones-to-the-rescue-as-cyclone-desmond-storms-mozambique-d7f501e40b0f

Richardson E, Mahle J (2020) Cyberattacks on the rise during the COVID – 19 pandemic. Cincinnati Business Courier. https://www.bizjournals.com/cincinnati/news/2020/06/01/cyberattacks-on-the-rise-during-covid-19.html

Saeed N, Bader A, Al-Naffouri TY, Alouini M (2020) When wireless communication responds to COVID-19: combating the pandemic and saving the economy. Front Commun Netw. https://doi.org/10.3389/frcmn.2020.566853

Saliola F, Islam AM (2020) How to harness the digital transformation of the covid era. Harv Bus Rev. https://hbr.org/2020/09/how-to-harness-the-digital-transformation-of-the-covid-era

Savvides L (2018) California's fires face a new, high-tech foe: drones. Cnet, 27 August. www.cnet.com/news/californias-fires-face-a-new-high-tech-foe-drones/

Shah D (2020) Class Central's MOOCReport; 2020. MOOCWatch 23: pandemic brings MOOCs back in the spotlight — Class central. https://www.classcentral.com/report/moocwatch-23-moocs-back-in-the-spotlight/

Silver L (2019) Smartphone ownership is growing rapidly around the world, but not always equally. https://www.pewresearch.org/global/2019/02/05/smartphone-ownership-is-growing-rapidly-around-the-world-but-not-always-equally/

Sinibaldi G (2020) COVID-19 is revolutionizing digital communications and testing providers' reliability and ability to innovate. Analysys Mason, April

Smiciklas J, Imran S (2018) Implementing ITU-T international standards to shape smart sustainable cities: the case of Moscow. International Telecommunication Union (ITU), Nov. https://www.itu.int/en/publications/Documents/tsb/2018-U4SSC-Case-of-Moscow/mobile/index.html#p=3

Sobaih EE, Hasanein AM, Abu Elnasr AE (2020) Responses to COVID-19 in higher education: social media usage for sustaining formal academic communication in developing countries. Sustainability 12(16):6520. https://doi.org/10.3390/su12166520

Statista (2020a) Global digital population as of October 2020. https://www.statista.com/statistics/617136/digital-population-worldwide/

Statista (2020b) Download growth rate of work from home apps in China among COVID -19 epidemic in 2020. https://www.statista.com/statistics/1105899/growth-rate-china-work-from-home-apps-download-coronavirus/

Statista (2020c) Statista Dossier on the global m-health industry and market. Available online at: https://www.statista.com/study/24501/mhealth-statista-dossier/. Accessed 4 May 2020

Tawil S (2020) Six months into a crisis: reflections on international efforts to harness technology to maintain the continuity of learning, ©UNESCO

Thompson D (2020) The workforce is about to change dramatically. The Atlantic. https://www.theatlantic.com/ideas/archive/2020/08/just-small-shift-remote-work-could-change-everything/614980/

UCLG, CGLU, Metropolis and UN Habitat (2020) Digital technologies and the COVID-19 pandemic, Briefing and Learning Note. https://www.uclg.org/sites/default/files/eng_briefing_technology_final_x.pdf

UN High Level Panel on Digital Cooperation (2019) The age of digital interdependence. https://www.un.org/en/pdfs/DigitalCooperation-report-for%20web.pdf

UN News (2020) COVID 19 has changed online shopping forever, survey shows, 8 October. https://news.un.org/en/story/2020/10/1074982

UNCTAD (2020) COVID-19 has changed online shopping forever, survey shows. Press Release UNCTAD/PRESS/PR/2020/029. https://unctad.org/press-material/covid-19-has-changed-online-shopping-forever-survey-shows

UNESCO (2020a) Digital innovation to combat COVID-19. https://en.unesco.org/covid19/communicationinformationresponse/digitalinnovation

UNESCO (2020b) UNESCO-UNICEF-World Bank survey on national education responses to COVID-19 school closures. http://tcg.uis.unesco.org/survey-education-covid-school-closure

UNFCCC (undated) Overview implementation of systematic observation. https://unfccc.int/topics/science/workstreams/RSO/overview

United Nations (2020a) The impact of digital technologies. https://www.un.org/en/un75/impact-digital-technologies

United Nations (2020b) Roadmap for digital cooperation. Report of the Secretary General, New York

Vegas E (2020) School closures, government responses and learning inequalities around the world during COVIV 19, Brooking Institutions. https://www.brookings.edu/research/school-closures-government-responses-and-learning-inequality-around-the-world-during-covid-19/

Vial G (2019) Understanding digital transformation: a review and a research agenda. J Strateg Inf Syst 28(2):118–144

Wakabayashi D, Nicas J, Lohr S, Isaac M (2020) Big tech could emerge from coronavirus crisis stronger than ever. New York Times, 23 March. https://www.nytimes.com/2020/03/23/technology/coronavirus-facebook-amazon-youtube.html

WARC, World Advertising Research Center (2020) Global online content consumption doubles in wake of COVID, 24 September. https://www.warc.com/news-andopinion/news/global-online-content-consumption-doubles-in-wake-of-covid/44130

Website Builder (2021) 30 truly fascinating app usage statistics to know in 2022. https://websitebuilder.org/blog/app-usage-statistics/. Accessed May 2022

Whitelaw S, Mamas MA, Topol E, Van Spall HGC (2020) Applications of digital technology in COVID-19 pandemic planning. Lancet. https://doi.org/10.1016/S2589-7500(20)30142-4

WHO (2020) COVID-19 and digital health: what can digital health offer for COVID-19? https://www.who.int/china/news/feature-stories/detail/covid-19-and-digital-health-what-can-digital-health-offer-for-covid-19

Williams CM, Chaturvedi R, Chakravarthy K (2020) Cybersecurity risks in a pandemic. J Med Internet Res 22(9):e23692. https://doi.org/10.2196/23692. PMID: 32897869

WMO (2022) Global observing system. https://public.wmo.int/en/programmes/global-observing-system

WMO and ITU (2017) Handbook on use of radio spectrum for meteorology: weather, water and climate monitoring and prediction. https://library.wmo.int/doc_num.php?explnum_id=3793

WMO, IOC-UNESCO, UNEP and ISC (2016) The global observing system for climate: implementation needs, Geneva

WMO, IOC-UNESCO, UNEP and ISC (2021) The Global Climate Observing System 2021, The GCOS Status Report, https://gcos.wmo.int/en/gcos-status-report-2021

Wong M (2020) Stanford research provides a snapshot of a new working-from-home economy. Stanford News. https://news.stanford.edu/2020/06/29/snapshot-new-working-home-economy/

Young L (2020) Facebook signs huge office lease at Vornado's Farley Building. New York Business Journal, 3 August. https://www.bizjournals.com/newyork/news/2020/08/03/facebook-leases-office-space-at-farley-building.html 86

Part III

Mega Risks, Urban Economic Development, and Sustainability

This part has explored the economic vulnerability of cities in the wake of mega risks. It highlights how the mega risks affected the economic functioning of cities and put them in a survival mode. It also discusses the lessons learned from the new adjustments that provide guidance for future development of city economies. In the post-pandemic growing climate-related disasters, the focus of the part is on the kind of thinking that is needed to design an urban development pattern, which should be more than quick fix or quick economic recovery. It means "learning from the past" and making development sustainable and resilient to reduce the likelihood of future shocks.

The impact of economic shock due to pandemic and its intensity varied and depended on cities' exposure; development status (developing or developed country cities), their economic composition, and labor market; crisis management responses related to social distancing, workplace management, and commuting possibilities; supply chain status; presence or absence of vulnerable groups; and support to business and industry groups by governments. Considering development status, cities of developing countries were worst affected. In terms of economic composition and labor market, cities with diverse economic structure were less vulnerable. Similarly, the cities and economies which relied on technology or the movement of information like San Francisco, Seattle, Hong Kong, and Singapore were comparatively less affected. Many cities to their predicament, in addition to COVID-19, also faced climate trouble, and in the wake of inaction or lack of required action, many more cities will fall into extreme crisis mode. The section highlights the economic costs of the pandemic and climate change and discusses short-term relief and recovery as well as long-term planning measures adopted and their efficacy. It discusses the measures needed for boosting the economy while making it sustainable.

The part also deals with energy use and sustainable development. While powering the engine of economic growth, energy use is a major cause of GHG emission and climate change. A special focus therefore is on how to promote urban energy use with climate resilience in mind. As such, the part highlights cities' major role in improving energy end use demand and reducing GHG emission. Case studies have been presented, and interdependent nature of urban systems has been emphasized as an opportunity to optimize the use of energy and minimize its waste in all forms through appropriate use of technology and integrated policies and strategies. The global climate change emergency and its related growing risk events are warning that the long-term economic and social damages could be far larger than those caused by COVID-19 in the short term. Together, they are calling for a change in development thinking not only to redesign post-pandemic economic recovery packages to "build back better" but also to make changes in the overall development paradigm leading to sustainability.

Mega Risks, Urban Economies, and Sustainability

5.1 Introduction

The twin mega risks coronavirus pandemic and climate crisis are similar in terms of their impacts on urban economies. They have both caused widespread devastation in cities' economies and markets, businesses, and industries, as well as across different sectors. Vulnerabilities to both have worsened in the wake of economic globalization and connectivity triggering multiplier effects. The COVID-19 crisis for one has demonstrated how quickly a natural calamity can cause the collapse of economies both at city and national level. Similarly, cities and countries with climate vulnerability to their predicament have already faced disaster risks frequently for a long time. In the wake of inaction and lack of concerted effort to significantly strengthen investment in mitigating and adapting to climate change, many more cities and countries will fall into crisis mode. In terms of disaster risk reduction, both mega risks require global-to-local response and political will for their resolution guided by science. The pandemic has given a deeper understanding of the ties, whether humanitarian or economic, that bind people from cities to a global scale. Hopefully this would help the world in grappling the equally big threat of the century—the climate change.

The pandemic has resulted in a severe economic recession. The severity of the pandemic's impact on global economic growth has been beyond anything experienced in nearly a century.

According to the US Congressional Research Service (2021), the virus reduced the global economic growth rate from −4.5% to −6.0% in 2020. United Nations (2021) estimated the contraction at 3.5%. The impacts of climate change have been more prolonged and have been estimated to reduce per capita GDP considerably in many cities and countries by 2100 (Kahn et al. 2019). In terms of risk to urban economy, the "Fourth National Climate Assessment" of the USA estimated that about $1 trillion in coastal real estate is threatened by rising sea level together with storms, floods, and erosion (National Oceanic and Atmospheric Administration 2018).

The outlook for global economic growth according to international organizations like the United Nations (2021), the International Monetary Fund (2021), and the World Bank (2021) appears promising. According to forecasts of all three organizations, a stronger economic recovery rate was projected and achieved in 2021, but it has been followed by increasingly gloomy developments in 2022 for several reasons (IMF 2022). The fiscal stimulus provided along with the vaccine rollouts across countries made the said organizations more optimistic for the progress of global economy in 2021. Nevertheless, some daunting challenges were pointed out to cause divergence in the speed of recovery between and within countries and cities due to the potential for continued economic damage from the crisis (IMF 2021; World Bank 2021; United Nations 2021). Since cities are the main contributor and driver of world economic growth, therefore, urban economies will play an extremely significant role in the future global development. It has been estimated that 65% of the world's future GDP growth is expected to come from a group of 600 cities. This becomes more important in light of the fact that the GDP per capita of the world's megacities is already 80% higher than the GDP of the rest of their respective national economies (Bloomberg and Mckinsey 2020).

Several studies have been conducted to evaluate the real-time impact of COVID-19 on cities' economy and society (Liu et al. 2020; Vandenberg et al. 2020; Ceylan et al. 2020; Lenzen et al. 2020). One such study in China (Liu et al. 2020) devised a Health Index of Cities (HIC) based on mobility data to assess the socioeconomic health of cities and regions. This study drew a parallel between mobility or flow of people in the city and the flow of energy in a system. It pointed out that the average HIC for Chinese cities from January to April (when the intensity of the pandemic was at its highest) was −28.6, representing a 28.6% reduction in urban health and vitality. Another study (Santos 2020) used input-output analysis, to model the impact of epidemic mitigation and control measures on the workforce in the USA. Yet another study used literature review to assess the impacts of the pandemic and its subsequent economic crisis on poor communities (Buheji et al. 2020). All these and similar studies indicate that economic impacts of the pandemic on cities and communities were colossal and affected a wide range of sectors and aspects of urban economy.

This chapter has been divided into five sections. The next section after this introduction deals with the impacts of the pandemic and climate change on cities. It is followed by policy response to COVID-19 in terms of short-term economic and fiscal relief measures including targeted actions meant for vulnerable groups. The section also highlights long-term measures for boosting the economy, restructuring it to build back better, making it sustainable and rejuvenating it by invigorating the lagging sectors badly affected by the pandemic. The following section

on climate action and urban economy takes up the case of cities as contributor and victims of climate change. It also presents the case for making low-carbon transition for 2030 and advocates for it because it is crucial in a build-better scenario. The findings of the chapter have been presented in the concluding section.

5.2   Impacts of Mega Risks on Urban Economy

The COVID-19 pandemic generated the third and greatest economic and financial shock in cities in the twenty-first century, after 9/11 and the global financial crisis of 2008–2009. It prompted a cessation of economic production in cities, cut their supply chains, saw a steep drop of mobility and consumption, and above all resulted in a collapse of confidence at least temporarily. Like the coronavirus pandemic, climate crisis is also causing great damage to cities' economy, which may increase enormously in business-as-usual scenario. Both highlight the importance of developing and implementing policies that can mitigate their impacts on the economy and society in the future.

5.2.1   Coronavirus and Urban Economy

The impacts on urban economy have been most severe in the wake of economic shutdowns during the pandemic. The consequences were complex and varied in scale across a wide range of sectors. Further, the recession in city economies due to the pandemic besides having direct impact on citizen's life also had spillover effects on the city budgets because of lower tax revenues and had direct impact on spending, etc. (Sharifi and Khavarian-Garmsir 2020). The reduction in tax revenues created budgetary shortfalls and also forced cities to defer infrastructure projects and capital investments and some other less important plans (Kunzmann 2020). This curbed job growth and local economic activity and impacted cities' abilities to deliver essential health and safety services (Vandenberg et al. 2020). It also severely affected citizens' incomes and businesses. The impact on vulnerable groups including 2 billion informal workers and 1 billion urban slum dwellers was particularly serious.

5.2.1.1 Intensity of the Impact

The intensity of economic shock due to the pandemic varied and depended on cities' exposure; their development status (developing or developed country cities), economic composition or diversification, and labor market; crisis management responses related to social distancing, workplace management, commuting possibilities; supply chain disruptions; presence of vulnerable groups; and support to business and industry groups by governments. Cities with diverse economic structure and those dependent on technology and the flow of information were less susceptible. Thus least negatively affected metropolitan areas in the USA were superstar technical centers such as San Jose, San Francisco, Seattle, Washington DC, and Boston. Similarly, comparatively less affected were university towns specialized in science and information technology like Trenton-Princeton, Ann Arbor, Durham-Chapel Hill, Boulder, and Ithaca. Tertiary industrial hubs including Elkhart-Goshen, Dalton, Battle Creek, and Columbus also fared better. None of these metros depend on tourism, leisure, or retailing, and none saw employment growth slip by more than 1.5% below the Bureau of Labour Statistics baseline (Muro and You 2020).

The worst affected cities were those specializing in vulnerable industries such as tourism (Earl and Vietnam 2020), leisure and hospitality, entertainment, retailing, travel arrangement and transportation, employment services, and mining related like oil- and gas-producing towns. The economic shock also brought chain reaction leading to a series of problems, such as unemployment, increasing urban poverty, and declining quality of life (Pineda and Corburn 2020), with impacts on urban resilience (Giannoccaro et al. 2018) and sustainability (Zhou et al. 2020). In the USA, Brooking Institution (Muro and You 2020), using the Bureau of Labor Statistics data, mapped

changes within the metro areas. They found that 20 most heavily impacted cities were those that had either location in the Sun Belt or were beach-based resorts and vacation centers such as Atlantic City and Ocean City, Las Vegas, Gulfport, Myrtle Beach, Orlando, and Cape Coral-Fort Myers. The group, according to the study may get their next 10-year employment growth sliced "anywhere between 2.5% (in Orlando and Cape Coral) to 3.3% or 3.6% (in Myrtle Beach, Las Vegas, Ocean City, and Atlantic City). These metro areas could brace for a sluggish recovery with potential stagnation for their vulnerable tourism-based businesses and waves of low-wage, low-stability workers" (Muro and You 2020).

One of the biggest impacts of COVID-19 in cities was on informal sector, which forms an extremely large and a very important component of urban economy in developing countries of global South (Bonnet et al. 2019). According to the International Labour Organization, it constitutes 90% of total employment in low-income countries, 67% in middle-income countries, and 18% in high-income countries and engages over 2 billion workers in the world (ILO 2020). The pervasive informal sector stretches in cities from Manila to Mumbai and Karachi in Asia to Cape Town in Africa and Bogota in South America. It engages diverse informal workers from vendors and domestic servants to couriers and traders and daily wage earners. The informal economy, despite its negative connotation, makes cities livable and more flexible and often offers an important venue of employment to large segments of the citizens (Martínez and Short 2021). More importantly, in post-COVID era, it could help a great deal in the economic reactivation and recovery in the cities of developing countries (Short 2020).

5.2.1.2 Impact on Informal Economy and Urban Poor

During the pandemic, the informal economy, which is predominant in developing countries' cities, was devastated in the wake of lockdown, quarantine, and social distancing orders. Living hand to mouth, without any savings, the fragility of informal workers became worse. Constituting a staggering 12.5% of the global population and 62% of all those working worldwide (ILO 2020), the pandemic's restrictions pushed them into poverty. On top of it, they were excluded, at least initially in cities and countries recovery strategies, whereby investments were largely targeted at formal employers and employees—ignoring this pervasive and growing group. It created big problems for workers and entrepreneurs in the informal economy, who did not have resources or means to overcome the impacts of the pandemic. Hence, city dwellers working informally, often in precarious jobs, increasingly became part of the group of "new urban poor" created by the COVID-19 crisis. ILO (2020) estimated that the pandemic effects increased the poverty of informal workers by 21%, 52%, and 56% in upper-middle, high-income, and lower-income countries, respectively.

Although the pandemic has affected all segments of urban population, and all cities of the world, the effect on urban poor and most vulnerable has been out of proportion. Estimates (CCSA 2021) on the increase in poverty show that COVID-19 pushed millions of people into poverty in 2020. The reasons for the same were economic stagnation, job losses, and decreased remittances (United Nations 2020). The findings of the study by Bottan et al. (2020) give preliminary estimates of the impact of the pandemic on the labor market and well-being in developing countries. It provides evidence that the coronavirus crisis exacerbated economic inequality. Short-term projections by the World Bank and the International Monetary Fund further indicated increase in extreme poverty and income inequality in low-income and emerging economies. Many other studies noted that while increasing poverty, the crisis sharpened and reinforced the inequalities in cities of both developed and developing countries. For example, studies by Adams-Prassl et al. (2020), Gonzalez et al. (2020), Martin et al. (2020), and Witteveen (2020) conducted in the developed countries highlight how the pandemic has exacerbated inequalities there. Similarly, studies by Ghosh et al. (2020), Boza-Kiss et al. (2021), Hill and

Narayan (2021), and Du et al. (2020) present the pandemic's impact in sharpening inequalities in developing countries. Temporarily closing the economy, shock waves were sent through cities—exposing the fault lines created by race, gender, age, and education in developed countries' cities. It also exacerbated the divide created by the social inequalities and unequal facilities available to slum and informal dwellers compared to non-slum dwellers in developing countries' cities. Overall, preexisting and current inequalities were worsened, weakening economic resilience and reinforcing disadvantage. The inequality aspect has been discussed in more details in Chap. 9 of this book.

5.2.1.3 Varied Sectoral Impacts

Within urban economy, the tourism, entertainment, and hospitality sectors were worst affected all over the world. Unmatched national travel bans and restrictions at global level created unprecedented challenges for cities relying on incomes from tourism and related activities. According to a report of the United Nations World Tourism Organization (UNWTO 2020), at one time, more than 217 countries and territories world over imposed measures that restricted or prevented people from entering their borders. It put millions of jobs and livelihoods at risk in tourist cities. It also led to flight suspensions and drops in hotel occupancy, causing airlines a shortfall of $391 billion in revenue in 2020 (Kulisch 2021). An UNCTAD (2021a) publication estimated that international tourism and its closely linked sectors suffered an estimated loss of $2.4 trillion in 2020 due to direct and indirect impacts of a steep drop in international tourist arrivals. Cities and communities of developing countries had the biggest impact, suffering the largest reductions in tourist arrivals estimated at between 60% and 80% in 2020—the most affected world regions included Northeast Asia, Southeast Asia, Oceania, North Africa, and South Asia. North America, Western Europe, and the Caribbean were least affected (UNCTAD 2021a). Altogether tourism industry recorded a decline of 74% in 2020 (CCSA 2021). Closely associated loss in air traffic estimated by the International

Civil Aviation Organization (ICAO) reveals that for the year 2020, global passenger traffic fell drastically—by 60% or 2.7 billion, compared with 4.5 billion in 2019. This reduced global air travel totals back to 2003 levels. Airline financial losses were estimated to have reached USD 370 billion (CCSA 2021).

Tourism also has linkages with other sectors which meant a decline in tourism sector caused a reduction in demand for inputs from other sectors down the supply chain. The net result was an indirect loss to other industries and businesses that supplied food, beverages, accommodation, and transport. It also affected retail trade, tourism-related sports, and hotel industry. Besides monetary losses, it also impacted jobs or employment. According to the State of the Hotel Industry 2021 Report (AHLA 2021), the USA alone suffered a loss of more than 670,000 jobs in hotel industry and some 4 million hospitality jobs in 2020 due to the pandemic. One study highlighted the impact's differential in the urban hotel market (Napierała et al. 2020). Sighting the case of Poland, the study pointed out that the pandemic affected more the large Polish cities, compared to smaller ones. The reason was that large cities primarily hosted international tourists, compared to the small cities, which mostly attracted the domestic tourists. Hence, the recovery of the hotel industry in comparatively more internationalized urban destinations is somewhat complicated and demands tackling the issue with due consideration of scale and the principle of sustainability.

5.2.1.4 Differential Impacts Within a City

Just like the pandemic had differential impact among cities, so it did within one city. For example, downtowns were worst affected within urban areas (Romei and Burn-Murdoch 2020). Thus, in megacities like London, Paris, New York, Toronto, Tokyo, Mexico City, and Delhi downtowns, and once crowded financial districts got deserted primarily because of switching to remote work – that emptied tall building blocks. It also grossly curtailed office worker visits to cafeterias and restaurants and resulted in abandonment of their leisurely shopping. The economic fallout

had a chain reaction. The core industries in cities that were negatively impacted by the COVID-19 recession like tourism, hospitality, and entertainment had broader and negative ripple effects on other economic sectors like retail, services, transportation, etc. The consequential economic downturn effected employment and citizens' income adversely. A group of United Nations and multilateral organizations have estimated that in terms of employment, 8.8% of global working hours were lost in 2020, equivalent to 255 million full-time jobs, an amount that is four times greater than the job losses during the 2008/2009 financial recession (CCSA 2021). The pandemic also exposed urban inequalities and made the marginal communities like ethnic minorities, slum dwellers, women, and migrant workers suffer more economically than others in terms of unemployment and income loss, pushing them into poverty.

5.2.2   Climate Change and Urban Economy

The number of cities and the number of people affected by climate change have been rising annually over time. According to UNCCRN (2018), almost all cities are at risk; however, some are more vulnerable than others. Majority of cities in the world have been dealing year after year with the devastating economic impacts of climate change along with the physical damage (UNCCRN 2018). The example of the USA is a case in point. During the past 40 years, the country's cities suffered from hundreds of climate-related disasters, the cost of which exceeded $1 billion each in terms of economic losses. The year 2020 alone saw 22 such billion-dollar disasters, breaking the previous annual record of 16 such events in 2017 and 2011. Among these events of 2020, 7 resulted from tropical cyclones, 13 from severe storms, and 1 each from drought and wildfire. Altogether, these events caused a cumulative damage of 95 billion dollars, almost double the cost suffered in 2019 (Smith 2021). The history of climate-related extreme events put the USA in the forefront. The country altogether faced 285 extreme climate-related billion dollar disasters from 1980 to 2020, including both years—the cumulative cost of which exceeded $1.875 trillion (NCEI 2021).

5.2.2.1 Damage Cost

Globally, the damage cost of weather and climate-related disasters to cities and communities in 2020 was $210 billion according to a report by reinsurance company Munich Re (2021). These monetary losses to cities increased by 26.5% compared to 2019's cost of $166 billion. Only $82 billion of the total damage in 2020 was insured, an increase of $25 billion from $57 billion insured in 2019. According to another report of Swiss Re (2021), climate change is likely to hit the world economy even more severely by 2050, in the business-as-usual scenario, as crop yields fall, diseases spread, and rising seas consume coastal cities. Unless the world succeeds in quickly slowing the use of fossil fuels, the effects of climate change are expected to cut 11–14% off global economic output by 2050 compared with economic growth without climate change (Swiss Re 2021). This, in absolute term, means reduced global economic output by US $23 trillion annually. However, if strengthened current efforts succeed in holding the global temperatures to less than 2°C above preindustrial levels (the goal set by the 2015 Paris agreement), economic losses by midcentury would be marginal. If this materializes, according to Swiss Re (2021), most countries' economies would be no more than 5% smaller than what would otherwise be the case. Nevertheless, the report warns that the levels of current emissions are far from attaining those targets—hence, based on current trajectories, global temperature is likely to increase as much as 2.6° by 2050. Such projections are important for the global, national, and cities' economy as they are likely to influence commercial investments—particularly by Swiss Re and other insurance companies, which collectively account for $30 trillion in assets.

5.2.2.2 Intensity of the Impact

Cities in general and metropolitan areas, in particular, account for the bulk of their national output and employment—hence, climate change-related damage to these cities may not only hurt them but also seriously affect the national economies of the country in which they are located. It is true for metropolitan areas in both developed and developing countries. For example, in developed countries, there are metropolitan areas that produce about one-half of the national GDP like Budapest in Hungary, Seoul in Korea, Copenhagen in Denmark, Dublin in Ireland, Helsinki in Finland, Brussels in Belgium, and Vienna in Austria (OECD 2008; UN Habitat 2011). Any damage to their economies will have direct bearing on their countries' economies. It is also valid for other metropolitan areas like Oslo in Norway, Auckland in New Zealand, Prague in Czechoslovakia, Tokyo in Japan, Stockholm in Sweden, London in the UK, Paris in France, and Brussels in Belgium which generate about one third of national GDP (OECD 2008). Among developing countries also, the central role of cities in national economies is very significant. For example, according to the UN Habitat (2011), Buenos Aires in Argentine generates 63.2% of national GDP, Nairobi in Kenya produces 20% of national GDP, Sao Paulo in Brazil generates 19.5% of GDP, and Dar es Salaam in Tanzania accounts for 14.9% of national GDP. Further, estimates of the cities' contribution to total GDP in India range from 60% to 80% (UN Habitat 2011). In addition, Kinshasa in Congo and Kabul in Afghanistan generate more than 500% higher GDP than their population share, while Hanoi in Vietnam generates more than 460% higher GDP than its population share. Among the cities that generate more than 200% higher GDP than their population share are Dhaka, Yangon, Chittagong, Khartoum, and Mumbai (UN Habitat 2011). It is, therefore, obvious that any damage to these cities' economies will have grave implications for their national economies. Overall impact of climate change on GDP was studied by Stanford University (2015), which found that there was a 51% chance that climate change would decrease the world's GDP by more than 20%. This can be compared to the Great Depression, during which GDP fell to −26.7%. The main difference between the two is that, in the case of climate change, the GDP reduction would be permanent (CRS 2010).

The concentration of economic activities in cities has also exposed their assets to climate-related risks. For example, a disproportionate number of megacities and their trillion dollars' worth assets are in coastal areas from Houston to New Orleans to Miami to Rotterdam to Alexandria to Dubai to Karachi to Mumbai and Guangzhou. Most of these cities are also playing a critical role in international trade and shipping as ports. In addition, they also house vital infrastructure. Climate change and sea level rise are expected to aggravate their risks to flooding and disrupt their trade and transport function. According to one estimate, (Hallegatte et al. 2013), average flood losses worldwide in 2005 were estimated at US$ 6 billion per year, and projected socioeconomic changes are likely to increase it to US$ 52 billion annually by 2050. Moreover, with expected climate change and subsidence, there will be a need to upgrade present protection of infrastructure to avoid unacceptable losses of US$ 1 trillion or more every year. Several cities currently face faster local sea level rise than the global average due to subsidence caused by sediment compaction and groundwater withdrawal (Syvitski et al. 2009). Among the largest cities in Europe, more than two thirds have areas that are not more than 10 meters above sea level (Kamal-Chaoui and Robert 2009). According to the same study, port cities most prone to coastal flooding are located both in the developed countries including Amsterdam, Miami, New York, Osaka, Rotterdam, and Tokyo as well as in developing countries such as Kolkata, Shanghai, and Guangzhou (Kamal-Chaoui and Robert 2009). Another study (OECD and Bloomberg Philanthropies 2014) includes most "at-risk" cities due to sea level rise (measured by annual average losses due to floods) in developed countries as the following: Miami, New York, New Orleans, Nagoya, Tampa-St. Petersburg, Boston, Osaka-Kobe, and Vancouver. In developing countries, Guangzhou, Mumbai, and Shenzhen were identified as the most at-risk cities.

5.2.2.3 Impacts on Port Cities

The issue of ports in relation to transport and shipping is vital in view of the significant role of the maritime traffic in the world economy. Studies on risks of disruptions in ports due to climate-related events (both sea level rise and discrete events like storms) suggest that the impacts on transport and their logistic chains would be substantial (Lam and Su 2015; Novati et al. 2015). It will not only affect the economy of the ports themselves but will also disrupt economies of the cities located in their hinterland which they serve through rails and roads. Additionally, it will also affect the economy of cities in their foreland or the network of ports or cities from which they receive cargo or to which they dispatch cargo. Christodoulou et al. (2019) conducted a study of European ports following an approach that allowed the quantification of the potential risk of disruption to them due to climate change. It also enabled them to evaluate the risks to the ports in their hinterland and foreland. Their study showed that in the high warming scenario of sea level increasing higher than 0.5 m, the largest part of the European coastline will be affected with regional variations—more than 1 m increase occurring in the North Sea, the Western part of the Baltic Sea, and parts of the British and French Atlantic coasts. The impacts are expected to be milder but much more frequent in the Black Sea and the Mediterranean (Christodoulou et al. 2019).

The potential disruptions noted in port cities (operations and economy) in the hinterlands of the European ports at regional level and on the foreland worldwide were also significant. Regarding the hinterland affected by sea level rise, its mapping suggested a much wider area to be affected than the immediate zone surrounding each port because ports have a wide catchment area that cover most of central Europe. Disruptions in the operations of a port were to directly affect road, rail, and inland waterways services connected to it and as a result were to hinder transport and trade for the goods as well as the origins—destinations served by the port in question. The analysis of the impacts on foreland enabled the study to identify European and overseas destinations to be mostly affected, based on their connections with European ports at risk. On the foreland side, disruption in the operations of even a limited number of European ports, due to sea level rise, were to affect the operations of ports in different geographical areas that are part of common supply chains. The impact noted was quite pronounced within Europe, where more than 60% of port traffic is dependent on connections with ports having a high sea level rise risk. Although, Mediterranean ports were not expected to suffer much from sea level rise, the indirect impacts on their operations, because of possible disruptions in Northern European ports, could be considerable. A similar impact—though at a much lower scale—was noted for the East coast of the USA, Brazil, and traffic through the Panama Canal.

5.2.2.4 Sectoral Impacts

Beyond maritime transportation and trade, climate-related events have wide-ranging impacts on many sectors of urban economy (Arent et al. 2014) particularly insurance and service sector, tourism, and recreation, as well as transport, water, and energy (particularly by affecting their infrastructure). Heat waves, for example, often have wider impacts from human health to those on real estate and tourism. Moreover, according to Corfee-Morlot et al. (2009), changes in temperature and the hydrological cycle also shorten the maintenance and replacement cycle for key infrastructure of energy production and transport, etc. They also influence their operational capacity in terms of brownout and blackouts, service interruptions, etc. (Corfee-Morlot et al. 2009). According to IPCC (Arent et al. 2014), more detailed economic data and the advancement of analytic methods and tools are needed to assess further the potential impacts of climate on key economic systems and sectors. Even in the absence of that, findings from the computable general equilibrium (CGE) model show that by 2030 some cities can become more attractive by curbing local pollution (e.g., Ankara, Auckland, Barcelona, Krakow, Lille, Melbourne, Montreal, Monterrey, and Toronto). In contrast, some other metroregions may lose their attractiveness if they do not strive to stop their ongoing pollution trends (Kamal-Chaoui and Robert 2009).

The degree of sectoral diversity of city's economy also has impact on its resilience to climate change. In other words, city's dependence on one or diverse local sources of income is an important facet in the magnitude of climate change impacts on its economy. Thus, impacts can be severe, if a city is reliant upon one sector or few industries for most of the local economic productivity. In extreme cases, in a city with low economic diversity, loss of a single industry leaves little options for its workers who lose their jobs. For example, in Venice, flood impacts on tourism and aquaculture make the city's future uncertain (UN Habitat 2011) at least in the short term.

5.3   Cities Response to Mega Risk

This section reviews cities' response in terms of adaption of economic policies and measures to tackle the twin mega risks—coronavirus pandemic and climate crisis. One major purpose is to see if the economic measures taken to deal with the mega risks represent a temporal adjustment, or do they provide a new mechanism toward adapting a fresh economic approach more effective in times of environmental and health crises? Another important intent is to assess whatever approach or measures used, whether these will (a) enable higher readiness of cities to face these and similar crises in future or (b) promote sustainable development as envisaged in Sustainable Development Agenda 2030.

The coronavirus pandemic came on top of ongoing climate emergency, which was exacerbating heat waves, storms, floods, and other weather and climate-related events. Moreover, several cities in both developed and developing countries faced such disasters during the ongoing pandemic when dealing with and recovering from new disasters was much more challenging. For example, the year 2020, well known for the pandemic also stands out for climate-related disasters, which, as discussed in Chap. 1, were responsible for the 389 recorded events that affected 98.4 million people and caused economic losses with estimates varying from US$ 171.3 billion (CRED and UNDRR, 2021) to over

$200 billion (Swiss Re). Among developed countries, the USA alone which was the hardest hit by the coronavirus also faced 22 climate-related billion dollar disasters costing together over $90 billion in 2020. Within developing countries, the Cyclone Amphan of May 2020 alone hit hard cities in Bangladesh and eastern India, which were still recovering from the impacts of Cyclone Bulbul of fall 2019 (Vandenberg et al. 2020). To cope with the grave situation, the quarantine centers had to be converted to cyclone shelters. However, maintaining social distancing was an uphill task given the huge number of evacuees roughly numbering 6 million. It clearly indicates the difficulties cities faced in dealing with simultaneous multiple risks.

5.3.1 Economic Response to Coronavirus Pandemic

The immediate response to the coronavirus was in the form of emergency relief, followed by efforts at recovery or revival of cities' economy. Surveillance was often accompanied by relief measures such as local service delivery, support to vulnerable groups and businesses, citizen engagements, and attention to their workplaces and commuting. Recovery strategies and planning and implementation for the future were often initiated in conjunction with one another. The response to economic shock due to the pandemic varied among cities and depended on the available resources, political will and awareness, available technology, economic composition, labor market, workplace management and commuting possibilities, supply chain management, presence of vulnerable groups, and support from city residents or communities' business, industry, and private sector. Since most cities were unable to face or fix the magnitude of the challenge on their own, the recovery outlay for dealing the crisis had to be shared with higher levels of government.

5.3.1.1 Rescue and Midterm Measures

The initial biggest challenge for cities was to help businesses survive and then help their economies bounce back as quickly as possible. However, financial weakness of cities' governments was the biggest hurdle to cope with this challenge and made them dependent on national governments' help. The objective of the initial rescue packages was to protect business environment and livelihoods in the face of unemployment and abrupt losses of income. McKinsey & Company carried out a survey (based on analysis of the economic responses of 54 of the world's largest economies, representing 93% of global GDP) to analyze these rescue or economic stimulus packages (Cassim et al. 2020). The size of these measures used in response to the COVID-19 crises up to only the end of May 2020 was ten trillion dollars—three times more than the response to the 2008/2009 financial crisis. Western European countries alone allocated close to $4 trillion, an amount almost 30 times larger than today's value of the Marshall Plan, and the crisis was still far from over. The measures used supported large parts of the economy in a very short time. The objective was to maintain financial stability, provide economic welfare to households, and help companies survive the crisis.

Household Measures Among *household measures*, most were to provide immediate relief to the most vulnerable, especially in countries without automatic stabilizers already in place. For example, by increasing pensions (Egypt), expanding unemployment insurance (several countries in South America), protecting ill or homeless and providing food (Indonesia), and providing coupons for use at night markets, shops, and restaurants (Taiwan). Some countries enacted broader income-distribution programs, primarily to support workers in the informal sector and the self-employed (Brazil and Morocco). Only around 20% of governments covered by survey took longer-term measures such as jobs redeployment and reskilling (Cassim et al. 2020).

Business Specific Measures In terms of *business-specific measures*, the initial steps in most countries focused on protecting vulnerable small- and medium-sized enterprises (SMEs) and companies within the most affected sectors. The most common approach (enacted by more than 80% of countries studied by McKinsey & Company) was in the form of debt restructuring and loan guarantees. There is a significant variation in how far countries went to protect companies' balance sheets. For example, Germany's loan guarantees amounted to 29% of its GDP, while the average was 4% for other G-20 countries. Equity injections were used by only around 10% of countries studied under the survey but may have become more common in recovery later, as opposed to relief measures (Cassim et al. 2020).

While national governments undertook measures to protect their economies from the economic fallout, city governments and local administrations also played a key role in developing and implementing recovery frameworks in the form of visions (Chicago for reopening economy); plans (Paris, Bilbao in Spain, Maringa and Paraty in Brazil); strategies ("Milan 2020: Adaptation Strategy," Hoboken in the USA, Sydney); recovery initiatives (Houston); guidelines (San Francisco); and projects (Barcelona Never Stops) (OECD 2020).

Financial and Technical Assistance Tools Cities also devised a variety of *financial and technical assistance tools* and measures for economic revival. Some of these included grants and loans for business support and jobs (Tokyo, Boston, Bilbao, and Maringa in Brazil), investment in infrastructure (Chicago), investment in businesses and public-private partnership (Paris), tax incentive (Madrid), subsidies (Tokyo and Milan), capacity building (Barcelona), promotion through communication and other means (Barcelona), allowing flexible use of public spaces (Hoboken in the USA and Yokohama), campaigns for business support (Frankfurt), and platforms for online delivery services (Yokohama and Dusseldorf) (OECD 2020; Landershauptstadt Dusseldorf 2020).

Tax and fiscal exemptions were also given by cities. For example, 63 million EURs were pro-

vided in tax break by Madrid City Council to commercial establishments, travel agencies, and department stores, subject to bans on layoffs until the end of year. Several other examples of cities helping businesses have been provided in an OECD (2020) report. In Montreal, for example, emergency financial support measures were undertaken to help businesses through postponement of municipal taxes and an automatic moratorium on capital and interest, as well as providing emergency financial assistance. Seattle in the USA waived taxes on receipts from the economic activities and the taxes for leisure, hospitality, and financial penalties for businesses that could not pay their taxes in time. Braga in Portugal exempted local businesses of city taxes for the occupation of public space or terraces or on advertisement for local business closed to the public. Braga city in Portugal also supported the business community under its InvestBraga program to meet their social security requirements and provided special assistance to tourist companies. Some cities gave exemption on commercial rents or froze rent (Lisbon, Viano do Castelo, and Porto in Portugal); others gave moratorium on road, terrace, and other municipal taxes for closed businesses and NGOs (Lisbon); still others deferred fees, increased scope of flexibility for homes and businesses, and reduced real estate fees (Reykjavík in Iceland).

Additionally, cities supported local products and their sales/distribution by developing marketplaces for local commerce or promoted local distribution like in Paris; and some developed supply chains in partnership with cooperatives, companies, distributors, restaurants, supermarkets (Vila Nova de Famãlicao in Portugal) (OECD 2020).

A widely used financial assistance was provision of loans. New York City, for example, in its local support for SMEs, included zero-interest loans repayable over 15–20 years to firms with under 100 employees, up to US\$ 75,000, if they showed a 25% decline in customer receipts. Tokyo provided one-time payment for SMEs that were taking steps to prevent the further spread of the virus by suspending the use of their facilities. In Buenos Aires (Argentina), the public bank Banco Ciudad launched an emergency loan program to provide funds to small- and medium-sized enterprises for the payment of their payrolls—a significant reduction was made in interest rates of all loan schemes (OECD 2020).

Businesses themselves also made efforts through *adjustment in work style* during the pandemic. Amid lockdowns or a series of shutdown orders to control the virus, businesses ranging from grocery stores and restaurants to multinational corporations changed their pattern of work. Grocery stores and restaurants, for example, relied more on curbside pick-up or deliveries. Corporates and big businesses or companies deployed more advanced technologies—digital products and tech talent—to stay afloat while speeding up innovation. Those who could not adjust faced major problems and bottlenecks. Most significant negative effects were on tourism and hospitality and big industries such as automotive, aerospace, and machine tools.

Generation of Funds Besides using their own resources to support these measures, cities also procured funds from national governments. Moreover, finances were raised through other mechanisms such as establishment of funds and floating of bonds. For example, the Mayor of Milan announced the establishment of a mutual aid fund to help those most in need and to support recovery of city activities. The fund, floated by the allocation of EUR 3 million by the government, was open to the economic participation of individual citizens, companies, and associations. EUR 800,000 were raised for the fund on the very first day—March 14, 2020 (AGI 2020). An example from the USA was King County which in cooperation with philanthropic organizations established a relief fund and designed a donations connector page for citizens—they could request on this page for what they needed and also give away things they wanted to give. Seattle in the USA also provided a relief fund to assist large tech companies. The city of Bilbao, on the other hand, floated bonds to raise funds for the purpose (OECD 2020).

Assistance to Lagging Sectors Cities' support to the lagging sectors like tourism and hospitality in their survival efforts was also significant. Vilnius, the Lithuanian capital, and a tourist city, for example, transformed itself (Serhan 2020) into an open-air café, where hundreds of restaurants and bars set up shop in its plazas, squares, and streets and served customers from a safe distance. It also briefly operated a drive-in movie theater at the city's idle airport, where people went in their cars to watch films on a giant screen. Vilnius is just one example among many. In a similar approach, the city of Saga in Japan used public pedestrian spaces as part of a new style of restaurants and bars under, "SAGA Night Terrace Challenge" collaborative program between Saga City and the local business association (Karamatsu 2020). To help the tourist destinations/cities, governments of countries and territories such as Barbados, Estonia, Georgia, Antigua and Barbuda, Aruba, and the Cayman Islands developed policies to issue new long-term permits for 12 months in those cities. The objective was to entice foreign visitors for bringing their virtual offices with them, in the new era of remote work to combine work and enjoyment. Despite the assistance of the governments, however, an UNCTAD (2021a) report, while talking about a rebound in international tourism in the second half of 2021, paints a gloomy picture. It still shows a loss of between $1.7 trillion and $2.4 trillion in 2021, compared with 2019 levels, and United Nations World Tourism Organization expects a return to pre-COVID-19 international tourist arrival levels only by 2023 or later (UNCTAD 2021b).

5.3.1.2 Long-Term Strategies

The above discussion so far has covered how cities responded to the pandemic initially and how they worked on recovery efforts. Many cities, particularly in developing world, are still in the first or second phase of the response cycle to the pandemic, but that has not prevented others from looking ahead in terms of future development planning. Based on lessons learnt from their initial and ongoing experiences, many cities have advanced further, bringing forward ideas on how to promote long-term strategies for cities economic and social development and make these sustainable and resilient. OECD (2020) conducted a study of 33 cities in this regard and put these urban strategies into three categories—inclusive, green, and smart. It was noticed that overall cities were seeing the recovery as an opportunity for far-reaching changes and make their economies more sustainable, equitable, and resilient. The change is triggered partly because (a) the pandemic clearly brought forward vulnerabilities and inequalities in cities and (b) the life experience during the control measures such as lockdown not only prepared the citizens to change their behaviors but also to accept some drastic changes. For example, Milan, in its Adaptation Plan 2020, used the exit from the crisis to tackle fundamental issues on the city scale for improving quality of life and well-being of its citizens.

The importance of resilience and sustainability in cities' development paradigm has also become obvious. Hence, several cities in the developed world such as Malmo, Milan, Paris, and Izmir are realizing the importance of a holistic and resilient urban development strategies. The crisis prompted these cities to reflect on how to update their strategies/plans to integrate the lessons learned from the pandemic and previous crises to better prepare and prevent future risks and emergencies (OECD 2020). These cities applied resilience lens to improve their overall strategies and/or sectoral plans and in their planning, development, and procurement processes. Cities are also integrating the COVID-19 recovery initiatives in updating their long-term development strategies toward sustainability. For example, Amsterdam, Barcelona, Tallinn (Estonia), and Vienna and Utrecht (Netherlands) are using the Sustainable Development Agenda 2030 of the United Nations, as a guide for their city development strategies. Although these strategies were already in place before the pandemic, the cities are currently using the global framework for sustainable development as a policy tool to guide the implementation of these strategies and building a "new normal" in cities while

reducing their economic, social, and environmental vulnerabilities (OECD 2020).

Cities while trying to fulfil their underlying governance and financing needs are also making efforts to build inclusive, greener, and smarter long-term strategies (Fig. 5.1). In targeting inclusion or making economy inclusive, they are supporting local businesses and employment, helping in affordable housing construction and renovation (Vienna, Mexico City, Liverpool), and providing rental subsidy (Mexico City, Bogota). They are also supporting vulnerable population—such as homeless and immigrants through provision of food and shelter, counselling, etc. (Nantes in France, Rotterdam, Milan, New Orleans in the USA, Istanbul, Madrid, and Vienna).

5.3.1.3 Greening of Recovery and Development

In terms of greening recovery and development, targeted measures at the city level are concentrating mainly on sustainable urban mobility and measures for improving energy efficiency and promoting its conservation. Measures undertaken include development of bicycle infrastructure (Bogota, Milan, Paris, Seattle, Medellin), promotion of electric scooters and bikes (Rome, Rotterdam, Middlesbrough in the UK, Dublin, Milan), enhancement of energy efficiency (Copenhagen, Lille), development of renewable energy (Seoul, solar panel; Swansea, marine energy), and subsidy for energy conservation (Lille in France). Commitment of international institution such as EU's pledge to spend 20% of its fiscal stimulus on accelerating the transition to a green economy (World Economic Forum 2021) has given additional boost to cities for greening their economy. Further, the central role of digitalization during the pandemic has encouraged many cities to capitalize on the use of smart city tools more permanently for information, participation, and management of services particularly through digital platforms for education and other means (Riga (Latvia), Bamberg (Germany), Istanbul, and Tirana in Albania (OECD 2020).

The extremely pressing global environmental challenges faced by cities of the world today such

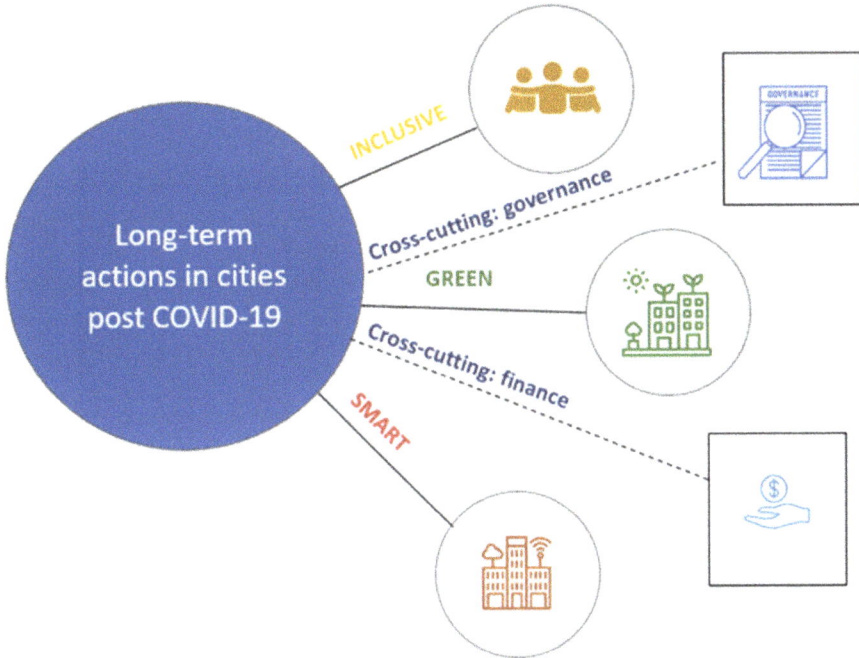

Fig. 5.1 Long-term actions in cities post COVID-19. (Source: OECD (2020))

as climate change, air pollution, and related health problems make it essential that the large funds mobilized by recovery plans lead to a transition toward more sustainable economies. Many countries have issued pledges to "build back better," and more than 130 nations (United Nations Climate Action 2021) have or are considering adopting a target of reducing emissions to net zero by mid-century. Since a very large amount of public funds are being injected into the economy, it is important to ensure that they are spent in an economically efficient and environmentally sustainable manner. Several tracking measures have been developed to track the sustainability of recovery packages, which have been discussed in Chap. 11. Out of these, OECD Green Recovery Database or OECD tracker on sustainability of development provides some important insights on the course various recovery measures are taking and their repercussions or implications for the environment (OECD 2021). These are given below:

- Although overall green measures increased in both number and budgetary spending, they

account for a small share of total recovery spending. Only around 21% or about one fifth of economic recovery spending in OECD, EU, and Key Partner countries is currently allocated to environmentally positive measures.
- The largest beneficiary of recovery measures are the energy and ground transport sectors.
- Current annual support to fossil fuels will likely outstrip all the one-off green recovery spending in just a few years, undermining efforts to meet the Paris climate goals.
- Green recovery measures need a stronger focus on skills for green jobs to ensure a "just transition"—current measures specifically targeting skills training represent only around 2% of the total.
- Recovery measures need to give more focus to innovation in green technologies and investment in R&D for the purpose. Current investment in R&D subsidies is only 8% of recorded measures.

OECD Green Recovery Tracker (OECD 2021) noted that one-off increase in public spending on green recovery measures is dwarfed by the continuing government support to fossil fuel producers and consumers. This means that while the USD 677 billions of identified green recovery budget will be spent over a variable number of years, government support measures for fossil fuels according to OECD-IEA estimates (OECD/IEA 2021) amounted to USD 345 billion just for the year 2020 in G20 and emerging economies. Although this indicates a decrease compared to 2019 levels (subsidies in 2019 amounted to USD 494 billion), the continued support provided by the governments to fossil fuels will undermine the efforts toward a green recovery.

According to OECD (2021), developing countries face different and even more difficult circumstances, where the COVID-19 crisis has compounded preexisting challenges. They are particularly constrained more because of low access to COVID-19 vaccines and their low fiscal capacity to deploy large rescue and recovery packages compared to high-income countries.

5.3.2 Economic Response to Climate Crisis

Many cities and metropolitan areas have been taking actions around the world on climate change using economic measures and instruments. Many have been doing this even in the absence of national policies and commitments. For example, in the USA, several cities undertook climate actions during the Trump administration. The cities are more agile on this issue because they recognize (a) the massive urban contributions to the crisis as well as risks, both physical and economic, posed to them from climate change and (b) the opportunities and the potential offered by promoting economic growth with environmental safeguard, and (c) climate action is a no-regret option that offers co-benefits in the form of increased efficiency and cost saving, public health betterment, and improved quality of life. Hence, the cities around the world are promoting policies, plans, and strategies to curb climate change. Currently, 6150 cities participating in the Global Covenant of Mayors and constituting one fifth of the urban population of the

world have developed climate action plans (CPI 2021a). The Global Covenant of Mayors for Climate & Energy is an alliance of the Compact of Mayors and the EU Covenant of Mayors. Bringing together world's two primary initiatives, it is helping cities/local governments in their transition to a low-carbon and climate resilient economy.

In the developed world, while some city governments have acted independently, others have benefited from guidance provided by networks of cities such as One Planet Charter of C40 Cities Climate Leadership Group and the Global Covenant of Mayors for Climate & Energy. The main objective of the Charter is to boost and accelerate the implementation of the Paris Agreement in metropolitan areas and cities throughout the world. Cities through this charter have committed to specific climate action that drives investments, sustainable public procurement, and policy decisions in renewable energy, energy efficiency, electric vehicles, and efforts for zero emission buildings and zero waste. The Charter reaffirms cities' commitment to promote sustainable and resilient infrastructure, products, and services while working closely with national governments and business sectors to mobilize global climate action. The Charter was built upon the achievements of the 23rd UN Climate Change Conference (COP 23), in particular the Bonn-Fiji Commitment of Local and Regional Governments to Deliver the Paris Agreement. It also promotes city networks' successful procurement and sectoral Initiatives/efforts already underway through ICLEI's Global Lead City Network on Sustainable Procurement, Sustainable Procurement Initiative, and 100% Renewable Energy Cities & Regions Campaign, as well as C40 Cities' Fossil Fuel Free Streets (C40 Cities 2017). According to the Vice Chair of the Global Covenant of Mayors, "The One Planet Charter demonstrates that when cities and local leaders embrace the opportunities that come with transitioning to a low-carbon world, both their constituents and their local economies benefit" (C40 Cities 2017). Urban climate actions have also emerged in response to national government mandates, such as Japan's Act on Promotion of Global

Warming Countermeasures, which requires local governments to formulate climate change action plans (Kamal-Chaoui and Robert 2009).

5.3.2.1 Policies and Plans

While large cities have developed landmark policies—notably Seoul, Stockholm, Toronto, Copenhagen, New York, London, and Tokyo—mid-sized and smaller cities have also created innovative climate policies, such as Mannheim and Freiburg in Germany, Toyama in Japan, Nantes in France, and Boulder in the USA (Kamal-Chaoui and Robert 2009). These climate policy packages, besides targeting various sectors of city economy like natural resources, energy, transport, waste, etc., have also used planning techniques such as land use zoning. They have also utilized economic instruments as well as existing and future investment potential for ameliorating climate change impacts—whether by maximizing GHG emissions cuts or enhancing their resilience to the adverse effects.

Natural Resource Policies Natural resource policies have been used mainly to reduce energy demand, enhance carbon sinks to absorb CO_2, and protect against adverse climate impacts. Tree plantation, for example, has been used in climate mitigation through carbon sequestration and offsets. It has also been effective in climate adaptation by planting mangroves to safeguard from flooding. According to Menéndez et al. (2020), in Miami (USA) and Cancun (Mexico), mangroves helped save more than \$US 500 million in avoided property damages every year, while "in Abidjan and Lagos in West Africa, Mumbai and Karachi in South Asia, Wenzhou in East Asia, and Cebu and Denpasar in South-east Asia existing mangroves protect more than 150,000 people from flooding every year." It is for the same reason that natural environment of the urban landscape is often included in climate plans as a means of absorbing CO_2 and reducing overall urban GHG emissions, as well as reducing potential urban heat island effects. The city of New York's PlaNYC Climate Plan has included a goal of planting 1 million additional trees between 2007 and 2030. The plan envisaged

planting 23,000 new trees every year (City of New York 2007). Likewise, Sejong, a new city in Korea, has planned to reduce average temperature of the city by 2.5 °C by allocating about half of the city area for parks, greenbelts, and waterfronts and operating a water circulation system based on natural water resources (Sejong City 2009). Tokyo also has a Greening Project that envisions tree-lined streets and green rooftops. The city of São Paulo, on the other hand, has developed linear parks along waterways that have minimized flooding effects, reduced water pollution, and contributed to the planting of more than half a million trees in about 4 years (OECD 2009).

Transportation Policy The transportation sector worldwide is responsible for 24% or about one fourth of direct CO_2 emissions from fuel combustion and road vehicles like cars, trucks, buses, and two- and three-wheelers. Majority of these are in cities, accounting for nearly three-quarters of transport CO_2 emissions (IEA 2020). Transportation has therefore been targeted as a major sector to reduce GHG emissions and adapt to expected climate change impacts. Cities' main effort in this direction has been to popularize the use of public transportation systems while discouraging the use of personal vehicles. For example, cities of Singapore and London have been discouraging use of personal vehicles by restricting these to enter certain zones at certain times. However, they have combined these restrictions with the provision of mass transit service and multi-modal linkages to maximize policies' effectiveness. Transportation agencies, in many cities such as in Stuttgart and Paris, have also implemented real-time signage systems to communicate expected arrival times to mass transit users. Cities, particularly in the developed world, are also giving incentives to promote no emission electric vehicles or less polluting alternate fuel vehicles. For example, as co-benefit, urban administrations in the USA have been allowing these vehicles to use carpool lanes. Use of non-motorized means of travel is also being encouraged in cities—the benefits of which apart from reduced GHG emissions include reduced congestion, increased walkability, and increased safety. Further, to facilitate bicycling, protected bicycle lanes have been promoted. Moreover, clear signage of bicycle routes has provided additional attractiveness for bike travel. Western European cities particularly Danish (Copenhagen) and Dutch (Amsterdam, Utrecht, and Rotterdam) are historically famous for promoting bicycle travel. During the pandemic, several cities world over established bike tracks. See Chap. 11 for details.

Waste Policy Cities have also reduced GHG emissions from the waste sector through reduced waste quantities and by increasing the energy efficiency of incinerators. The quantity of waste has been reduced by recycling and composting services and imposing fee to discourage waste. The city of San Francisco's recycling and food composting efforts resulted in diversion of 70% of its waste going to landfills as far back as the first decade of this century (OECD 2008). The city of Zurich restricts the amount of waste that residents can generate and sets fees for additional amounts (OECD 2008). Waste quantities have also been reduced through education campaigns, which are already common in many urban areas even in developing countries' cities such as Bangkok. The EU Landfill Directive requires reductions in the volume of biodegradable municipal waste cities send to landfills. Cities are also capturing methane gas from landfills to be used as a source of energy. The city of Monterrey, Mexico, which has been active in generating electricity by harvesting methane, constructed a 7-megawatt energy plant that captures and converts enough landfill gas into electricity to power the city's light-rail transit system and its streetlights (OECD 2008). In China, the city of Guangzhou in Guangdong province, one of the largest landfill energy capture projects, generates more than 50 Gigawatt of electricity, enough for 30,000 households. Other cities investing in landfill methane gas capture include Amman in Jordan and Christchurch and Nelson in New Zealand.

Land Use Policy Land use zoning has also been used as a means to promote climate change mitigation and adaptation. Cities, for example, have promoted higher densities and mixing of land use to reduce travel distance and frequency while developing mass transit linkages to promote compact growth and increase complementarities (Paris, London, New York). Managing compact growth and controlling outward sprawl or expansion of cities has become more popular and was made part of long-term growth plans in several metropolitan areas of the Netherlands, the UK, and Japan which initiated the concept of "Eco-Compact City" in their policies (Kamal-Chaoui and Robert 2009). Through these policies, targeting the higher residential densities in tandem with mixing of land uses in urban neighborhoods has considerably reduced travel distances between home, work, and other activities (shopping, leisure etc.) and promoted non-motorized travel with the most direct effect on GHG emissions.

Land use planning has also been used to create "eco-neighborhoods" or "sustainable neighborhoods" by integrating various measures for land use, transportation, natural resource, building, waste, and water for responding to climate change while also reducing the environmental footprint in cities. Such noteworthy "eco-neighborhoods" exist in some Western and Northern European countries. For example, in Sweden, these include Bo01 and Augustenborg in Malmö and Hammarby Sjöstad in Stockholm. Among others, they include Viiki (Helsinki) in Finland, Vauban and Rieselfeld (Freiburg) and Kronsberg (Hanover) in Germany, Vesterbro (Copenhagen) in Denmark, and Leidsche Rijn (Utrecht) in the Netherlands (Kamal-Chaoui and Robert 2009). In newly emerging economies and developing countries, eco-cities along similar lines have been developed in Korea, China, and Abu Dhabi. Development principles in these neighborhood go beyond residential density and include promotion of energy efficiency, use of sustainable building materials, and restricted personal vehicle use (cars were restricted or prohibited such as in Vauban city in France). Waste collection and management rules are also more restrictive in these neighborhoods compared to other parts of the city. The emerging concept of "15-Minute City" has been developed along similar lines, which became quite popular during the coronavirus pandemic (Moreno et al. 2021, C40 Cities 2020). It has been discussed in detail in Chap. 11 of this book.

Other Related Policies Overall, in terms of mitigation and adaptation policies, each city has its own specific needs and characteristics, based on its geographic milieu, population density, and other local settings. Cities around the world are currently at various stages of planning and development of these policies. A summary of these is available in the report "Cities on the Route Towards 2030: Building a Zero Emissions, Resilient Planet for All" which is released by CDP (2021) based on data reported by 812 cities in the world and highlights ongoing climate action of cities during 2020. 57% of cities have adaptation plan, while 43%, representing a projected population of over 400 million people by 2030, do not have an adaptation plan to tackle climate risk. City governments, in many of these adaptation plans, are making use of their jurisdiction over environmental features within their boundaries to protect cities' built environment from potential climate change impacts. Thus, in some coastal cities, engineering solutions with public investment have been sought as a primary adaptation tool for flood protection. Venice in Italy, New Orleans in the USA, Helsinki in Finland, and Rotterdam in the Netherlands are examples. These investments are not without controversy as they have affected the ecological resources to protect the built environment. Moreover, in terms of constraints, environmental zones often do not fall within city boundaries, therefore adaptation planning and management often require horizontal coordination with multiple local governments within the same region as well as vertical coordination with regional and national governments.

Parks and natural spaces have also been used as an adaptation measure—by planning new parks in areas that are most vulnerable to flooding like New York. Several cities like the city of Dresden in Germany have implemented adaptation programs to prevent flooding and minimize and manage rainwater and stormwater. Likewise, through national hydraulic engineering and forestry legislations, the Swiss federal government is providing funding at the canton level for protective measures against climate hazards, which is matched by funding from cantons, municipalities, and infrastructure owners.

Among other measures, according to CDP (2021) report, a vast majority of cities (87%) are incorporating sustainability into master planning or are intending to do so soon. Cities that have incorporated sustainability into their master planning have identified more than twice as many opportunities by addressing climate change as cities that have not done it so far. Cities also identified over 1000 climate projects seeking finance, worth US$ 72 billion. An encouraging sign is that more than three fourths of cities are collaborating with businesses on sustainability projects or are intending to do so in the coming 2 years.

5.3.2.2 Fiscal Instruments and Urban Climate Financing

Cities have already used fiscal and economic instruments for abating climate change. Additional measures on mitigation to cut GHG emissions and adapt to possible climate change impacts require many other actions. These require financing through both restructuring of existing funds and exploration of additional avenues such as investment financing.

Fiscal Instruments Among instruments used included congestion charge (Energy Foundation 2014), caps-and-trade mechanisms, carbon tax, etc. (Kamal-Chaoui and Robert 2009) The congestion charge has been applied in several cities and proved effective in reducing congestion and reducing CO_2 emissions from transport (reductions between 10% and 20% in London, Stockholm, and Milan). Development charges and value-added taxes have been used in Miami, Milan, and Bogotá, and transport-related fuel taxes, congestion charges, and parking fees have been used to relieve traffic congestion and greenhouse gas emissions there.

European cities have operated a cap-and-trade program since 2005. Similarly, several Chinese cities have carbon caps since 2013. Mexico initiated a pilot cap-and-trade program from January 1, 2020 (C2ES 2021). Some cities participate in the cap- and-trade mechanisms for emission trading (e.g., the Chicago Climate Change Exchange). In some cases, cities have set up their own urban cap-and-trade mechanisms (Los Angeles, Chicago, Santiago, and metropolitan Tokyo in 2010). Other cities such as London have explicitly defined emissions trading as a business opportunity to increase their metropolitan competitiveness. Carbon taxes and climate change levies have also been used for climate mitigation. Boulder in the USA was the first city to introduce carbon tax. Funding from clean development mechanisms (CDM) and joint implementation mechanisms—the two main Kyoto carbon offset instruments—are also helping reduce cities' carbon emissions by providing carbon offsets for urban projects such as mass transit expansion. However, some of these mechanisms are complex and were rarely used by cities, Bogotá and Sao Paulo being a notable exception. There is a need to use these mechanisms effectively in other cities and promote innovation in other existing fiscal instruments such as taxes, fee, bonds, etc. There is also a need for raising funds from carbon markets.

C40 (undated) cities network has recommended six most effective avenues that cities can use to ensure funding of projects, programs, and investments set out in their climate action plans as follows:

- Mainstream climate considerations into the city's budget and investment decisions
- Develop city officials' capacity to prepare climate projects

- Introduce policies to improve and expand revenue collection, targeting the biggest emitters
- Lobby national governments for improved access to finance and regulatory environment
- Work with partners, especially the private sector, to encourage investment
- Divest investments and pension funds from fossil fuel companies and increase investments in sustainable projects and companies

Investment Mobilization Despite the momentum on climate action, cities, particularly in developing countries, face serious constraints in mobilizing investment financing for transformational climate action. Many of the barriers to financing include the lack of technical and financial capacity, control over resources, and workable funding models (CPI 2021a). International Finance Corporation (IFC 2018) pointing to the investment barriers faced by cities lists creditworthiness, bankability, and the lack of a viable project pipeline, as well as limit on what cities can do on their own as obstacle to attracting private finance. The coronavirus pandemic has further increased financial difficulty of cities. It has brought the ambitions of cities for climate action to a critical juncture where they need partnership with national governments, international organizations, civil society, and the private sector. The rapidly growing cities in Africa and South Asia are the worst affected (CPI 2021a).

Urban climate finance (resources directed to activities limiting city-induced GHG emissions or aiming to address climate-related risks faced by cities, contributing to cities' low-carbon development or resilience) flows reached an estimated USD 384 billion annually on average in 2017/2018, far short of urban climate finance needs estimated at USD 4.5–5.4 trillion annually (CPI 2021a). Investment trends during the pandemic and beyond are highly uncertain with both positive and negative factors at play. On the positive side, while development banks have increased their climate commitments and many cities have adopted green recovery packages, simultaneously, consumers' investment is increasing toward promoting green consumption—for example, investments in electric vehicles have continued to show a rising trend. On the negative side, many cities are delaying or reducing non-essential capital expenditures, while electric vehicle subsidies at the national level have been decreasing (IEA 2020). From the last available data on climate financial flows for cities, of the US$ 384 billion financing in 2017/2018, US$ 147 billion accounted for expenditures on urban green transport and USD 161 billion for urban green buildings and appliances. This has been discussed in more detail in Chap. 11 of this book.

It is important to emphasize that urban financing constitutes a lifeline in building urban resilience strategies and achieving mitigation targets. Further, International Finance Corporation (IFC) has estimated that despite barriers that constrain investment in climate-smart urban infrastructure, cities in emerging markets around the globe have the potential to attract more than $29.4 trillion in cumulative climate-related investments in six key sectors (waste, renewable energy, public transportation, water, electric vehicles, and green buildings) by 2030 (IFC 2018). The biggest share of the opportunity, according to estimates, was in green buildings ($24.7 trillion), covering both new constructions and retrofits, in the wake of cities' efforts to accommodate soaring populations. Improvements in low-carbon mobility solutions, driven by public transport infrastructure and the expected surge in electric vehicles, are other potential areas that account for $1 trillion and $1.6 trillion, respectively. The IFC (2018) report, however, mentions regional variations in the size of the investment opportunity by sector based on both the range in the ambitions of targets set by cities and the differing costs in terms of technologies to be used and implementation mechanisms. It is extremely important for developing countries' cities and nations to tap this source, where the infrastructure projects tend to be funded solely by the public sector. The sheer scale of the required investment in these projects necessitates the unlocking of private sources of financing.

Within urban climate adaptation context, activities that aim to maintain or increase the resilience of cities and urban communities, in response to climate-related risks affecting the city, directly constitute "unban climate adaptation finance." Climate adaptation finance for cities is largely inadequate and requires urgent review to protect people from future disasters. According to Climate Policy Initiative report on climate adaptation finance (CPI 2021b), less than $4 billion was invested annually in 2017–2018 in urban adaptation finance projects. This is far shorter than the World Bank's estimate amounting to $11–20 billion that will be needed by 2050 on an annual basis to protect global urban infrastructure from climate risks. Water and wastewater management projects addressing urban climate risk received the most finance of any sector, followed by disaster risk management. The report (CPI 2021b) noted the significant gap between the assessed financing need to adequately address urban climate risk and tracked flows of finance. It also identifies significant opportunities to increase finance for climate adaptation. The opportunities mentioned 12 examples of the use of financial instruments in the urban adaptation context, which among others included resilience bonds, disaster risk insurance pooling, public-private partnerships, and catastrophe bonds.

In terms of financing, the connection between financial flows and climate risk reduction and its advanced tracking is extremely important. It can help development finance institutions (DFIs) and national and international policy makers to target capital flows to locations with highest need. It can also permit development of evaluation metrics, thereby easing investor hesitancy.

5.4 Conclusion

The mega risks COVID-19 pandemic and climate crisis had compounding effects on the economy of cities—the severity of their impact has been beyond anything experienced in nearly a century. Together, they have generated a threat, the effects of which have spread across all sectors of economy and all areas from local to global. The worst impact has been on cities and metropolitan areas, which drive the global economy by generating 80% of the global GDP and where 65% of the global economic growth is expected in the next few years. Their impact as risk multiplier has been devastating—worsening vulnerabilities and causing failures in economic systems. Significant losses were incurred in cities' tax revenues and citizens' income, while disproportionate effects were felt on certain sectors (such as tourism and hospitality) and businesses such as small- and medium-sized enterprises. Disruptions were experienced in the urban supply chains, and a new normal lifestyle had to be adopted to maintain the economic functioning of cities. The supply chain and logistical disruptions had serious impacts on industrial production with such ramification as semiconductor shortages, which were difficult to predict. Finally, both crises have been "economically regressive" in that they have mainly affected the most vulnerable communities—urban poor and downtrodden daily wage earners.

The impact of economic shock due to the pandemic and its intensity varied and depended on cities' exposure; development status (developing or developed country cities) and their economic composition and labor market; crisis management responses related to social distancing, workplace management, and commuting possibilities; supply chain status; presence or absence of vulnerable groups; and support to business and industry groups by governments. Considering development status, cities of developing countries were worst affected. In terms of economic composition and labor market, cities with diverse economic structure were less vulnerable. So were the cities whose economies relied on the core industries or businesses dependent on technology or the movement of information like San Francisco, Seattle, Hong Kong, and Singapore.

The informal economy, which has a lion's share in the developing cities' economy, suffered from a catastrophic effect in the wake of lockdown measures. Living hand to mouth, without any savings, the fragility of informal workers became worse. Constituting a staggering 12.5% of the global population and 62% of all those working worldwide, the pandemic's restrictions pushed them into poverty. The World Bank estimated that over a million people fell under poverty in 2020, given the economic stagnation, job losses, and the decrease of about 40% remittances to low- and middle-income countries. With increasing poverty, the crisis sharpened and reinforced the inequalities in cities of both developed and developing countries. The pandemic economic shock magnified the existing fault lines in developed countries' cities created by race, gender, age, and education, while it exacerbated the divide created by the social inequalities and dwelling conditions (slum and informal vs non-slum dwellers) in developing countries' cities. All in all, preexisting and current inequalities were worsened in world cities, weakening economic resiliency, and reinforcing disadvantage. It is therefore imperative to promote inclusive measures in post-pandemic development paradigm.

Cities and countries facing climate crisis to their predicament have already faced disaster risks frequently for a long time, and in the wake of inaction or proper action, many more cities and countries are falling into crisis mode. Globally, the damage cost of weather and climate-related disasters to cities and communities in 2020 has been estimated to be between $166 and $210 billion. These monetary losses to cities were up 26.5% compared to 2019's cost. Only $82 billion of the total damage in 2020 was insured. Climate change is likely to hit the world economy even more significantly by 2050, in the business-as-usual scenario, as risk events increase, diseases spread, and rising seas consume parts of coastal cities. Unless the world succeeds in quickly slowing the use of fossil fuels, the effects of climate change are expected to cut 11–14% off global economic output by 2050 compared with economic growth without climate change. This means reduced global economic output by US $23 trillion every year. However, if strengthened current efforts succeed in holding the global temperatures to less than 2°C above preindustrial levels (the goal set by the 2015 Paris agreement), economic damage will be marginal. Therefore, climate action has become an urgent necessity.

The immediate economic response to coronavirus was in the form of emergency relief, followed by rescue packages as well as recovery or revival of city economy. The concentration is now also on the planning for the future. The objective of the rescue packages was to protect business environment and livelihoods in the face of unemployment and abrupt losses of income. The size of these measures used in response to the COVID-19 crisis up to only the end of May 2020 was ten trillion dollars—three times more than the response to the 2008–2009 financial crisis. Western European countries alone allocated close to $4 trillion, an amount almost 30 times larger than today's value of the Marshall Plan, and the crisis is far from over. While national governments undertook measures to protect their economies from the fallout, city governments also played a key role in developing and implementing recovery frameworks in their own jurisdiction in the form of visions, plans, strategies, recovery initiatives, guidelines, etc. They also devised a variety of financial and technical assistance tools and measures for economic revival. Some of these tools included grants and loans for business support and jobs, investment in infrastructure and businesses, and development of public-private partnerships. Economic instruments such as tax incentive and subsidies have also been used. Based on lessons learnt from their initial and ongoing experiences, many cities have advanced further, bringing forward ideas on how to promote long-term strategies for cities' economic and social development. It is important to make these sustainable and resilient by selecting the right trade-offs for sustainability. With massive stimulus packages being undertaken in cities and nations to promote the economy around the world, governments, businesses, and societies have both a responsibility and self-interest to not only look for near-term measures to shore up livelihoods and employment but also to reflect on the political and economic driving forces leading to the current crisis and take concrete steps not to repeat the same mistakes in the future.

Many cities and metropolitan areas have also been taking actions around the world on climate change. Several have been doing this even in the absence of national policies and commitments. Currently, 6150 cities participating in the Global Covenant of Mayors and constituting one fifth of the urban population of the world have developed climate action plans. While large cities—notably Seoul, Stockholm, Toronto, Copenhagen, New York, London, and Tokyo—have developed landmark climate policies, mid-sized and smaller cities have also created innovative policies, such as Mannheim and Freiburg in Germany, Toyama in Japan, Nantes in France, and Boulder in the USA. In developing these policy packages, cities have used measures for the management of various sectors of city economy—including natural resources, energy, transport, waste, etc. They have utilized planning techniques such as land use zoning to develop sustainable neighborhoods. They have also applied economic instruments as well as existing and future investment potential for ameliorating climate change impacts—whether by maximizing GHG emissions cuts or enhancing their resilience to the adverse effects through various adaptation measures. It is extremely important to implement these policies and orient them toward achieving the target of Paris Agreement.

Despite the momentum on climate action, cities, particularly in developing countries, face serious constraints in mobilizing finances. Among others, barriers to financing include the lack of technical and financial capacity, control over resources, and workable funding models. The International Finance Corporation pointing to the investment barriers faced by cities lists creditworthiness, bankability, and the lack of a viable project pipeline, as well as limit on what cities can do on their own, as obstacles to attracting private finance. The pandemic has brought the ambitions of cities for climate action to a critical juncture where they need partnership with national governments, international organizations, civil society, and the private sector. The rapidly growing cities in Africa and South Asia are the worst affected. Even in the existing barriers that constrain investment in climate-smart urban infrastructure, however, IFC estimates that cities in emerging markets around the globe have the potential to attract more than $29.4 trillion in cumulative climate-related investments in six key sectors (waste, renewable energy, public transportation, water, electric vehicles, and green buildings) by 2030 (IFC 2018). The biggest share of the opportunity, according to estimates, is in green buildings ($24.7 trillion), covering both new constructions and retrofits.

Finally, the Agenda for Sustainable Development 2030 demands a "paradigm shift" to move away from perceiving growth and sustainability as an either/or proposition to defining the two policy objectives as linked and mutually reinforcing. This paradigm shift is happening across local and national governments albeit slowly. An issue particularly in developing countries is whether national and city/local governments accept this premise both generally and in the context of mega risks. Most importantly, it is important to analyze whether within government authorities responsible for issues related to risks in particular mega risks and economic development fully understand the challenge and the need for change. In fact, tackling the challenge lies in coherence in developing and implementing cross-sectoral approaches. The ones that would make cities economically vibrant and socially inclusive, environmentally safe, as well as resilient to risks.

References

Adams-Prassl A, Boneva T, Golin M, Rauh C (2020) Inequality in the impact of the coronavirus shock: evidence from real time surveys. J Public Econ 189:104245

AGI Agenzia Italia (2020) A Milano sembra di essere tornati agli anni di piombo dice Beppe Sala. https://www.agi.it/cronaca/news/2020-03-04/coronavirus-sala-milano-rilancio-7325067/. Accessed 26 Jul 2021

AHLA American Hotel and Lodging Association (2021) State of the hotel industry. https://www.ahla.com/sites/default/files/2021_state_of_the_industry_0.pdf

Arent DJ, Tol RSJ, Faust E, Hella JP, Kumar S, Strzepek KM, Tóth FL, Yan D (2014) Key economic sectors and services. In: Climate change 2014: impacts, adaptation, and vulnerability. Part A: global and sectoral aspects. contribution of working group II to the fifth assessment report of the intergovernmental panel on climate change. Cambridge University Press, Cambridge, United Kingdom and New York, pp 659–708

Bloomberg New Economy Forum and McKinsey & Company (2020) NEF spotlight: a pandemic reboot for cities. https://www.mckinsey.com/business-functions/strategy-and-corporate-finance/our-insights/nef-spotlight-a-pandemic-reboot-for-cities

Bonnet F, Vanek J, Chen M (2019) Women and men in the informal economy: a statistical brief. International Labour Office, Geneva. http://www.wiego.org/sites/default/files/publications/files/Women%20and%20Men%20in%20the%20Informal,20

Bottan N, Hoffmann B, Vera-Cossio D (2020) The unequal impact of the coronavirus pandemic: evidence from seventeen developing countries. PLoS One 15(10):e0239797. https://doi.org/10.1371/journal.pone.0239797

Boza-Kiss B, Pachauri S, Zimm C (2021) Deprivations and inequities in cities viewed through a pandemic lens. Front Sustainable Cities 3:645914. https://doi.org/10.3389/frsc.2021.645914

Buheji M, da Costa CK, Beka G et al (2020) The extent of covid-19 pandemic socio-economic impact on global poverty, a global integrative multidisciplinary review. Am J Econ 10(4):213–224

C2ES (2021) Cap and trade basics. https://www.c2es.org/content/cap-and-trade-basics/

C40 Cities (2020) How to build back better with a 15-Minute City. https://www.c40knowledgehub.org/s/article/How-to-build-back-better-with-a-15-minute-city?language=en_US

C40 CITIES (undated) Climate emergency: urban opportunity. https://www.c40knowledgehub.org/s/article/Climate-Emergency-Urban-Opportunity-How-national-governments-can-secure-economic-prosperity-and-avert-climate-catastrophe-by-transforming-cities?language=en_US

C40 CITIES and ICLEI (2017) Global covenant of mayors, and C40 announce one planet charter to accelerate local implementation of the Paris agreement, https://www.c40.org/press_releases/iclei-global-covenant-of-mayors-and-c40-announce-one-planet-charter-to-accelerate-local-implementation-of-the-paris-agreement

Cassim Z, Handjiski B, Schubert J, Zouaoui Y (2020) The $10 trillion rescue: how governments can deliver impact. https://www.mckinsey.com/industries/public-and-social-sector/our-insights/the-10-trillion-dollar-rescue-how-governments-can-deliver-impact

CCSA The Committee for the Coordination of Statistical Activities (2021) How COVID-19 is changing the world, a statistical perspective, Vol III. https://reliefweb.int/sites/reliefweb.int/files/resources/CCSA_COVID19_Volume-III.pdf

CDP (2021) Cities on the route towards 2030: building a zero emissions, resilient planet for all. https://cdn.cdp.net/cdp-production/cms/reports/documents/000/005/759/original/CDP_Cities_on_the_Route_to_2030.pdf?1621329680

Ceylan RF, Ozkan B, Mulazimogullari E (2020) Historical evidence for economic effects of COVID-19. Eur J Health Econ 21(6):817–823

Christodoulou A, Christidis P, Demirel H (2019) Sea-level rise in ports: a wider focus on impacts. Marit Econ Logist 21:482–496. https://doi.org/10.1057/s41278-018-0114-z

City of New York (2007) PlaNYC – a greener, greater New York. City of New York

Congressional Research Service (2021) Global economic effects of COVID-19. https://fas.org/sgp/crs/row/R46270.pdf

Corfee-Morlot J, Kamal-Chaoui L, Donovan MG, Cochran I, Robert A, Teasdale PJ (2009) Cities, climate change and multilevel governance, OECD environment working paper, Paris

CPI Climate Policy Initiative (2021a) The state of cities climate finance. https://www.climatepolicyinitiative.org/publication/the-state-of-cities-climate-finance/

CPI Climate Policy Initiative (2021b) An analysis of urban climate adaptation finance. https://www.climatepolicyinitiative.org/publication/an-analysis-of-urban-climate-adaptation-finance/

CRED and UNDRR (2021) 2020 The non-COVID year in disasters, Brussels

CRS, Congressional Research Service (2010) The 2007–2009 recession: similarities to and differences from the past. https://crsreports.congress.gov/product/pdf/R/R40198/10

Du J, King R, Chanchani R (2020) Tackling inequalities in cities is essential for fighting COVID-19. World Resources Institute

Earl C, Vietnam R (2020) Living with authoritarianism: ho chi minh city during COVID-19 lockdown. City Soc 32(2):12306. https://doi.org/10.1111/ciso.12306

Energy Foundation (2014) International best practices for congestion charge and low emissions zone. https://www.efchina.org/Attachments/Report/reports-20140812-en/reports-20140812-en

Ghosh S, Seth P, Tiwary H (2020) How does Covid-19 aggravate the multidimensional vulnerability of slums in India? A commentary. Soc Sci Human Open 2:100068. https://doi.org/10.1016/j.ssaho.2020.100068

Giannoccaro I, Albino V, Nair A (2018) Advances on the resilience of complex networks. Complexity 2018:8756418

Gonzalez D, Karpman M, Kenney GM, Zuckerman S (2020) Hispanic adults in families with noncitizens disproportionately feel the economic fallout from COVID-19. https://www.urban.org/sites/default/files/publication/102170/hispanic-adults-in-families-with-noncitizens-disproportionately-feel-the-economic-fallout-from-covid-19_2.pdf

Hallegatte S, Green C, Nicholls RJ, Corfee-Morlot J (2013) Future flood losses in major coastal cities. Nat Clim Chang 3:802–806

Hill R, Narayan A (2021) What COVID-19 can mean for long-term inequality in developing countries. World Bank, https://blogs.worldbank.org/voices/what-covid-19-can-mean-long-term-inequality-developing-countries

IEA International Energy Agency (2020) Transport. https://www.iea.org/topics/transport

IFC International Finance Corporation (2018) Climate investment opportunities in cities. https://www.ifc.org/wps/wcm/connect/875afb8f-de49-460e-a66a-dd2664452840/201811-CIOC-IFC-Analysis.pdf?MOD=AJPERES&CVID=mthPzYg, Washington, DC

ILO International Labour Organization (2020) COVID-19 crisis and the informal economy immediate responses and policy challenges. ILO brief May. https://www.ilo.org/wcmsp5/groups/public/%2D%2D-ed_protect/%2D%2D-protrav/%2D%2D-travail/documents/briefingnote/wcms_743623.pdf

IMF International Monetary Fund (2021) World economic outlook. https://www.imf.org/en/Publications/WEO

IMF (2022) World Economic Outlook. https://www.imf.org/en/Publications/WEO

Kahn ME et al (2019) Long-term macroeconomic effects of climate change: a cross-country analysis. Federal Reserve Bank of Dallas, https://www.dallasfed.org/~/media/documents/institute/wpapers/2019/0365.pdf

Kamal-Chaoui L, Robert A (eds) (2009) Competitive cities and climate change, OECD regional development working papers N° 2 2009. OECD Publishing, © OECD

Karamatsu N (2020) Demonstration experiment of avoiding "3 Dense" in saga prefecture with terrace seats on the sidewalk at night. https://project.nikkeibp.co.jp/atclppp/PPP/news/052801568/. Accessed 5 Dec 2020

Kulisch E (2021) Airline absorb $ 391 billion revenue shortfall in 2020. https://www.freightwaves.com/news/airlines-absorb-391-billion-revenue-shortfall-in-2020

Kunzmann KR (2020) Smart cities after covid-19: ten narratives. disP – the Planning Review 56(2):20–31

Lam JSL, Su S (2015) Disruption risks and mitigation strategies: an analysis of Asian ports. Marit Policy Manag 42(5):415–435. https://doi.org/10.1080/03088839.2015.1016560

Landershaupstadt Dusseldorf (2020). https://corona.duesseldorf.de/

Lenzen M, Li M, Malik A et al (2020) Global socio-economic losses and environmental gains from the coronavirus pandemic. PLoS One 15(7):e0235654

Liu H, Fang C, Gao Q (2020) Evaluating the real-time impact of COVID-19 on cities: China as a case study. Complexity 2020:1–11. https://doi.org/10.1155/2020/8855521

Martin A, Markhvida M, Hallegatte S, Walsh B (2020) Socio-economic impacts of COVID-19 on household consumption and poverty. Econ Disaster Clim Chang 4(3):453–479. https://doi.org/10.1007/s41885-020-00070-3

Martínez L, Short JR (2021) The Pandemic City: urban issues in the time of COVID-19. Sustainability 13:3295. https://doi.org/10.3390/su13063295

Menéndez P, Losada IJ, Torres-Ortega S et al (2020) The global flood protection benefits of mangroves. Sci Rep 10:4404. https://doi.org/10.1038/s41598-020-61136-6

Moreno C, Allam Z, Chabaud D, Gall C, Pratlong F (2021) Introducing the "15-Minute City": sustainability, resilience and place identity in future post-pandemic cities. Smart Cities 4:93–111. https://doi.org/10.3390/smartcities4010006

Munich Re (2021) Record hurricane season and major wildfires – the natural disasters figures for 2020. https://www.munichre.com/en/company/media-relations/media-information-and-corporate-news/media-information/2021/2020-natural-disasters-balance.html

Muro M, You Y (2020) In some cities the pandemics economic pain may continue for a decade. Brookings Institution, Washington, DC

Napierała T, Leśniewska-Napierała K, Burski R (2020) Impact of geographic distribution of COVID-19 cases on hotels' performances: case of Polish cities. Sustainability 12(11):4697

National Oceanic and Atmospheric Administration (2018) Fourth national climate assessment, Chapter 8, coastal effects, 83. https://nca2018.globalchange.gov/chapter/coastal

NCEI NOAA's National Center for Environmental Information (2021) 2020 US billion-dollar weather and climate disasters in historical context. https://www.climate.gov/disasters2020

Novati M, Achurra-Gonzalez P, Foulser-Piggott R, Bowman G, Bell MGH, Angeloudis P (2015) Modelling the effects of port disruption: assessment of disaster impacts using a cost based container flow assignment in liner shipping networks. Transportation Research Board 94th Annual Meeting, Washington, DC, USA

OECD (2009) Green cities: new approaches to confronting climate change, OECD workshop proceedings, 11 June 2009, Las Palmas de Gran Canaria, Spain

OECD (2020) OECD policy responses to coronavirus: cities policy response. https://www.oecd.org/coronavirus/policy-responses/cities-policy-responses-fd1053ff/

OECD (2021) Key findings from the update of the OECD green recovery database. https://www.oecd.org/coronavirus/policy-responses/key-findings-from-the-update-of-the-oecd-green-recovery-database-55b8abba/

OECD and Bloomberg Philanthropies (2014) Cities and climate change. https://www.oecd.org/env/cc/Cities-and-climate-change-2014-Policy-Perspectives-Final-web.pdf

OECD Organization of Economic Cooperation and Development (2008) OECD environmental outlook to 2030, Paris

OECD/IEA (2021) Update on recent progress in reform of inefficient fossil-fuel subsidies that encourage wasteful consumption. https://www.oecd.org/g20/topics/climate-sustainability-and-energy/OECD-IEA-G20-Fossil-Fuel-Subsidies-Reform-Update-2021.pdf. Accessed 14 Nov 2021

Pineda VS, Corburn J (2020) Disability, urban health equity, and the coronavirus pandemic: promoting cities for all. Journal of Urban Health 97(3):336–341

Romei V, Burn-Murdoch J (2020) From peak city to ghost town: the urban centers hit hardest by COVID-19. Financial Times, London

Santos J (2020) Using input-output analysis to model the impact of pandemic mitigation and suppression measures on the workforce. Sustainable Prod Consumption 23:249–255

Sejong (Korea) (2009) Green city Sejong. Multifunctional Administration City Construction Agency, Yeongi-gun

Serhan Y (2020) Vilnius shows how the pandemic is already remaking cities. The Atlantic. https://www.theatlantic.com/international/archive/2020/06/coronavirus-pandemic-urban-suburbs-cities/612760/

Sharifi A, Khavarian-Garmsir AR (2020) The COVID-19 pandemic: Impacts on cities and major lessons for urban planning, design, and management. Sci Total Environ 749:142391. https://doi.org/10.1016/j.scitotenv.2020.142391

Short JR (2020) Street vendors make cities livelier, safer and fairer—here's why they belong on the post-COVID-19 urban scene. https://theconversation.com/street-vendors-make-cities-livelier-safer-and-fairer-heres-why-they-belong-on-the-post-covid-19-urban-scene-141675

Smith AB (2021) 2020, U.S. billion-dollar weather and climate disasters in historical context. https://www.climate.gov/news-features/blogs/beyond-data/2020-us-billion-dollar-weather-and-climate-disasters-historical

Swiss Re (2021) The economics of climate change, climate change poses the biggest long-term risk to the global economy. https://www.swissre.com/institute/research/topics-and-risk-dialogues/climate-and-natural-catastrophe-risk/expertise-publication-economics-of-climate-change.html

Syvitski JPM, Kettner AJ, Overeem I, Hutton EWH, Hannon MT, Brakenridge GR, Day J, Vorosmarty C, Saito Y, Giosan L, Nicholls RJ (2009) Sinking deltas due to human activities. Nat Geosci 2(10):681–686

UN Habitat (2011) Global report on human settlements: cities and climate change. Earthscan, London

UNCCRN-Urban Climate Change Research Network (2018) The future we don't want: how climate change could impact the world's greatest cities, Technical Report, February, in collaboration with C40 Cities, Global Covenant of Mayors for Climate and Energy and Acclimatize

UNCTAD United Nations Conference on Trade and Development (2021a) COVID-19 and tourism – an update: assessing the economic consequences, Geneva. https://unctad.org/system/files/official-document/ditcinf2021d3_en_0.pdf

UNCTAD United Nations Conference on Trade and Development (2021b) Global economy could lose over $4 trillion due to covid-19 impact on tourism. https://unctad.org/news/global-economy-could-lose-over-4-trillion-due-covid-19-impact-tourism

United Nations (2020) The sustainable development goals report 2020. https://unstats.un.org/sdgs/report/2020/

United Nations (2021) World economic situation and prospects as of mid-2021. Department of Economic and Social Affairs, New York

United Nations, climate action (2021) For a livable climate: net zero commitments must be backed by credible action. https://www.un.org/en/climatechange/net-zero-coalition

United Nations Habitat (2011) The economic role of cities, Nairobi

United Nations World Tourism Organization (2020) 100% of global destinations now have COVID-19 travel restrictions. https://www.unwto.org/news/covid-19-travel-restrictions

Vandenberg R, Lauren NS, Arnoud M, Rafael T (2020) Equitable cities during COVID-19. https://www.wri.org/insights/building-climate-resilient-and-equitable-cities-during-covid-19

Witteveen D (2020) Sociodemographic inequality in exposure to COVID-19-induced economic hardship in the United Kingdom. Res Soc Stratif Mobil 69:100551

World Bank (2021) Global economic prospects, Washington, DC

World Economic Forum (2021) Chief economists outlook, Geneva

Zhou L, Mu H, Wang B, Yuan B, Dang X (2020) Evaluating urban community sustainability by integrating housing, ecosystem services, and landscape configuration. Complexity 2020:3460962

Mega Risks, Urban Energy Use, and Sustainable Development

6.1 Introduction

Energy is a basic need that plays a crucial role in economic and social development of cities. Over 70 percent of the global energy is consumed by cities providing lifeline for their functioning (United Nations 2022). As the engine of economic growth, energy is vital for all sector of urban economy including transport and communication, industries, residential sector, business, and commerce, trade, tourism, etc. An uninterrupted supply of energy is also needed for heating and cooling of buildings; for operation of urban infrastructure, utilities, and services such as water distribution, waste management, health, and education; and for running city administration itself. As the biggest user of energy worldwide, cities account for a mammoth carbon footprint, releasing an increasingly large proportion of GHG emissions, a major cause of climate change mega risk.

The chapter has been divided into seven parts. After the introduction, the next section discusses the nature of energy use and consumption in the world and its cities. It is followed by examining the impact of COVID-19 pandemic and resulting economic crisis on the energy production and its use as well as on GHG emission patterns. The objective is to highlight the lessons that can be derived from it for handling climate crisis in cities. The next section examines carbon footprint of cities. It is followed by a section on urban policies on energy use and GHG emissions. It dis-

cusses the potential contribution of local/urban policy solutions and instruments in reducing energy demand through the trade-offs between economic and environmental objectives. Case studies on the use of several direct and indirect tools including regulatory measures, economic instruments, green innovation, etc., for increasing energy efficiency and reducing GHG emissions in various sectors, have been presented. The section following it covers the constraints and shortcomings in the implementation of policy measures. The findings of the chapter are highlighted in the concluding section.

6.2 Energy Supply/Consumption

In the absence of actual consumption data, urban energy supply has been used as proxy for consumption, which is important in addressing the challenges to sustainability. The amount and type of energy cities consume, or use, is extremely important as it not only affects the economy, the environment, and the well-being of its citizens but also has a bearing on mega risk like climate change. Lack of urban energy data, however, is a big constraint especially in developing countries. Some studies (Pranab et al. 2020; Kennedy et al. 2015; Tong et al. 2021; Office of Energy Efficiency and Renewable Energy 2015) have utilized/suggested methods for monitoring and quantifying urban energy use patterns. For example, Pranab et al. (2020) used the relationship between the settlement types and the correspond-

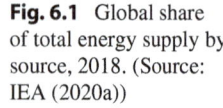 **Fig. 6.1** Global share of total energy supply by source, 2018. (Source: IEA (2020a))

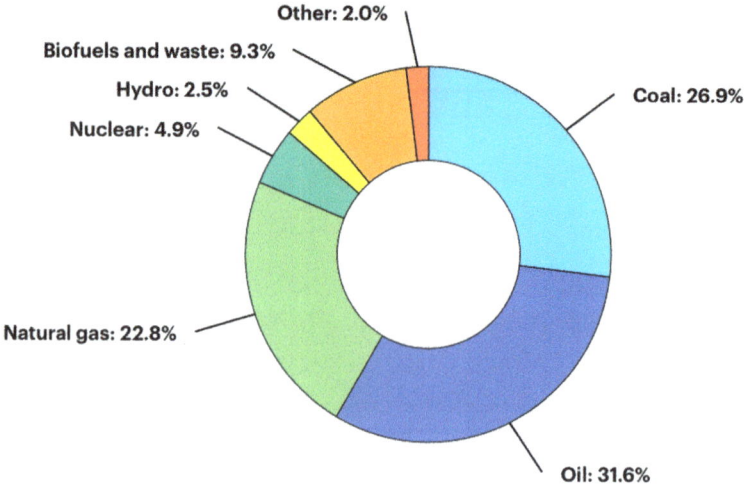

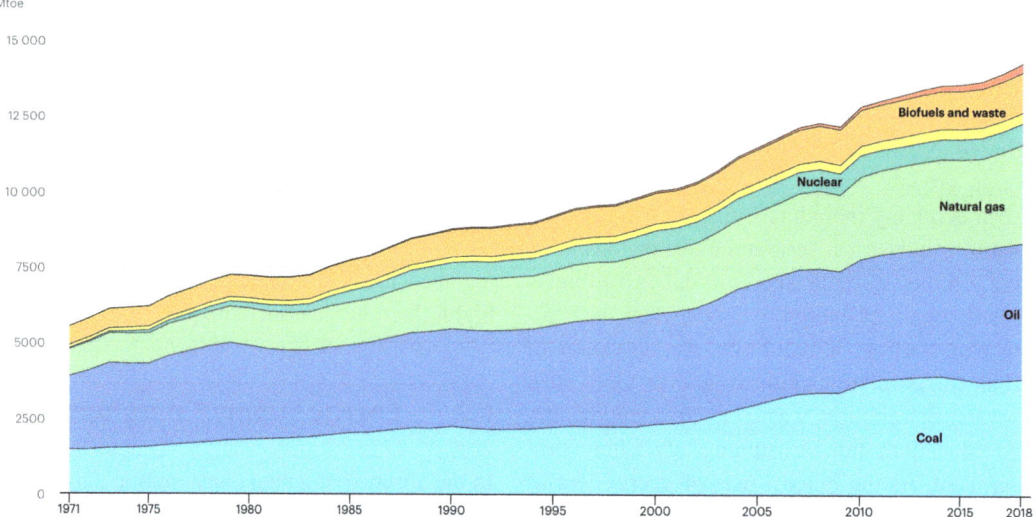

Fig. 6.2 World's total energy supply by source, 1971–2018. (Yellow color is for hydroelectricity and brown on top for other sources. Source: IEA (2020a))

ing nighttime light emission (using space images), as a proxy of electricity consumption to assess the differential electricity consumption patterns. Most of these studies, however, targeted an individual or a group of cities only, avoiding global coverage of urban areas. Hence, the final total energy supply data in the world is generally utilized to grasp the general pattern of energy consumption (Fig. 6.1).

It shows that almost 80% of energy supply/consumption in 2018 came from fossil fuels including oil, coal, and gas. This widespread use of fossil fuels although reduced with time

(Fig. 6.2) carries several challenges in terms of rising greenhouse gas emissions and pollution.

It is clear from Fig. 6.2 that energy consumption has increased almost every year for more than 45 years. The only exceptions were the early 1980s and 2009 following the global financial crises. Another dip in energy consumption occurred during the coronavirus pandemic, which will be discussed in the next section. Developed countries including Iceland, Norway, Canada, and the USA and wealthy oil-producing countries in the Middle East such as Oman, Saudi Arabia, and Qatar and their cities are the largest energy consumers in the world. The average urbanite in these countries consumes several times more energy than the average person in some of the poorest countries (Ritchie and Roser 2020). In terms of energy mix, according to the International Energy Agency, the year-on-year demand of global oil, coal, gas, and nuclear reduced by −9.12, −7.73, −4.99, and −2.52 percent respectively in 2020 compared to 2019, while for renewable energy, it was estimated to increase by 0.79% (IEA 2020b).

Like countries, cities are also currently dependent on fossil fuels as their main source of energy with their negative consequences. This makes them suffer too in the form of risks and disasters from climate change, heat island effect, and air pollution. These issues are likely to become more serious because the energy consumption in cities is likely to increase with the growing urban population (Fig. 6.3). Within cities three major consumers of energy are residential, industrial, and transportation sectors. In comparatively rich cities of the developed world, buildings—both residential and commercial—are the biggest user of energy for heating, lighting, cleaning, etc. Transport is the next biggest user of energy followed by industry. Metro areas like Berlin, Bologna (Italy), London, New York, Seoul, Singapore, and Tokyo, for example, consume

more than half of their energy in residential and commercial buildings, while the transport sector consumes 25 to 38 percent of their energy (UN Habitat 2008). The only exception in transport energy consumption in this group is Singapore, which has successfully made urban mobility more energy efficient. Berlin, London, and Tokyo are among the metros where industry consumes less than 10 percent of energy, mainly because economic activities in these cities have moved to service industry. Industry sector in Singapore however is a big consumer of energy (UN Habitat 2008).

Cities in the developing world are different from those of the developed world in their energy consumption, and even within the developing world, their energy consumption pattern varies by their development phase. For example, in metropolitan areas of comparatively fast-growing economies of China and India like Shanghai, Beijing, and Kolkata, industries constitute the leading sector in energy use, consuming more than half of the total end use energy. Some Chinese cities are huge consumers of energy, for example, Shanghai's industrial sector consumes as much as 80 percent of energy, compared to 10 percent consumed in transport. Transport sector is the biggest consumer of energy in metros of middle-income countries such as Bangkok,

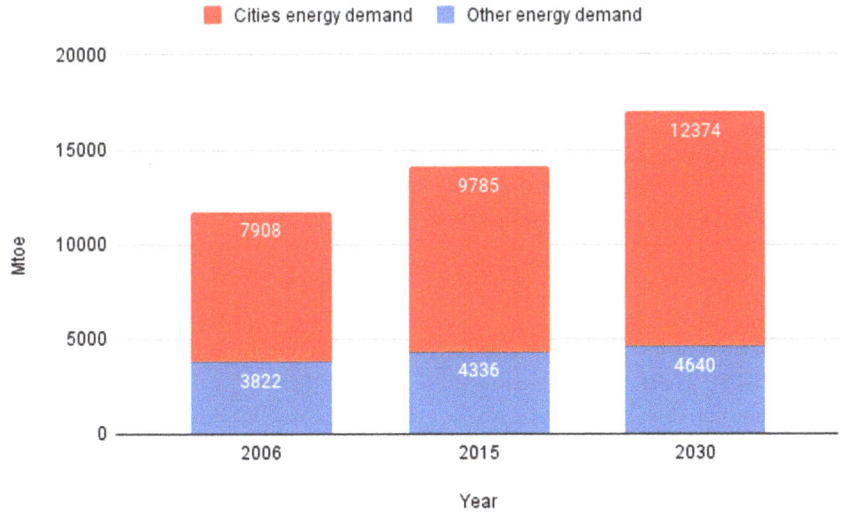

Fig. 6.3 Cities' energy demand, 2006–2030. (Source: Data taken from OECD (2016))

Buenos Aires, Cape Town, and Mexico City accounting for almost half or more of the total energy consumption followed by industrial or residential sector (UN Habitat 2008).

In a study of energy and material flows in 27 megacities of the world, Kennedy et al. (2015) noted that these megacities, according to conservative estimates (excluding air and marine transport energy, etc.), consumed 9.3% of global electricity and 9.9% of global gasoline. The energy consumption varied in these cities—ranging from 78 petajoule (PJ) for Kolkata (population 16.3 million) to 2824 PJ for the New York metropolitan area (population 22.2 million). They observed that although Tokyo is much larger than New York, its energy consumption is lower than New York because the latter city had higher consumption of transport fuel (47 gigajoule (GJ) per capita vs. 18 GJ per capita in Tokyo) and heating/industrial fuels (56 GJ per capita vs. 29 GJ per capita in Tokyo). Among others nine megacities—Moscow, Seoul, Los Angeles, Shanghai, Guangzhou, Osaka, Tehran, Mexico City, and London—consumed more than 1000 PJ/year. To put these figures in perspective, an oil supertanker can hold about 12.2 PJ of oil; New York consumed the energy equivalent of one supertanker approximately every 1.5 days.

The study (Kennedy et al. 2015) concluded that rates between the lowest- and highest-consuming megacities differed by a factor of 28 for energy use per capita, but the situation was not straightforward—not only the economies of cities had a bearing on cities' use of resources but heat degrees days (HDD), urban form, and growth rates also affected energy use. Moreover, while some of the wealthier megacities needed to decrease their level of energy consumption to reduce environmental risks, others in countries of developing world, for example, South Asia, needed to increase energy flows and provide access to this basic resource to all citizens (Kennedy et al. 2015). There are differences in energy consumption between communities even in the same nation, for example, in India, about a third of the population that live in cities consume 87 percent of the nation's electricity and in China urbanites use 40% more energy compared to vil-

lage dwellers (Sawin and Hughes 2007). Also, there are big variations in the consumption rates of energy and carbon emissions between different income groups within urban areas. Thus, the higher-income groups have seven times higher per capita CO_2 emissions from household energy consumption than the low-income group (living on $1.9 consumption a day) in India (Lee et al. 2021).

In terms of future growth in urban energy use, a study by Creutzig et al. (2015) of 274 cities representing all city sizes and regions worldwide showed that if current trends in urban expansion continue, energy use in cities will increase more than threefold, from 240 EJ (exajoule) in 2005 to 730 EJ in 2050. Their analysis using a model, however, showed that urban planning and transport policies can limit the future increase in urban energy use to 540 EJ in 2050 and contribute to mitigating climate change. The results of the study also showed that, for affluent and mature cities, higher gasoline prices together with compact urban form can contribute to savings in both residential and transport energy use. In contrast, cities with emerging or nascent infrastructures in developing countries or rapidly urbanizing Asia, Africa, and the Middle East, developing compact urban form, and appropriate transport planning can help avoid high-carbon emissions (Creutzig et al. 2015).

6.3   Urban Energy and Coronavirus

The coronavirus pandemic has affected almost every sector of urban economy and energy sector is no exception. This section reviews the impacts of coronavirus pandemic on energy demand and consumption in cities and highlights lessons learned and opportunities created in the energy sector. According to the International Energy Agency, the energy demand in 2020 according to estimates declined by 6% compared to 2019—the largest in the last 70 years and a fall seven times greater than the 2008/2009 financial crisis (IEA 2020b). In terms of regional impact, the electricity generation in 16 European countries dropped in April 2020 by 9% compared to the mean generation value from 2015 to 2019. Fossil and nuclear energy generation declined by 28% (24 gigawatt (GW)) and 14% (11 GW) respectively, while renewables increased by 15% (15 GW) (Werth et al. 2020).

6.3.1 Pandemic Impacts on Urban Energy

The available literature on the impact of coronavirus crisis on energy sector covers various aspects. The work on energy security or demand and supply and energy systems includes those of Jiang et al. (2021a), Memmott et al. (2021), IIASA and ISC (2021), Zhong et al. (2020), Abu-Rayash and Dincer (2020); and WEF (2020), while some examples of work on energy equity are those of Mastropietro et al. (2020a, b), Brosemer et al. (2020), and Boza-Kiss et al. (2021). The work on energy in relation to environmental sustainability includes those of Gillingham et al. (2020), Klemeš et al. (2020), Mofijur et al. (2021), Norouzi et al. (2020); Rugani and Caro (2020); and Mulvaney et al. (2020). The impact of the pandemic on specific cities has also been carried out—for example, Ningbo City in China was selected as a case study to explore the impacts of COVID-19 on household energy use (Cheshmehzangi 2020). Likewise, another study was conducted on impact of the COVID-19 on the energy sector in commercial tourist city of Macao.

A review of these studies shows that the pandemic's impact on urban energy use was as follows:

- In the short term, the energy demand declined particularly during the early implementation of lockdowns but showed gradual recovery later.
- Industrial and commercial demand plummeted but household energy use enhanced.
- Fossil fuel demand declined, while renewable energy demand increased.
- The dynamics of peak time for electricity demand changed in terms of days and hours.

- Transportation use of energy by both public and private vehicles declined in the short term. However, private transport use increased later after lifting the lockdown to avoid public transportation in which social distancing was difficult to observe.
- The pandemic while reducing energy demand also created structural changes by creating extra energy demand from multiple pathways in some sectors such as household or residential sector.
- The district-level buildings' thermal energy demand declined, but overall electricity demand in buildings increased.
- Energy intensity increased in cities of China, the USA, and Japan but also depicted some variation by geographic regions.
- Short-term air pollution and carbon emissions reduced.
- Energy inequality worsened (income losses, increased unaffordability to pay energy bills).
- Fossil fuel prices fell resulting in job losses and lower income.
- Digitalization and use of online platforms increased positive lifestyle changes and reduced transportation energy.

The lockdowns curtailed mobility and economic activities particularly construction and manufacturing, caused a fall in the urban energy demand, and in certain cases even damaged the energy industry. For example, the pandemic caused bankruptcy affected at least 19 energy companies in the USA (Crider 2020) alone. July 2020 data on the peak reduction rates (weather corrected) for electricity consumption showed that these rates went down by more than 10% in six European countries—France, Germany, Italy, Spain, the UK, China, and India—due to lockdowns (IEA 2020c). Further, the pandemic not only changed the spatial and temporal patterns of energy use in the short term at local and urban scale but also shifted the peaks in electricity consumption. For example, a case study in Toronto Province of Canada revealed that the pre-pandemic peak time occurred from Wednesday to Friday compared to the in-pandemic peak which was brought forward—Monday to Tuesday. The reduced demand for electricity also affected the supply of fossil fuels like coal. The trend of their replacement by cheaper renewables was also observed (Henry et al. 2020; Watts and Ambrose 2020).

The structural changes through extra energy demand by household or residential sector came due to confinement measures and changed lifestyle—remote work from home, teleconferencing, mail order shopping, telemedicine, etc. The measures adopted for disease prevention and management such as development and operation of medical facilities, preparation and use of disinfectants, and manufacture of personal protective equipment increased the energy demand related to health sector. A major impact of confinement measures was considerably increased time at home with a major implication for energy use in urban residential sector. The thermal energy demand alone has been estimated to have increased in Global North and South by about 18 and 6 percentage point, respectively, in 2020 compared to 2019 (Kikstra et al. 2021). Changes in electricity demand were also enhanced even in developing country cities due to use of ICT, cooking, cooling, heating, lighting, etc. For example, Indian cities saw a 26% increase, while Bhutanese cities experienced 6.7% increase in electricity consumption in 2020 compared to 2019 (Chhetri 2020). Home confinement for long periods also enhanced video watching and streaming, as a mean for entertainment in both developed and developing countries' cities with implications for energy use. On face value, it appears very small but a study by Save on Energy (2020) showed that energy generated from 80 million views of the Netflix thriller Bird Box was equivalent to driving 146 million miles (equal to driving from London to Istanbul and back 38,879 times) and emitting just about 66 million kg of CO_2 (Save on Energy 2020). In terms of equity, higher energy use also translated into higher energy costs (Boza-Kiss et al. 2021) adding to a vicious cycle of poverty that had trapped low-income households, living in low-energy standard buildings (Weinsziehr et al. 2017) whose income had already reduced by economic recession due to the pandemic.

In health sector, Klemeš et al. (2020) have highlighted energy implications and environmental footprints of hospitalization, production of personal protective equipment, working shifts, disinfections, massive testing, and supply chains. Their preliminary estimate on the annual extra energy used to produce ethanol for disinfectants, some PPE, and fast test was estimated at 236.5 PJ. The additional energy consumption for sample collection and storage and their refrigeration evaluated by Schatz Energy Research Center (2020) and use of energy by health-related technologies (Jiang et al. 2021b) was also substantial.

Regarding energy equity, urban energy poverty had affected millions of households across the world even before the pandemic, which were sharpened by the spread of virus and its economic consequences. Therefore, measures had to be placed by governments (Mastropietro et al. 2020b; Goyens 2020; IMF 2020) to mitigate these negative impacts on marginalized households and businesses. During the pandemic, disconnection from energy supply was prohibited in case of nonpayments. These bans were often linked to deferral plans, payment extension solutions, and/or zero-interest rate loan solutions (Mastropietro et al. 2020b). For example, in Mali and Togo, special energy tariffs were created for poor and vulnerable households (IMF 2020). Some countries (Spain, Italy, and Ukraine) extended the validity and eligibility of these social tariffs (IMF 2020; Mastropietro et al. 2020a). Efforts to address energy poverty were also initiated by organizations—such as EU-funded STEP IN project (https://www.step-in-project.eu), which is a consortium "that brings together a wide expertise in the area of energy poverty—research institutes, universities, municipalities, energy providers, charities, consumer associations and regulatory authorities."

In terms of energy efficiency improvements, according to the International Energy Agency (IEA 2020d), the COVID-19 crisis added new uncertainties due to three factors: (a) the economic recession threatened to delay investments by business and households in more efficient technologies; (b) the crisis triggered changes in lifestyles such as increased rates of remote working, online shopping, teleconferencing, etc., which changed transportation patterns in cities and that may have reduced energy intensity in some instances but increased it in others; and (c) government policy responses, for better or worse, were also to affect energy efficiency progress, as witnessed in the past—whereby stimulus packages sometimes resulted in continuation of aging and energy-inefficient industrial or business facilities for longer period. Lack of consideration of the energy system in the design and implementation of stimulus packages may have yielded similar results. Nevertheless, IEA also noted growing recognition of the socioeconomic benefits of energy efficiency by governments in "building back better" from the crisis.

Comprehending the full impact of the pandemic may take years, but the crisis clearly provided both risks and opportunities for global energy efficiency according to IEA (2020d). In terms of energy intensity, the projections varied among countries. For example, China, the USA, and Japan presented differences in the degrees of elevation. Based on the two forecasted scenarios, the USA was projected to have a high change rate of energy intensity (+29.3%), followed by Japan (+7.8%), while China presented no significant change (+2.8%). The energy intensity in the EU is projected to have a slight elevation at +1.03% (Jiang et al. 2021a).

6.3.2 Environmental Gains from Energy Use Reduction

The pandemic-triggered restrictions not only resulted in reduced energy use in industry and transport but also caused a decline in emissions, resulting in cleaner air in cities from Los Angeles to Delhi and Beijing. A study by Kumari and Toshniwal (2020) built on the ground-based station air pollution data from 162 monitoring station in 12 cities in the countries seriously affected by COVID-19 showed a major decline in pollutants such as $PM_{2.5}$, PM_{10}, and NO_2. Cities of Beijing, Bengaluru, Delhi, Lima, Mumbai,

Rome, and Wuhan recorded a notable decline in air pollution in 2020 during the lockdown compared to the same period in 2019. Individual pollutants like $PM_{2.5}$ recorded a drop in the range of 20.2–34.3%; PM_{10} concentration reduced by 23.7–47.3%; while NO_2 witnessed a much higher decline—between 31.6% and 64.5%. There was a big reduction in GHG emissions too. A GHG emission study by Le Quéré et al. (2020) revealed that during the height of the pandemic-related lockdowns in Asia, Europe, and North America, estimated daily global carbon dioxide emission went down by 17% compared to the same period in 2019. The situation changed after lifting of strict lockdown but still total carbon dioxide emission for 2020 fell by 7 percent compared to 2019 (Le Quéré et al. 2021), which in absolute term amounts to 2.6 billion tons of carbon dioxide (Future Earth's Global Carbon Project estimates it at 2.4 billion tons)—something that has never happened before. It is more than what is needed to limit warming as proposed in the Paris Agreement—"to reduce carbon emissions by 1 or 2 billion tons every year" (Chow 2021). The biggest decrease in emissions was in the transport sector—with emission reduction from surface transportation (cars, trucks, etc.) by 10 percent and aviation by 40 percent. With return to normality, however, emissions are likely to bounce back. IEA (2021) projected that increasing global energy demands in 2021 could drive emissions above pre-pandemic levels.

Although at least some lifestyle changes will stick around, resistance to restrictions such as face mask and social distancing in the pandemic have shown the difficulties in relying upon behavioral changes. Nevertheless, commitments by countries and cities and groups and coalition of cities have also signaled that their intention is to pursue aggressive action to combat global warming. For example, lawmakers in the European Union are finalizing an accord that would include the goal of achieving net-zero emissions by 2050. Such pledges will encourage countries and provide them incentives to implement emission targets as part of the Paris Agreement (Chow 2021). This has been discussed further in Chap. 11.

6.3.3 Energy Lessons and Opportunities from the Pandemic

The coronavirus crisis, while causing disruptions in urban energy sector, also presented lessons and opportunities toward energy management and meeting the Sustainable Development Agenda 2030 (Goal 7 and 11 on Energy and Cities) as well as the pledges of the Paris Agreement on climate change. An important lesson highlighted by the pandemic is the importance of interconnection in the world—the way a crisis can become global and cause devastation across all parts of society. It is an important future signal for governments that they cannot ignore the emerging climate crisis and the relationship between energy use, climate change, and sustainable development. If they do, it will be at their own peril. In terms of energy management, the pandemic has brought forward four important opportunities as follows:

• Energy resilience enhancement and promotion of circular economy.
• Exploration of opportunities for renewables.
• Energy storage in the situation of drop in demand.
• Exploration of avenues for energy efficiency and conservation.

In terms of resilience, the virus outbreak prompted disruptions in some type of energy transportation and other supply chains, which makes it urgent to develop more reliable and resilient supply chains—with rationale to prevent the exhaustion of supplies in emergencies. Sharma et al. (2020) based on important factors built a conceptual framework to enhance the survivability of green and sustainable supply chains in post-COVID-19 pandemic. Notwithstanding supply chain problems, the resilience of energy system by itself is important. Although the pandemic has exposed vulnerabilities in the power systems, it has also brought forward the need and opportunity to safeguard sustainable and reliable energy supply and resilient power sector planning (Jiang et al. 2021a). A promising solution

for resilience improvement post-pandemic is the promotion of circular economy, which "in the climate context, involves: reducing the carbon that has to be managed in the first place through efficiency, fuel substitution, renewables, nuclear, hydrogen, and so forth; reusing carbon to create feedstocks and fuels (e.g., methanol, ethanol, fertilizer, curing concrete); recycling carbon through the natural carbon cycle (including via bioenergy); and removing excess carbon and storing it (e.g., direct air capture, biologic storage, geologic storage). All these zero-carbon alternatives need to be pursued to make progress on decarbonization" (Center on Global Energy Policy 2020).

The investments in renewables created more jobs than those in fossil fuels (UNSG 2020) during the pandemic. In terms of energy mix, therefore, the growth rate of the renewables should increase in total energy supply. However, high renewable energy outputs sometimes—as in Germany—in good-weather days also set a record of negative wholesale power prices (Amelang 2020) which is a lesson and motivation for energy producers to design more flexible systems for renewables. The pandemic has also unleashed the importance of energy storage industry, although the storage industry suffered a short-term decline during the crisis due to economic recession and some of its related projects were stalled even though energy storage not only helps mitigate demand variations but also enhances the flexibility of energy systems. In the long-term, however, the crisis could trigger energy transitions offering potential development/opportunities of novel energy storage technologies. In addition, experiences of other crises in history, including the Spanish flu as well as IEA (2020e) projections, show that energy demand will not only recover but bounce back triggering the energy storage needs for a sustainable supply. The need for energy conservation in health sector also became apparent in the pandemic to meet the high-energy consumption due to enhanced hospitalization, medical equipment usage for disease control, sample storage, research, vaccine development, etc.

However, the positive environmental benefits of the pandemic may not be replicated for non-pandemic times (El Zowalaty et al. 2020). For example, global carbon dioxide (CO_2) emissions declined in 2020 due to the pandemic and its resulting economic contraction. These were neither sustainable gains nor produced by deliberate structural transformation or planning. Moreover, "lower economic growth does not produce a cleaner energy world and is a poor low-emissions strategy. Energy efficiency suffers the most with low energy prices, and clean energy investments suffer too" (Center on Global Energy Policy 2020). Nevertheless, the gains made have shown that some solutions previously thought to be out of reach are more possible than expected (IIASA and ISC 2021). An example of positive outcome of the crisis is the digitalization of physical activities in the form of remote work, online shopping, remote education, teleconferencing, and other online activities. Besides short-term lifestyle changes, it has enabled to cut transportation and avoid travel, offered to use resources more efficiently, made consumption more sustainable and reduced carbon footprint. There is a need to capitalize on such examples in a post-pandemic world with the evaluation of pros and cons. Further, in terms of reducing carbon footprint, suggestions have also been floated toward redesigning cities and reinventing urban spaces (Moreno et al. 2021) to increase accessibility and reduce mobility thereby cutting energy use. It has also been proposed to increase urban density while creating more parks and green spaces as carbon sinks. These proposals have been discussed in detail in Chap. 11 of this book, which deals with the urban design aspects in relation to mega risks.

In terms of clean energy transition, whether the coronavirus crisis will be a catalyst is an open question. The pandemic may have caused some increase in renewables and some decline of fossil fuels, but the real test of this transition will depend on the policies that city/local and national governments implement in response to the economic recession caused by the crisis.

6.4   Energy's Carbon Footprint and Climate Change

In the wake of increasing use of energy to maintain rapid economic development, the biggest challenge for cities is to mitigate climate change impacts. This section attempts to explore the energy carbon footprints in cities in terms of challenges and opportunities. It is well documented that the cities are the major emitters of carbon dioxide and other GHG emissions; therefore, they also have a big responsibility for its mitigation. This makes it important and essential to have accurate and consistent carbon inventories to identify the main sources of emissions and target them for action. It is through this process in time that carbon reduction progress can be monitored locally as well as globally (Ting et al. 2021). Several inventory methods have been developed to enable cities comprehend the emission contributions of various activities and control these (ICLEI, WRI and C40 2014). Efforts have also been made to standardize these. Despite the establishment of a recommended international standard for the accounting of community-scale GHG emissions (GPC 2014), inventory methods used in making assessment for cities' emissions significantly vary, making it somewhat difficult to compare both the inventories and progress of emission mitigation over time and space (Ting et al. 2021).

Moran et al. (2018) made a major effort to map the carbon footprint of 13,000 cities of the world for the first time using a consistent methodology—Gridded Global Model of City Footprints. They found that "a small number of large and/ or affluent cities drive a significant share of national total emissions, which means that a concerted action by a small number of local mayors and governments has the potential to significantly reduce national total carbon footprints" (NASA 2019). Five cities with the largest footprints in the world were Seoul, Guangzhou, New York, Hong Kong, and Los Angeles, respectively. The study (Moran et al. 2018) compared the carbon footprints to three variables: the citi-

Fig. 6.4 Population—carbon footprint relationship in selected cities. (Source: NASA (2019))

zens' income, the city population, and the density of development. The cities with the largest per capita carbon footprints tended to be exceptionally wealthy including Hong Kong, Abu Dhabi, Dubai, Singapore, and Hulun Buir (China). The city's population was also important (Fig. 6.4), but there were a few exceptions on this factor. For example, the metropolitan Cairo, Jakarta, and Tokyo each had a population exceeding 30 million, but they had comparatively low-carbon footprints (below 150 million tons). In contrast, affluent Hong Kong, Singapore, Chicago, New York, and Los Angeles, even though they had less than 20 million people, each had a high-carbon footprint (over 150 million tons).

In the USA, suburban parts of cities with low residential density had larger carbon footprints. The top 100 cities with the largest carbon footprint accounted for 18 percent of the global carbon emissions. In both rich and poor, developed or developing countries, the top three largest cities had more than one-fourth share in their national emissions. Further, carbon footprints are highly concentrated in cities which hold big share in global GDP. This strategic combination "augurs well for future development of innovative strategies to reduce (and forecast) carbon footprints" (Moran et al. 2018). In other words, it means that targeted measures in these few affluent cities by responsible authorities and selected coalitions are likely to yield strong and favorable results in reducing carbon footprint.

It is important to mention that some studies (Sudmant et al. 2018; Chen et al. 2020; Ting et al. 2021; Sudmant et al. 2018), while embarking on the inventories of CO_2 emissions or estimating carbon footprints, not only have considered production-based emissions taking territorial jurisdiction of cities into account but also used inflows and outflows or carbon trade between cities/urban territories and outside areas as well. This is considered important because of cities' high consumption-based demands for goods and services from other regions (OECD 2020), which contributes to higher emissions. This also necessitates assessment of sectoral energy use and consumption. Sudmant et al. (2018) found that consumption-based emissions increase more strongly compared to production-based emissions as the city's per capita income and population density rise. One advantage of calculating consumption-based emissions according to OECD (2020) is that it provides additional opportunities for reducing emissions and supporting a more circular economy.

Ting et al. (2021) conducted inventories of 167 major cities across the world located in countries at different developmental stages by sector (Fig. 6.5). While comparing their differences in emission, the study also analyzed the progress

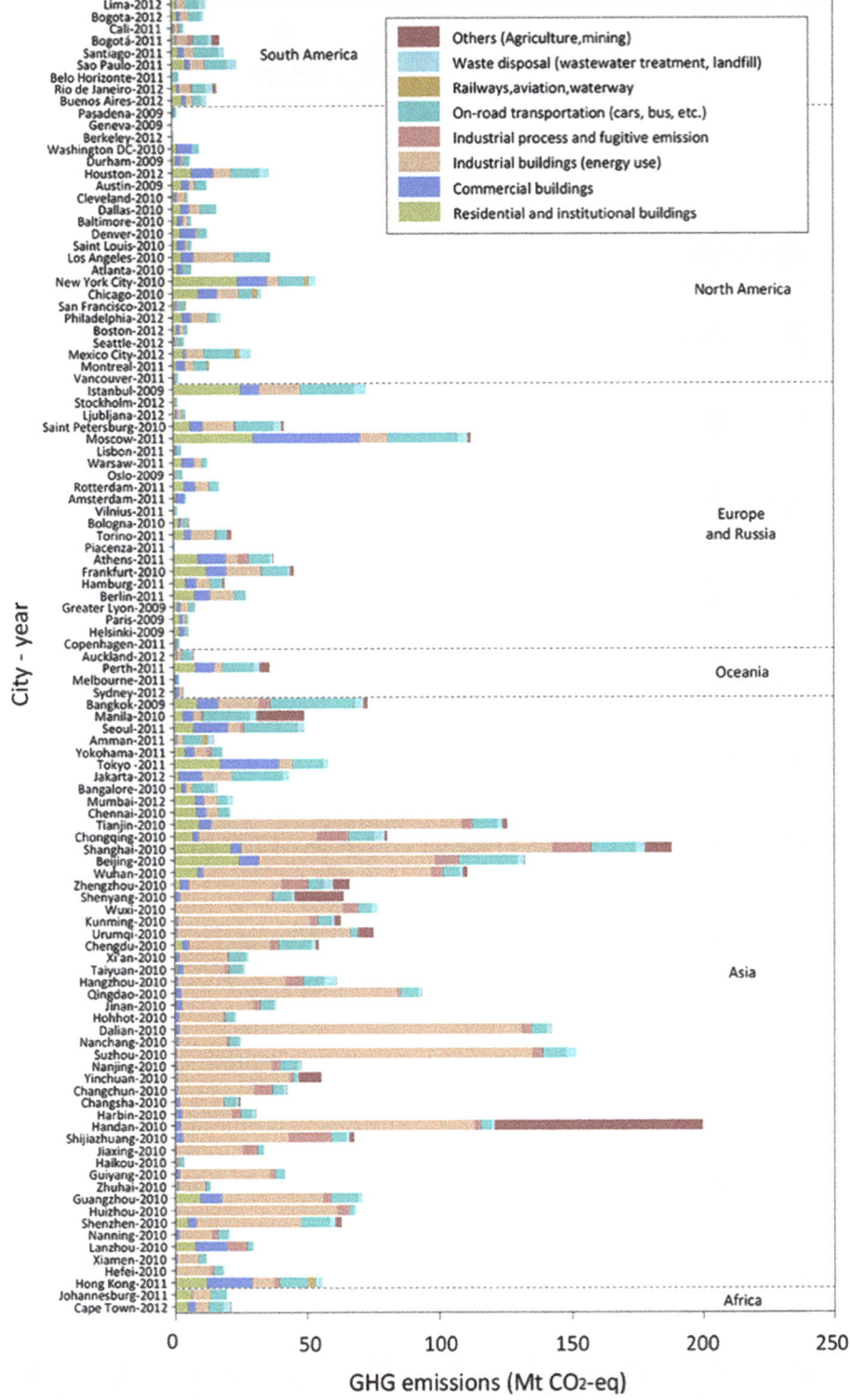

Fig. 6.5 Sector-wise contribution of GHG emissions in selected cities of the world. (Source: Ting et al. (2021))

made by these cities in their carbon reduction over time. It was noted that energy use by stationary objects such as residential and institutional buildings, as well as commercial and industrial buildings, were the largest contributors to urban GHG emissions. In nearly half of over 100 cities studied, energy emissions from various types of buildings had a major share (70%) in total GHG emissions. Hence, significant GHG reduction can be achieved if this disproportionate energy emissions can be controlled. In terms of regional differences in the world, Chinese cities like Shanghai, Suzhou, Dalian, Handan, and Tianjin (Fig. 6.4) had the largest energy emissions from this source (mainly from industrial buildings). In contrast, some South American and other Asian cities (outside China) had a relatively small proportion of energy emissions from buildings. For example, Belo Horizonte in Brazil accounted for only a quarter of GHG emissions from use of energy by this source, while the proportions for Amman and Manila were 24% and 22%, respectively. Contribution of emission from this source in Europe and North American cities ranged from 60 to 80% (e.g., this share in New York City was 74%). The work showed that in addition to buildings, transport sector was another major user of urban energy and producer of GHG emissions in most cities. In about one-third of the cities studied, more than 30% of total GHG emissions came from road transportation. The average GHG emissions from road transportation in cities of developed region were about two times that of the developing region (Ting et al. 2021). Moreover, it was noted that consumers in richer cities of the developed world often buy carbon-intensive products manufactured in cities of China or other upstream regions—if this was counted, then their share will rise substantially in both the energy consumption and GHG emissions (Sudmant et al. 2020; Chen et al. 2020). It is therefore argued that the control of GHG emissions worldwide should promote a supplementary mitigation strategy based on international trade.

6.5   Urban Energy Policies/ Strategies and Climate Change

Urban energy policies and strategies are extremely important to promote development that is sustainable. Since many countries are aiming to achieve net-zero greenhouse gas emissions, their cities will play a key role in transforming energy use in their building stock, mobility systems, enterprises and industries, and urban infrastructure. Management of urban energy supply and its efficient use can create resilience to climate change. Billions of dollars already committed in green stimulus packages, and pledges for hydrogen energy, batteries, electric vehicles, and more have risen the hope for sustainability of energy development and climate change mitigation. "Governments around the world are pledging to increase clean energy R&D investments. There is talk from China that the next five-year plan will be much more focused on green energy and digital innovation. The policies may not be as substantial as the climate crisis demands, but government planning and large spending coming out of the COVID pandemic are raising the hopes that they may accelerate an energy transition going forward" (Center on Global Energy Policy 2020).

6.5.1   Policy Framing and Setting Targets

The first step in framing and implementing an energy policy in a city is to have an energy consumption and GHG inventory. This enables to set goals, target appropriate sectors, identify actions as needed, and track the progress. For example, GHG inventories in Flagstaff and Knoxville in the USA showed that commercial and industrial buildings contributed most heavily to community-wide GHG emissions followed by transportation and enabled the cities to take initiatives accordingly (Aznar et al. 2015). In terms of setting targets, it is encouraging to note that many cities have set carbon emission and circular economy targets, often more ambitious in scope and time horizon than their national equivalents (OECD 2020). Some examples of these are Copenhagen, Adelaide, Glasgow, Amsterdam, and several other cities in Europe and the USA. Some Chinese cities are also taking lead in improving their performance in energy use and/or cutting emissions drastically. Copenhagen aims to become the world's first carbon-neutral capital city by 2025. The climate plan 2025 of the city is based on four pillars—energy use, energy production, mobility, and municipal initiatives (OECD 2020). Adelaide's carbon-neutral strategy 2015–2025 has set a net carbon zero-emission target for 2025 and to have 50% renewables in energy mix by 2025. Glasgow is committed to become a net-zero-carbon city by 2030. The city of Amsterdam aims to make its economy fully circular by 2050. The city is building universal indicators to measure progress toward circularity (OECD 2020). The city of Helsinki has targeted to become carbon neutral by 2035. Sydney's operations became carbon neutral in 2007 and it was the first city government in Australia certified as such in 2011. The city of Melbourne became a certified carbon neutral a little later. 100+ cities committed to net-zero emission by 2050 at the Climate Action Summit held in New York (IISD 2019). Information on several cities striving to become carbon neutral can be obtained from the Carbon Neutral Cities Alliance website (CNCA undated). Rizhao is the first city in China that has pledged to become carbon neutral (Biello 2008). Among developing cities, Cocody City located in the north of Abidjan in Ivory Coast announced to reduce carbon emissions by 70% up to 2030. Initiatives include using solar energy to power large public buildings, installing solar lampposts and traffic lights, and supplying households with PV power kits (WEF 2021a).

In terms of tracking progress toward urban CO_2 reduction, Ting et al. (2021) studied the annual change data of per capita GHG emissions over the period 2005–2016 for 42 cities and found that 30 cities, mostly from Americas and Europe, reduced their annual GHG emissions—Seattle, Oslo, Bogotá, and Houston were the top four cities with the largest per capita emission reduction (Bogotá was also the second largest city in terms of total reduction). European cities were heading for achieving their climate mitigation targets from a territorial perspective. From 2008 to 2016, the EU Emissions Trading System regulated 50% of EU GHG emissions and reduced more than one billion tons of CO_2 from 2008 to 2016 (Bayer and Aklin 2020). It is important to note in this connection that some cities from the developed world in the past moved a certain number of high-carbon industries to developing country cities in Asia and Latin America (Ting et al. 2021). Such transfers of industrial activities between cities meant an ineffective GHG emission reduction for developed cities as the net result was not positive. Therefore, appropriate accounting methods need to be applied in carbon inventory taking all factors into consideration while identifying the climate responsibility of cities or giving credit to them for carbon reduction (Chen et al. 2020).

Despite reduction in emissions noted above, the growing frequency and severity of climate-related disasters and increased public concern are the two indicators that reflect that the progress toward energy transition has been too slow so far. The urgency of the matter demands that the present decade should be one of diligent and strong actions if there is to be hope for meeting the Paris Agreement goals (Center on Global Energy Policy 2020). Governmental policies both at local and national level will play a key role in this regard. Market forces may be important but effectiveness of even these forces need government support. For example, market, technological, and policy forces led coal demand to peak in 2014, but global coal consumption since then has declined by only a few percentage points and this fossil fuel is still the leading source of emission in the world (Center on Global Energy Policy 2020). Likewise, whether oil demand peaks soon or not, supply, demand, and emissions may not meaningfully change without policy support from the government. Similar is the case in promoting renewables. Even with great progress in development and deployment of renewables, the share of this clean energy globally has not changed much. This amply demonstrates that an energy-resilient path and a tipping point in sustainable energy cannot be reached without strong government investment and policy support (Center on Global Energy Policy 2020).

6.5.2   Policy-Based Solutions in Urban Energy Management

Good energy policies lead to a win-win situation, whereby both energy consumption and GHG emission can be reduced while enhancing cities' economic performance. Although cities' energy policies operate within national and international frameworks, local or city governments are better placed in addressing local-level energy issues. This is firstly because cities are better connected to their citizens and secondly, their influence on entrepreneurs is stronger in promoting the most sustainable energy solutions. Moreover, in the long run, cities' success in local-level energy-related endeavors can also promote bottom-up approach by attracting the attention of national government and even international institutions for replicating these efforts. In addition, cities' political clout and procurement power and their role as birthplaces of innovation and experimentation are undeniable. For example, Amsterdam has set a target of 50% circular procurement by 2025 for itself (Amsterdam Smart City, undated). Further, cities around the world are learning that through the development of effective partnerships between municipal leaders and other key stakeholders (e.g., in business, industry, financial institutions, citizen groups), they can leverage funding and maximize opportunities for using climate action plans to stimulate new economic development. For example, a group of pioneering cities such as Frieberg, Germany, and Philadelphia have capitalized on a "first-mover" advantage and witnessed substantial growth in their renewable energy industries (OECD 2008).

City governments, in implementing energy use and climate-related policies, have used both direct and indirect measures and solutions. Direct measures by city governments have included investments in climate-friendly technologies and materials and municipal support for investment in energy-efficient and low-carbon infrastructure as well as climate-friendly purchases or procurements. Moreover, many policy options for reducing energy consumption and decreasing CO_2 emissions fall under the authority of local governments, including direct energy consumption linked to the provision of public services (sanitation, water, public transport, and local public buildings), capital work programs, and infrastructure development. Overall, urban local governments' energy consumption varies ranging from 1% to 5% in New Zealand and France to as high as 16% in Austria (IEA 2008).

Local governments have also used effective indirect solutions and measures to reduce energy use and GHG emissions. These have included regulatory measures, economic instruments, technological solutions, institutional capacity building, educational, and information-based as well as participatory and partnership building measures (Table 6.1). Among indirect measures or solutions, cities' regulatory powers (energy efficiency requirements in building codes, land use, and urban planning rules) have proved very important.

Cost-effective strategies to tackle energy efficiency and climate change have also used economic instruments that put a price on greenhouse gas emissions. Cap-and-trade schemes, or emission taxes, have proved cost-effective because they induce firms to look for abatement options where they are cheapest and boost incentives to scale up climate-friendly R&D. These economic instruments have often been complemented by other measures to seek market and information support, for example, developing partnership with stakeholder and voluntary agreements with businesses. Communication of information to consumers, for example, on energy efficiency performance of electrical appliances and light bulbs, has proved quite effective in optimizing energy consumption. Use of energy efficiency standards, and ecolabeling, and communication with citizens through websites and devices have also been effective for the purpose. For example,

Table 6.1 Policy-based instruments and solutions used by cities for sustainable use of energy and reducing carbon footprint of cities

Action category	Action subcategories
Direct government action	Procurement Investment in sustainable energy project
Use of regulatory (command and control) instruments	Laws, regulations, and standards (land use and buildings) Energy use codes and standards on energy performance and emissions, product standards, mandatory rulings like ban on motorized transport in certain areas
Use of economic instruments	Disincentives (taxes, e.g., carbon tax, tax on operation of private vehicles in certain areas at certain times, etc. User charges, product charges, effluent or emission charges, tradeable rights and permits) Incentives-subsidies-fuel substitution, renewables
Promoting technology for	Reducing building energy use, clean energy promotion—mandatory installments of solar water heater, etc. Sustainable urban transport—mass transit, electric vehicles Improving energy efficiency, standards for production process
Institutional capacity building	R&D, training
Education and information	Mandatory ecolabeling, consumer advice, knowledge campaigns, websites, and dashboards (to display goals, targets, and actions)
Stakeholders' participation	Voluntary agreements Networking Participatory planning
Participation in networks of cities	Exchange of ideas and best practices Undertaking bilateral and multilateral cooperation programs and projects
Urban design and spatial planning	Land use planning and urban density, mixed neighborhoods Climate-neutral and circular mobility system—facilitating non-motorized transport bicycle lanes, pedestrian areas Building and waste management

the city of Newcastle created the world's first and only device, known as a greenhouse gas "speedometer" and accessible online at climatecam.

com, that monitors and reports the city's consumption of electricity, gas, liquid fuels, waste to landfill, water consumed, trees planted, and the resultant equivalent in tons of carbon dioxide expended. The electronic billboard includes a 500-megawatt electricity meter in the town square, updated hourly from data sourced directly from the energy providers in the 15 electrical zone substations that power the city (OECD 2008). Cities' policy-based interventions on energy and climate front are numerous and diverse in various arenas and economic sectors, some of these have been covered in the following account.

6.5.2.1 Building and Construction

Governments particularly the city or local ones are typically responsible for setting land use and planning rules and standards that influence city design, layout, and other planning measures including low-emission zones, mixed-use zones, and densification of local areas. Among other things, their enforcement of energy use and CO_2 emissions in building codes is particularly effective. For example, stringent building codes/specifications in Tokyo have enabled 30% decrease in building CO_2 emissions. The building and construction sector, as the largest consumer of urban energy, has the biggest potential of any sector to cut both energy use and GHG emissions. This sector in 2018 accounted for 36% of final energy use and 39% of energy- and process-related carbon dioxide emissions across the world—11% of which came from producing building materials and products (IEA 2019). According to UNEP and IEA (2017), the built environment's energy intensity will need to improve by 30% up to 2030 to meet the goals of the Paris Climate Agreement. The segment, however, is off track in terms of level of investment and action necessary toward a zero-emission, energy-efficient, and circular building sector (OECD 2020). According to Building Performance Institute of Europe (BPIE 2018), as much as 85% of existing buildings in the EU will still be in use in 2050. Moreover, not even 1% of these buildings are net-zero-carbon today (WRI 2019). This shows that the main policy challenge in high-income country cities is the decarbonization of existing building stock through retrofits and deep renovations.

Some cities of the world such as London, Singapore, Toronto, and Vancouver have already taken the first step toward carbon-neutral or zero-carbon buildings (ZCB). Among developing cities, Kochi in India has announced a public commitment toward a zero-carbon building. World Resources Institute of India has been working with the Kochi Municipal Corporation on developing a citywide roadmap to help it transition to ZCB (Malaviya 2020). The initiative of the city of Toronto in instituting mortgages that encourage green building is also an important step to which the market responded positively by creating its own instruments. Soon after, builders in the city started observing the highest environmental standards as a routine (OECD 2008). Toronto's effort toward energy retrofits on its 2000 concrete-slab apartment towers constructed in the 1960s and the 1970s was also a significant step in this regard. The Tower Renewal Project in the city lowered the total energy expenditure by 5%, created better living conditions in socially disadvantaged neighborhoods, and generated new jobs in the building retrofit industry (OECD 2008).

City policies in many countries support energy transitions in built environment through introduction or improvement in building codes, subsidy schemes, and other incentives. Cities are also promoting energy efficiency approaches in the built environment. Support to retrofitting of existing buildings in New York is an example. The city's Climate Mobilization Act of April 2019 requires buildings above 25,000 sq. ft. to reduce their emissions by 40% up to 2030 compared to the base level of 2005 and up to 80% by 2050 compared to the same base level. Buildings larger than 25,000 square feet (applicable to 50,000 buildings across the city) are required to meet greenhouse gas emission caps, beginning in 2024. Moreover, the city has already been mandating the disclosure of energy demand of all its buildings over 10,000 sq. ft. The Act is expected to reduce New York City's overall emissions by one-tenth up to 2030, which in absolute term means eliminating six million tons of greenhouse gas emissions. It will also create 26,700 green jobs and is expected to prevent 50 to 130 premature deaths annually by 2030 (OECD 2020).

World Economic Forum (2021b) has highlighted two interesting case studies of Lidl Distribution Center, Finland, and EDGE Olympic Building, Netherlands. The former, a warehouse completed in 2019, covers an area of 60,000 square meters and operates on 100% renewable energy producing more energy than it consumes throughout the year. To achieve the highest efficiency possible, the digitally enabled warehouse employs a microgrid along with a comprehensive building management system. Microgrid and EcoStruxure Building Operation solution enable energy cost savings up to 70%. The latter, an office building redeveloped from an old post office, is shaped on a holistic four-pillar approach of well-being, sustainability, design, and technology. The building embraces circular, zero-waste design principles and smart digital infrastructure to optimize energy use. It has 15,000 sensors to constantly measure indoor climate performance, is energy neutral, and consumes 70% less energy than the average nonresidential buildings (WEF 2021b).

6.5.2.2 Transport Interventions

Besides buildings, local/city governments have played a crucial role in energy savings while reducing GHG emissions and pollution from transport sector. A diverse range of policy measures and instruments have been used including promotion of public transport system while extending its reach, significantly making walking to and from hubs and stops easier, and encouraging modal shifts from private vehicles toward public transit and to nonmotorized transport such as cycling and walking—through development of infrastructures as bicycle lanes and lanes for pedestrians as well as mixed-land use developments. Regulatory measures were also used, for example, city of Boston implemented Federal Clean Air Act (1970) by enforcing a freeze on the construction of new public parking facilities to discourage private vehicle use. The city tried new policy tools based on demand management, by installing parking meters with varied rates considering the timings of the day and congestion levels. The city also created a compact, walkable environment through participation—by coordinating with a wide range of stakeholders, from community-based groups to federal government agencies. Citizen's advocacy groups are very active and regularly participate in development project public reviews and through town meetings and other fora, help promote pedestrian accessibility and vibrant streetscapes. The effectiveness of the policy measures is evident from city's inventory of GHG, which shows that emissions from transport sector are only about one-fifth of the total GHG emissions compared to as much as 50% in other American cities (OECD 2020).

The city of Vienna launched an ambitious Climate Protection Program (KLiP), with 36 climate protection initiatives, that included 241 actions covering electricity and heat generation, district heating, housing, business, administration, and transport. Urban mobility issues in this package have targeted traffic reduction, promoted ecologically sound transport alternatives, and raised the efficiency of motor vehicles. The main objective of the Transport Master Plan of Vienna, adopted by the City Council in 2003, was to increase the share of public transport to 40% by 2020 and to double the share of cyclists from 4% to 8%. This strategy has already worked, and the use of public transport increased from 29% in 1993 to 35% in 2008 and the use of private cars declined from 40% to 34% (OECD 2008). The city government also improved the energy efficiency of its own vehicle fleet (public transport vehicles included), simultaneously launching initiatives with other groups, such as taxi operators numbering in thousands (OECD 2008).

Regulatory and technological improvements to vehicles along with increased energy efficiency have also been used as the key measures for improving energy efficiency and reduction of greenhouse gases in urban transport sector. These are, however, being reinforced with other measures and instruments. Among economic instruments used, some interesting examples include the congestion fee, user charges, emission charges, congestion fees, tradeable rights, and market permits. Congestion fee, for example, in Seoul and London, is estimated to have contributed to reduce emissions between 10% and 20% (OECD 2020). They have also yielded successes in Stockholm. Among regulatory instruments, Singapore provides an example of the Vehicle Quota System, which the government uses to control the number of cars on the road. This is reviewed regularly, and the quota gets changed every month, based on road conditions and the number of cars taken off the road in that month (UN Habitat 2012). New avenues such as digital-based ride-sharing are also being explored. For example, a recent modeling study of ride-sharing for the Dublin metropolitan area in Ireland, based on its daily mobility patterns, by International Transport Forum (ITF 2020) suggests that the number of vehicles, traffic, CO_2 emissions, and congestion would be reduced by up to 98%, 38%, 31%, and 37%, respectively, by ride-sharing. Broadly, similar results have been obtained for other cities, such as Lyon, Auckland, Helsinki, and Lisbon. Further benefits could include substantially lower pollution and freeing up space occupied by parked cars. In addition, emission reductions are larger if the shared vehicle fleet is electric (ITF 2020).

6.5.2.3 Technological Solutions

Governments' contribution to innovations has been most effective in integrated energy production technology—as it has moved the biggest energy-using sectors in cities—buildings and mobility, to the electricity vector while supporting the development of renewables. A very good example of the government coordinated planning for innovative solutions for cleaner energy infrastructure and near-zero-carbon development is the city of Meishan in China. The near-zero-carbon zone technology projects in Meishan have included a smart, near-zero-carbon port, with gradually electrifying container trucks, improved mass transit coordination, and high-voltage on-dock power charging stations. Innovative technology solutions are also being used for renovation of old communities in improving roads, water and power supply, information, logistics, flood control, and wastewater treatment facilities. Once a small fishing village, Meishan currently is a flourishing industrial port city with warehouses, high-tech industries, tourism, commerce, and other services (WEF 2021b).

Another good example of innovation is the use of renewable energy in district heating and cooling systems. International Renewable Energy Agency (IRENA 2017), for example, has highlighted 21 case studies on this including three from cities of Munich (solar), Copenhagen (biomass), and Hohhot (wind power). From Munich, it provides an example of a local solar heating system for 320 apartments with rooftop solar thermal collectors connected to hot water tank for seasonal storage—coupled by heat pump to city district heating network. In Copenhagen, the district heating system of the city has the most efficient combined heat and power (CHP) plants in the world. These plants are being modified to permit a larger proportion of biomass co-combustion or a complete switch to biomass to achieve the city's goal to become carbon-neutral by 2025 (IRENA 2017). In Inner Mongolian region of China, where Hohhot City is located, high amounts of wind power generated, and limited transmission capacities, created very high curtailment rates. To overcome this deficiency, the Chinese National Energy Administration promoted the utilization of excess wind power for heating in the city of Hohhot, the main urban center in Inner Mongolian region, which had a developed district heating network but was fed by relatively inefficient isolated coal boiler. As such, the city became the first test bed for large-scale implementation of this power-to-heat scheme in the region (IRENA 2017).

A big potential technology-related innovation in urban mobility is electrification of vehicles and investment in related infrastructure. Electric car deployment has grown rapidly over the last decade, with the global stock of electric cars increasing from 17,000 in 2010 to 7.2 million in 2019. The largest share of electric cars on the road in 2019 were in China. Among other countries, Canada, Finland, France, India, Japan, Mexico, the Netherlands, Norway, Sweden, and the UK were significant. Seventeen countries have already announced 100% zero-emission vehicle targets or the phase-out of internal combustion engine vehicles through 2050. France was the first country to put this intention into law, with a 2040 timeframe (IEA 2020e).

Many cities support the use of electric vehicles, and some have already announced specific goals concerning electric vehicles. For example, the city of London has a goal to sell 70,000 ultralow-emission vehicles by 2020 and 250,000 by 2025; the city of Los Angeles is planning to have 10% of its vehicle stock to be electric by 2025 and 25% by 2035; New York envisages to have 20% electric vehicle sales share by 2025; Amsterdam and Oslo are to have zero-emission transport within the city by 2025 and 2030, respectively; and Shenzhen and Tianjin will have 120,000 and 30,000 new energy vehicles sold by 2020, respectively (Hall et al. 2017). Cities while promoting the use of electric vehicles more actively are also improving the infrastructure required particularly by investing in public charging infrastructure and making it highly visible and easily accessible to drivers. Cities in the Netherlands (Amsterdam, Rotterdam-Utrecht), Norway (Oslo), and China (Beijing, Shanghai) have the highest concentration of public charging points (Hall et al. 2017). Fiscal incentives for the vehicle purchase, as well as complementary measures (e.g., road toll rebates and use of low-emission zones and carpool lanes), have also been conducive to attract consumers and businesses to choose the electric vehicle option. Cities are also investing directly in transit bus electrification as in Kolkata (India), Shenzhen (China), Santiago (Chile), and Helsinki (Finland). Case studies on these have been featured by IEA (2020e) that can provide lessons to policymakers and stakeholders across cities of the world for considering the electric vehicle adoption.

6.5.2.4 Renewable Adaptation and Innovation

Cities have gradually become active in this arena realizing the potential of renewables in creating clean and sustainable cities. Over 800 cities in the world had adopted a total of 1088 renewable energy targets in at least one sector by the end of 2020 (REN 2021). Out of a total of 1088 targets adopted, 653 are for 100% renewable energy use either in municipal operations or to cover the entire city operations. The municipal targets for renewable energy are more common in cities of developed world (Europe and the USA) but less so in developing cities of Asia, Latin America and the Caribbean, and sub-Saharan Africa. These targets are mostly in the power sector, but interventions in heating and cooling and transport sectors are also increasingly attracting attention. Several cities have adopted detailed plans to transition to renewables in their heating systems, and by the end of 2020, at least 67 cities had e-mobility targets, up from 54 in mid-2019, creating opportunities for wider use of renewables in transport (REN 2021).

Among the cities that have taken the lead, three in developed countries—Adelaide in Australia, Malmö in Sweden, and Seoul in Korea—and one in a developing country, Cape Town in South Africa, are significant. In Adelaide, the municipal operations are powered entirely by renewable energy since July 2020. Wind and solar farms are the main supplier of energy to the city which has commitment to reach carbon neutrality by 2025. The city has promoted initiatives for energy-efficient buildings, has schemes that support the uptake of hybrid and electric vehicles, and invested in energy storage technologies by establishing a power reserve at Hornsdale (WEF 2021a). In the city of Malmö (Sweden), Western Harbor District is being operated on 100% renewable energy since 2012, while the industrial area of Augustenborg has solar thermal panels connected to a central heating system. The city has planned to switch from 43% renewable in 2020 to totally renewable by 2030. A geothermal deep-heat plant is under construction and expected to be operational in 2022; four more of these geothermal plants are in the pipeline and expected to be completed by 2028 (WEF 2021a). In Seoul, the Solar City project has generated 237,805 MWh in annual energy while reducing 109 tons of CO_2 and 27.6 tons of fine particulate matter. The city's total installed rooftop photovoltaic (PV) capacity in 2019 is to be raised from

around 200 MW to approximately 1 GW by the end of 2022. The project will supply one million households with mini solar power stations and install solar power in every single public site where installation is possible (Bellini 2019). Cape Town in South Africa had a surge in photovoltaic (PV) panel installations in the past decade and had the highest intensity of registered rooftop solar PV systems in the country in 2019. The solar power has been targeted for greater use in water heating systems in low-income areas. Within transport sector, the city is exploring biofuels' use and has a pilot program on locally made electric buses (WEF 2021a).

6.5.2.5 Smart Energy Infrastructure

Energy infrastructure is quite elaborate; it involves energy generation, transmission, distribution, and consumption and power generation and distribution storage and waste wires, heating and cooling networks, meters, charging, and everything else that encompasses the "grid." One major problem is that in most cities energy infrastructure was developed years ago and was catered through a centralized power system, based primarily on fossil fuels. It has caused major losses and inefficiencies in energy use and contributed to GHG emissions. In contrast, the modern efficient or smart energy system calls for an increasingly decentralized and digitalized system based on high volumes of renewables. The term "smart" does not have a single connotation in scientific literature and has been subjected to considerable debate, which will be discussed in more detail in Chap. 11. In this section and for this chapter, the term is used to depict new technologies that are making energy utility and its communities greener and more efficient, for example, by deployments of smart meters, sensors, and data analytics using ICT that allows cities to make their energy consumption and the critical energy infrastructure that serves the city efficient. Two case studies are presented here—one example depicts how existing grid can be made more efficient and sustainable while the other is a model of new energy infrastructure development with cutting-edge technology.

In the first case, one of the largest energy providers in Michigan, USA, is transforming its infrastructure to make it efficient and climate friendly. The company initially relied on power plants that used fossil fuels for 74% of the electricity it provided to 6.7 million of the state's ten million residents. The company has now planned to have more than half of its electricity from renewables (solar and wind) by 2040 and phase-out fossil fuels. Simultaneously, it decided to roll out smart thermostats and smart meters to remote track and manage energy consumption. The company has already retired seven of its coal-fired power plants and the remaining five are to be phased out by 2040. The low-cost renewables and efficiency of the system in the long run, as noted by World Economic Forum (WEF 2021b), will eventually boost return for shareholders despite high cost of transition. The second case relates to "Vila Olímpia" in the São Paulo metropolitan area, originally a marginal district but now known as city's Silicon Valley (WEF 2021b). This first NETWORK DIGITAL TWIN of South America was developed by an Italian company Enel in partnership with local authorities and National Electricity Agency "ANEEL." This Twin model simulates the grid based on approximately 5000 sensors installed on it, each communicating information on the status in real time to the energy distributor and local stakeholders using ICT and network automation. An important new feature of the system is that it has moved from corrective maintenance or repairs to preventive maintenance. It can now spot a risk beforehand and act in advance to control it. The project is also a kind of laboratory to test the interaction of the grid, electric vehicles, charging points, solar rooftop generation, and energy service providers and consumers.

6.5.2.6 Circular Economy

Adoption of circular economy is also helping in transforming urban energy consumption and reduction of emissions from waste, the living examples of which are the cities of Seoul in Korea and Lahti in Finland. At one time, Seoul, like many other cities in the developing world, suffered from a spate of waste overloads and power shortages due to rapid growth. As a solution, the city decided to develop Gangnam Resource Recovery Facility (RRF). The facility after sorting out recycling materials transforms waste into energy (Fig. 6.6). This helps reduce landfill on the one hand and power neighborhood heating on the other. The approach has also led to reduce fossil fuel use and carbon emissions, contributing to climate mitigation (Kang 2020).

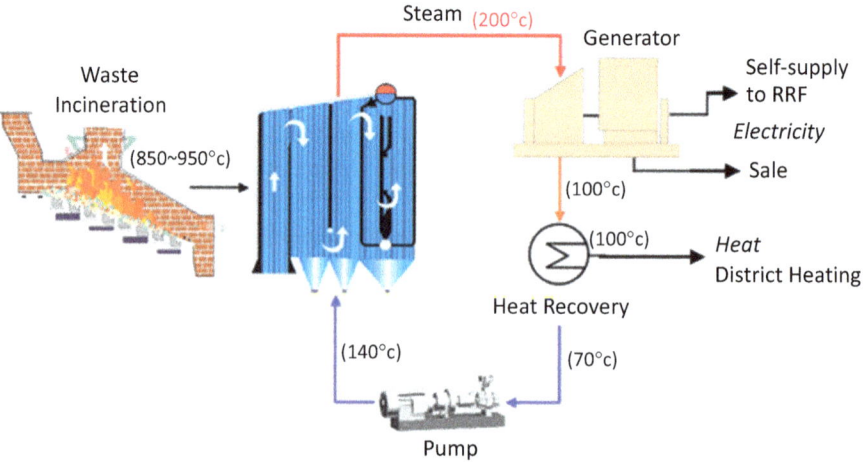

Fig. 6.6 Gangnam Resource Recovery Facility in Seoul transforms waste into electricity and heat. (Source: Kang (2020))

The city of Lahti in Finland has set targets to become carbon neutral by 2025, curbing over-consumption and becoming a zero-waste city by 2040. Lahti experienced great progress in recycling as early as the 1990s by creating a special regional waste sorting and treatment culture, and today the circular economy has become a specific focal theme in the city. In 2019, Lahti was the first city to launch a carbon trade program for citizens' personal carbon emissions. Eight different types of waste are collected from households and treated differently to promote a circular business approach among local businesses and citizens in Lahti (OECD 2020). The city of Toronto has also achieved reductions in greenhouse gas emissions by capturing methane from its landfill and using it to generate electricity since long (OECD 2008).

6.5.2.7 Solutions Through Networking

In addition to their individual efforts to save energy and mitigate GHGs, cities have also undertaken coordinated actions through various networks such as ICLEI, C40 Cities, and the like. Through these networks, they have not only been successful in exchange of information and best practices but achieved significant savings by coordinated actions—like pooling of purchasing power. For example, the C40 energy-efficient buildings initiative has led to several agreements for the collective purchase of energy-efficient products. At the second Summit of the C40 Cities in 2007, 16 cities signed a USD 5 billion program to improve the energy efficiency of buildings, starting with public buildings (OECD 2008).

6.6 Problems and Constraints in Undertaking Policy Initiatives

Despite several initiatives undertaken by cities toward energy and climate actions, some shortcomings are apparent. Among these, the first and most important is that only a relatively small proportion of world cities are active in pursuing reduction in energy consumption and CO_2 mitigation policies. Second, most cities particularly in developing countries have not fully mainstreamed climate change into their policies and plans, even in some developed country cities, it has not been mainstreamed in day-to-day action. Further, often one action is taken without being fully embedded in broader government planning and operations. In addition, several policy solutions have not been fully exploited due to lack of political will, resources, and technology. The use of urban and land use planning to address climate change is one example of lack of political will where local governments could make progress.

Moreover, there are a range of obstacles external to the local government such as clear national-level guidance on climate change for local or city governments, as well as the lack of energy and climate change policy competence delegated to them. Regarding reduction in urban energy consumption, cities of the world are facing a dilemma; while there is an imperative of decarbonization, there is also the desires of millions of city dwellers in emerging economies to move into a more energy-intensive lifestyle. In addition, there is an urgent need to provide energy accessibility to the billions in cities of developing countries and alleviate their energy poverty. Further, in many cities such as those of South Asia, power cuts, resulting from load shedding when the demand is higher than supply, are common. The problem can get worse as economic growth leads to increased demand for electricity (UN Habitat 2012). Increasing new capacity is not always the best solution because it is costly and needs time. There are other options such as lifestyle changes, smart appliances, co-generation, and reduction in the transmission losses that need to be explored in developing country cities. Overall, challenges for achieving sustainable energy system demand energy sufficiency but not extravagancy as well as energy conservation and energy efficiency. In addition, it requires clean energy preferably renewable energy systems and appropriate technologies. Promoting green energy also requires policies that cater for capacity building, knowledge transfer, and financial support mechanisms, market stimulation, and mechanisms for sensitizing the population as well (UN Habitat 2012).

One of the main obstacles preventing political leaders from moving ahead with actions to respond to climate change is a perception that such actions force inevitable trade-offs against the goals of economic growth. Cities around the world are learning, however, that through the development of effective partnerships between municipal leaders and other key stakeholders (e.g., in business, industry, financial institutions, citizens groups), they can leverage funding and maximize opportunities for using climate action plans to stimulate new economic development. This may include linking strategies that connect climate change response goals to business profitability. This may also mean training for populations struggling with high unemployment rates, or transforming strategies that use climate change goals to help "green" existing businesses so they can remain profitable or expand into new markets. This could also include leapfrogging strategies that attempt to create an entirely new sector in a green technology area. For example, a group of pioneering cities have capitalized on a "first-mover" advantage in renewable energy. Thus, Freiburg in Germany developed a citywide strategy as early as 1986 with environmental guidelines that served as a basis for its economic specialization in the solar energy industry. This included such policy measures as building city-owned solar projects, instituting a local ordinance requiring that 10% of the city's electricity be obtained from renewable sources by a certain date, creating public subsidies, and proactive research and development support. Besides other advantages, these efforts led to the creation of thousands of jobs. Along the same line, Philadelphia has specialized in renewables including solar and wind energy, whereas Pittsburg specialized in retrofitting of buildings and now ranks fifth in the USA for the number of its green buildings. Cities are also capitalizing upon "mega events" (such as Olympic Games, International Expos, World Cup tournaments, and policy summits) as opportunities for simultaneously advancing goals for economic development and climate change mitigation.

For policy purposes, the complexity of a zero-carbon transition is immense, but as stated earlier, the challenge appears different in emerging markets and developing countries as compared to the developed world. It also looks different in varied urban sectors—for example, the power sector vis-à-vis heavy industry. Adoption of methods for cleaner energy solutions in terms of higher proportion of renewables in energy mix or energy efficiency, etc. may also vary. Innovation and management, however, are necessary across the board. Innovation is needed, "to adapt to a low-carbon energy transition and for promoting circular economy" (Center on Global Energy Policy 2020). Management means use of low-carbon energy and energy technologies on the supply side, and lowered consumption on the demand side as well as efficient and fair (inclusive and accessible) distributional infrastructure.

A low-carbon, energy pathway will not necessarily cost much; however, it will require a shift in the intervention type. For example, according to the report of the Global Commission on the Economy and Climate (2014), just strengthening climate action between 2018 and 2030 in the world could by 2030 generate over 65 million new low-carbon jobs and deliver at least $26 trillion in net global economic benefits. Another study by Hepburn et al. (2020), based on a survey of central bank officials, finance ministry officials, and other economic experts, assessed the performance of 25 major fiscal recovery packages and concluded with five recommendations for recovery—two of these were in the field of energy as follows: (a) recovery investment in clean physical infrastructure should be in the form of renewable energy assets, storage (including hydrogen storage), grid modernization, and carbon capture and storage (CCS) technology and (b) recovery spending should be in building efficiency for renovations and retrofits, including improved insulation, heating, and domestic energy storage systems.

6.7 Conclusion

The amount and type of energy cities consume, or use, is extremely important not only because it affects their economy, the environment, and the well-being of their citizens but also has a bearing on mega risks like climate change. The cities, as an overwhelming consumer of energy, have a large carbon footprint, which is likely to increase with the growing urban population. Mapping the carbon footprint of 13,000 cities of the world shows that a small number of large and/or affluent cities drive a significant share of total emissions, which means that a concerted action by a small number of local mayors and governments has the potential to significantly reduce total and national carbon footprints. This is further confirmed by another study of 27 megacities which consume 9.3% of global electricity and 9.9% of global gasoline. The rates of consumption of energy even within the group of large or megacities vary by a factor of 28 for energy use per capita. The situation, however, is not straightforward as the use of energy in cities is affected not only by their economies but also by many other factors such as per capita wealth, urban form, and growth rates as well as heat degree days (HDD). This brings forward the fact that there is no one solution applicable to all. Moreover, situation varies between developed and developing countries. While there is no doubt that some of the wealthier megacities consume excess energy, others in countries of developing world suffer from energy shortages. They need to provide access to this basic resource to their energy poverty-ridden citizens.

Overall, as an overwhelming consumer of energy, however, cities have a big responsibility for mitigation of GHG. Indeed, many of them were making efforts toward it when the world was struck by the coronavirus pandemic. The lockdowns curtailed mobility and economic activities caused a fall in the urban energy demand. According to the International Energy Agency (IEA), the energy demand in 2020 was to decline by 6% compared to 2019—the largest in the last 70 years and a fall seven times greater than the 2008/2009 financial crisis. The pandemic, however, while causing disruptions in energy sector also presents lessons and opportunities toward management of urban energy. It has brought forward the interconnection in the world—the way a crisis can become global and

cause devastation is an important future signal for governments that they cannot ignore the emerging climate crisis and the relationship between energy use and climate change. If they do, it will be at their own peril. While exposing vulnerabilities in the energy and power systems, the pandemic has also brought forward the need and opportunity to safeguard reliable and sustainable energy supply and resilient power sector planning. In terms of energy management, the pandemic has brought forward four important opportunities—energy resilience enhancement and promotion of circular economy, exploration of opportunities for renewables, importance of energy storage in the situation of drop in demand, and exploration of avenues for energy efficiency and conservation. The economic recession due to crisis no doubt threatens to delay investments by business and households in more efficient technologies; however, governments' attitude in the wake of rising to the challenge of "building back better" from the crisis has raised hopes for the future.

In terms of clean energy transition, whether the coronavirus crisis will be a catalyst is an open question. The pandemic may have caused some increase in renewables and some decline of fossil fuels, but the real test of this transition will depend on the policies that city/local and national governments implement in response to the economic recession caused by the crisis. Some positive environmental benefits of the pandemic may not be copied or replicated. For example, global carbon dioxide (CO_2) emissions, which declined due to economic contraction, are not sustainable. Lower economic growth does not necessarily produce a cleaner energy world and is a poor low-emission strategy. However, some gains made during the pandemic have shown that certain solutions previously thought to be out of reach are more possible than expected. For example, one positive outcome of the crisis in the digitalization of physical activities in the form of remote work, teleconferencing, and other online activities has cut transportation by avoiding unnecessary travel, offered to use resources more efficiently, made consumption more sustainable, and reduced carbon footprint. This needs to be capitalized in a post-pandemic world with the evaluation of pros and cons.

Good energy policies lead to a win-win situation, whereby both energy consumption and GHG emission can be reduced while enhancing cities' economic performance. A low-carbon energy pathway will not necessarily cost much; however, it does require a shift in the intervention. For example, interventions that lead to the decarbonization of the electricity system, like renewable generation and storage, while making the power infrastructure more flexible can also improve resilience to future shocks. Moreover, mini-grid renewable systems or off-grids may also carry additional benefits for employment generation, health improvements, etc. Apart from reducing network infrastructure costs, they curtail the environmental footprint and increase affordability.

Although cities' energy policies operate within national and international frameworks, local or city governments are better placed in addressing local-level energy issues. This is firstly because cities are better connected to their citizens and secondly, their influence on entrepreneurs is stronger in promoting the most sustainable energy solutions. In the long run, cities' success toward local-level energy-related endeavors can also promote bottom-up approach by attracting the attention of national governments and international institutions for replicating these efforts. In addition, cities' political clout and procurement power and their role as birthplaces of innovation and experimentation are also important. Above all, many policy options for reducing energy consumption and decreasing CO_2 emissions fall directly under the authority of local governments, including energy consumption linked to the provision of public services (sanitation, water, public transport, and local public buildings), capital work programs, and infrastructure development. The city energy consumed by local government over which they have direct control has ranged from between 1% and 16% of total city energy use. Local governments can also use effective indirect solutions and measures to reduce energy use and GHG emissions.

Such solutions include regulatory measures, economic instruments, as well as support to technological innovations, institutional capacity building, cooperation with stakeholders, as well as education and information. All these measures are already in use within cities, but they need to be applied universally.

References

Abu-Rayash A, Dincer I (2020) Analysis of the electricity demand trends amidst the COVID-19 coronavirus pandemic. Energy Res Soc Sci 68:101682

Amelang S (2020) Negative electricity prices: lockdown's demand slump exposes inflexibility of German power. https://energypost.eu/negative-electricity-prices-lockdowns-demand-slump-exposes-inflexibility-of-german-power/. Accessed 22 July 2021

Amsterdam Smart City (undated) Roadmap circular land tendering. Amsterdam Smart City. https://amsterdamsmartcity.com/projects/roadmap-circular-land-tendering. Accessed 22 July 2021

Aznar A, Day M, Doris E, Mathur S, Donohoo-Vallett P (2015) City-level energy decision making: data use in energy planning, implementation, and evaluation in U.S. cities. National Renewable Energy Laboratory, U.S. Department of Commerce, Alexandria

Bayer P, Aklin M (2020) The European Union emissions trading system reduced CO_2 emissions despite low prices. Proc Natl Acad Sci U S A 117:8804–8812. https://doi.org/10.1073/pnas.1918128117

Bellini E (2019) Seoul launches 1 GW rooftop solar plan. PV Magazine. https://www.pv-magazine.com/2019/11/18/seoul-launches-1-gw-rooftop-solar-plan/

Biello D (2008) Sunrise on China's first carbon neutral city. Scientific American. https://www.scientificamerican.com/article/sunrise-on-chinas-first-carbo-neutral-city/

Boza-Kiss B, Pachauri S, Zimm C (2021) Deprivations and inequities in cities viewed through a pandemic lens. Front Sustain Cities 3. https://doi.org/10.3389/frsc.2021.645914

Brosemer K, Schelly C, Gagnon V, Arola KL, Pearce JM, Bessette D (2020) The energy crises revealed by COVID: intersections of indigeneity, inequity, and health. Energy Res Soc Sci 68:101661

Building Performance Institute Europe (BPIE) (2018) Towards a decarbonized EU building stock: expert views on the issues and challenges facing the transition - factsheet. Buildings Performance Institute Europe, Brussels. http://bpie.eu/wp-content/uploads/2018/10/NZE2050-factsheet_03.pdf. Accessed 25 Mar 2020

Carbon Neutral Cities Alliance (CNCA) (undated) Carbon neutral cities alliance members. https://carbonneutral-cities.org/cities/

Center on Global Energy Policy (CGEC) (2020) Upheavals in global energy in 2020, COVID, climate change and energy transitions. The Aspen Institute

Chen SQ, Long H, Chen B, Feng K, Hubacek K (2020) Urban carbon footprints across scale: important considerations for choosing system boundaries. Appl Energy 259:114201. https://doi.org/10.1016/j.apenergy.2019.114201

Cheshmehzangi A (2020) COVID-19 and household energy implications: what are the main impacts on energy use? Heliyon 6(10):e05202

Chhetri R (2020) Effects of COVID-19 pandemic on household energy consumption at college of science and technology. Int J Sci Res Eng Dev 3:1383–1387

Chow D (2021) Did the pandemic show us that we can cut carbon emissions. CNBC. https://www.nbcnews.com/science/environment/pandemic-show-us-can-cut-carbon-emissions-sort-rcna715

Creutzig F, Baiocchi G, Bierkandt R, Pichler P, Seto KC (2015) Global typology of urban energy use and potentials for an urbanization mitigation wedge. PNAS 112(20):6283–6288. https://doi.org/10.1073/pnas.1315545112

Crider J (2020) COVID-19 Bankrupts 19 Energy (Oil & Gas) Companies. https://cleantechnica.com/2020/08/05/covid-19-bankrupts-19-energy-oil-gas-companies. Accessed 30 Oct 2020

El Zowalaty ME et al (2020) Environmental impact of the COVID-19 pandemic - a lesson for the future. Infect Ecol Epidemiol 10(1):1768023. https://doi.org/10.1080/20008686.2020.1768023

Gillingham KT, Knittel CR, Li J, Ovaere M, Reguant M (2020, June) The short-run and long-run effects of covid-19 on energy and the environment. Joule 4:1337–1341

Global Commission on the Economy and Climate (GCEC) (2014) Better growth, better climate: the new climate economy report. Washington, DC. https://sustainabledevelopment.un.org/content/documents/1595TheNewClimateEconomyReport.pdf

Goyens M (2020) COVID-19 means tackling energy poverty is more urgent than ever. Euractiv. Available online at https://www.euractiv.com/section/energy/opinion/covid-19-means-tackling-energy-poverty-is-more-urgent-than-ever/. Accessed 21 Dec 2020

GPC (2014) Global protocol for community-scale greenhouse gas emission inventories: an accounting and reporting standard for cities. Greenhouse Gas Protocol

Hall D, Cui H, Lutsey N (2017) Electric vehicle capitals of the world: what markets are leading the transition to electric? The International Council of Clean Transportation, http://www.theicct.org/EV-capitals-of-the-world. Accessed 22 May 2020

Henry MS, Bazilian MD, Markuson C (2020) Just transitions: histories and futures in a post-COVID world. Energy Res Soc Sci 68:101668

Hepburn C, O'Callaghan B, Stern N, Stiglitz J, Zenghelis D (2020) Will COVID-19 fiscal recovery packages accelerate or retard progress on climate change? Smith School Working Paper 20-02. Smith School of Enterprise and the Environment, University of Oxford

ICLEI, WRI and C40 (2014) Global protocol for community-scale GHG emissions. http://c40-production-images.s3.amazonaws.com/other_uploads/images/143_GHGP_GPC_1.0.original.pdf?1426866613

IISA International Institute for Applied System Analysis and ISC International Science Council (2021) Rethinking energy solutions. https://council.science/wp-content/uploads/2020/06/IIASA-ISC-Reports-Energy.pdf

IISD (2019) 77 Countries 100+ cities commit to become carbon neutral by 2050 at climate summit. https://sdg.iisd.org/news/77-countries-100-cities-commit-to-net-zero-carbon-emissions-by-2050-at-climate-summit/. Accessed 28 July 2021

IMF (2020) COVID-19 Policy Tracker. Policy responses to COVID-19. International Monetary Fund. Available online at https://www.imf.org/en/Topics/imf-and-covid19/Policy-Responses-to-COVID-19#S. Accessed 22 Dec 2020

International Energy Agency (IEA) (2008) World energy outlook 2008. Paris

International Energy Agency (IEA) (2019) Global energy & CO2 status report 2019. International Energy Agency, Paris. https://www.iea.org/reports/global-energy-co2-status-report-2019/emissions, All Rights Reserved. Accessed 25 July 2021

International Energy Agency (IEA) (2020a) Key world energy statistics 2020. https://www.iea.org/reports/key-world-energy-statistics-2020/final-consumption, All Rights Reserved. Accessed 26 Oct 2020

International Energy Agency (IEA) (2020b) Global energy review 2020: the impacts of the Covid-19 crisis on global energy demand and CO2 emissions. Paris. https://www.iea.org/reports/global-energy-review-2020, All Rights Reserved. Accessed 26 Oct 2020

International Energy Agency (IEA) (2020c) Covid-19 impact on electricity. International Energy Agency (IEA), Paris. https://www.iea.org/reports/covid-19-impact-on-electricity, All Rights Reserved. Accessed 26 Oct 2020

International Energy Agency (IEA) (2020d) COVID-19 and energy efficiency. https://www.iea.org/reports/energy-efficiency-2020/covid-19-and-energy-efficiency, All Rights Reserved. Accessed 16 July 2020

International Energy Agency (IEA) (2020e) Global EV Outlook 2020. International Energy Agency, Paris. https://www.iea.org/reports/global-ev-outlook-2020, Paris All Rights Reserved. Accessed 26 July 2020

International Energy Agency (IEA) (2021) Global Energy Review, 2021. https://www.iea.org/reports/global-energy-review-2021, All Rights Reserved

International Renewable Energy Agency (IRENA) (2017) Renewable energy in district heating and cooling. IRENA Innovation and Technology Center, Bonn

International Transport Forum (ITF) (2020) Shared mobility simulations for Lyon, International Transport Forum Policy Papers, No. 74. OECD Publishing, Paris. https://doi.org/10.1787/031951c3-en

Jiang P, Fan YV, Klemeš JJ (2021a) Impacts of COVID-19 on energy demand and consumption: challenges, lessons, and emerging opportunities. Appl Energy 285:116441. https://doi.org/10.1016/j.apenergy.116441

Jiang P, Klemeš JJ, Fan YV, Fu X, Bee YM (2021b) More is not enough: a deeper understanding of the COVID-19 impacts on healthcare, energy and environment is crucial. Int J Environ Res Public Health 18(2):684

Kang M (2020) How is Seoul, Korea transforming into a smart city? World Bank Blogs. https://blogs.worldbank.org/sustainablecities/how-seoul-korea-transforming-smart-city

Kennedy CA, Stewart I, Facchini A et al (2015) Energy and material flows of megacities. Proc Natl Acad Sci U S A 112(19):5985–5990. https://doi.org/10.1073/pnas.1504315112

Kikstra J, Vinca A, Lovat F, Boza-Kiss B, van Ruijven B, Wilson C et al (2021) Maintaining energy demand shifts induced by COVID-19 reduces mitigation challenges. Nat Energy [Preprint]. https://doi.org/10.21203/rs.3.rs-155224/v1

Klemeš JJ, Fan V, Jiang P (2020) The energy and environmental footprints of COVID-19 fighting measures – PPE, disinfection, supply chains. Energy 211:118701

Kumari P, Toshniwal D (2020) Impact of lockdown on air quality over major cities across the globe during COVID-19 pandemic. Urban Clim 34:100719. https://doi.org/10.1016/j.uclim.2020.100719

Le Quéré C, Jackson RB, Jones MW et al (2020) Temporary reduction in daily global CO2emissions during the COVID-19 forced confinement. Nat Clim Change 10:647–653. https://doi.org/10.1038/s41558-020-0797-x

Le Quéré C, Peters GP, Friedlingstein P et al (2021) Fossil CO2 emissions in the post-COVID-19 era. Nat Clim Change 11:197–199. https://doi.org/10.1038/s41558-021-01001-0

Lee J, Taherzadeh O, Kanemoto K (2021) The scale and drivers of carbon footprints in households, cities, and regions across India. Glob Environ Change 66:102205, ISSN: 0959-3780

Malaviya S (2020) We need zero carbon building for green cities, business India. https://businessindia.co/climatechange/we-need-zero-carbon-building-for-green-cities

Mastropietro P, Rodilla P, Batlle C (2020a) Measures to tackle the Covid-19 outbreak impact on energy poverty. FSR Blog. Available online at https://fsr.eui.eu/measures-to-tackle-the-covid-19-outbreak-impact-on-energy-poverty/. Accessed 23 Dec 2020

Mastropietro P, Rodilla P, Batlle C (2020b) Emergency measures to protect energy consumers during the Covid-19 pandemic: a global review and critical analysis. Energy Res Soc Sci 68:101678. https://doi.org/10.1016/j.erss.2020.101678

Memmott T, Carley S, Graff M et al (2021) Sociodemographic disparities in energy insecurity among low-income households before and during the COVID-19 pandemic. Nat Energy 6:186–193. https://doi.org/10.1038/s41560-020-00763-9

Mofijur M, Fattah IMR, Alam MA et al (2021) Impact of COVID-19 on the social, economic, environmental and energy domains: lessons learnt from a global pandemic. Sustain Prod Consum 26:343–359. https://doi.org/10.1016/j.spc.2020.10.016

Moran D, Kanemoto K, Jiborn M, Wood R, Többen J, Seto KC (2018) Carbon footprint of 13000 cities. Environ Res Lett 13:064041

Moreno C, Allam Z, Chabaud D, Gall C, Pratlong F (2021) Introducing the "15-minute city": sustainability, resilience and place identity in future post-pandemic cities. Smart Cities 4:93–111. https://doi.org/10.3390/smartcities4010006

Mulvaney D, Busby J, Bazilian MD (2020) Pandemic disruptions in energy and the environment. Elementa 8(1):052. https://doi.org/10.1525/elementa.052

NASA (2019) Sizing up the carbon footprints of cities. https://earthobservatory.nasa.gov/images/144807/sizing-up-the-carbon-footprint-of-cities

Norouzi N, de Rubens GZ, Choubanpishehzafar S, Enevoldsen P (2020) When pandemics impact economies and climate change: exploring the impacts of COVID-19 on oil and electricity demand in China. Energy Res Soc Sci 68(2020):10165

OECD (2008) Competitive cities and climate change. https://www.oecd.org/cfe/regionaldevelopment/50594939.pdf

OECD (2016) Energy and resilient cities. https://doi.org/10.1787/5jlwj0rl3745-en

OECD (2020) Managing environmental and energy transitions for regions and cities. OECD Publishing, Paris. https://doi.org/10.1787/f0c6621f-en. Accessed10 Dec 2021

Office of Energy Efficiency and Renewable Energy (2015) Energy analysis data and tools for state and local energy planning. https://www.energy.gov/eere/analysis/energy-analysis-data-and-tools-state-and-local-energy-planning

Pranab K. R. C, Jeanette E. W, Eric M. W, Dalton L, St. Thomas M. L, Amy N. R, Budhendra L. B (2020) Electricity consumption patterns within cities: application of a data-driven settlement characterization method. International Journal of Digital Earth 13(1):119–135. https://doi.org/10.1080/17538947.2018.1556355

REN21, Renewable in Cities (2021) Global status report. https://www.ren21.net/wp-content/uploads/2019/05/REC_2021_full-report_en.pdf

Ritchie H, Roser M (2020) Energy, our world in data. https://ourworldindata.org/energy

Rugani B, Caro D (2020) Impact of COVID-19 outbreak measures of lockdown on the Italian carbon footprint. Sci Total Environ 29:139806

Save on Energy (2020) Does online video streaming harm the environment? Available on https://www.saveonenergy.com/uk/does-online-video-streaming-harm-the-environment/. Accessed 12 July 2021

Sawin JL, Hughes K (2007) Energizing Cities. In: Starke L (ed) State of the world, our urban future. World Watch Institute, Norton & Co, New York

Schatz Energy Research Center (2020) Energy requirements of the screening sites in a COVID-19 hub and spoke testing approach. https://www.lightingglobal.org/wp-content/uploads/2020/07/Covid-19-Screening-Energy-Requirements.pdf

Sharma M, Luthra S, Joshi S, Kumar A (2020) Developing a framework for enhancing survivability of sustainable supply chains during and post-COVID-19 pandemic. Int J Logist Res Appl. https://doi.org/10.1080/136755 67.2020.1810213

Sudmant A, Gouldson A, Millward-Hopkins J, Scott K, Barrett J (2018) Producer cities and consumer cities: Using production- and consumption-based carbon accounts to guide climate action in China, the UK, and the US. Journal of Cleaner Production 176:654–662. https://doi.org/10.1016/j.jclepro.2017.12.139

Ting W, Junliang W, Shaoqing C (2021) Keeping track of greenhouse gas emission reduction Progress and targets in 167 cities worldwide. Front Sustain Cities 3. https://doi.org/10.3389/frsc.2021.696381

Tong K, Nagpure AS, Ramaswami A (2021) All urban areas' energy use data across 640 districts in India for the year 2011. Sci Data 8:104. https://doi.org/10.1038/s41597-021-00853-7

UN Habitat (2008) State of the world cities 2008/2009. Earthscan, London

UN Habitat (2012) Sustainable urban energy: a sourcebook for Asia, HS/084/12. Nairobi. https://unhabitat.org/sites/default/files/2020/10/3378_alt.pdf

UNEP and IEA (2017) Towards a zero-emission, efficient, and resilient buildings and construction sector: global status report 2017. United Nations Environment Program

United Nations (2022) Climate action, generating power. https://www.un.org/en/climatechange/climate-solutions/cities-pollution

United Nations Secretary-General (UNSG) (2020) Secretary-General's remarks to 19th Darbari Seth memorial lecture "The rise of renewables: shining a light on a sustainable future"

Watts J, Ambrose L (2020) Coal industry will never recover after coronavirus pandemic, say experts. https://www.theguardian.com/environment/2020/may/17/coal-industry-will-never-recover-after-coronavirus-pandemic-say-experts. Accessed 31 Oct 2020

Weinsziehr T, Grossmann K, Gröger M, Bruckner T (2017) Building retrofit in shrinking and ageing cities: a case-based investigation. Build Res Inf 45:278–292. https://doi.org/10.1080/09613218.2016.1152833

Werth A, Gravino P, Prevedello G (2020) Impact analysis of COVID-19 responses on energy grid dynamics in Europe. Appl Energy 281:116045

World Economic Forum (WEF) (2020) Here's why energy security is a vital tool in tackling a pandemic. https://www.weforum.org/agenda/2020/04/pandemic-energy-access-coronavirus/. Accessed 14 July 2021

World Economic Forum (WEF) (2021a) These 5 global cities are leading the charge to a renewable future. https://www.weforum.org/agenda/2021/04/renewable-energy-urban-city-emissions/

World Economic Forum (WEF) (2021b) Net zero carbon cities: an integrated approach. https://www3.weforum.org/docs/WEF_Net_Zero_Carbon_Cities_An_Integrated_Approach_2021.pdf

World Resources Institute (WRI) (2019) Accelerating building decarbonization: eight attainable policy pathways to net zero carbon buildings for all. https://www.wri.org/publication/accelerating-building-decarbonization. Accessed 27 July 2021

Zhong H, Tan Z, He Y, Xie L, Kang C (2020) Implications of COVID-19 for the electricity industry: a comprehensive review. CSEE J Power Energy Sys 6(3):489–495

This part evaluates the social vulnerabilities of cities in the wake of mega risks. The pandemic and climate-related risk events in exacerbating poverty and inequality have clearly brought out the interdependence between social variables such as poverty, health, education, social welfare, and human well-being. The concept of social welfare and human well-being incorporates a broader spectrum of basic facilities that stretch from food, health, nutrition and safe shelter to education, and social protection. In cities, it means their availability across all spheres or units—from family or household to community and neighborhood. While highlighting the impacts of mega risks on health, education, and social welfare, the section stresses the need and urgency for promoting a socially inclusive approach to sustainable development.

The part has three chapters. The chapter on health covers the coronavirus pandemic and climate crisis in relation to the health of city residents and management of their health systems. The promotion of sustainable development in a city cannot be imagined without a healthy population. The chapter critically assesses cities' preparedness and response, containment and control measures adopted, pressures experienced by urban health system, and lessons learned in resilience. Like healthy population, an educated, knowledgeable, and well-informed public is critical in a resilient and sustainable city. The chapter on education examines the impact of the pandemic on urban education. It examines preparedness of cities and their response to education delivery in the wake of mega risks. The chapter on social protection concentrates on well-being and social protection as hallmark of human development and builds a case for making cities inclusive with equal opportunities for all. A special focus is placed on the poor, the vulnerable, and slum dwellers, whose preponderance can make sustainable development an elusive goal. All chapters include case studies to highlight the lessons learned and how these can be applied in future.

Both national and city governments were ill-equipped and/or unprepared to deal with the mammoth socioeconomic impacts of the COVID-19 pandemic and growing risks of climate change in 2020. The austerity measures adapted in the aftermath of the 2008/2009 financial crisis resulted in severely underfunded healthcare systems and public services and decline in global labor income shares and made cities less resilient to face shocks. In the wake of the rapid spread of coronavirus disease coupled with climate change-related disasters and public services still under stress, the temporary relief measures and limited cash transfers proved only to be a "drop in the ocean" for people living in poverty, whether in developed, developing, or least developed countries. Thousands of social protection measures were adopted by hundreds of jurisdictions to cushion the shock during the pandemic. However, these constituted only an emergency response to a situation and were in no way a substitute for social protection as a set of permanent entitlements making individuals as rights-holders and public authorities as duty-bearers. The part therefore advocates that it is important to learn from the mega risks and make cities socially resilient, which demands universal health coverage, quality education for all, and a set of social protection measures to meet the need of the time.

7.1   Introduction

The pandemic caused by the coronavirus, as stated earlier in this book, is not the first of its kind. Many momentous epidemics and pandemics as well as climate-related disasters have occurred over time and decimated cities and societies, altered the course of human history, and killed millions of people. Nevertheless, they are also witness to the fact that mankind has survived these epidemics/pandemics and disasters. Although causing great human suffering, they have also cleared the way for innovations and advances in sciences (including medicine and public health), economy, and political systems (Cohn 2002).

This chapter has been divided into four parts. After this introduction, the second part discusses the impact of coronavirus pandemic on health of city residents as well as their health system and documents the lessons learned from this experience. It has been divided into four sections. Highlighting the vulnerabilities and response of cities to the coronavirus crises first, it focuses in its second section on the pressure that has been exerted on the urban healthcare system and health costs of the pandemic. The third section discusses the actual preparedness of the cities to face the pandemic and lessons learned. The final section traces the ebb and flow of events in the pandemic in terms of various waves, emerging new strains of virus, and when and how the pandemic will be tamed. It highlights the vaccination efforts and its shortcomings which caused the virus to spread out of control in various cities and parts of the planet. The third part of the chapter highlights the impact of climate change on health and its management. It has two sections. The first of these concentrates on the impacts of climate change on the health of urban residents, while the second section discusses the management aspect of climate change impacts on the human health in cities. The chapter ends with a conclusion discussing the findings of the chapter.

7.2   Coronavirus Pandemic: Impacts and Response

Scientific investigations of epidemiologists revealed that the virus possibly came from an animal, most likely a bat. The new virus caused a severe acute respiratory syndrome (CDC 2020a). It was like SARS-CoV, so it was named SARS-CoV-2 and the disease caused by the virus was named COVID-19 by WHO referring to the coronavirus that caused it and the year of its origin. Since its first appearance in December 2019, in Wuhan (China), the virus quickly spread to most cities of the world and became an international crisis. The World Health Organization (WHO) declared the COVID-19 an "international emergency" on January 30, 2020, and a pandemic on March 11, 2020. While it seemingly ended in China, where it was first reported, it still caused the second and more waves in cities of Europe, the USA, Asia, and Africa and in other parts of the world, including in many low-income and middle-income countries.

The impact of the COVID-19 pandemic when it initially emerged varied among cities and even within different parts of the same city. Several studies (Chowdhury and Jomo 2020; Singu et al. 2020; Kochańczyk and Lipniacki 2021; United Nations 2020) have pointed out factors that were responsible for this variation in impacts. In general, vulnerability to infection and response capacity has been attributed to factors such as preparedness of healthcare system, education, leadership experience, and ability to manage challenges posed. Governments' response depended upon system capacity and capabilities—e.g., authorities' ability to speedily trace, isolate, and treat the infected population and mobilize resources to quickly enhance testing capacity and secure personal protective equipment. Socioeconomic and demographic factors, such as poverty, economic status of a community, where in a city a person lives and works, race, ethnicity, and age, also had impact and so did deep-rooted inequalities, including development status of the city. Sociopolitical factors including greater "trust" in the public authority, less individualistic and narcissistic cultures, and the use of experience of past outbreaks such as SARS also proved important in quick implementation of control mechanisms. The United Nations (2020), pointing out some other factors, stated that there is evidence that cities with high level of crime and violence, poor infrastructure, weak local governance, ill-equipped resourced frontline workers, and limited access to healthcare were more vulnerable and tackling COVID-19 in them were more challenging. The disparities observed in different cities of countries as diverse as Canada, Chile, Korea, Taiwan, and the UK as well as within one large metropolitan area like New York have been explained due to these factors.

The most important aspect in the origin and diffusion of virus in various cities was how the first "clusters" of cases originated and spread. In many instances, large cities, with their strong international links due to increased globalization—including business travel, tourism, marketing, etc.—were the entry points for the virus.

Contagion spread more quickly in large urban areas, due to proximity, particularly when preventive, protective, or containment measures were not introduced early enough. However, a clear correlation could not be established between density and incidence of the disease as some very densely populated Asian cities, such as Hong Kong (7.5 million), Seoul (9.8 million), Singapore (5.6 million), and Tokyo (9.3 million) saw limited diffusion of COVID-19, thanks to early and very proactive measures, mask wearing and extensive testing (OECD 2020a).

7.2.1 Cities' Vulnerabilities and Impacts

The cities became the major battlegrounds of the crisis facing once-in-a-century test to global resilience (Acuto et al. 2020). According to the United Nations (2020), 90% of COVID-19 cases took place in urban settlements, with cities affected in thousands worldwide. The plight of large cities can be explained from the fact that in Chile, Metropolitan Santiago accounted for 70% of cases as of November 2020 (Ministry of Health of Chile 2020). In Brazil, São Paulo registered 25% of cases at the same time (Government of Brazil 2020). In Russia, Moscow represented 24% of total cases as of November 2020 (OECD 2020a).

7.2.1.1 Spatial and Temporal Impacts
The pandemic brought very serious problem to cities in both the developed and developing countries. For many developing country cities which were highly underresourced and unable to serve the basic needs of growing populations, lockdowns meant locking poor communities into cycles of poverty, poor health, and low productivity (van den Berg et al. 2020). In slums such as Dharavi in Mumbai, India, where 67% of households relied on community toilets, and where soap and clean water was scarce, physical distancing was impossible. Not surprisingly, by mid-July, 40–60% of people living in Mumbai's slums were infected (Bai et al. 2020). Slums and shanty towns in developing countries cities cur-

rently accommodate more than 1 billion people with lack of access to affordable and secure housing. Moreover nearly 3 billion city residents lack access to basic handwashing facilities in their homes. In addition, up to 70% of the urban population lacks access to core services and infrastructure, relying instead on informal or alternate arrangements for water, sanitation, transportation, and energy. These were indeed massive problems during a pandemic of an extremely infectious disease (van den Berg et al. 2020).

Even in developed countries, some deprived areas or relatively poor cities were more seriously affected by the pandemic. For example, municipalities in the lower quartile of the national income distribution in France had higher mortality rates (twice as high) compared to municipalities in upper quartiles. This heterogeneity may be explained by municipal differences in housing conditions and occupational exposure (OECD 2020a). Likewise, in the USA, poor counties recorded more deaths (60 COVID-19 deaths per 100,000 people) than wealthier ones (48 COVID-19 deaths per 100,000). In the first income quintile of per capita GDP, new COVID-19 deaths were significantly higher than in other quintiles between August and October 2020. However, overcrowding associated with poor housing conditions played a role in the spread of the virus. It is important to note that mortality rates were also determined by the health system capacity, and preexisting health conditions (like high blood pressure, obesity, and diabetes), which themselves tend to be correlated to income and education (OECD 2020a). Limited access to healthcare basic services and adequate housing and/or public space further undermined COVID-19 response and enhanced mortality.

It is also important to note that virus situation has varied during various waves. During the first wave of the pandemic, after Wuhan and other cities of China, the big cities on America's east coast were hit hard. However, most other states in the USA had lockdown quick enough to prevent major outbreaks in their cities. Nevertheless, mortality remained high across much of the country's large cities. Some of the Midwest cities in the USA suffered large spikes in death rates

during the second wave in November and December, after experiencing few fatalities during the first wave. While COVID-19 was devastating New York, cities in Western Europe particularly those of Britain, Spain, Italy, Belgium, and Portugal were also suffering large outbreaks during the first wave. After a sharp fall in COVID-19 deaths during the summer of 2020, a second wave in autumn and winter caused excess mortality to rise again across the cities of Europe (The Economist 2021). Russian cities were affected in both the first and second wave in 2020.

In 2021, while the virus started receding in developed countries with vaccination campaigns, it spiked in cities of developing countries like India and South Africa which showed the grimmest picture during the third wave. They recorded a huge spike in fatalities in 2021, as a strongly infectious variant of the virus spread through cities of these countries. It also swamped some cities of Southeast Asian countries like Thailand and Cambodia that had largely kept the virus at bay early on. In Latin America, most cities experienced a devastating first wave from April to July 2020, when Bolivia and Ecuador were hit particularly hard. A second wave surged through the region in late 2020, as Mexico, Peru, and Brazil all recorded higher peaks of mortality than at any previous point during the pandemic.

7.2.1.2 Pandemic Pressure on the Healthcare System

Public healthcare systems and services in most developing countries were under tremendous pressure in dealing with the pandemic. At the peak of the crisis, they often suffered from lack of equipment required to take care of COVID-19 patients, such as personal protective equipment, bedside oxygen supply, pulse oximeters, ventilators, and ICU beds. Insufficient infection control training of healthcare workers was also a constraint (Levison 2020). Regarding equipment, fewer than 2000 working ventilators were available to serve hundreds of millions of people in public hospitals across 41 African countries (Maclean and Marks 2020). There were also chronic shortages of more basic supplies like face masks. These challenges were exacerbated by the prevalence of tropical parasitic diseases, malaria, HIV/AIDS, tuberculosis, and cholera in these countries. Data from South Africa's Western Cape province indicated that people living with HIV or tuberculosis had developed more than twofold increased risk of death from COVID-19 (Saltzman 2020).

India, which was the second after USA in terms of cases due to COVID-19 in November 2020, had a total of 713,986 government hospital beds available according to research of Brookings Institution (Singh et al. 2020). This amounted to 0.55 beds per 1000 population, out of which 5–8% were ICU beds (35,699–57,119 ICU beds). If 50% of these ICU beds had ventilators, the estimated number came to 17,850–25,556 ventilators in the country (Singh et al. 2020). Even in the best-case scenario where all ICU beds were equipped with ventilators, a maximum of 57,000 ventilators would be available to cater to a growing number of COVID-19 patients. Despite ongoing efforts to manufacture locally, the growing demand for ventilators outstripped the limited supply. The situation in other South Asian countries was the same if not worse.

Wuhan in China felt shortage of healthcare facilities in the initial stage. After the lockdown on January 23, 2020, the number of seriously ill COVID-19 patients there continued to rise, exceeding local hospitalization and ICU capacities for at least a month (Li et al. 2020). Few large hospitals there were designed to isolate hundreds of people with infectious diseases. For example, only 3 of Wuhan's 90 hospitals were initially designated as suitable for people with COVID-19. However, government acted fast and by April 2020 their number increased to 65. Mobile and large temporary wards were quickly constructed to treat thousands of more people (Bai et al. 2020).

Even the richest cities in most developed countries of the world lacked medical facilities that were needed to cope with epidemics. In the US cities, for example, with growing cases of the virus infection, plans were urgently needed to mitigate the effect of COVID-19 outbreaks for the local healthcare, but they did not come in

time. New York City is a case in point where a Wuhan-like outbreak happened, but in the absence of preparedness, the healthcare resources were overstretched beyond any one's imagination. What to say of hospital beds and ventilators shortage was experienced even in basic items of protection for healthcare workers such as N95 masks and face shields for intubating patients. According to an emergency room (ER) doctor, masks which were worn for observing a single patient were so short in supply that they had to be used at first for multiple patients, then for the whole single shift, and then for the multiple shifts (Ouyang 2020). Risk to doctors because of quickly rising outbreak and shortage of resources reached a level… the gravity of which can be guessed from the following paragraph of diary of the same ER doctor in *The New York Times*:

"I've taken part in humanitarian relief missions in more than 20 countries, in settings as resource-poor as - mobile clinics in South Sudan immediately after its secession, refugee camps in Kenya, an abandoned war hospital in Liberia, medical facilities in Somalia. Never have I personally felt unsafe, like I didn't have enough protection for myself. People are now referring to ours as "a third-world country," but in terms of personal protective equipment in this pandemic, it's actually worse than those overseas hospitals" (Ouyang 2020).

Speaking of overcrowded conditions of his New York hospital, the emergency room doctor stated: "Patients are now triple-bunked into single-person spaces, curtains pushed aside. In one room, three men, who appear to be in their 80s or so, are side by side in their stretchers, each one pulling at his oxygen mask, confused, their frail limbs swinging in the air. Some have sat in their own feces for a day… I dodged patients, colleagues and stretchers to get around—forget six feet of separation; we're not able to maintain six inches" (Ouyang 2020).

Due to high pressure on ICU and ER beds, because of rapidly rising cases, the patients had to be released sooner than they should have been. Referring to this the same emergency room (ER) doctor pointed out: "Everyone in medicine knows that one of the most heart-dropping phrases you can hear is: 'You know that patient you saw the other day? Well, he came back and. …' I think of all the doctors who sent their patients home because they looked well or were young or didn't have medical problems, and they came back to the ER needing a breathing tube. I'm sure these patients all looked OK a few days ago" (Ouyang 2020). When pressure increased further, makeshift hospitals opened around the city to take some of the load off. Initially, it was decided that the Javits Center and the Navy hospital ship Comfort will care only for non-COVID patients. A few days after it opened, the Javits temporary hospital changed its admission policy to take in COVID-19 patients; the Comfort did the same the following week (Ouyang 2020).

This is rather sad that this type of situation occurred in a city of the USA that has more critical care hospital beds per capita (34.7 per 100,000 population) than most other countries—almost ten times as many as China (3.6) and about triple the rate in Italy (12.5). Still, as demonstrated by the case of New York, the number of ICU beds and mechanical ventilators even in the US hospitals constituted only a fraction of what was needed (Boyles 2020). The situation like New York seems to have been encountered by cities in some other developed countries too, which saw similar outbreaks, like in cities of Lombardy region in Italy and Madrid in Spain, which at one time accounted for half of all the coronavirus cases in Spain. Situations in some developing country cities were also more if not equally worse like São Paulo in Brazil, Santiago in Chile, and Delhi in India.

The resources of healthcare including ICU beds, support equipment (including ventilators), and the critical care team are essential components of a patient management system during a pandemic. According to the Society of Critical Care Medicine (SCCM 2020), simply adding more of one resource element without considering the interconnectedness of the healthcare system's other assets is unwise and potentially unsafe in planning for or managing patients during a pandemic like COVID-19. To deal with the shortage of the first component, hospital/ICU beds, the strategies followed by cities included

repurposing decommissioned facilities and converting other use facilities such as convention centers and sports stadiums to temporary field hospitals and/or to build makeshift hospitals. In Chicago, for example, Metro South Hospital in Blue Island and Westlake Community Hospital in Melrose Park were revamped for use. Metro South Hospital was a 400-bed hospital that reopened with 574 beds; Westlake Community Hospital reopened with 230. A third facility Sherman Hospital in Elgin became a 274-bed treatment center (JLL 2020). In New York City during the month of April, which saw the city's largest number of cases, the Javits Convention Center was converted into a temporary medical facility which treated more than 1000 patients. In Poland, the government announced in October 2020 to use National Stadium in Warsaw as temporary medical facility to handle surging cases of coronavirus—built for the Euro 2012 football championships, the 60,000-seat stadium had conference rooms under its stands, which accommodated patients (Gera 2020). In Wuhan, China, where the outbreak began, officials built 16 temporary hospitals out of sports stadiums, gymnasiums, and exhibition centers. They closed down on March 10, showing the flexibility of fast-care solutions (JLL 2020). Likewise, according to *The Guardian* newspaper (Wainwright 2020), when faced with a soaring number of patients in Italy, a network of architects and engineers joined hands to convert 20-foot-long shipping containers into two-bed intensive care units for a hospital in Milan. Dubbed CURA (Connected Units for Respiratory Ailments), each container had all the features of an ICU plus extractors to create negative air pressure. Tents were also used to treat as the cases surged. Two Navy ships in the USA, Mercy in Los Angeles and Comfort in New York, were also used to take care of patients.

Regarding support equipment, a national service provision assessment (SPA) was conducted from representative surveys of hospitals (within the past 5 years) in five low-income countries—Afghanistan, Democratic Republic of Congo, Haiti, Nepal, and Tanzania (Demographic and Health Surveys quoted in McMahon et al. 2020). The assessment and analysis of hospital general clinics confirmed limited availability of PPE, with only 24–51% of hospitals reporting any type of face mask, 22–92% medical gowns, and 3–22% eye protection. Sanitation supplies were also scarce, with 52–87% of hospitals recording soap plus running water and 38–56% alcohol-based hand sanitizer (McMahon et al. 2020). Further gaps were found in ability to provide care for respiratory conditions, demonstrating under-investment in hospital-based services (Rajbhandari et al. 2020). The hospitals analyzed had a lack of pulse oximeters (12–48% available), oxygen tanks (10–82%), and bag-masks necessary for basic resuscitation (28–45%). As has been noted by prior studies, advanced respiratory support such as intensive care units (ICUs) and ventilators were even more scarce (Murthy et al. 2015).

Even rich nations such as the UK and the USA reported dangerously low supplies of PPE in March 2020. In Italian cities, the shortages contributed to the high burden of infections and deaths among hospital staff (Burki 2020). At the end of March, WHO Director-General stated that "the chronic global shortage of personal protective equipment is now one of the most urgent threats to our collective ability to save lives" (Burki 2020). Nearly 90% of the mayors who responded to a national survey conducted from March 20 to March 24, 2020, in 213 US cities on coronavirus said they lacked sufficient tests kits, face masks, and other protective equipment for their emergency responders and medical workers, while 85% said they did not have enough ventilators for their hospitals (Hart 2020). Before the pandemic, China was responsible for half the world's supply of surgical masks and was the only place capable of mass-producing clinical gowns. So, the severe shortages that characterized the early stages of the pandemic were probably unavoidable. The things became somewhat better later, but the situation was not absolutely resolved (Burki 2020) all over the world. To meet the shortages of ventilators in the USA, companies such as Tesla and General Motors repurposed their factories to make emergency ventilators. In addition, SpaceX helped Medtronic, the largest ventilator producer, to meet surging demand of ventilators by preparing a proportional solenoid (PSOL) valve, a highly complex piece of machinery that controls the flow of air and oxygen inside the ventilator.

The critical care team of doctors and nurses, the third important component of healthcare service, was also under great pressure. Its link with other components can be judged from the fact that although new emergency healthcare facilities were being developed in cities and more beds were being added, the healthcare workers—doctors, nurses, and technicians—were not enough to staff those beds even in cities of rich countries like the USA. The COVID-19 patients, according to some healthcare workers, were the sickest people they had ever served. They required twice as much attention as a typical intensive-care-unit patient and three times longer time than the normal length of a patient's stay (Yong 2020). Speaking on the availability of healthcare workers and working of the health system in the wake of critical patient surge, one nurse talking to a journalist said: "Last Monday we had 25 patients waiting in the emergency department. They had been admitted but there was no one to take care of them." When she was asked how much slack the system left, she replied: "There was none" (Yong 2020). An infectious disease doctor at the University of Iowa, seeing the surging cases of disease said: "The health-care system in Iowa is going to collapse, no question. In the imminent future, patients would die because there simply aren't enough people to care for them" (Yong 2020).

7.2.1.3 Pressure on Non-pandemic Healthcare Services

In terms of non-pandemic health services, a WHO (2020b) survey of the impact of COVID-19 on health system in 105 countries highlighted the problems in the present systems and pointed out new strategies to improve healthcare provision during the pandemic and beyond. The survey noted that countries on average experienced disruptions in 50% of a set of 25 tracer services. The most frequently disrupted areas reported were routine immunization—outreach services (70%) and facility-based services (61%)—noncommunicable disease diagnosis and treatment (69%), family planning and contraception (68%), treatment for mental health disorders (61%), and cancer diagnosis and treatment (55%). Countries also reported disruptions in malaria diagnosis and treatment (46%), tuberculosis case detection and treatment (42%), and antiretroviral treatment (32%). While some areas of healthcare, such as dental care and rehabilitation, may have been deliberately suspended in line with government protocols, the disruption of many of the other services had harmful effects on health of the population in the short, medium, and long term (WHO 2020b).

Potentially lifesaving emergency services, according to the survey, were disrupted in almost a quarter of responding countries. Disruptions to 24-h emergency room services, for example, were affected in 22% of countries, urgent blood transfusions were disrupted in 23% of countries, and emergency surgery was affected in 19% of the countries. In terms of outpatient care service, 76% of countries reported reductions in this service due to lower demand and other factors such as lockdowns and financial difficulties. The most reported factor on the supply side was cancellation of elective services (66%). Other factors reported by countries included staff redeployment to provide COVID-19 relief, unavailability of services due to closings, and interruptions in the supply of medical equipment and health products. Many countries started to implement some of the WHO recommended strategies to mitigate service disruptions, such as triaging to identify priorities, shifting to telehealth or online patient consultations, and changes to prescribing practices and supply chain and public health information strategies.

7.2.1.4 Health Costs of Pandemic

The health risk losses associated with COVID-19 have been monetized like the cost of disruptions to the economy. The mortality and morbidity effects of COVID-19 were very real and among others have been attributed to cause substantial economic losses. Viscusi (2020) worked on worldwide COVID-19 costs using data for over 100 countries. The calculation was based on income-adjusted value of statistical life (VSL) estimates for the selected countries. The total global mortality cost through July 2, 2020, was $3.5 trillion; the USA alone accounted for 1.4 trillion or 41% of the world mortality cost. Morbidity effects of COVID-19 also affected many more patients than did the disease's mortality risk. Consideration of the morbidity effects, according to author (Viscusi 2020), increases the expected health losses associated with COVID-19 illnesses by 10–40%. The study did not include mental illness due to anxiety or long-term health impairment which if included enhances the costs further considerably. For example, according to Cutler and Summers (2020), health costs (losses from premature death, long-term health impairment, and long-term mental impairment) in the USA alone have been estimated at over 8 trillion dollars (Cutler and Summers 2020). It may be noted that even when only the costs during the acute infection and not those of follow-up care after infection are considered, the direct medical costs of a symptomatic COVID-19 case tend to be substantially higher compared to the costs for other common infectious diseases (Bartsch et al. 2020).

Besides direct human health costs, healthcare institutions also faced serious financial problems. For example, international hospitals and healthcare facilities faced catastrophic financial challenges related to the COVID-19 pandemic. The American Hospital Association (AHA 2020) estimated a financial impact of $202.6 billion in lost revenue between March 1, 2020, and June 30, 2020 for America's hospitals and healthcare systems—an average of $50.7 billion per month. Furthermore, it may have costed low- and middle-income countries US$52 billion (equivalent to US$ 8.60 per person) each 4 weeks to provide an effective healthcare response to COVID-19 (Kaye et al. 2020).

7.2.2 Response to the Health Impacts

As with the other disease outbreaks and pandemics before it, control of COVID-19 depended on the detection and containment of clusters of infection and the interruption of community transmission. During the plague outbreak that affected fourteenth-century Europe, isolation of affected communities and restriction of population movement were used to avoid further spread (Tognotti 2013). These public health response measures remain valid even today and were applied with reinforcement by the latest technology to contain and control the coronavirus pandemic.

7.2.2.1 Public/Governmental Response

The governmental measures included promotion of social distancing, use of face mask, staying at home and working from home, quarantine, closure of educational institutions and undertaking online teaching, closure of nonessential business, promotion of telehealth and stopping nonessential surgeries. Other measures included enhancing testing capabilities and tracking of disease and its clusters, communicating information on personal hygiene and care such as handwashing, and avoiding crowded spaces.

At the time of declaration of COVID-19 pandemic in March, more than 90% of cases were in cities of four countries—China, Iran, Italy, and South Korea; the new infections were on the decline in China and South Korea, 81 countries had no cases, and 57 reported 10 cases or less (WHO 2020b). The WHO was hoping that the pandemic could still be checked by mobilizing resources to detect, test, isolate, and trace those infected and putting them in quarantine. This strategy, could not be applied. Only a few East and Southeast Asian countries and state of Kerala in southwest India acted early, urgently, and adequately, thus avoiding highly disruptive total lockdowns and associated human and economic costs (Chowdhury and Jomo 2020). Most others were slow to respond, or complacent, with some hoping or expecting the virus would bypass them or believing that "herd immunity" would protect them—hence, they became most exposed to the virus. "A few influential government leaders refused to acknowledge the severity of the COVID-19 threat, distracting many with conspiracy theories and 'blame games', instead of

quickly learning from and correcting policy errors as new knowledge became available" (Chowdhury and Jomo 2020). Thus, new infections and deaths quickly rose exponentially as the epidemic rapidly spread to cities worldwide, especially in Western countries, better connected by passenger air traffic. The main strategy adopted by majority of governments was to "flatten the curve." It was followed to allow the cities' health systems to cope with new infections by tracing, testing, isolating, and treating those infected until an approved vaccine or "cure" was available to all. It was, however, easier said than done (Chowdhury and Jomo 2020).

The variation in effective application of some or all the public health response measures between cities and even within communities of the same city depended on several factors including the available resources, technology, and living conditions and culture. As such they produced different results. For example, use of intervention measures backed by firm governance decision and strong community solidarity and behavior yielded positive results in cities of China, Japan, Republic of Korea, Taiwan, and New Zealand. Supported by extensive use of emerging technologies particularly apps for disease tracking and identifying hotspots and clusters along with medical/healthcare treatment made the response more effective and reduced the risk of the spread of the disease. Although the pandemic was a global one, its responses and effectiveness were to a large extent local, depending on the local governance, socioeconomic factors, and cultural context (Shaw et al. 2020). For example, societies with close physical human contact as an intrinsic element of lifestyle—as in cities of Brazil, France, Italy, and Spain—were exposed much more to the disease compared to cities of Japan and Republic of Korea and countries of Scandinavia, where people naturally maintain social distance.

The implementation of response measures like social distancing was often more difficult in cities of developing countries, where people usually live in crowded, multigenerational households (Levison 2020). In addition, the communities there often lack access to water—

what to say of running water for handwashing. They also lack adequate sanitation and have little or no savings to back up a loss of income (Gibson and Rush 2020). Even basic supplies that are taken for granted in developed countries, like soap and disinfectants, are unavailable, short in supply, or are unaffordable due to poverty. In terms of technology, the major deficiency was absence of even simple digital devices, which along with no or poor Internet connection proved to be a major handicap in online teaching, tele-healthcare, work from home, and effective communication by the governments during the pandemic.

Deep-rooted inequalities, even in developed countries, made COVID-19 disproportionately affect the urban poor and certain minority ethnic groups particularly because of difficulties in implementing social distancing and other control measures. According to CDC (2020b) and APM Research Lab (2020), evidence showed that some racial and ethnic minority groups in cities of the USA were excessively affected by the pandemic. Increased rates were recorded among Black or African Americans, and in Hispanic or Latino communities compared to white and Asian American groups. For example, African Americans constitute 13% of the US population but accounted for 20% of COVID-19 cases and more than 22% of COVID-19 deaths, up to July 22, 2020. Likewise, Hispanic, constituting 18% of the population, accounted for almost 33% of new cases nationwide at the same time (CDC 2020b). Nearly 20% of US counties are disproportionately Black, and these counties accounted for more than half of COVID-19 cases and almost 60% of deaths nationally in 2020 (Millett et al. 2020). Prevalence of disease in Latino populations was 30–40%, in the city of Chelsea in Massachusetts (Saltzman 2020). Hospitalization rates of American Indian and Alaskan Native groups for COVID-19 were five times higher than those of white, non-Hispanic groups (Bai et al. 2020). Likewise, disease incidence rate among the Indians in Navajo reservations in New Mexico surpassed that of New York State (Silverman et al. 2020). Another vulnerable group in the US cities was homeless population.

According to the government estimates, 1.4 million people live in homeless shelters each year—a population with the potential for widespread transmission of disease—often crowded, with no room for social distancing. Moreover, "many persons experiencing homelessness are older or have underlying medical conditions, placing them at higher risk for severe COVID-19 associated illness" (Fertel 2020). The Howard Center for Investigative Journalism tracked at least 153 homeless deaths in just six areas with large homeless populations in the early period of the pandemic—San Francisco; Los Angeles; New York City; Washington, D.C.; Seattle; and Phoenix.

Migrant workers have also been badly affected by the pandemic. Research conducted prior to the pandemic (Nagendra et al. 2018) pointed out that the governments rarely consider them in their planning, and they are invisible in censuses and surveys and therefore information for them is limited. Whatever information is available is enough to show their plight in the pandemic. For example, in Singapore, coronavirus infections spread quickly among its migrant workers, many of whom were living in cramped, unsanitary spaces, and unlike other citizens, they were not allotted face masks at the outset (Bai et al. 2020). According to Ministry of Health Singapore (2020), over 300,000 unskilled migrant workers in the country lived in packed dormitories that housed up to 20 people per room, and they accounted for more than 90% of all the COVID-19 cases in the city-state. Outbreaks also took place in a meatpacking plant in Germany staffed largely by migrant workers from eastern Europe—Teonnie's meatpacking plant in Rheda-Wiedenbrueck closed for a month after about 1500 workers tested positive for coronavirus in June 2020. "Germany's meatpacking sector has been criticized for the widespread use of sub-contracted migrant workers from eastern Europe, whose cramped accommodation and poor oversight were blamed for accelerating coronavirus outbreaks" (Reuters 2020). Across India, Africa, and Latin America, millions of migrants were excluded from aid during lockdowns (Bai et al. 2020). Overall, migrants, refugees, and the homeless living in encampments on the outskirts or even in the centers of cities like Los Angeles, Seattle, New York, Oakland, Paris, and London were highly vulnerable because under their living conditions, physical distancing was difficult or even impossible.

7.2.2.2 Nongovernmental Response

Coronavirus pandemic put tremendous pressure on governments' limited resources at both national and local levels that brought them to crippling point. On the brighter side, this however led to stepping in of communities and other bodies at city level. Bai et al. (2020) mention good practices from developing countries such as Brazil, India, and Vietnam. G10 Favelas, a network of community leaders across the ten largest informal settlements in Brazil, for example, provided medical supplies and raised donations to hire private ambulances and health staff for residents. By late June 2020, women in São Paulo's largest slum, Paraisópolis, had sewn more than 50,000 masks. In India, when Mumbai's private health system collapsed under the strain of COVID-19 infections, collaboration between the municipality and local low-cost private health practitioners capped the virus's spread in the Dharavi slum. Likewise, NGOs, charities, and philanthropists provided food and personal protective equipment to clinical staff and city residents in Bangalore, Chennai, Mumbai, and Delhi in India. Vietnamese cities also forged connections with local clinicians and businesses. Within a month of the first cases, local manufacturers were producing affordable COVID-19 testing kits, allowing the number of testing stations to rise from 3 in January 2020 to 112 by the end of April 2020 nationwide in Vietnam (Bai et al. 2020).

AlKhidmat Foundation in Pakistan, one of the largest charity organizations, distributed ration, enough for 15 days, among the families in the low-income areas, in addition to soap, sanitizer, and face masks across the country. A total of 52 small and big hospitals of this organization across the country were designated to tackle the coronavirus outbreak, of which two were taken over later by the government trained staff. The Edhi charity, which runs a nationwide ambulance ser-

vice in Pakistan comprising 1500 vehicles, made a bulk of its ambulances available to transport the suspected virus patients to the hospitals and quarantine centers across the country. In terms of charity, a remarkable scene unfolded during the early period of crises, outside grocery stores in Karachi. Instead of rushing home after shopping to avoid being exposed to coronavirus there, many people were pausing outside to offer food, money, or other charity to the people on the street with no "place" (Imtiaz 2020).

Connections between cities were another source of strength. Many cities in China received masks and protective gear from sister cities in South Korea, Japan, and other countries. Once the Chinese cities had recovered, they reciprocated the assistance. People also crossed borders for treatment, for example, between cities in Guangdong and Liaoning provinces in China, and from Italy to Germany.

7.2.2.3 Response Effectiveness

As discussed above, the cities' response to COVID-19 pandemic has been mixed but tells a lot about what works and what does not (Muggah and Katz 2020). Well-run cities such as Hong Kong, Taipei, Singapore, and Hanoi in Vietnam were able to avoid major outbreaks at the outset by rapidly restricting travel and scaling up testing, tracking, and quarantining. Singapore, Hong Kong, and Taipei all had direct flights to and from Wuhan, the first outbreak center in China, but the vigilant monitoring systems in these cities helped a lot as they started screenings of passengers from Wuhan in late December even before Beijing admitted that the coronavirus was spreading between humans. These systems were built over years, after their failures to stop an earlier dangerous outbreak of SARS, 17 years ago (Barron 2020). In contrast, the USA disbanded its pandemic response unit in 2018 (Beech 2020) and its cities suffered the worst outbreaks.

One good example of a city quoted by Bai et al. (2020) is Thiruvananthapuram (Trivandrum), the capital of the Indian state of Kerala. With an estimated population of more than 2.5 million, it had one of the lowest infections and death rates in India by August 1, 2020—just over 4000 confirmed cases and 12 deaths—despite having thousands of returning international students and workers. By contrast, places such as São Paulo in Brazil, Delhi in India, and New York City in the USA that reacted slowly or ineffectively ended up with overstretched hospital facilities and hundred times more deaths (Bai et al. 2020).

Some aspects of response in Chinese cities are also worth mentioning. In Wuhan, after the chaos of the first few weeks, people who suspected they had the virus were not directed to hospitals but sent to temporary drive-through testing sites to avoid sending them unprotected into the hospitals. They had to wait for their test results and on being positive were either isolated in special makeshift hospitals for minor symptoms or were sent to ICUs if serious (Dollar 2020). Measures used by authorities in Beijing and Shanghai and major mainland cities included stricter controls on the movement of residents and vehicles, compulsory mask wearing, and shutting down leisure and other nonessential community services (Jiangtao 2020). Beijing also made mandatory for anyone returning to the city from elsewhere to go into 2-week quarantine in a hotel—they could not shelter at home where they could infect family members. The lockdown-style measures were aimed at controlling possible community transmission of the virus. In addition, in cities where cases emerged, steps were taken for rigorous contact tracing with testing and, if necessary, isolation of the contacts. However even in China, a comparison between Wuhan and Guangzhou illustrated that an early intervention in Guangzhou led to lower outbreak size and peak compared to Wuhan (Li et al. 2020).

South Korea used measures like China with extensive testing and contact tracing primarily using cell phones and electronic media, isolation, and quarantine. These measures, however, were less restrictive and more hi-tech than those of China but did make it possible to control the virus quickly. In contrast to well-executed measure in Chinese and Korean cities, there were other urban areas which laxed in implementing control measures initially such as Milan, New York City, Barcelona, and São Paulo. As a result, they had to

go through serious agony due to the pandemic. For example, as of mid-May 2020, there were 50 times more deaths in the USA than in Korea, relative to population because the basic measures of testing, contact tracing, isolation, and quarantine were not applied (Dollar 2020) with the same zeal in the US cities as in China or South Korea.

The control strategy in a country or area was later changed with time and made more specific to a particular city or cities to suit a place-based or city-sensitive approach. This meant that the control measures such as mask wearing, school and facility closures, and full lockdowns were adopted for specific localities, cities, or territories based on local conditions, rather than applied universally in a country or area. For example, in England, the government had set three tiers of local COVID alert levels (medium, high, and very high) to ensure that the right measures are taken in the right places to manage the outbreaks (UK Government 2020). In places with very high alert, stringent measures were applied and locally reinforced with additional measures in consultation with local authorities. Cross-boundary decisions were implemented at local authority level. Liverpool City and Greater Manchester regions, for example, were under tier 3 alert in October, while London, Essex, Elmbridge, and York remained under tier 2 (OECD 2020b). However, testing and tracing were still at the heart of all coronavirus management strategies, as recommended by the WHO. In fact, effective testing strategies, together with social distancing, had become the norm to limit the large costs of confinements and lockdowns (OECD 2020b).

The WHO (2020a) recommended massive testing to fight the coronavirus by identifying infected people and isolating contagious contact-cases before symptoms developed. Testing and contact tracing, as stated earlier, were at the core of Korea's successful strategy to manage the first wave of infections and prevent a second one. Local governments were responsible for COVID-19 screening stations allowing for quick and safe testing and monitoring of those in self-quarantine. European countries also considerably increased their capacities and generalized testing for suspicious cases between April and October 2020 (OECD 2020a). The official data shows that in Europe more than 6 million RT-PCR tests were performed every week in October compared to 1.5 million in April (ECDC 2020).

7.2.2.4 Preparedness Versus Response: Lessons Learned

All countries are required by the International Health Regulations (WHO 2005) to have core capacity to ensure national preparedness for infectious hazards that have the potential to spread internationally. The emergency management plans in many cases were no doubt developed and agreed upon, but they remained static and were rarely revisited although they should have been more frequently practiced becoming flexible living documents that align with evolving operations (Cheney 2020). Moreover, research and development of new methods and technologies to strengthen the core capacities should have been practiced—which often transpire with actual experience during outbreaks when innovation becomes an absolute necessity (Budd et al. 2020). One major weakness in emergency plans was lack of emphasis on contagion response probably because the 2003 SARS, 2009 H1N1, and 2014 Ebola viruses, which were expected to have a dramatic effect on public health and the healthcare system, but caused less severe effects than anticipated. The medical experts were therefore trapped into a false sense of security (Cheney 2020). In addition, hospitals' and health systems' annual vulnerability threat assessments related to disasters and major risks focused upon varied local hazards like floods, forest fires, etc. A global pandemic was not high on any hospital's threat agenda.

The above shortcomings and experiences during the coronavirus pandemic have taught several lessons for preparedness (Cheney 2020). Among these, the first highlights that the emergency management should form an integral part of operational strategy and incident command systems as a part of regular work. The second – it is important to have vigilance whereby health systems and hospitals need to be alert and maintain essential capabilities to occurrence and resur-gence of novel infectious diseases, i.e., those which have been relatively common over the past two decades as well as to the outbreak of new pathogen. Finally, although after an epidemic or pandemic is over, it may not be necessary to always maintain incident command readiness, but some aspects of command systems and monitoring need to remain a permanent part of daily operations. This would ensure to maintain infectious outbreaks as a continuous and essential feature of operational strategy.

In terms of healthcare resources, before the pandemic, many health systems and hospitals functioned with "just-in-time" supply chains. It has now become important that healthcare organizations should take time to identify and activate supply streams, review stockpile levels, and rotate supplies to avoid expiring items. Health systems and hospitals should also look closely at longer-term and renewable personal protective equipment (PPE) such as powered air-purifying respirators and N95 respirator masks as well as methods to extend the life of PPE (Cheney 2020). The pandemic has taught us that it is crucial to involve all departments and decision-makers in emergency preparedness planning and recovery efforts. Involving frontline clinicians in planning improves understanding of their experience during an emergency, helps with engagement and adoption of new processes, and gives medical professionals a vested interest in emergency preparedness (Cheney 2020). Finally, health systems and hospitals should work to improve their technological infrastructure such as establishing robust telemedicine and remote work capabilities (Budd et al. 2020). These investments ensure business and operational continuity during a crisis.

In terms of management of the pandemic, the first important lesson is never to be complacent, and once the first cases are detected, it becomes urgent to mobilize resources to detect, test, isolate, and trace those infected and put them in quarantine. Simultaneously communication channels with public should be established and strengthened continuously to provide important information while controlling spread of wrong or misinformation on the crisis. Misinformation, in infectious disease epidemics or pandemics, in the worst-case scenario, could lead to psychosocial implications with detrimental mental affects including panic, fear, anxiety, uncertainty, depression, and other adverse psychological issues. Moreover, community best practices should be promoted—including those on wearing high-quality masks, social distancing, and how to carry out day-to-day activities related to work, shopping, health, and education and how to break the social isolation. Additionally, vulnerable population should be identified and equitably protected, while continued provision of essential services should be ensured including emergency medical and surgical services, sexual and reproductive health services, drug and alcohol misuse services, psychological and mental health services, as well as infrastructure services related to transport, water and sanitation, garbage disposal, energy, etc.

7.2.2.5 Pandemic's Future: Possible Scenarios

The rollout of several highly effective SARS-CoV-2 (COVID-19) vaccines in cities and nations, since the beginning of 2021, raised the hopes that the virus will vanish. However, there are three factors that temper the general expectations: firstly, the patchy vaccine coverage and inequalities in global access to vaccines; secondly, the hesitancy to vaccination; and thirdly, the mutation of the virus and development of its new strains or variants. No doubt, the experts have time and again voiced and agree that increased vaccinations will help reduce the number of severe cases, hospitalizations, and deaths, but vaccine may not always block virus transmission. Additionally, although massive vaccine deployment may lead to the end of the pandemic, but the end of the pandemic does not necessarily mean the end of SARS-CoV-2 (Telenti et al. 2021). Medical experts, including White House chief medical advisor Dr. Anthony Fauci, agree that coronavirus may never completely vanish—it may stay as an endemic such as influenza.

Telenti et al. (2021) have visualized three possible future scenarios for COVID-19. According to them, in the most worrisome scenario, it may

not be possible to gain rapid control on this pandemic, and thus the future may carry ongoing manifestations of severe disease combined with high levels of infection that, in turn, could foster further evolution of the virus. However, vaccinations and previous infection could lead to achieving a long-term herd immunity. Therefore there will still be a need for a very broad application of vaccines worldwide combined with comprehensive disease surveillance using accurate and readily available diagnostic devices (Aschwanden 2021).

Another more likely scenario, also predicted by other experts, is the transition of the pandemic to an epidemic or seasonal disease such as influenza. This scenario is like flu, whereby the virus constantly circulates in the background, rising and falling seasonally each year, yet people get vaccinated regularly against it, helping them build up multiple layers of immunity over many years (Stieg 2021). Effective therapies that prevent progression of COVID-19 disease (e.g., monoclonal antibodies that reduce hospitalization and death by 70–85%) may bring the burden of SARS-CoV-2 infection to levels that are equivalent or even lower than influenza. Nevertheless, it is important to note that the estimated annual mortality burden of influenza, even in non-pandemic years, ranges between 250,000 and 500,000 deaths comprising around 2% of all annual respiratory deaths (Paget et al. 2019). Still, this is a relatively "optimistic" outlook of the COVID-19 pandemic (Telenti et al. 2021).

A third scenario is the transition to an endemic disease like other human coronavirus infections that have a much lower disease impact than influenza or SARS-CoV-2. There is, however, limited data on the global burden of disease by common human coronaviruses (Gilca et al. 2021). Also, it is not possible to predict with confidence whether further adaptations of SARS-CoV-2 to humans will increase or decrease its intrinsic virulence (Telenti et al. 2021). Immaterial of the scenarios, the pandemic has presented both the importance of individual initiatives in cities and nations and the necessity of global cooperation for pandemic control in an interdependent world. "Global health leaders will need to be vigilant with respect to the trajectory of SARS-CoV-2 in the near future while assessing the strategies and approaches used in the pandemic to develop more effective structures and processes to ensure a more effective and equitable response in the future" (Telenti et al. 2021).

7.2.2.6 Vaccination and the Pandemic

Vaccines provide the most effective public health measures against infectious disease (Andre et al. 2008). The track record of vaccines has given hope that SARS-CoV-2 may soon get under control (Anderson et al. 2020). The quick and concerted actions and research in the pandemic led to development of vaccines in an unexpected manner and in an unprecedented timeframe. According to WHO (2021), 76 vaccine candidates based on several different platforms are being currently evaluated in human clinical trials, while another 182 candidates are under investigation in preclinical models. Four well-known SARS-CoV-2 vaccines have already been licensed by the regulatory agencies including the two developed by Moderna (mRNA-1273) and Pfizer/BioNTech (BNT162b2) pharmaceutical companies, and the other two developed by Oxford University/AstraZeneca (AZD1222) and Janssen (Ad26.COV2-S) pharmaceutical company. Israel was the first country to demonstrate that vaccines can bend the curve of infections. It was a leader in early vaccinations, where more than 84% of people aged over 70 received two doses by February 2021. As a result, cases declined there rapidly. This pattern of vaccination and recovery was repeated across dozens of other cities and nations.

However, in terms of vaccination against COVID-19, there are many concerns. One potential concern is the low rate of vaccine production and administration (Our World in Data 2021; Bloomberg Vaccine Tracker 2021). Currently, the biggest vaccination campaign in history is underway. More than 8.9 billion doses have been administered across cities and communities of 184 countries, according to data collected by Bloomberg Vaccine Tracker (2021). The latest rate was roughly 36.8 million doses per day. However, it is not the administration of vaccine doses; it is rather unequal access to vaccines as well as varying degrees of efficiency in getting people vaccinated that is the cause of concern. For example, before March 2021, few African nations had received even a single shipment of shots. In contrast, 151 doses had been administered for every 100 people in the USA (Bloomberg Vaccine Tracker 2021). Delivering billions of vaccines to stop the spread of COVID-19 worldwide is posing one of the greatest logistical challenges today. "At the current pace of 10.5 million people getting their first shots each day, the goal of halting the pandemic remains elusive. Manufacturing capacity, however, is increasing, thanks to new vaccines and added capacity from existing drug makers" (Bloomberg Vaccine Tracker 2021).

The emergence of COVID-19 (SARS-CoV-2) variants in different areas of the globe is another big cause of concern to both human health and vaccination. Khan et al. (2021) have listed some of the SARS-CoV-2 variants—B.1.1.7, Alpha; B.1.351, Beta; P.1, Gamma; B.1.617, Delta; B.1.617.2, Delta-plus; B.1.525, Eta; B.1.429, Epsilon; etc. A new variant recently discovered in South Africa is Omicron. All the variants have some common characteristics of a higher transmissibility, becoming dominant within populations in a short time span, and an accumulation of a high number of mutations in the spike (S) protein—with implications on virus infection rates and reinfection frequency (Gómez et al. 2021). Major efforts are ongoing for the control of the emerging variants by different nations and institutions. Among institutions World Health Organization (WHO), Coalition for Epidemic Preparedness Innovations (CEPI), Bill and Melinda Gates Foundation, Global Alliance for Vaccines and Immunizations (GAVI), and others are striving both for the provision of universal access to vaccines and to control virus infection. The general indication is that current and incoming vaccines will cope with the control of variants (Mahase 2021) and work for the potential eradication of the virus as well. However, it will be only through the detailed understanding of the virus structure, biology, and vaccine developments that the control of the virus infections is likely to be achieved (Gómez et al. 2021).

The virus and its variants have so far triggered a multi-wave pandemic that has infected 278 million people globally and killed over 5.38 million people as of December 22, 2021. The first wave of the pandemic reached its peak between early spring and early summer of 2020, when most cities introduced strict lockdown measures to increase physical distancing. A significant reduction of COVID-19 cases in late summer and early fall of 2020 led many cities and nations to believe that the pandemic was under control and could be defeated. However, the quick rebounding infection rates in October–December 2020 induced a second COVID-19 wave in late fall and winter 2020/2021 with even more serious health impacts compared to the first wave particularly in the USA and Brazil, the cities of which showed high peaks. The start of vaccination campaigns in early 2021 was widely interpreted as "light at the end of the corona-tunnel" and relaxation of lockdown measures. At the same time, variants of the virus spread out; among these were British, South African, and Brazilian variants, which brought a much higher infection rate and a third wave with rapidly increasing infection rates in some countries like India (highest peak in the third wave), France, and Germany, while countries like the USA and the UK applied successful vaccination strategies or vaccine protectionism by maximizing vaccination of their population. In early December 2021, a new virus variant Omicron was discovered in South Africa, which brought the beginning of the fourth wave in that country. This highly contagious virus has already spread to several countries and according to CNN (2021) became the dominant strain in a host of countries including the USA, Denmark, Portugal, and the UK and has raised urgent need to vaccinate people.

"Vaccine hesitancy" or the reluctance of people to receive safe and recommended available vaccines is also considered as a hurdle in achieving sufficient immunization coverage to end the pandemic. Vaccination, in general, was a growing concern even before the COVID-19 pandemic (Macdonald 2015). However, promoting the uptake of vaccines (particularly those against COVID-19) demands understanding of certain issues including (a) whether people are willing to be vaccinated, (b) the reasons why they are willing or unwilling to do so, and (c) the most trusted sources of information in their decision-making to get vaccinated. Solís Arce et al. (2021) investigated these issues using a common set of survey items deployed between June 2020 and January 2021, across 15 studies carried out in Africa, South Asia, Latin America, Russia, and the USA.

One important result the study brought out was the fact that the vaccination acceptance rate in every low- and middle-income countries (LMIC) sample was higher than those from the USA (64.6%, CI 61.8–67.3%) and Russia (30.4%, CI 29.1–31.7%). In LMIC reported acceptance was lowest in Burkina Faso (66.5%, CI 63.5–69.5%) and Pakistan (survey 2; 66.5%, CI 64.1–68.9%). According to the study, Pakistan's relatively low acceptance rate could be linked to negative historical experiences with foreign-led vaccination campaigns, and in Burkina Faso, it might reflect general vaccine hesitancy (Solís Arce et al. 2021). The most important reasons for vaccine acceptance were personal or family protection—protecting one's community did not feature prominently among stated reasons for acceptance. The main reason for nonacceptance of vaccine in both developed countries (the USA and Russia) and LMIC was side effects. In terms of most trusted source of guidance on decision-making identified for taking a COVID-19 vaccine or not in both developed and LMIC was healthcare system and health workers. The next most trusted source mentioned was family and friends. The respondents in the study rarely cited conspiracy theories or ulterior motives on the part of corporations, politicians, or the pharmaceutical industry. However, some other studies have identified fears related to these issues as a cause in nonacceptance of vaccines in higher-income countries (Loomba et al. 2021).

Hence, messaging about vaccine efficacy is likely to play the most important role in alleviating hesitancy or persuading people to accept vaccine in the future. Therefore, the need and urgency for proactive messaging before large-scale vaccination campaign rollout can hardly be overemphasized. Media particularly need to play responsible role in stressing the high efficacy rates of the COVID-19 vaccines in reducing or eliminating disease, hospitalizations, and death. It also needs to responsibly communicate accurate information on potential side effects, including the rarity of severe adverse events that may have been blown out of proportion by some in media coverage (Puri et al. 2020). It is also important to inform the unvaccinated people worldwide that they are more at risk than ever and that the pandemic has now become the "pandemic of the unvaccinated."

7.3 Climate Change and Health: Impacts and Response

Along with coronavirus, climate change is a "mega health threat" to cities in the twenty-first century. Due to the complexity of the processes involved, the exact magnitude to which health may be influenced by the changing climate remains unknown. However, like other natural and human-made factors, climate change will influence human health and disease in numerous ways—various existing health threats will intensify, and new health risks are likely to emerge.

7.3.1 Impacts on Health

Based on current knowledge and future projections, the climate is changing and that health in general (and urban health in particular) is and will continue to be affected by this change (PAHO and WHO 2017). The impacts of climate change, nevertheless, will vary across cities—depending on their development status, demographic character, socioeconomic condition, infrastructural and institutional capacity, and health system resiliency. Certain populations, such as children, pregnant women, older adults, and people with feeble and poor health, will face increased risks (EPA 2017). Thus, there will be variation in terms of the strength and severity of

associated health impacts. It is, however, extremely important to monitor and collect health risk information due to climate change for putting early warning systems in place and undertaking appropriate interventions through adaptation planning.

There are various ways in which climate change may affect human health in cities (Fig. 7.1). Firstly, it may exacerbate the intensity or severity of a health problem or disease as well as its frequency of occurrence. Secondly, it may create a new health threat or generate an unprecedented or unanticipated diseases or health problems that may not have occurred previously. Broadly, climate change-related risks can be classified into direct or indirect impacts as follows:

Direct
- Mortality and morbidity due to temperature extremes particularly extreme heat
- Injuries, mortality, and morbidity due to extreme climate-related events or severe weather

Indirect
- Food and waterborne diseases
- Food and water supply insecurity—hunger and malnutrition and unsafe water consumption
- Vector-borne diseases
- Impacts on air quality
- Increase in allergens
- Occupational health
- Stress and psychological problems

7.3.1.1 Temperature-Related Impacts
Temperature is rising in cities with adverse effects. There are three dangers to human health from rising temperatures: (a) higher risk of heat strokes as well as cardiovascular, respiratory, and cerebrovascular disease particularly during heat waves, (b) some illness worsens like asthma or diabetes where patients may need oxygen or dialysis, and (c) outdoor workers such as those involved in construction and farmers, who may not be able to tolerate extended exposure and may suffer from cardiac failure and strokes. Heat

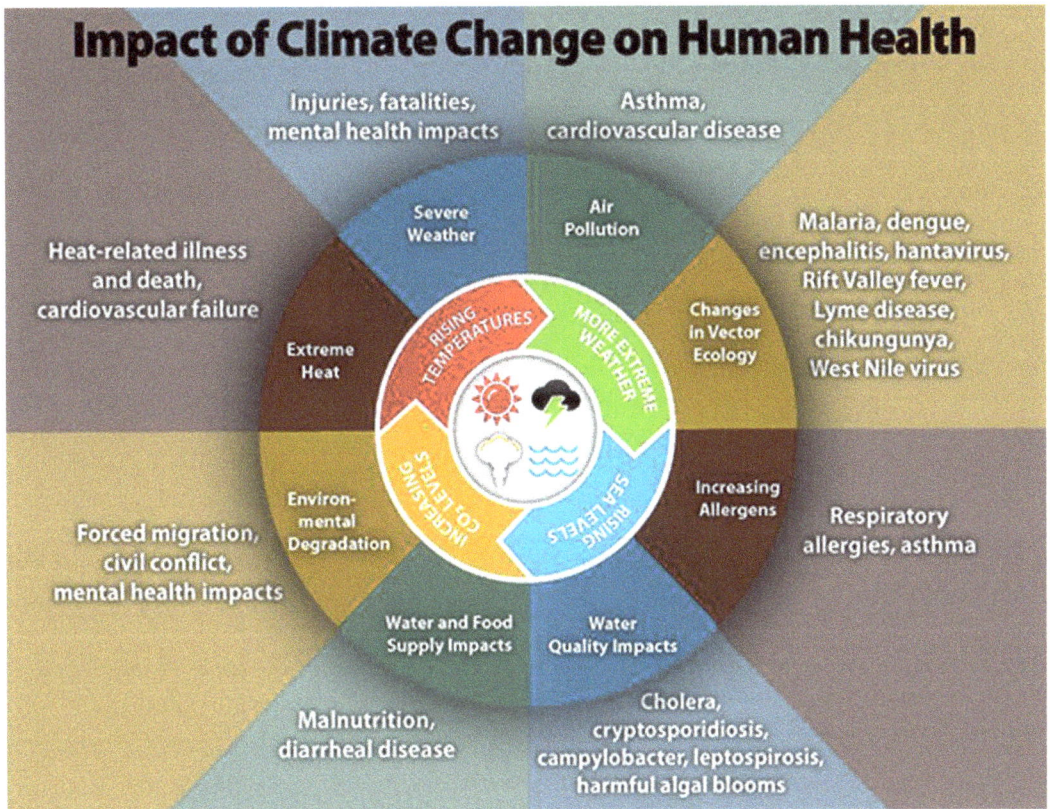

Fig. 7.1 Direct and indirect impact of climate change on human health. (Source: CDC (2021))

waves until recently have not been treated like a state of emergency but more like a seasonal nuisance (Naumova 2018). However, in 2003, a heat wave in Europe resulted in over 70,000 deaths (Robine et al. 2008). Similarly, a heat wave in Chicago in 1995 had adverse impacts (Fig. 7.2). Indirect impacts of high temperatures are less obvious or occur over longer periods. For example, in 2003 a heat-related electricity blackout in the northeastern USA led to failures of emergency generators in hospitals, untreated sewage, and food contamination from losses of refrigeration, resulting in additional deaths and illnesses in New York City (NCEI 2017).

7.3.1.2 Extreme Climate-Related Events

Climate change has resulted in changes in the frequency, intensity, and geographic distribution of extreme climatic events, and it will continue to be a key factor of change in the future. Some of these events include heat waves, droughts, wildfires, dust storms, rainfall regimes and intensity, coastal flooding, storm surges, and hurricanes. The pathways connecting extreme events to health outcomes are diverse and complex. The difficulty in predicting these relationships depends on the local societal and environmental factors that affect disease burden. Bell et al. (2018) have given a description of observed and projected changes in extreme events and their general health effects, providing a concept of observed associations and potential mechanisms, and exemplified these by specific events. The impacts of extreme climate events can be direct or indirect and either short or long term. Direct and short-term impacts can be death, injury, or disease. However, it could also result in trauma after disasters that can last longer

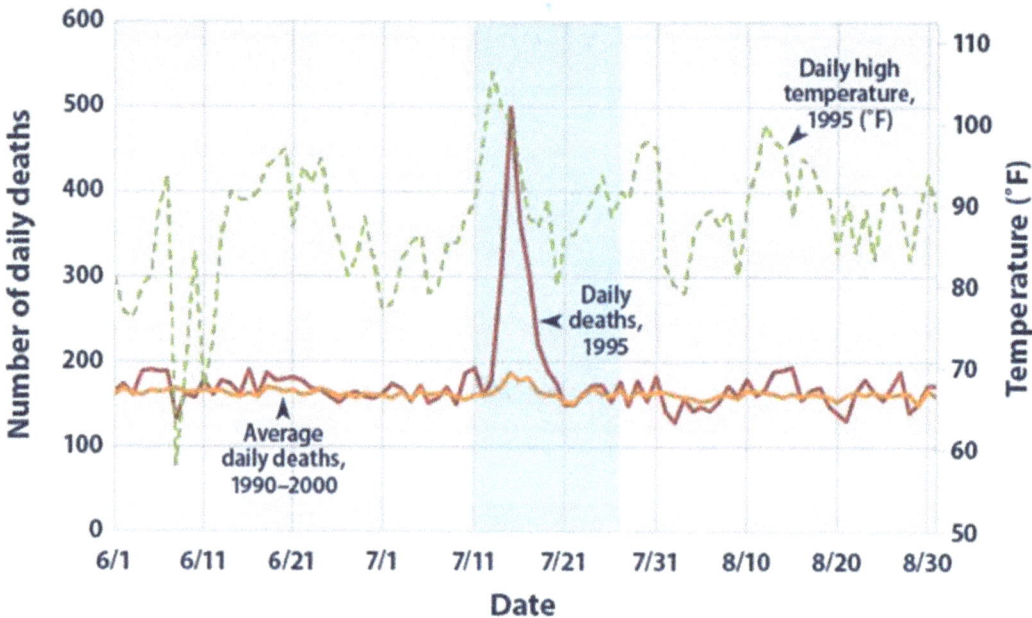

Fig. 7.2 Chicago metropolitan statistical area: the relationship between high temperatures and deaths observed during the 1995 Chicago heat wave. (Source: EPA (2017))

and lead to severe stress and mental illness— flooding in High River, Alberta, in 2013, resulted, for example, in several adverse mental health outcomes (Sahni et al. 2013).

7.3.1.3 Foodborne and Waterborne Diseases

Climate change can exacerbate food and water safety risks in a number of ways. For instance, illnesses from pathogens such as *Salmonella* and *Campylobacter* are generally more common when temperatures are higher. Rapid snowmelt or extreme rainfall events can also cause sporadic increases in streamflow rates; along with changes in water treatment, these increases have been linked to an increase in waterborne pathogens (Luber et al. 2014).

7.3.1.4 Food and Water Supply Insecurity

Climate-related disruptions in the food production system can indirectly impact human health by diminishing food security and through pathways of nutrition (Schnitter and Berry 2019). The

World Health Organization (WHO) estimates that without adaptation, climate change will result in 7.5 million more cases of stunted children by 2030 and 10.1 million by 2050 (WHO 2014). Furthermore, an additional 4.8 million cases of undernutrition in children under 5 years of age is projected by 2050 due to climate change (Martinez and Berry 2018). Currently, world hunger is on the rise and climate-related shocks have exacerbated this trend (FAO. IFAD. UNICEF. WFP. WH 2017).

Climate change also poses the most significant challenge to urban water security, sanitation, and health, which move in tandem. Risks to health-related water supply emerge from damage to water and sanitation infrastructure from flooding, loss of water sources due to declining and changing rainfall pattern, and sea level rise. Alderman et al. (2012) found infectious disease epidemics tended to occur more during mass population displacement by floods and that there was good evidence of increased water-related diseases after floods. Leptospirosis was identified as causing epidemics during floods and as a key post-flood pathogen with cholera, hepatitis A and E, and pathogenic *E. coli*. Carlton et al. (2014) found that rainfall was associated with diarrhea incidence and water treatment-reduced incidence in Ecuador.

7.3.1.5 Vector-Borne Diseases

Climate change can affect the transmission dynamics, geographic spread, and re-emergence of vector-borne diseases through multiple pathways, including direct effects on the pathogen, the vector, nonhuman hosts, and humans. In addition to having direct effects on individual species, climate change can alter entire habitats (including urban habitats), in which vectors or nonhuman hosts may thrive or fail. Regional and local impacts of climate change on vector-borne diseases point to the need for vigilance (Rocklöv and Dubrow 2020). There are many signals to indicate that climate change has already affected or is likely to affect vector-borne disease transmission or spread. A Wildlife Conservation Society (2020) report listed 12 pathogens that could spread into new regions because of climate change, with potential impacts to both human and wildlife health. Referred to as deadly dozen, these include bird flu or avian influenza, babesiosis, cholera, Ebola, intestinal and external parasites, Lyme disease, plague, Rift Valley fever, sleeping sickness, tuberculosis and yellow fever, and toxic algal blooms or red tides. The 2019 Lancet Countdown report (Watts et al. 2019) emphasizes the continued upward trend of climate suitability for transmission of other diseases like dengue—with 9 of the 10 most suitable years occurring since the year 2000 (Fig. 7.3). Malaria suitability also continues to increase in highland areas of Africa according to the report. Siraj et al. (2014) also provide evidence for an increase in the altitude of malaria distribution in warmer years, which implies that climate change will, without mitigation, result in an increase of the malaria burden in the towns or areas located on highlands of Africa and South America.

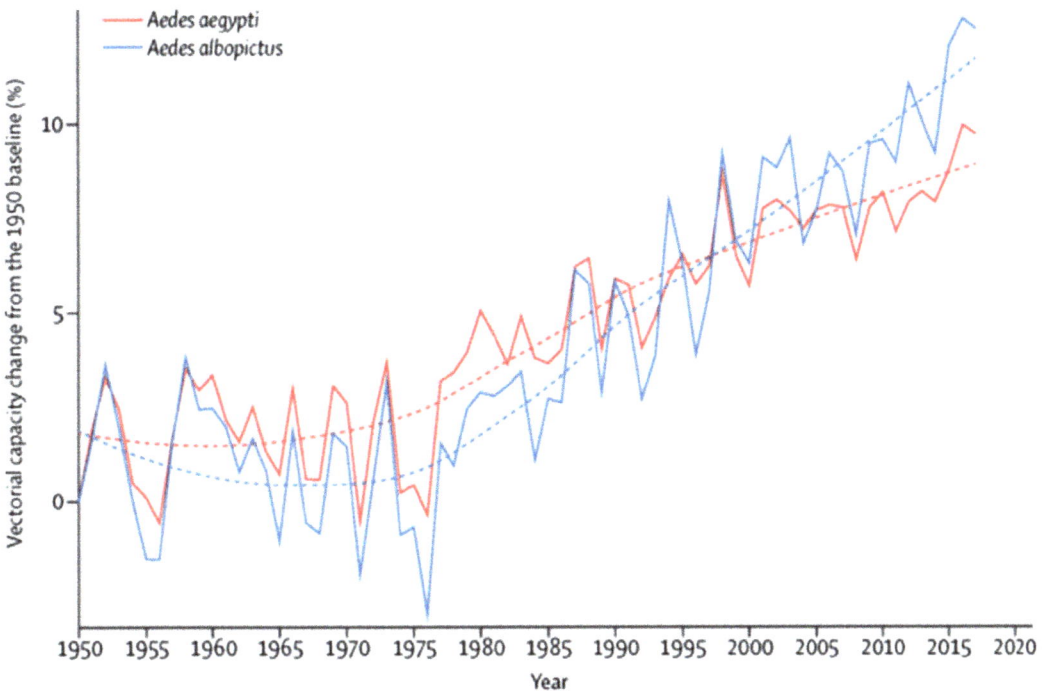

Fig. 7.3 Changes in global vectorial capacity for the dengue virus vectors *Aedes aegypti* and *Aedes albopictus* since 1950. (Source: Watts et al. (2019))

7.3.1.6 Climate Change Impacts on Air Quality and Aeroallergens

Changes in the climate affect the air quality in three ways—outdoor air pollution, aeroallergens, and indoor air pollution. The changing climate has modified weather patterns, which in turn have influenced the levels and location of outdoor air pollutants such as ground-level ozone, fine particulate matter, and carbon dioxide. For example, increased future wildfire risks have been predicted for Russia, the western USA, and the European Mediterranean area (Spracklen et al. 2009). Westerling and Bryant (2008) predicted a 10–35% increase in large fire risk by midcentury in California and Nevada, depending on the greenhouse gas emission scenario and Global Change Model (GCM) used. Moreover, increasing carbon dioxide levels, which are triggering climate change, promote the growth of plants that release airborne pollen and allergens (aeroallergens). Increased aeroallergens and ambient air pollutants (including ozone, nitrogen dioxide, sulfur dioxide, and fine particulate matter) may increase health risks for vulnerable population such as aged and people already suffering from chronic diseases like asthma and heart problems. Finally, these changes to outdoor air quality and aeroallergens also affect indoor air quality as both pollutants and aeroallergens infiltrate homes, schools, and other buildings (Fann et al. 2016).

7.3.1.7 Occupational Health

Outdoor workers from occupational sectors in urban settings that include construction, transportation, oil production, landscaping, firefighting, and other emergency response operations are often among the first to be affected by climate change. Individuals exposed to hot indoor work environments, such as steel mills, dry cleaners, manufacturing facilities, commercial kitchens, and warehouses, are also at risk for climate change impacts, including extreme heat exposure and indoor air pollution (MISPH 2017).

A number of studies have been conducted on the impacts of climate change on workers' health in both developing and developed countries—Ghana, South Africa, Saudi Arabia, Germany, Australia, the USA, Italy, and India. An analysis (Moda et al. 2019) of 32 of these papers published between 2002 and 2019 on the influence of climate change and heat stress, workplace injury, and work productivity could be grouped as climate change impacts on outdoor workers' safety and health. These studies covered urban heat islands and their impacts on outdoor workers, heat stress and outdoor workers' performance, and occupational health hazards and effects related to climate change. The impacts of climate change on workers' health was the major commonality, while their major differences were in the interventions considered. Broad findings from the studies revealed that exposure to extreme heat due to climate change is associated with negative health impacts and possible decreases in productivity.

7.3.1.8 Stress and Psychological Problems

Climate change-related impacts on mental health can be both direct and indirect. Direct or immediate and short-term effects could be due to heat waves, and indirect in the short term could be due to extreme climatic events such as floods, wildfire, or hurricanes. Indirect long-term impact could result from prolonged droughts, persistent urban heat island, increase in the sea level, forced migration, etc. On the basis of review of several papers and reports, Paolo et al. (2020) concluded that "all these events affect the mental health of a population, with the appearance of psychiatric conditions such as Posttraumatic Stress Disorder (PTSD), mood disorders such as depression, anxiety, increased suicide rate and substance use, as well as increased aggressive behavior." Climate change according to them will also exert the greatest impact on groups of vulnerable populations that have an increased probability of developing psychopathologies. The groups with higher

On reducing emissions:

- Assess the health system's emissions
- Reduce emissions of health-care facilities
- Monitor the hospitals' and health systems' climate footprint
- Identify environmentally friendly suppliers
- Implement green purchasing policies

On strengthening health system resilience:

- Develop vulnerability and adaptation assessment plans
- Map climate risks and vulnerable populations
- Prepare Early Warning Systems
- Implement emergency preparedness and disaster risk management plans

Fig. 7.4 Mitigation and adaptation actions for health. (Source: PAHO and WHO (2017))

risk include women, the elderly, children, people with previous psychiatric illnesses who can consequently worsen their mental condition, and people with low income or poor social network, as well as indigenous and native communities. Extreme weather events, according to another study (Torres and Casey 2017), also have the power to destroy social ties.

7.3.2 Response to Climate Change Impacts on Health

7.3.2.1 Mitigation and Adaptation Actions

Increasing risks of climate change-related diseases and injuries in cities demand that urban health systems should prepare for these exacerbated risks through mitigation and adaptation actions (Fig. 7.4). Climate risk information and early warning systems for adverse health outcomes are necessary to enable appropriate interventions and prepare mitigation and health adaptation plans. Improving basic public health and healthcare services, developing and implementing early warning systems, and training citizens' groups in disaster preparedness, recovery, and resilience constitute effective adaptation measures in the short term. Likewise, conducting vulnerability, capacity, and adaptation assessments and developing the health component of city or regional adaptation plans are equally important in the long run. "A vulnerability, capacity, and adaptation assessment provides evidence-based information on current associations between weather and climate and health outcomes, identifies vulnerable populations, projects future risks, and identifies opportunities to build resilient communities and health systems" (Ebi and Hess 2020). The assessments also help in determining program priorities to address the health risks of climate change in adaptation plans. Progress on inclusion of health and climate change in adaptation to date has been slow. It has been affected by negligible funding allocated to health adaptation (which totals less than 0.5% of international climate adaptation finance), leading the 2018 Adaptation Gap Report to conclude that "there is a significant global adaptation gap in health, as efforts are well below the level required to minimize negative health outcomes" (UN DTU partnership 2018).

The scientific literature on adaptation to climate change has gradually increased, but investment "in specific health protection activities is growing less rapidly" (Smith et al. 2014). There is a need to increase this investment to strengthen urban health resilience. Further, the public health sector, municipal governments, and the climate change community should work together to integrate health as a key goal in the policies, plans, and programs of all cities. Climate change and health interactions should also be made clear to public health practitioners, city planners, policy-

Fig. 7.5 Building
Resilience Against
Climate Effects:
framework of the US
CDC initiative. (Source:
CDC (2019))

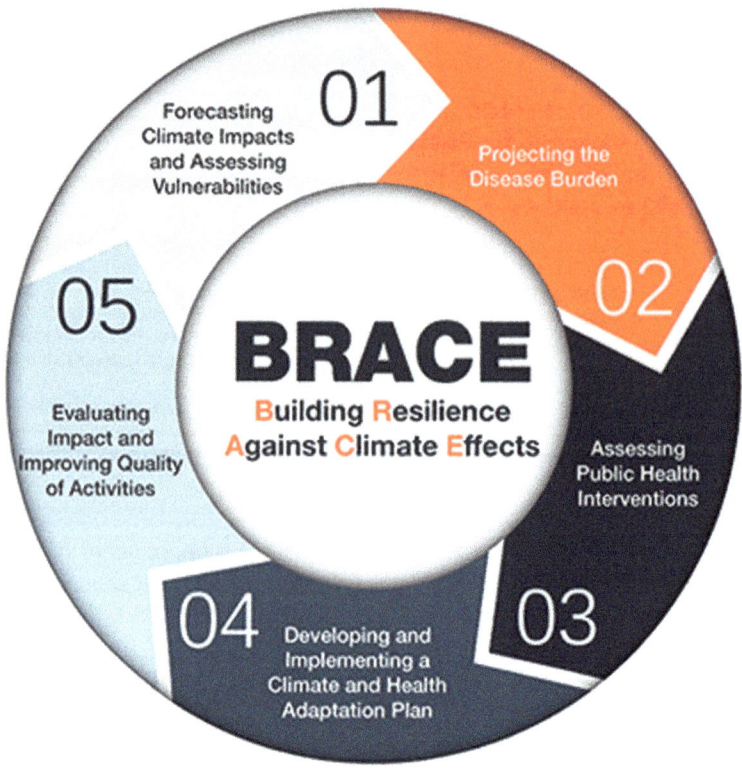

makers, and the general public (PAHO and WHO 2017).

7.3.2.2 Building Urban Climate Resilience

Public health expertise also need to be used effectively to prepare for and respond to the health effects that a changing climate may bring to their communities. This is already being done in the USA where the Centers for Disease Control and Prevention (CDC) has started Climate-Ready States and Cities Initiative (CRSCI), which is helping a network of 16 states and two cities to use the five-step Building Resilience Against Climate Effects (BRACE) framework (Fig. 7.5) to identify likely climate impacts in their communities, potential health effects associated with these impacts, and their most at-risk populations and locations. The BRACE framework is helping cities and states develop and implement health adaptation plans and identify and address gaps in critical public health functions and services.

7.4   Conclusion

Despite considerable progress made in the medical sciences, the emergence and re-emergence of novel viruses causing epidemic and pandemic continue to remain a persisting problem to human health in the twenty-first century. Although many pandemics have occurred in human history with serious repercussion on human life, recent emergence of COVID-19 pandemic and re-emergence of its waves provided a clear example of complications that a pandemic can afflict on human health and well-being even in the modern world. Since its first appearance in December 2019, in Wuhan (China), a city of over 10 million people, the virus quickly spread to most cities of the world and became an international crisis—by December 23, 2021, it had affected 278 million people and resulted in over 5.38 million deaths worldwide.

As centers of large gatherings and concentration of population, the cities offered ideal settings for the spread of the virus and became the major battlegrounds of the crisis facing once-in-a-century test to global resilience. The most important aspect in the origin and diffusion of virus in cities was how the first "clusters" of cases originated and spread. In many instances, large cities, with their strong international links due to increased globalization—including business travel, tourism, marketing, etc.—were the entry points for the virus. The impact of the COVID-19 varied among cities and even within different parts of the same city depending mainly on response measures. Some cities such as Hong Kong, Taipei, Singapore, and Hanoi in Vietnam and cities of South Korea, Japan, and New Zealand were able to avoid major outbreaks at the outset by rapidly restricting travel and scaling up testing, tracking, surveillance, and quarantining. Preparedness system (Muggah and Katz 2020) was a great help to some—for example, Hong Kong, Taipei, and Singapore had preparedness systems, which were built over years, after their failures to stop an earlier dangerous outbreak of SARS, 17 years ago. In contrast, the USA disbanded its pandemic response unit in 2018 and its cities suffered the worst outbreaks.

The implementation of response measures was often more difficult in cities of developing countries, where crowded living conditions made social distancing difficult if not impossible. Hygienic practices were also difficult there in the wake of lack of access to basic utilities such as water, sanitation, and supplies that are taken for granted in developed countries, like soap and disinfectants. However, even in developing countries, there are good examples of the cities (e.g., in the state of Kerala within India—the second largest country affected by COVID-19 after the USA) that had one of the lowest infections and death rates due to early implementation of effective social distancing and quarantine measures. Deep-rooted inequalities, even in cities of developed countries like the USA and Europe, made the pandemic affect disproportionately the urban poor and certain minority ethnic and migrant groups living in crowded and unhealthy conditions. Cities in many developed countries also suffered because they were slow to respond or complacent, with some expecting that the virus would bypass them or believing that "herd immunity" would protect them. As a result, they became most exposed to the virus and its effects. A few influential government leaders, as in the USA and Brazil, refused to acknowledge the severity of the COVID-19 threat, distracting many with conspiracy theories and "blame games." The result was a big problem in terms of spread of infections and many deaths—a price paid by the cities in the USA and Brazil.

The pandemic has taught several lessons for the management of healthcare system. The biggest lesson for the system is to have adequate preparedness with the emergency management as an integrated part of operational strategy. It has also highlighted the need of vigilance—meaning alertness of health systems and hospitals and maintaining essential capabilities to occurrence and resurgence of novel infectious diseases. In terms of management, the most important lesson is never to be complacent. Resources should be mobilized immediately and urgently with the detection of first case(s) to test, isolate, and trace the infected population and to put them in quarantine. At the same time, communication channels with public need to be established and strengthened continuously to provide important information while controlling spread of misinformation. Simultaneously, dissemination of community best practices should be promoted—such as wearing high-quality masks, social distancing, and on carrying out day-to-day activities related to work, shopping, health, education, as well as overcoming the social isolation.

The coronavirus crisis led to the rollout of several highly effective SARS-CoV-2 vaccines in cities and nations, since the beginning of 2021, which raised the hopes that the virus will vanish. However, three factors tempered the general expectations: firstly, the patchy vaccine coverage and inequalities in global access to vaccines; secondly, the hesitancy to vaccination; and thirdly, the emergence of COVID-19 variants in different areas of the globe. The virus and its variants triggered three pandemic waves and the fourth was in making at the end of 2021 due to Omicron variant since its discovery in South Africa. Regarding the first problem, delivering billions of vaccines to stop the spread of COVID-19 worldwide is posing one of the biggest logistical challenges. At the current pace of 10.5 million people getting their first shots each day, the goal of halting the pandemic remains elusive. Manufacturing capacity, however, is increasing, thanks to new vaccines and added capacity from existing drug makers. Major efforts are also ongoing for the control of the emerging variants by different nations and international institutions. They are striving both for the provision of universal access to vaccines and to control virus infection. Regarding the variants, there is a general indication that the current and incoming vaccines will cope with the control of variants and work for the potential eradication of the virus as well. However, it will be only through the detailed understanding of the virus structure, biology, and vaccine developments that the control of the disease is likely to be achieved. As far as the vaccine hesitancy, messaging particularly through media and leaders is likely to play the most important role in persuading those who may be hesitant. Since the most important concern of those who are hesitant to get the vaccine are side effects and vaccine efficacy, the urgency of proactive positive messaging before large-scale vaccination campaign rollout, can hardly be overemphasized.

Experts generally believe that the pandemic outbreaks will be tamed once the bulk of global population (90–95%) develop a degree of immunity either by vaccination or previous infection. However, as its previous outbreaks have indicated, the pandemic is likely to end at different times in different places. While some countries are going for zero COVID-19 cases, the world is unlikely to eradicate the virus completely (Cortez 2021). Medical experts, including White House chief medical advisor Dr. Anthony Fauci, agree that coronavirus may never completely vanish. Among countries, Denmark and Singapore have managed to keep cases relatively contained, and they are already moving toward a post-pandemic future with fewer safety restrictions. Meanwhile, China, Hong Kong, and New Zealand have vowed to keep vigilantly working to eliminate the virus locally. As a result, they are likely to be among the last places to leave behind the disruption wrought by walling out the pandemic (Cortez 2021).

Along with the COVID-19 pandemic, climate change is also a "mega health threat" to cities. Climate change is affecting and will influence human health in numerous ways—by exacerbating the intensity or severity of a health problem or disease as well as its frequency of occurrence. It may create a new health threat or generate an unprecedented or unanticipated disease or health problems that may not have occurred previously. A Wildlife Conservation Society (2020) report listed 12 pathogens that could spread into new regions due to climate change, with potential impacts to both human and wildlife health.

An effective public health response to climate change is essential to preventing injuries and illnesses, enhancing public health preparedness, and reducing risk. The scientific literature on adaptation to climate change is on the increase, but investment in specific health protection activities is growing less rapidly. There is a need to increase investments to strengthen urban health resilience. The public health sector, municipal governments, and the climate change community also need to work together to integrate health as a key goal in the policies, plans, and programs for city development. Connections between climate change and health should also be made clear to public health practitioners, city planners, policymakers, and the public.

References

Acuto M, Larcom S, Keil R et al (2020) Seeing COVID-19 through an urban lens. Nat Sustain 3(12):977–978. https://doi.org/10.1038/s41893-020-00620-3

Alderman K, Turner LR, Tong S (2012) Floods and human health: a systematic review. Environ Int 47:37–47. PMID: 22750033. https://doi.org/10.1016/j.envint.2012.06.003

American Hospital Association (AHA) (2020) Hospitals and health systems face unprecedented financial pressures due to COVID-19. https://www.aha.org/guidesreports/2020-05-05-hospitals-and-health-systems-face-unprecedented-financial-pressures-due

Anderson RM, Vegvari C, Truscott J, Collyer BS (2020) Challenges in creating herd immunity to SARS-CoV-2 infection by mass vaccination. Lancet 396:1614–1616

Andre FE et al (2008) Vaccination greatly reduces disease, disability, death, and inequity worldwide. Bull World Health Organ 86:140–146

APM (2020) The colour of coronavirus. go.nature.com/349uvhd

Aschwanden C (2021) Five reasons why COVID herd immunity is probably impossible. Nature 591:520–522

Bai X, Nagendra H, Shi P, Liu H (2020) Cities build networks and share plans to emerge stronger from COVID-19. Nature 584:517–520

Barron L (2020) What We Can Learn from Singapore, Taiwan and Hong Kong about Handling Coronavirus, Time Magazine, March. https://time.com/5802293/coronavirus-covid19-singapore-hong-kong-taiwan/

Bartsch SM, Ferguson MC, McKinnell JA, O'shea KJ, Wedlock PT et al (2020) The potential health care costs and resource use associated with COVID-19 in the United States. Health Affairs 39(6):927–935. https://doi.org/10.1377/hlthaff.2020.00426

Beech H (2020) Tracking the coronavirus, how crowded Asian cities tackled an Endemic. The New York Times. https://www.nytimes.com/2020/03/17/world/asia/coronavirus-singapore-hong-kong-taiwan.html

Bell JE, Brown CL, Conlon K, Herring S, Kunkel KE, Lawrimore J, Luber G, Schreck C, Smith A, Uejio C (2018) Changes in extreme events and the potential impacts on human health. J Air Waste Manage Assoc 68(4):265–287. https://doi.org/10.1080/10962247.2017.1401017

Bloomberg Vaccine Tracker (2021). https://www.bloomberg.com/graphics/covid-vaccine-tracker-global-distribution/

Boyles S (2020) SCCM: too few ventilators, ICU beds available for worst-case COVID-19 scenario, Society of Critical Care Medicine (SCCM). https://www.medpagetoday.com/infectiousdisease/covid19/85462

Budd J, Miller BS, Manning EM et al (2020) Digital technologies in the public-health response to COVID-19. Nat Med 26:1183–1192. https://doi.org/10.1038/s41591-020-1011-4

Burki T (2020) Global shortage of personal protective equipment. Lancet Infect Dis 20(7):785–786. https://doi.org/10.1016/S1473-3099(20)30501-6

Carlton EJ, Eisenberg JNS, Goldstick J, Cevallos W, Trostle J, Kevy K (2014) Heavy rainfall events and diarrhea incidence: the role of social and environmental factors. Am J Epidemiol 173(3):344–352. https://doi.org/10.1093/aje/kwt279

CDC (2019) CDC's building resilience against climate effects (BRACE) framework. https://www.cdc.gov/climateandhealth/BRACE.htm

CDC (2020a) Identifying the source of the outbreak. https://www.cdc.gov/coronavirus/2019-ncov/cases-updates/about-epidemiology/identifying-source-outbreak.html

CDC (2020b) Health equity considerations and racial and ethnic minority groups. go.nature.com/37fffny

CDC, Center for Disease Control and Prevention, USA (2021) Climate effects on health. https://www.cdc.gov/climateandhealth/effects/default.htm

Cheney C (2020) How to improve emergency preparedness for pandemics. HealthLeaders. https://www.healthleadersmedia.com/clinical-care/how-improve-emergency-preparedness-pandemics

Chowdhury AZ, Jomo KS (2020) Responding to the COVID-19 pandemic in developing countries: lessons from selected countries of the global south. Development 63:162–171. https://doi.org/10.1057/s41301-020-00256-y

CNN (2021) The latest on the coronavirus pandemic and the Omicron variant. https://www.cnn.com/world/live-news/omicron-variant-coronavirus-news-12-21-21-intl/index.html

Cohn SK (2002) The Black Death transformed: disease and culture in early Renaissance Europe. Arnold, London

Cortez MF (2021) Here's what the next six months of the pandemic will bring. Bloomberg. https://www.bloomberg.com/news/features/2021-09-12/6-month-covid-outlook-2021

Cutler DM, Summers LH (2020) The COVID-19 pandemic and the $16 trillion virus. JAMA 324(15):1495–1496. https://doi.org/10.1001/jama.2020.19759

Dollar D (2020) China recovers first – with what lessons. In: Allen JR, West DM (eds) Reopening the world. Brooking Institutions, https://www.brookings.edu/wp-content/uploads/2020/06/Brookings-Reopening-the-World-FINAL.pdf

Ebi KL, Hess JJ (2020) Health risks due to climate change: inequity in causes and consequences. Health Affaire 39(12):2056–2062. https://doi.org/10.1377/hlthaff.2020.01125

ECDC (European Center for Disease Prevention and Control) (2020) COVID-19 data platform. https://www.ecdc.europa.eu/en/covid-19/data

EPA, US Environmental Protection Agency (2017) Climate impacts on human health. https://archive.epa.gov/epa/climate-impacts/climate-impacts-human-health.html

Fann N, Brennan T, Dolwick P, Gamble JL, Ilacqua V, Kolb L, Nolte CG, Spero TL, Ziska L (2016) Chapter 3: air quality impacts. In: The impacts of climate change on human health in the United States: a scientific assessment. U.S. Global Change Research Program, Washington, DC, pp 69–98. https://doi.org/10.7930/J0GQ6VP6

FAO. IFAD. UNICEF. WFP. WH (2017) The state of food security and nutrition in the world: building resilience for peace and food security. Food and Agriculture Organization of the United Nations, Rome, Italy

Fertel I (2020) False perception of COVID-19's impact on the homeless. https://www.factcheck.org/2020/05/false-perception-of-covid-19s-impact-on-the-homeless/

Gera V (2020) Poland turning national stadium into COVID 19 field hospital. ABC News. https://abcnews.

go.com/Health/wireStory/poland-turning-national-stadium-covid-19-field-hospital-73691186

Gibson L, Rush D (2020) Novel coronavirus in Cape Town informal settlements: feasibility of using informal dwelling outlines to identify high risk areas for COVID-19 transmission from a social distancing perspective. JMIR Public Health Surveill 6(2):e18844. https://doi.org/10.2196/18844

Gilca R, Carazo S, Amini R, Charest H, De Serres G (2021) Relative severity of common human coronaviruses and influenza in patients hospitalized with acute respiratory infection: results from 8-year hospital-based surveillance in Quebec, Canada. J Infect Dis 223:1078–1087

Gómez CE, Perdiguero B, Esteban M (2021) Emerging SARS-CoV-2 variants and impact in global vaccination programs against SARS-CoV-2/COVID-19. Vaccines 9:243. https://doi.org/10.3390/vaccines9030243

Government of Brazil (2020) Covid-19 data platform, "Coronavírus Brasil", painel coronavírus COVID19. https://covid.saude.gov.br/

Hart K (2020) Cities face severe supply shortages amid coronavirus outbreak. https://www.axios.com/cities-supply-shortages-coronavirus-04c60c13-2dd5-4232-baab-4a9fa3e88aaf.html

Imtiaz A (2020) The law of generosity combating coronavirus in pakistan. BBC. http://www.bbc.com/travel/story/20200331-the-law-of-generosity-combatting-coronavirus-in-pakistan

Jiangtao S (2020) Beijing and Shanghai impose new controls on residents as China battles to contain coronavirus. South China Morning Post

JLL (2020) How cities mobilized to convert spaces into COVID-19 care facilities. https://www.us.jll.com/en/trends-and-insights/cities/how-cities-mobilized-to-convert-spaces-into-covid19-care-facilities

Kaye AD, Okeagu CN, Pham AD et al (2020) Economic impact of COVID-19 pandemic on healthcare facilities and systems: international perspectives [published online ahead of print, Nov 17]. Best Pract Res Clin Anaesthesiol 35(3):293–306. https://doi.org/10.1016/j.bpa.2020.11.009

Khan WH, Hashmi Z, Goel A, Ahmad R, Gupta K, Khan N, Alam I, Ahmed F, Ansari MA (2021) COVID-19 pandemic and vaccines update on challenges and resolutions. Front Cell Infect Microbiol 11:690621. https://doi.org/10.3389/fcimb.2021.690621

Kochańczyk M, Lipniacki T (2021) Pareto-based evaluation of national responses to COVID-19 pandemic shows that saving lives and protecting economy are non-trade-off objectives. Sci Rep 11:2425. https://doi.org/10.1038/s41598-021-81869-2

Levison ME (2020) COVID-19 challenges in developing countries. Merck Manual. https://www.merck-manuals.com/home/news/editorial/2020/07/08/20/55/covid-19-challenges-in-the-developing-world

Li R, Rivers C, Tan Q, Murray MB, Toner E, Lipsitch M (2020) The demand for inpatient and ICU beds for COVID-19 in the US: lessons from Chinese cities. medRxiv 2020. https://doi.org/10.1101/2020.03.09.20033241

Loomba S, de Figueiredo A, Piatek SJ, de Graaf K, Larson HJ (2021) Measuring the impact of COVID-19 vaccine misinformation on vaccination intent in the UK and USA. Nat Human Behav 5:337–348

Luber G, Knowlton K, Balbus J, Frumkin H, Hayden M, Hess J, McGeehin M, Sheats N, Backer L, Beard CB, Ebi KL, Maibach E, Ostfeld RS, Wiedinmyer C, Zielinski-Gutiérrez E, Ziska L (2014) Chapter 9: human health. In: Melillo JM, Richmond T, Yohe GW (eds) Climate change impacts in the United States: the third national climate assessment. US Global Change Research Program, pp 220–256. https://doi.org/10.7930/J0PN93H5

MacDonald NE; SAGE Working Group on Vaccine Hesitancy (2015) Vaccine hesitancy: definition, scope, and determinants. Vaccine 33(34):4161–4164. https://doi.org/10.1016/j.vaccine.2015.04.036

Maclean R, Marks S (2020) Ten African countries have no ventilators: that is only part of the problem. New York Times Apr 18, 2020. Updated May 17, 2020. https://www.nytimes.com/2020/04/18/world/africa/africa-coronavirus-ventilators.html

Mahase E (2021) Covid-19: where are we on vaccines and variants? BMJ 372:n597. https://doi.org/10.1136/bmj.n597

Martinez GS, Berry P (2018) The adaptation gap report. United Nations environment programme; Nairobi, Kenya: 2018. The Adaptation Health Gap: A Global Overview

McMahon DE, Peters GA, Ivers LC, Freeman EE (2020) Global resource shortages during COVID-19: bad news for low-income countries. PLoS Negl Trop Dis 14(7):e0008412. https://doi.org/10.1371/journal.pntd.0008412

Millett GA, Jones AT, Benkeser D et al (2020) Assessing differential impacts of COVID-19 on black communities. Ann Epidemiol 47:37–44

Ministry of Health, Chile (2020) Action plan coronavirus COVID-19, Confirmed cases in Chile COVID-19. https://www.minsal.cl/nuevo-coronavirus-2019-ncov/casos-confirmados-en-chile-Covid-19/

Ministry of Health Singapore (2020) Updates on covid-19 local situation. https://www.moh.gov.sg/docs/librariesprovider5/local-situation-report/situation-report%2D%2D23-june-2020.pdf

MISPH Milken Institute School of Public Health (2017) Hazard zone: the impact of climate change on occupational health. https://onlinepublichealth.gwu.edu/resources/impact-of-climate-change-on-occupational-health/

Moda HM, Filho WL, Minhas A (2019) Impacts of climate change on outdoor workers and their safety: some research priorities. Int J Environ Res Public Health 16:3458. https://doi.org/10.3390/ijerph16183458

Muggah R, Katz R (2020) How cities around the world are handling COVID-19 – and why we need to measure their preparedness. World Economic Forum. https://

www.weforum.org/agenda/2020/03/how-should-cities-prepare-for-coronavirus-pandemics/

Murthy S, Leligdowicz A, Adhikari NK (2015) Intensive care unit capacity in low-income countries: a systematic review. PLoS One 10(1):e0116949. Epub 2015/01/27. pmid:25617837; PubMed Central PMCID: PMC4305307

Nagendra H, Bai X, Brondizio ES et al (2018) The urban south and the predicament of global sustainability. Nat Sustain 1:341–349. https://doi.org/10.1038/s41893-018-0101-5

Naumova EN (2018) 3 dangers of rising temperatures that could affect your health now. The Conversation. https://theconversation.com/3-dangers-of-rising-temperatures-that-could-affect-your-health-now-105028

NCEI – National Center for Environmental Information (2017) Changing extremes and human health. https://www.ncei.noaa.gov/news/changing-extremes-and-human-health

OECD (2020a) The territorial impact of COVID-19: managing the crisis across levels of government. https://www.oecd.org/coronavirus/policy-responses/the-territorial-impact-of-covid-19-managing-the-crisis-across-levels-of-government-d3e314e1/

OECD (2020b) Cities policy responses. http://www.oecd.org/coronavirus/policy-responses/cities-policy-responses-fd1053ff/

Our World in Data (2021) Coronavirus vaccinations. https://ourworldindata.org/covid-vaccinations

Ouyang H (2020) I'm an E.R. Doctor in New York. None of us will ever be the same. New York Times. Published April 14, updated May 27. https://www.nytimes.com/2020/04/14/magazine/coronavirus-er-doctor-diary-new-york-city.html

Paget J et al (2019) Global mortality associated with seasonal influenza epidemics: new burden estimates and predictors from the GLaMOR Project. J Glob Health 9:020421

PAHO – Pan American Health Organization and WHO (2017) Climate change and health. https://www.paho.org/salud-en-las-americas-2017/?tag=health-effect-of-climate-change

Paolo C, Betrò S, Luigi J (2020) The impact of climate change on mental health: a systematic descriptive review. Front Psychiatry 11:74. https://doi.org/10.3389/fpsyt.2020.00074

Puri N, Coomes EA, Haghbayan H, Gunaratne K (2020) Social media and vaccine hesitancy: new updates for the era of COVID-19 and globalized infectious diseases. Hum Vaccin Immunother 16:2586–2593

Rajbhandari R, McMahon DE, Rhatigan JJ, Farmer PE (2020) The neglected hospital—the district hospital's central role in global health care delivery. N Engl J Med 382(5):397–400. Epub 2020/01/30. pmid:31995684

Reuter (2020) German meatpacking plant told to shut after coronavirus outbreak. https://www.reuters.com/article/health-coronavirus-germany-meat-idUSKBN26T0ZF

Robine JM, Cheung SL, Le Roy S, Van Oyen H, Griffiths C, Michel JP, Herrmann FR (2008) Death toll exceeded 70,000 in Europe during the summer of 2003. C R Biol 331:171–178. https://doi.org/10.1016/j.crvi.2007.12.001

Rocklöv J, Dubrow R (2020) Climate change: an enduring challenge for vector-borne disease prevention and control. Nat Immunol 21:479–483. https://doi.org/10.1038/s41590-020-0648-y

Sahni V, Scott AN, Beliveau M, Varughese M, Dover DC (2013) Talbot J. Public health surveillance response following the southern Alberta floods. Can J Public Health 107:142–148. https://doi.org/10.17269/cjph.107.5188

Saltzman J (2020) Nearly a third of 200 blood samples taken in Chelsea Show exposure to coronavirus. Boston Globe. https://www.bostonglobe.com/2020/04/17/business/nearly-third-200-blood-samples-taken-chelsea-show-exposure-coronavirus/

SCCM (2020) United states resource availability for COVID 19. https://sccm.org/Blog/March-2020/United-States-Resource-Availability-for-COVID-19

Schnitter R, Berry P (2019) The climate change, food security and human health nexus in Canada: a framework to protect population health. Int J Environ Res Public Health 16(14):2531. https://doi.org/10.3390/ijerph16142531

Shaw R, Kim YK, Hua J (2020) Governance, technology and citizen behavior in pandemic: lessons from COVID-19 in East Asia. Prog Disaster Sci 6:100090. https://doi.org/10.1016/j.pdisas.2020.100090

Silverman H, Toropin K, Sidner S, Perrot L (2020) Navajo Nation surpasses New York state for the highest Covid-19 infection rate in the US. CNN.com. https://www.cnn.com/2020/05/18/us/navajo-nation-infection-rate-trnd/index.html. Accessed 6 Jul 2020

Singh P, Ravi S, Chakraborty S (2020) COVID-19 Is India's health infrastructure equipped to handle an epidemic? Brooking Institutions, https://www.brookings.edu/blog/up-front/2020/03/24/is-indias-health-infrastructure-equipped-to-handle-an-epidemic/

Singu S, Acharya A, Challagundla K, Byrareddy SN (2020) Impact of social determinants of health on the emerging COVID-19 pandemic in the United States. Front Public Health 8:406. https://doi.org/10.3389/fpubh.2020.00406

Siraj AS, Santos-Vega M, Bouma MJ, Yadeta D, Ruiz Carrascal D, Pascual M et al (2014) Altitudinal changes in malaria incidence in highlands of Ethiopia and Colombia. Science 343(6175):1154–1158

Smith KR, Woodward A, Campbell-Lendrum D, Chadee DD, Honda Y, Liu Q, Olwoch JM, Revich B, Sauerborn R (2014) Human health: impacts, adaptation, and co-benefits. In: Climate change 2014: impacts, adaptation, and vulnerability. https://www.ipcc.ch/site/assets/uploads/2018/02/WGIIAR5-Chap11_FINAL.pdf

Solís Arce JS, Warren SS, Meriggi NF et al (2021) COVID-19 vaccine acceptance and hesitancy in low- and middle-income countries. Nat Med 27:1385–1394. https://doi.org/10.1038/s41591-021-01454-y

Spracklen DV, Mickley LJ, Logan JA, Hudman RC, Yevich R, Flannigan MD, Westerling AL (2009) Impacts of climate change from 2000 to 2050 on wildfire activity and carbonaceous aerosol concentrations in the western United States. J Geophys Res 114(D20). https://doi.org/10.1029/2008JD010966

Stieg C (2021) The covid pandemic could end next year experts say – here's what that looks like and how the us could get there. https://www.cnbc.com/2021/12/09/bill-gates-how-covid-pandemic-ends-and-becomes-endemic-with-omicron.htm

Telenti A, Arvin A, Corey L et al (2021) After the pandemic: perspectives on the future trajectory of COVID-19. Nature 596:495–504. https://doi.org/10.1038/s41586-021-03792-w

The Economist (2021) Tracking COVID-19 excess deaths across countries. https://www.economist.com/graphic-detail/coronavirus-excess-deaths-tracker

Tognotti E (2013) Lessons from the history of quarantine, from plague to influenza A. Emerg Infect Dis 19:254–259

Torres JM, Casey JA (2017) The centrality of social ties to climate migration and mental health. BMC Public Health 17:1–10. https://doi.org/10.1186/s12889-017-4508-0

UK Government (2020) Local COVID alert levels: what you need to know. https://www.gov.uk/guidance/local-covid-alert-levels-what-you-need-to-know

UN Environment DTU Partnership (2018) Adaptation gap report [Internet]. United Nations Environment Programme, Nairobi; [cited 2020 Nov 10]. Available from: https://www.unenvironment.org/resources/adaptation-gap-report

United Nations (2020) COVID 19 in an urban world. https://www.un.org/en/coronavirus/covid-19-urban-world

van den Berg R, Sorkin LN, Molenaar A, Tuts R (2020) Building climate resilient and equitable cities during COVID-19. WRI. https://www.wri.org/news/building--climate-resilient-and-equitable-cities-during-covid-19

Viscusi WK (2020) Pricing the global health risks of the COVID-19 pandemic. J Risk Uncertain 61:101–128. https://doi.org/10.1007/s11166-020-09337-2

Wainwright O (2020) Architect in Italy turns shipping containers into hospitals for treating Covid-19. Guardian. https://www.theguardian.com/artanddesign/2020/mar/27/architect-in-italy-turns-shipping-containers-into-hospitals-for-treating-covid-19

Watts N, Amann M, Arnell N, Ayeb-Karlsson S, Belesova K, Boykoff M et al (2019) The 2019 report of The Lancet Countdown on health and climate change: ensuring that the health of a child born today is not defined by a changing climate. 394(10211):1836–1878. https://doi.org/10.1016/S0140-6736(19)32596-6

Westerling A, Bryant B (2008) Climate change and wildfire in California. Climate Change 87:S231–S249

WHO (2005) International health regulations (2005) third edition. https://www.who.int/publications/i/item/9789241580496 (2016)

WHO (2014) Quantitative risk assessment of the effects of climate change on selected causes of death, 2030s and 2050s. World Health Organization, Geneva

WHO (World Health Organization) (2020a) Laboratory testing strategy recommendations for COVID-19. https://apps.who.int/iris/bitstream/handle/10665/331509/WHO-COVID-19-lab_testing-2020.1-eng.pdf

WHO (World Health Organization) (2020b) WHO director-general's opening remarks at the media briefing on COVID-19. https://www.who.int/dg/speeches/detail/who-director-general-s-opening-remarks-at-the-media-briefing-on-covid-19%2D%2D-11-march-2020

WHO (World Health Organization) (2021) COVID-19 vaccine tacker and landscape. https://www.who.int/publications/m/item/draft-landscape-of-covid-19-candidate-vaccines

Wildlife Conservation Society (2020) The deadly dozen: wildlife diseases in the age of climate change released at the IUCN World Conservation Conference, Barcelona

Yong E (2020) No one is listening to us. The Atlantic. https://www.theatlantic.com/health/archive/2020/11/third-surge-breaking-healthcare-workers/617091/

Mega Risks, Education, Knowledge, and Sustainability

8.1 Introduction

Education is a strong driver of human development, and one of the most powerful tools in reducing poverty and improving social well-being of individuals as well as communities and society. Education creates opportunities and employment and serves as a mean for both income generation and returns on income. Globally, there is a 9% increase in hourly earnings for every extra year of schooling. Moreover, social returns to schooling remain high, above 10% at the secondary and higher education levels (Psacharopoulos and Patrinos 2018). Investments in education are critical for developing the human capital, ending extreme poverty, and making progress toward the achievement of Sustainable Development Goals. However, education systems across the world are going through one of the worst crises in human history. The quick spread of COVID-19 pandemic in early months of 2020 impacted on schooling which affected at least 1.6 billion children and youth. Since then, education systems around the world are continuously in the grip of the complex decisions of how to cope with the changing day-to-day situation. Likewise, world over, 450 cities each with a population over 1 million face recurring climate-related extreme events—cyclones, typhoons, and hurricanes, the deadliest and costliest of disasters. Droughts and desertification also affect 250 million people and threaten 1.2 billion in 110 countries (Patel 2008). These climate-related events also seriously affect education systems through denied schooling time, lost instruction, and drop in quality of education.

This chapter on mega risks, education, knowledge, and sustainability has been divided into four parts. The introduction is followed by two parts which discuss the impacts of two mega risks—coronavirus and climate change, respectively, on urban education and associated policy response. The second part dealing with the coronavirus pandemic has been divided into three sections. The first section traces the impact of the pandemic on education in cities which led to the school closure, created uncertain conditions in education, and led to the most severe crisis in years. The educational authorities and educators were caught by such a surprise that it took them some time to formulate response strategies. These strategies constituting the next section focus on the paradigm shift in education delivery—with the big-time use of online learning, distance, and continuing education, which became a panacea in the new normal. However, they also posed great challenges to both educators and the learners particularly in developing countries. The biggest among these were quality of remote learning and its access, which varied greatly not only among cities particularly between developed and developing ones but even within different areas of the same city. Marginalized and vulnerable students were in either case had least likelihood to access decent remote learning opportunities. The crisis has in many ways widened pre-COVID gaps in education, which is the focus of the next section that deals with the learning recovery and equity post-pandemic that is crucial for children returning to school. The third part of this chapter is on climate change and education. It has two sections. The first deals with the climate change impacts on education and the second is on the related response strategies for managing education The findings of the chapter have been highlighted in the conclusion.

8.2   Impact of COVID-19 on Education

The coronavirus pandemic has caused the largest disruption in history in education systems of the world. It has affected nearly 1.6 billion students in about 190 countries (United Nations 2020). The lockdowns and associated social distancing

policies in cities brought closures of schools and other learning institutions impacting 94% of the world's student population—up to 99% in low- and lower-middle income countries (United Nations 2020). A recent World Bank, UNESCO, and UNICEF (2021) report provides evidence on the severity of the learning losses incurred during school closures (Fig. 8.1). Globally, from late February 2020 until early August 2021, education systems were on average fully closed for 121 instructional days and partially closed for 103 days—the world's poorest children were disproportionately affected (UNESCO 2021a). While some countries quickly reopened schools, many kept all schools closed for exceptionally long periods or reopened, but only partially (Fig. 8.1). For example, some education systems reopened but offered access to face-to-face schooling only in certain areas, to certain grades, or to all students on a part-time basis, adopting a

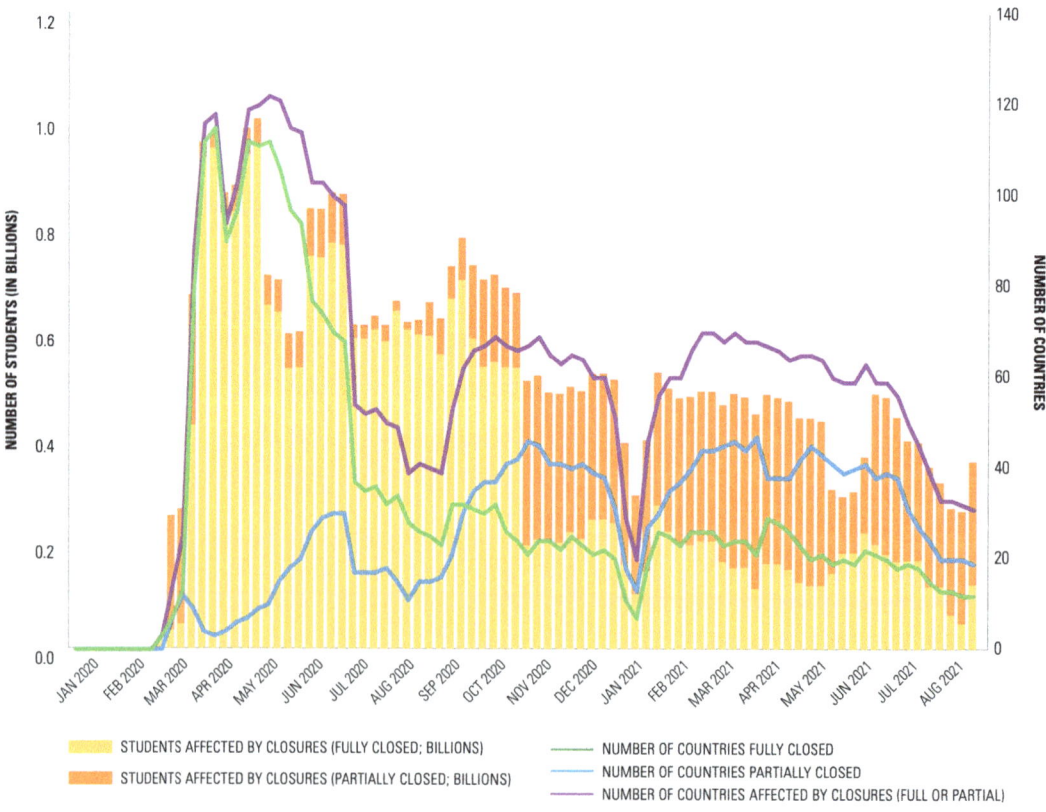

Fig. 8.1 Billions of students in low- and middle-income countries were affected by full and partial school closures since the beginning of the pandemic. (Source: The World Bank, UNESCO, and UNICEF (2021))

hybrid model where students rotated in receiving in-person instruction. The concern is that even up to the end of 2021, 21 months later schools remained closed for millions of children and youth, while millions more are at risk of never returning to education (World Bank, UNESCO, and UNICEF 2021). What to say of developing countries, according to a RAND Corporation report (Schwartz et al. 2021) even in the USA, two-thirds of school districts lost some enrollment in 2020–2021 from the year before; their average decline was 5%. This is twice as many districts and twice as large a decline as the year before. While the situation was improving, the advent of Omicron virus and its variants has made the situation uncertain again.

8.2.1 Learning Losses

Simulations of various scenarios at the end of 2020 by the World Bank (World Bank, UNESCO, and UNICEF 2021) showed that closures of school lasting 7 school months may have costed enormously in monetary terms. In intermediate scenario, the current generation of students may have lost an estimated $10 trillion in lifetime earnings in present value. Under a pessimistic scenario, however, losses were worse, whereby this generation of students could have lost $17 trillion in lifetime earnings at present value (Azevedo et al. 2021a). This projected loss is equivalent to 14% of today's global GDP. The new simulations indicate a shift in the distribution of learning losses by income groupings—they show a larger share of losses in middle-income countries, where the reported school closures were longer than those in high- and low-income countries (World Bank, UNESCO, and UNICEF 2021).

In terms of learning-adjusted years of schooling (LAYS), World Bank simulations in 2020 expected to see a global loss of 0.9 LAYS, driving the global average down from 7.8 LAYS to 6.9 LAYS (Azevedo et al. 2021b). Under this scenario, a typical student would lose $25,000 in lifetime earnings in present value, and with that duration of school closures, learning poverty may have increased by 10% points (Fig. 8.2), reaching 63% (Azevedo 2020). The new 2021 simulation, however, depicts an increase in learning poverty to as much as 70% in low- and middle-income countries (Fig. 8.2) under the pessimistic scenario (Azevedo et al. 2021b).

Learning poverty is a measure developed jointly by the World Bank and UNESCO's Institute of Statistics, to assess the number of children particularly in low- and middle-income countries, who cannot read and understand a simple story by the end of primary school or age 10. High levels of learning poverty ring a warning bell that all global educational goals and other related Sustainable Development Goals are in jeopardy (World Bank 2021a). Even before the pandemic in low- and middle-income countries, the share of children living in learning poverty had already reached over 50% (Fig. 8.2), which is expected to increase sharply, potentially up to 70%, given the long school closures and the varying quality and effectiveness of remote learning. Learning losses were significant even in high-income countries, which were able to switch quickly to online teaching. Based on several studies, the World Bank, UNESCO, and UNICEF (2021) report narrated substantial learning losses in both Europe and the USA up to 20% and above.

8.2.2 Other Related Losses

Besides education and learning, schools also provide critical services and offer safe spaces for protection. In fact, school closures deprive opportunities for growth and development to many children and youth. For example, the impact was disproportionate on underprivileged students (UNESCO 2021b). A case in point is child nutrition—more than 370 million children missed out on school meals in 150 countries due to worldwide school closures during the pandemic (Borokowski et al. 2021). It is sad for these children, as the only reliable source of food and daily nutrition for them was school meals. According to the World Bank, UNESCO, and UNICEF (2021), low- and middle-income nations have

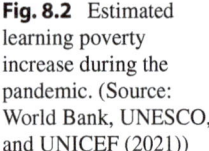

Fig. 8.2 Estimated learning poverty increase during the pandemic. (Source: World Bank, UNESCO, and UNICEF (2021))

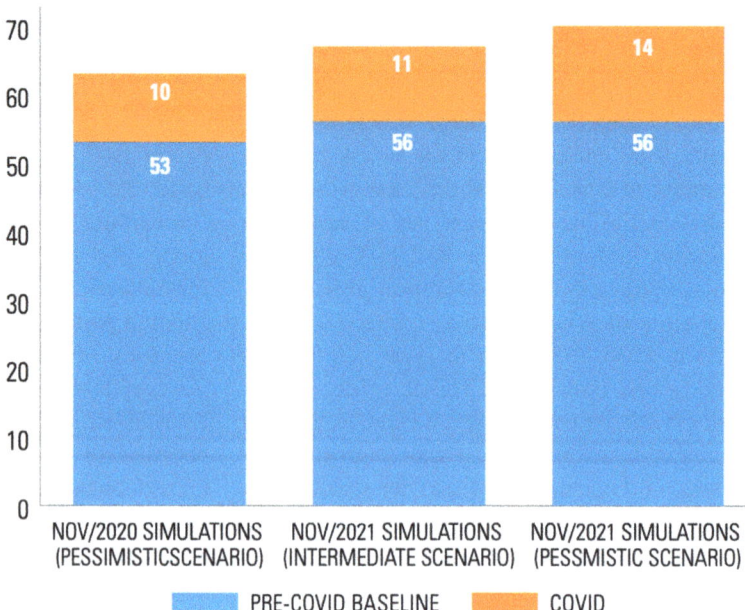

already seen about 30% reduction in essential nutrition services such as school meal programs, iron and folic acid supplementation, and deworming. This kind of hunger and malnutrition seriously affected children's immediate and long-term growth and development, as well as learning capacities (UNICEF 2019a). Estimates under the most pessimistic scenario show that because of inadequate food access and nutrition, 40 million more children may suffer from wasting, while an additional 7 million may be stunted by 2030 (UNICEF, FAO, IFAD, WFP, and WHO 2021). Further, it is expected that the school closures are likely to have differential impacts on cities and countries depending on their income levels. For example, while undernutrition is expected to deepen in poor countries and cities, threatening years of global progress, there are concerns that pediatric obesity may increase in middle- and high-income countries (Zemrani et al. 2021).

School closure jeopardized many children's health and safety too, with increased child labor and domestic violence. According to UNICEF and ILO (2021), 9 million additional children face the risk of being pushed into child labor by the end of 2022 because of rising poverty triggered by the pandemic. A past study (Edmonds

and Schady 2012) showed that children engaged in paid employment face additional challenges in returning to formal education and are likely to remain out of school. The evidence on the consequences of the pandemic on violence against children are growing too (Cappa and Jijon 2021; Bourgault et al. 2021). Lost employment and reduced household income often increase the risk of violence against women and children (Bourgault et al. 2021). A review by Cappa and Jijon (2021) documented 48 studies published in 2020 (from March to December 2020) highlighting the impact of the pandemic on violence against children. Despite increasing numbers of these studies which explored risk factors and recent trends in violence against children, little attention has been given to violence prevention policy or programs (Bhatia et al. 2021). It is important to mention here that schools and educators play an important role in detecting and reporting cases where children experience domestic violence, suggesting that in some cases, instances of violence may be underreported (World Bank, UNESCO, and UNICEF 2021).

In addition to education, schools promote children's well-being by delivering health services and vaccinations, allowing children to develop their socioemotional skills, and provide psychosocial support, hence offering additional incentives to parents for sending their children to school. COVID-19 and previous health crises have shown enough evidence that school closures reduce children's access to many of these critical services and safe spaces with detrimental effects on child protection and their well-being (Villegas et al. 2021). Most underprivileged children, for example, tend to have fewer educational and other opportunities and social contacts beyond school. The schools as such provide opportunities to these students to build positive relationships with their peers and teachers, which can foster inclusion with positive impact on mental health and well-being. School closures around the world made these children experience social isolation, caused disruption in their daily routines, and created stress often associated with parental unemployment, and feelings of uncertainty about their future. Several studies showed high rates of anxiety and depression in children and youth due to COVID-19, with some studies finding that girls, adolescents, and some others were disproportionately affected (Bignardi et al. 2021; Lee 2020; Sharma et al. 2021). Global estimates suggest that mental health challenges doubled because of the pandemic (Racine et al. 2021). An effective way of supporting children under stress and adversity is play, which can improve equity for children (Solis et al. 2020), but school closures cut or limited access to even playful spaces.

Students with disabilities faced more barriers to engaging in remote learning compared to others because remote learning modalities were more often not adequately tailored for sign language interpretation, closed captioning, braille, etc. (World Bank 2020). Only 33% of low-income countries had taken steps to support learners with disabilities (World Bank, UNESCO, and UNICEF 2021). In a global survey of learners with disabilities, their teachers and parents or caregivers provided information on a wide range of environmental, educational, and health-related challenges that learners with disabilities faced at the onset of the pandemic and during school closures. Most caregivers (60%) showed concern that learners with disabilities were falling behind in learning due to inaccessible distance learning modalities. Among teachers, only 19% responded positively indicating that students with disabilities were learning during school closures, while more than twice as many pointed otherwise—that they were falling behind in learning (World Bank 2021b).

School closures and strict lockdown and containment measures made more families to rely upon technology and digital solutions—to keep their children engaged in learning and entertainment. However, all children had neither the necessary knowledge nor skills, and resources to keep themselves safe online (Pokheral and Chethhri 2021). Children's vulnerability to online exploitation also increased in the wake of dependence of an extremely large number of students on virtual platforms. Many children were potentially exposed to harmful and violent content as well as faced greater risk of cyberbullying due to increased but unstructured time spent due to online learning.

Progress in gender equality came under threat too, with school closures placing an estimated 10 million more girls at risk of early marriage in the next decade (UNICEF 2021b). Education offers protection to girls from child marriage and provides reproductive health services that prevent unplanned pregnancy. Previous epidemics like the Ebola crisis in West Africa showed the interrelationship between poverty, school dropout, early marriage, and early pregnancy (Villegas et al. 2021). School closures and economic hardships provide incentives to families for marrying their daughters early to offset financial burden, especially when families give limited immediate value to education (Save the Children 2021). Pregnancy during COVID-19 school closures also increasingly threatened to keep many adolescent girls barred from education even when they reopen (Bhengu 2021; Mavhunga 2021; Yusuf 2020). Although drivers of adolescent pregnancy varied, economic hardships led some girls to engage in transactional sex, increasing their risk of sexual exploitation and pregnancy (Oulo et al. 2020). Intersecting inequalities also put displaced and migrant girls at a heightened risk of never returning to school (Nyamweya 2020).

Schools reopening after long closures pose a major challenge to ensure children and youth continue their education. There is an increased risk of student's dropouts, particularly those who fail to continue some form of learning during school closure and/or those who got occupied in some income-generating activities. A UNESCO (2020) Projection on the COVID-19 crisis predicted that 23.8 million children, adolescents, and youth (from pre-primary to tertiary education) globally will be at risk of not returning to care centers, schools, or universities in 2020 alone. About 11 million or roughly half of the estimated number belonged to primary and secondary level students. This is in addition to the 258 million children and youth of primary and secondary school age who were already out of school prior to the pandemic (UNESCO 2020). Hence, it becomes critically important that in the recovery phase the dropout rate is contained, and the level of current student participation rate is maintained.

8.2.3   Teachers' Predicament due to School Closures

Besides students, the closure of schools also created confusion and stress for teachers especially because the institutions closed unexpectedly and for unknown durations. As a result, teachers were often unsure of their obligations and had problems in maintaining connections with students to support their learning. Transitions to distance learning platforms tended to be chaotic and frustrating, even in the best circumstances. One study from Pakistan (Bhamani et al. 2020) highlighted the predicament of teachers as follows:

- Distance learning is problematic when teachers are themselves not trained for it.
- Teaching is moving online on an untested and unprecedented scale. Student assessments are also moving online, with a lot of trial and error and uncertainty for everyone.

In many instances, therefore, school closures led to furloughs or separations of teachers.

Further, demand for distance learning skyrocketed due to school closures and often overwhelmed existing portals for remote education. Moving learning from classrooms to homes at unprecedented scale and in a hurry presented enormous challenges, both human and technical to teachers. From the beginning of the COVID-19 pandemic, teachers started accumulating adverse psychological symptoms due to school closures and the need to adapt to different teaching modalities. Based on papers published on the prevalence of depression, anxiety, stress, and burnout in teachers from December 1, 2019, to June 15, 2021, Ozamiz-Etxebarria et al. (2021) showed that among teachers reported level of anxiety was 17%, depression 19%, and stress 30%. Asia, according to them, experienced more anxiety compared to other continents. In terms of level of education, anxiety has been higher among schoolteachers compared to those teaching in universities. In contrast, stress levels have been higher among teachers in universities compared to schools. However, statistically, no significant differences were noted by gender and age in any of the symptoms. Overall, the results suggested that teachers at different educational levels were experiencing adverse psychological symptoms during the pandemic, but anxiety levels varied between countries.

8.2.4   Pressure on Parents

Pressure was generated on parents as well due to school closures. For example, when schools closed, parents were often obligated or asked to facilitate the learning of children at home. Many of them, however, struggled to perform this task especially those with limited education and resources. Further, working parents, in the absence of alternatives, had no options, but often to leave their children alone when schools closed with high probability of risky behaviors, including increased peer pressure and substance abuse. In addition, some parents especially women missed work when schools closed to take care of their children with negative impacts on family income and productivity. In terms of remote learning, while most parents felt helpless in keeping their children engaged, many tried to learn technology to grapple with online learning. Those already familiar with technology and online tools were far more effective in creating a routine of learning at home with their children. They also felt comfortable with activities and home assignments given by the schools and using more online resources for reading and home-based activities. Many mothers used Facebook groups to interact with the community on how to keep their children engaged with pot painting, indoor gardening, simple games, worksheets, and activities. Quite a few parents were able to use free services that have become available for pleasure reading, including audiobooks, e-books, and flip-books (Bhamani et al. 2020).

8.2.5 Inequality of Impacts

The impacts of school and educational closures varied by developing status of the cities and nations, by urbanization status (whether urban or rural district), and by structure of communities such as race, ethnicity, and income. Globally, full and partial school closures lasted an average of 224 days although in cities of low- and middle-income countries, where most of the school-age children live, school closures often lasted longer than in cities of high-income countries. Response also varied by urban and rural districts. For example, in the USA, in February 2021, only 17% of urban districts were offering fully in-person instruction to students, compared with 42% of rural districts (Schwartz et al. 2021). School timings were also shortened. For example, in the USA, more school districts shortened school time in 2020–2021 than have lengthened it—especially those offering remote instructions. Thus, more than 33% of such districts had shortened the school day, and one-quarter had reduced instructional minutes (Schwartz et al. 2021).

The response effectiveness also varied. For example, in low- and middle-income nations, it was less effective because teachers in many of their cities and countries received limited professional development support to transition to

Table 8.1 Comparison of intensity of learning losses by dimensions of inequality in low- and middle-income countries

Dimension of inequality	Intensity of loss
Area	Bigger losses in disadvantaged and rural areas
Age/grade	Bigger losses for students in earlier grades
Gender	Bigger losses to girls compared to boys
Public/private	Bigger losses to students in government schools compared to private schools
Socioeconomic status	Bigger losses to low-income households

Source: Compiled from UNESCO, UNICEF, World Bank, and OECD (2021)

remote learning, hence leaving them unprepared or partially prepared to engage with learners. The households' or parents' ability to respond to the shock also varied depending on their income and educational level. Most importantly, children from disadvantaged households and groups were often at a greater loss (Table 8.1) and less likely to benefit from remote learning than their peers primarily due to a lack of electricity, connectivity, devices, and caregiver support. In terms of age and fitness, the youngest students and those with disabilities were largely left out of cities and countries' policy responses because remote learning was rarely designed in a way that met their developmental needs. In terms of gender, girls faced compounding barriers to learning amidst school closures—as social norms, limited digital skills, and lack of access to devices constrained their ability to keep learning.

8.2.6 Impacts on Higher Education

In higher education, COVID-19 disruptions affected more than 220 million students globally (UNESCO 2021c). Review of peer-reviewed papers (Khan 2021) and relevant literature, primarily from academic databases and government/nongovernment, credible organizations (Treve 2021) and surveys and findings of international organizations (World Bank, UNESCO, and

UNICEF 2021), show several effects of the pandemic on higher education and institutions of higher learning. Firstly, there has been a large increase in online learning for two reasons: (a) worldwide institutions for higher education shifted to online, distance, or hybrid modes of instruction and (b) due to international travel restrictions – the pandemic drastically cut the international mobility of students and faculty. One major problem in online or remote teaching was to appropriately design the learning content not only for the purpose of delivery but also to foster the creative thinking in students considering their capabilities. Moreover, there were significant concerns of teachers on online teaching even before the pandemic, some of which included loss of control over students, lack of concentration compared to the traditional mode, difficulty in using online platforms, and felt "left in the dark" because they could not observe students (Islam et al. 2015). In addition, a global analysis of responses by higher education systems revealed that many universities had neither the resources nor academic capabilities to transition to an online delivery system (Treve 2021). They were temporarily adopting a short-term approach that was not suitable in the long run (Crawford et al. 2020)

Students' mobility including cross-border movement of students is an important aspect of the global higher education landscape (Treve 2021). International student mobility has been increasing at an average annual growth rate of 10% during the last two decades. Some 5 million students travel to different nations every year for tertiary education. The number was predicted to increase to 8 million by 2025 in the business-as-usual scenario according to OECD (BizED 2020). However, the trend was severely affected by the pandemic—around 95% of higher education institutions in Europe and 91% in America reported that their students' mobility was affected due to COVID-19 (International Association of Universities 2020).

In terms of students' enrollment in universities, a UNESCO survey showed that most countries did not experience any significant difference in enrollment in higher education institutions. Nevertheless, 14 countries experienced up to 20% decreases in enrollment, whereas three countries (Armenia, Venezuela, and Hungary) reported decreases in enrollment between 21% and 40% (UNESCO 2021d). In the USA, undergraduate enrollment fell by 6% in the spring 2021 compared to 2020—community colleges were worst affected, with a decrease of 10% in the fall and 11% in the spring (NSC Research Center 2020). The Brazilian National Examination of Upper Secondary (Examen Nacional do Ensino Médio, ENEM) for university entry had the lowest number of applicants since 2007. Only 4 million students registered to take the exam compared to over 6 million in 2020. The share of black, brown, or indigenous students declined considerably in the total number of applicants—with the risk of enhancing existing inequalities in university admissions (World Bank, UNESCO, and UNICEF 2021). Enrollment in Technical and Vocational Education Training institutions also declined, for example, in Colombia it decreased by 50% between 2019 and 2020 (Ministerio de Educación Nacional de Colombia 2021).

8.3   Policy Response

The matter of most immediate concern, in the wake of intensity of shock to education, was to ensure that learning continues through some actions in the short term followed quickly by suitable mechanisms for recovery. The crisis, nevertheless, provided an opportunity for adopting long-term policies, programs, and approaches to address the deep-seated existing problems. It also opened the way to transform education to the need of the time while examining the practicalities of resource availability and professional capacity. The response measures and policies have been reviewed in the following account under short-term recovery measures as well as long-term outlook.

8.3.1 Immediate Response: Introduction of Remote Learning

Governments and education authorities' immediate response to the school closure was adoption of remote learning. The measures utilized varied by delivery channels—from digital tools to TV-/radio-based teaching and take-home packages to allow the students to learn continuously. A coalition of three international organizations UNESCO-UNICEF-World Bank conducted a Survey on National Education Responses to COVID-19 School Closures to collect critical information on how ministries of education in more than 110 countries continued to provide learning opportunities during school closures. The Survey (UNESCO-UNICEF-World Bank Joint Survey 2020) results showed that for each level of education, most countries developed policies using digital (Internet-based) or broadcast (TV- or radio-based) remote learning. The digital instruction was used by 42% of countries for pre-primary education, 74% countries for primary education, and 77% countries for upper secondary education (Fig. 8.3). It was found that many countries also developed broadcast curricula, especially for use in primary and lower secondary education.

Overall, 94% of countries' ministries of education studied formulated policies to provide at least one form of remote learning that involved digital and/or broadcast instruction—although only 60% used this policy for the pre-primary education. This shortcoming is serious because estimates show that every dollar invested in increasing enrollment in pre-primary education returns $9 in benefits to society in the form of reduced repetition and dropouts in primary and secondary school as well as increased lifetime earnings for individuals (UNICEF 2020b). Therefore, it is important to recognize the importance of education at this level.

The choice of remote learning technology, whether for a city or country, was influenced by their economic status or their position within countries of an income group (UNESCO, UNICEF, World Bank, and OECD 2021;

UNICEF 2020a). For example, radio-based instruction was more common in nations belonging to low-income or lower-middle-income group (Fig. 8.4)—over 80%—compared to upper-middle-income group, 59%, or high-income group, 25%. In contrast, online platforms and TV-based instruction were more common in larger groups of lower-middle-income countries (both above 95%) and upper-middle-income countries (90% and 95%) compared to low-income countries (56% and 83%, respectively). Higher-income countries relied principally on online platforms (96%).

8.3.2 Disparity in Access to Devices and Means for Remote Learning

During the shift to remote learning, a huge number of students were left behind due to digital divide or lack of access to devices. For example, the cities and nations that adopted technological solutions digital divide for one made it a "remote learning paradox." Globally, online platforms were the most common response for remote learning (UNESCO, UNICEF, World Bank, and OECD 2021), but 1.3 billion school-age children did not have access to the Internet at home. Although situations in cities were better than villages, access to the Internet was low even there in less developed countries compared to developed ones (UNICEF and International Telecommunication Union 2020). Digital divides between urban and rural communities were greatest in Eastern and Southern Africa, East Asia and the Pacific, and Latin America and the Caribbean (World Bank, UNESCO, and UNICEF 2021). Even in cities of developed countries, digital divide emerged by household incomes. For example, in ten European countries, fourth grade students from low socioeconomic status backgrounds were half as likely to have Internet access at home compared to their more affluent peers (European Commission 2020). In West and Central Africa, Eastern and Southern Africa, and South Asia, Internet access for children from the bottom wealth quintile is nearly nonexistent

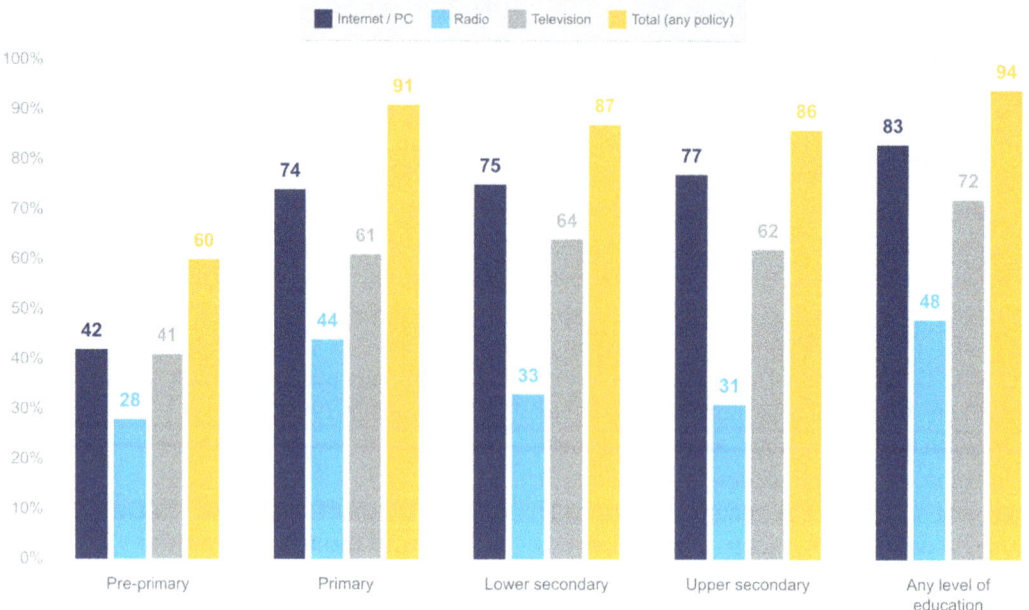

Fig. 8.3 Share of countries that implemented remote learning policies by digital/broadcast methods used at various levels of education. (Source: UNICEF (2020a))

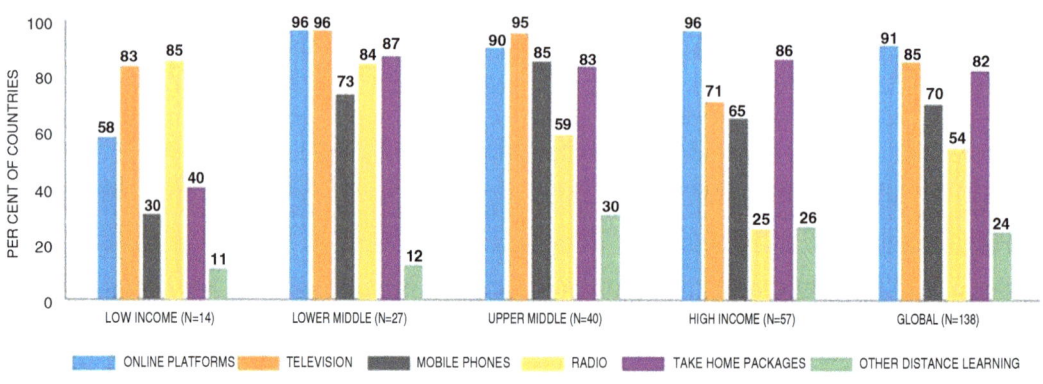

Fig. 8.4 Choice of remote learning technology by countries in various income groups. (Source: World Bank, UNESCO, and UNICEF 2021)

(Wang et al. 2021). Countries with longer durations of school closures tended to have lower rates of school-age children with Internet connection at home (UNICEF 2021a), which affected adversely access to remote learning if other modalities were not available or utilized.

Many cities and communities with low rates of Internet connectivity utilized other remote learning modalities, but their access and use still varied considerably. A vast majority of low-income countries used broadcast media (television and radio) as a remote learning modality to reach learners (UNESCO, UNICEF, World Bank, and OECD 2021). Dearth or lack of access to devices, however, formed the biggest barrier to learning through broadcast media, though it was somewhat better in cities compared to rural communities. For example, in West and Central Africa, broadcast media proved pivotal in remote learning, yet 27% of households in urban areas still lacked it. In South Asia, situation was a little better where 50% of households, on average, owned a television (The World Bank, UNESCO, and UNICEF 2021). Although mobile phones have widespread prevalence globally, they were the second-least-used mode to deliver remote learning. This mismatch of available infrastructure to remote learning mechanism in cities and nations needs further exploration for the possibility of their more extensive use in the future (UNESCO, UNICEF, World Bank, and OECD 2021).

Besides Internet connectivity and lack of available devices, nonavailability of electricity was another constraint that impeded equitable access to digital learning (Brossard et al. 2021). Many countries adopted take-home learning materials to reach households without access to electricity or technology. However, even this mechanism had challenges related to distribution (UNICEF 2021b).

8.3.3 Remote Learning Problems

The deployment, application, and effectiveness of remote learning programs in cities and communities varied greatly in most countries and offered an inadequate substitute for in-person learning for several reasons (Table 8.2). The key obstacles in a shift to distance learning during COVID-19, noted by various studies (Treve 2021; Dhawan 2020; Reimers and Marmolejo 2022), included lack of technical resources, differential access to education, difficulties in adaptation, and resistance to change. Despite its many problems and challenges, however, these studies also depicted some beneficial side of the pandemic—mainly as a springboard for future promotion of alternatives such as digitalization, promotion of 5G technologies, and public-private educational collaborations.

Undeniably, the abrupt changes disrupted the general "comfort zone" of teachers who had been teaching students in-person for a long time. Further, the situation appeared to be in disarray especially in developing countries because of the technological problems faced in conducting

Table 8.2 Challenges in remote learning

Challenges	Associated problems
Online pedagogy	Accessing technical infrastructure Quality of the network Computers' availability Teacher-student interaction Low affordability in low-income countries
Teaching and learning competencies	Student motivation Student concentration Student output
Resistance to change	Difficulties in shifting to technology Moving to something unusual Worries—feelings of getting lost Dealing with changed situation
Studies requiring technical equipment	Laboratory access Art and design equipment
Sharpening inequalities	Shortage of technical infrastructure Inadequate access to online networking Low access to the Internet Low access to devices
Assessing competencies of students	Possibility of cheating Possibility of plagiarism
Impact on students' health	Balancing digital- and screen-free activities Prolonged classes and health Issues—blood circulation, oxygen supply, computer vision syndrome

online classes, which were more visible compared to the benefits. Hopefully, the right technological elements and tools as well as experience in the long run will help create a harmonious mix of online and off-line educational programs in the future. The digital divide, however, will need to be mitigated by developing cooperation between developed and developing nations through technical assistance. If the world can collaborate in fighting the pandemic, it can continue the same in making positive changes in the years to come. This is extremely important because the right infrastructure and e-learning together provide the best way to the developing nations to permanently transition into the global technological framework and society. e-learning can also help marginalized ethnic groups in both developing and developed cities to participate in economic activities and social discourse in a meaningful way. Therefore, despite some negative experiences of the shift, opportunities ahead should not be denied or ignored in the long run (Treve 2021). For example, the pandemic has proved to be a catalyst for educational institutions to explore and absorb innovation via technological solutions. Thus, students in Hong Kong started home learning via interactive apps, while China facilitated approximately 120 million students' access to learning material through live television broadcasts (World Economic Forum 2020). The crises also facilitated educational institutions in connecting students globally through such means as Zoom, Microsoft Teams, and interactive webinars.

Even before the pandemic, the educational innovation and solutions had been attracting the attention of not only typical government-funded or non-profit-backed social project but also far greater interest and investments from the private sector. The initiatives involved diverse companies like Microsoft and Google from the USA to Samsung in Korea and to Tencent, Ping An, and Alibaba in China. Most initial initiatives in that direction were limited in scope, and relatively isolated. The pandemic, however, paved the way toward the formulation of much larger-scale, cross-industry coalitions around the common educational goal. The crisis promoted the development of learning consortiums and coalitions among a variety of stakeholders, including educators, publishers, technology providers, and network operators. Such free tech sector coalitions are likely to play a notable role in developing new cloud-based online learning platforms to upgrade the existing infrastructure in the emerging nations. Since governments in these nations play an important role in imparting education, they can reap considerable benefits in this regard through public-private partnerships. In China, the Ministry of Education has already taken the initiative of assembling a group of diverse constituents to develop a new cloud-based, online learning and broadcasting platform as well as to upgrade a suite of education infrastructure, led by the Education Ministry and Ministry of Industry and Information Technology.

While carrying some critical lessons for short-term action, the pandemic also highlighted the need for the transformation of education in medium-term recovery as well as long-term development of education and learning.

8.3.4   Recovery Actions

The COVID-19 Global Education Recovery Tracker (2021), sourced by teams across the World Bank, Johns Hopkins, and UNICEF, monitors recovery planning efforts in more than 200 countries and territories. The latest data shows that most countries have already opened schools partially (parts of the country) and mostly with hybrid (combining remote and in-person) approaches. Even by May 2021, schools in 141 countries had reopened for some in-person instruction (World Bank 2021c), indicating the dawn of recovery phase. Still two-thirds of these countries' students were not fully back to school (World Bank 2021c). A return to in-person operations in the middle of a global health pandemic is by no means easy. It requires actions to serve the best interests of students and teachers while also safeguarding citizens' health. The immediate tasks for the educational authorities along with safe operations were (a) stemming learning losses; (b) introducing accelerated learning recovery programs based on proven techniques and evidence-based strategies such as consolidating the curriculum, extending instructional time, and making learning more efficient through targeted instruction, structured pedagogy, small-group tutoring, and self-guided learning programs; and (c) encouraging reenrollment and giving the most marginalized children priority attention.

Obviously, school operations could not be undertaken as they were prior to COVID-19. In the new situation, schools needed to consider social distancing measures, which required planning of classroom sizes as well as teacher capacities to serve multiple classes—straining the financial, physical, and human resources. The

People's Republic of China provided a case of addressing some of these issues—when schools reopened there after the first wave. The country had already witnessed an enormous surge of student participation in online learning through public and private education platforms—thanks to many providers of online education as well as concerted efforts by the government. So, by the time of school reopening, all grade levels across different schools started offering online learning. Using the opportunity, school leaders decided to continue it even after schools reopened. Hence, a blended learning approach was adopted during the first wave of school reopening. Beijing Middle School 101 offered an innovation for meeting social distancing requirements— whereby classes were split into two concurrent sessions taking place in two adjacent classrooms. Teaching was managed in both rooms simultaneously using a large-screen intelligent monitor. Teacher and students across both classrooms were thus able to have interaction concurrently.

Jordan also followed two-pronged teaching modalities after school reopening: in-person education and hybrid education. To catch up on content or learning losses, the first 2 weeks of school were devoted to a continuation of the remedial education programs, and then the year's curriculum was introduced. While preparing for school reopening, the country prioritized teacher vaccinations and reported that over 90% of teachers were fully vaccinated. Those who were not needed a PCR test required every Sunday and Thursday. Saudi Arabia also followed both in-person and hybrid modalities. To recover learning losses, the county's annual school session was switched to a trimester or three-term system instead of two semesters, with each term consisting of 13 weeks (with 12 holidays during the school year). This strategy enabled the development of curricula and educational plans in response to development requirements and Vision 2030. This also allowed an increase in the actual school days, and the number of study hours for primary and intermediate levels. In-person return was initially limited to intermediate and secondary school students. Kindergarten and primary

school students were to continue with distance education until 70% of the population was to be fully vaccinated or October 30, 2021—whichever was to come first.

To promote reenrollment, Chicago Public Schools, in partnership with the University of Chicago, developed a student prioritization index (SPI) that identified students at highest risk of unfinished learning and dropping out of school. The consolidated index was based on a combination of socioeconomic variables related to academics, attendance, socioemotional, and community vulnerability factors. The district also reached out to all students with a back-to-school marketing campaign and targeted more vulnerable students with additional support. Schools also arranged home visits in partnership with community-based organizations. In addition, recognizing that many students were able to find paid work during the pandemic, they offered various paid summer opportunities to reduce the trade-offs students may have to make between summer school and summer jobs. The district tracked and monitored the results to learn which tactics worked better (Dorn et al. 2021).

Among international organizations, UNICEF, UNESCO, the World Bank, the World Food Programme, and the UN Refugee Agency (UNHCR) published a joint framework for reopening schools and curbing learning losses. The framework provides high-level guidance on schools' safe operations, stemming learning losses, ensuring the well-being of students and teachers, etc. The Asian Development Bank also issued a guidance note for the purpose (ADB 2021).

8.3.5 Long-Term Actions

The long-term actions are very important because they will help education systems innovate as well as set new norms and practices, based on lessons learned from the pandemic. Moreover, they will allow to make appropriate changes in policies and practices. Basically, two sets of policies are needed for the future: the first set to promote actions that are likely to make education systems more resilient and capable of responding to similar crises in the future and the second set to promote innovative actions, practices, and models that either were implemented during the period of ongoing crisis or were developed as a potential coping mechanism to support students and teachers. Rather than returning to business-as-usual scenario, school operations should be based on more deliberate thinking by policymakers and educators so that there is a conscious effort to address both deficiencies systematically while synchronizing education with job opportunities.

Education systems in several countries had shifted to online learning as an emergency measure, and as discussed above, many have already indicated that they are unlikely to roll back completely after schools' reopening. Hence, a hybrid system is likely to prevail in many countries. Likewise, several new digitally powered initiatives introduced during the pandemic may continue and be mainstreamed. However, it may require additional financial and human resources. The crisis has amplified several existing dysfunctions in education systems—such as learning poverty, shortcomings in student assessments/examinations, and teachers' professional development. These are some of the areas that could benefit from new digital solutions. However, governments alone cannot take the agenda forward to raise the quality of education through innovation particularly in developing countries. Therefore, more partnerships and joint initiatives are required with the private sector including EdTech companies and corporations as well as civil society organizations in developing state-of-the-art practices. It is also important to ensure that these models and prototypes must take into considerations the interests of the poor and disadvantaged. Adequate financial and technical resources will need to be put in place to adopt new technology options and training programs.

In the immediate future, new normal digitalization is expected in education at a faster pace. Some nations are already poised to transform their digital solutions implemented during COVID-19 into long-term education delivery operations. Indonesia is one example, which has ramped up investments in education technology for online and distance learning. Thus, the government there has designated seven e-learning platforms to enable students to continue learning at home including (a) Zenius Education, which is providing open access to over 80,000 learning videos for elementary and senior high school students; (b) Rumah Belajar Kemendikbud, with digital learning materials and provision for digital classes and virtual laboratories; (c) G Suite Education, to help students and teachers conduct remote classes in areas with limited Internet access; (d) Microsoft Teams, with all accompanying features; (e) Quipper School, offering digital materials and examination preparation exercises; (f) Ruangguru, providing live teaching sessions and teacher training; and (g) Sekolahmu, providing online and off-line digital learning programs.

In addition, TV Edukasi—an educational television station owned by the Ministry of Education and Culture, in Indonesia—is airing live education programs for students and teachers (ADB 2021). Overall, its aim is to expand online education and enable open access to quality education to citizens with greater flexibility and opportunities for learning. For assuring the quality of online education, the government established Indonesia Cyber Education (ICE) Institute. ICE is responsible for assessing and assuring quality of online education in the country while acting as a marketplace for online education (Pannen et al. 2021). Actions have been undertaken to allow universities to offer a mix of online and off-line courses so that students have greater flexibility in learning and obtaining credits and degrees. These measures will provide digital learning experiences for students with greater readiness for future job markets. Simultaneously, it will substantially expand the provision of higher education, which probably would not be possible through enrollments in brick-and-mortar universities in the short term.

While promoting digital platforms and e-learning, however, it is important for the governments and educational institutions to keep in mind that the shift toward online learning must also take into consideration the fact that many students are unable to access digital platforms. Moreover, they should also take note of the fact that learning through online platforms needs raising of its effectiveness. Additionally, at least in the short term, non-digital resources will continue to be important even where digital solutions are effective, irrespective of the fact—whether they are widely available to students or not. Finally, as hybrid and new learning techniques and approaches in the delivery of education system get underway and enhance, their appropriate governance, real-time monitoring, and regular stocktaking become crucial. This would require involvement of teachers, communities, and other local stakeholders, and the use of real-time data tools to aid decision-making. A review of roles and responsibilities of the education stakeholders, as well as the governance structure in place, may also be needed to monitor feedback and improve future response (ADB 2021).

As the school education system is striving to move toward placing greater emphasis on learning rather than schooling, more activity-based materials will also be needed to replace or supplement standard textbooks. These materials can be produced in both print and digital formats for self-paced learning by students. Further, the long-term education reform agenda could also promote greater decentralization of governance authority to districts and cities or municipal level. This would incorporate some localized curriculum over and above the core national curriculum. Cities/municipalities provide promising opportunities to develop innovative schools that could also support other schools through networking. More importantly, a radical change from an instructive approach in teaching to one that is more student-centric must be a key component of school education reform (ADB 2021).

A major lesson from the pandemic and its response is development of a more resilient system—capable of withstanding future disruptions—from opening schools safely in the short term to building teachers' capabilities and digital capacity in the medium term and realizing a hybrid system of learning in the long term. The pandemic provided an opportunity to all those involved in education with the motivation and impetus to speed up reform to prepare future generations to be both productive and resilient. The key to building forward better is building system-wide resilience so that lessons from the current pandemic response will help to build crisis response systems against all future disruptions be that from natural disasters, conflicts, closures of public facilities and strikes—not just health-related disruptions.

8.4 Impact of Climate Change on Education

Climate change affects education in several ways—for example, cyclones, floods, and wildfires can severely damage or destroy schools and related educational infrastructure directly which can lead to delayed or reduced learning time. In addition, changing weather patterns can also affect education indirectly by impacting food production, nutrition, health, and livelihoods, which may cause physical and financial stress to students and/or their families resulting in cutting students' learning time or resulting in absenteeism or abandonment of school. The details of direct and immediate impacts by rapid-onset hazards (floods, storms, and wildfire) and indirect and delayed impacts of slow-onset hazards (droughts, rising temperatures, and rising sea levels) have been discussed below.

8.4.1 Direct Impacts

Direct impacts of climate change on education are increasing for two reasons:

- Increasing frequency and magnitude of climate-related hazards such as storms, floods, rainfall intensity, wildfires, etc.
- Increasing exposure and vulnerability of school because of their location in hazard-prone areas (Fig. 8.5)

The second factor suggests that people are making an economic judgment to establish lives and businesses in hazard-prone areas despite the

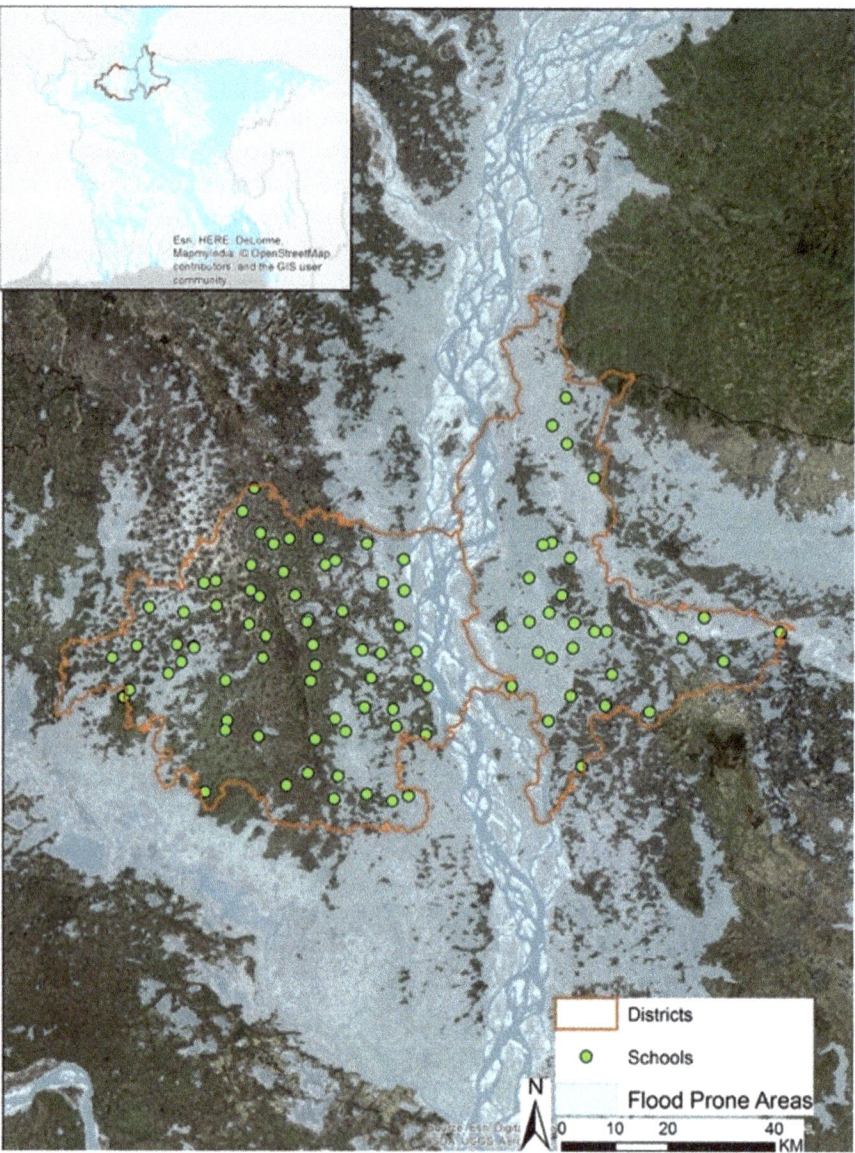

Fig. 8.5 Bangladesh: school location in flood-prone areas liable to direct damage. (Source: Chuang et al. (2018))

inherent risks. Figure 8.5 shows two districts in Bangladesh where 45 out of 96 secondary schools are in high flood risk areas (Chuang et al. 2018).

High intensity of climate-related extreme events can cause exceptional direct damage as occurred in heavy monsoon devastating floods in Pakistan in 2010—resulting in the destruction of 11,000 schools in the country. Similarly, heavy monsoon floods in Nepal, India, and Bangladesh destroyed or damaged 18,000 schools (OCHA 2017). Likewise, in cities of sub-Saharan Africa, tens of thousands of children were forced out of school at the very start of their academic year (when the schools were to reopen after long closure due to COVID-19) whereby heavy floods destroyed their homes and classrooms (Save the Children 2020).

The details of direct impact of climate change have been presented in Fig. 8.6. The first arc of blue circles illustrate three major climate-related

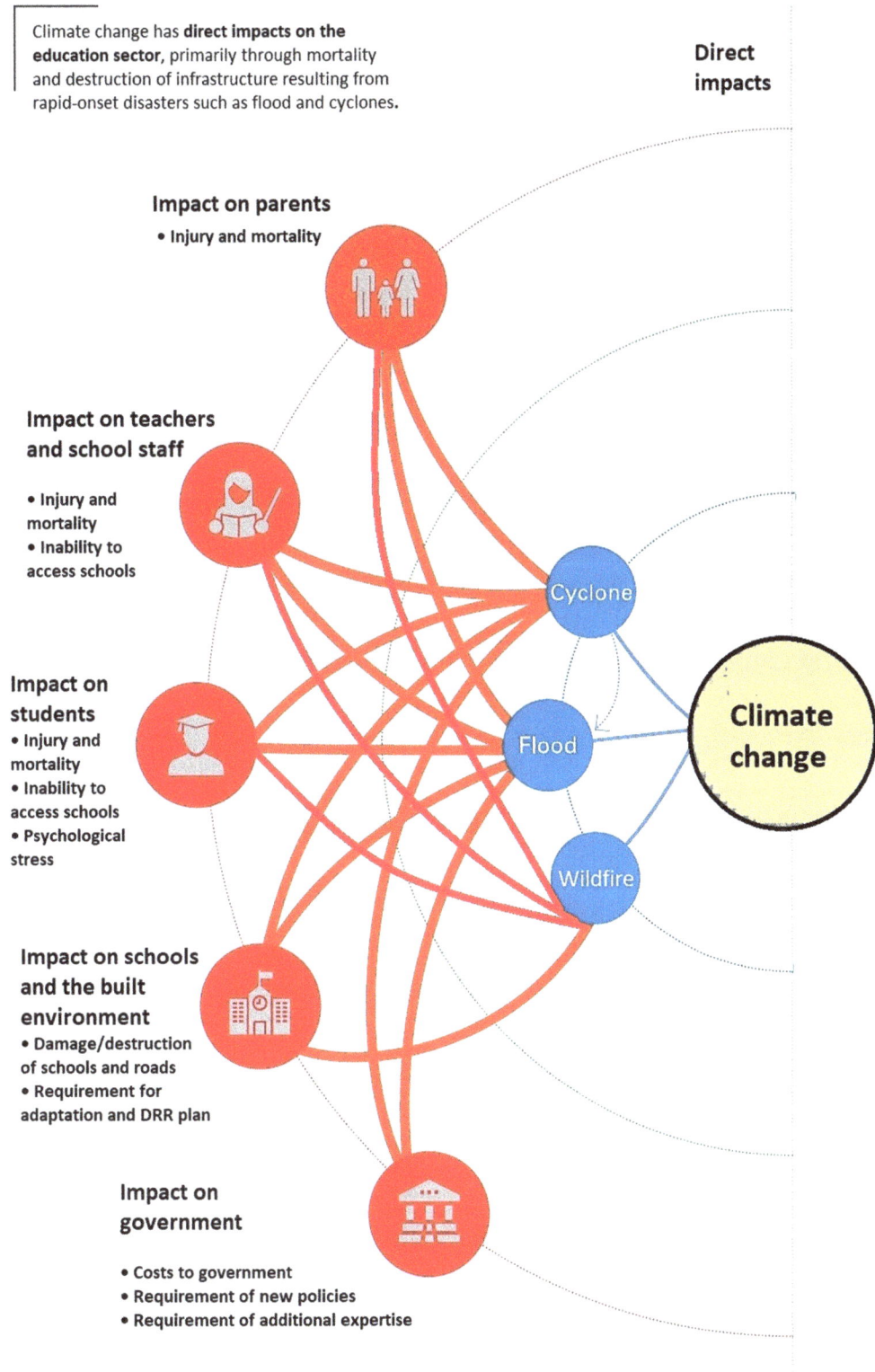

Fig. 8.6 Direct impacts of climate change on education. (Source: UNICEF (2019b))

risks (cyclones, floods, and wildfire). The second arc of red circles on the fringe of the diagram shows varied stakeholders experiencing the impact. These include parents, teachers, students, the built environment such as school buildings, and the government including education authorities. The red lines connecting the blue circles and the red circles indicate the direct pathways through which each of the stakeholders may be affected, and the impacts are listed with each stakeholder.

Climate-related extreme events—storms, floods, strong winds, and rains associated with cyclones, as well as wildfire—can all damage educational (schools) and transport infrastructure such as roads and bridges (Kousky 2016; Benson 1997; Arndt et al. 2015). Such destruction can disrupt education for days and even weeks—and in countries with limited alternative education modalities during disasters, this can lead to missed classes and lower academic performance. Table 8.3 provides some examples of extreme climate events that destroyed or caused damage to schools in selected countries and suspended learning for weeks.

These extreme events not only damage schools and the built environment but can also result in losses of textbooks and classroom materials such as desks and chairs, display boards, etc. For example, over 40,000 textbooks were lost in the aftermath of the 2000 floods in Cambodia (UNICEF 2019b). Wildfires also cause considerable damage to education particularly in Australia and in California State of the USA. A snapshot of damage to schools and education in California in the twenty-first century has been presented in Fig. 8.7.

The damage to schools and their educational infrastructure including furniture, books, etc. interrupts children's education in all cities and communities, but their impact is more devastating in nations where access to educational resources is either already under strain or limited. Following damage, repairs to schools and infrastructure tend to be slow and delayed in such areas.

Additionally, climate-related extreme events can cause injury to parents, who may not be able to take their young children to school, whereas injury to students and teachers also results in

Table 8.3 Climate-related hazards damaging or destroying schools in developing countries

Country	Climate event/ date	School closure/ damage
Cambodia	Destructive floods, 2000	155 schools closed for 9 weeks in Tonle Sap Basin
Indonesia	Flood event, 2015	351 schools submerged and closed for 3–14 days in North Jakarta
Mongolia	Multiyear 1999–2002 dzuds	50 schools reported heavy damage to roofs and schools
Myanmar	Cyclone Komen, 2015	4116 schools damaged and 608 were destroyed completely
Philippines	Typhoon Koppu, 2015	803 schools closed for about 14 days
Vanuatu	Cyclone Pam, 2015	50% of primary and secondary schools closed for about 1 month
Vietnam	Cyclone Damrey, 2017	Over 325 schools destroyed in four provinces—closed for 9 days
Sudan and Niger	Floods, August– September 2020	116 schools damaged
Bangladesh, India, Nepal	Monsoon floods, 2017	18,000 schools destroyed or damaged
Mozambique	Floods, 2000	More than 500 schools destroyed
Bangladesh	Cyclone Sidr, 2007	849 schools destroyed and over 3700 damaged
Myanmar	Cyclone Nargis, 2008	2250 schools completely collapsed and 750 damaged severely
Nicaragua, Guatemala	Hurricane Mitch, 1998	367 schools destroyed and 286 damaged
Philippines	Cyclone Durian, 2006	50–100% damage to school in five cities—worth US$20 million
Zambia	Floods, 2009	67 schools severely damaged
US Gulf states	Hurricane Katrina, 2005	56 schools destroyed and 1162 damaged

Sources: UNICEF (2019b), Save the Children (2020), OCHA (2017), and Patel (2008)

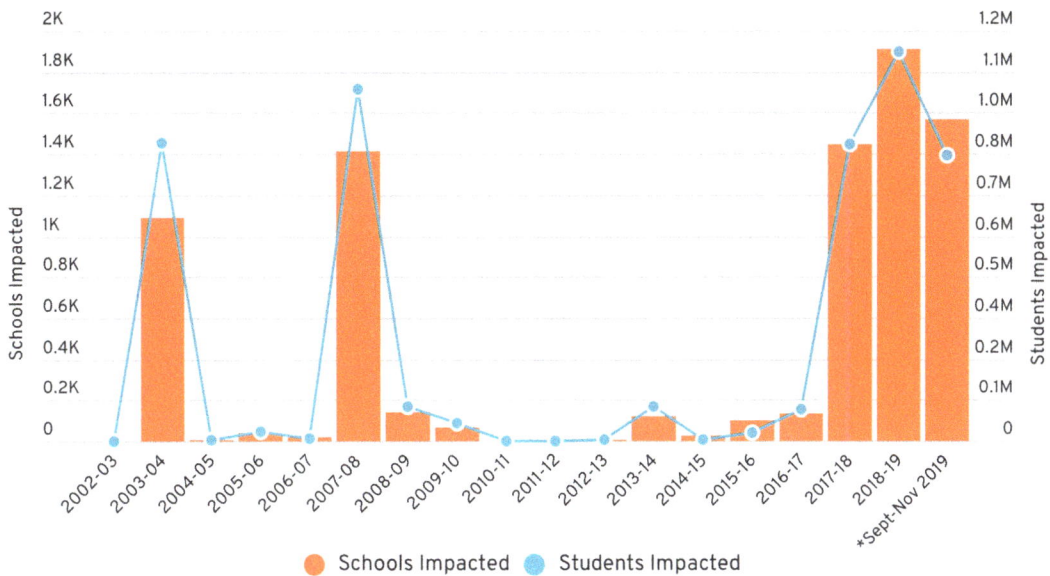

Fig. 8.7 Schools and students impacted by wildfires in California (USA) in the twenty-first century. (Source: Cano (2020))

absenteeism (UNICEF 2019b). In extreme cases, mortality may occur with devastating consequences—evidence shows that children and adolescents who lose a classmate, a teacher, or a parent after a major disaster are more likely to be traumatized. Further, they are likely to concentrate less during class, thereby compromising their academic performance (McMillen et al. 2002).

The consequences for educational authorities and the government or public sector can be substantial too. For example, significant amount of funds may be needed to support affected communities, teachers, and students and repairing infrastructure damaged by climate-related events. Moreover, governments and educational authorities may also need skills and additional resources for promoting resilience in the long term by developing policies and guidelines to better manage climate-related risks.

8.4.2   Indirect Impacts

The indirect pathways through which climate change and climate-related hazards impact education are shown in Fig. 8.8. The inner arc of the blue circles highlights seven major climate-related risks including rapid-onset hazards such as cyclones and floods, multi-causal disasters (dzuds) and slow-onset events like droughts, sea level rise, glacier melt, and rising temperatures. The second arc of green circles in the middle illustrates socioeconomic and environmental effects resulting from the climate-related hazards given in the first arc. Six socioeconomic and environmental conditions have been presented that are to be affected by climate change and which have an indirect impact on education and learning. These include food security and nutrition, livelihood and income, air pollution, water quality and availability, climate-sensitive diseases and health effects, and energy security. The third arc of red circles on the fringe shows various stakeholders that may be impacted—parents, teachers, students, schools and the built environment, and the government or public sector including educational authorities.

Akresh (2016) has added economic dislocation and forced migration to six socioeconomic conditions listed in the diagram which are basically related to livelihood and income. Most importantly, climate change disrupts basic requirements for health—clean water, clean air,

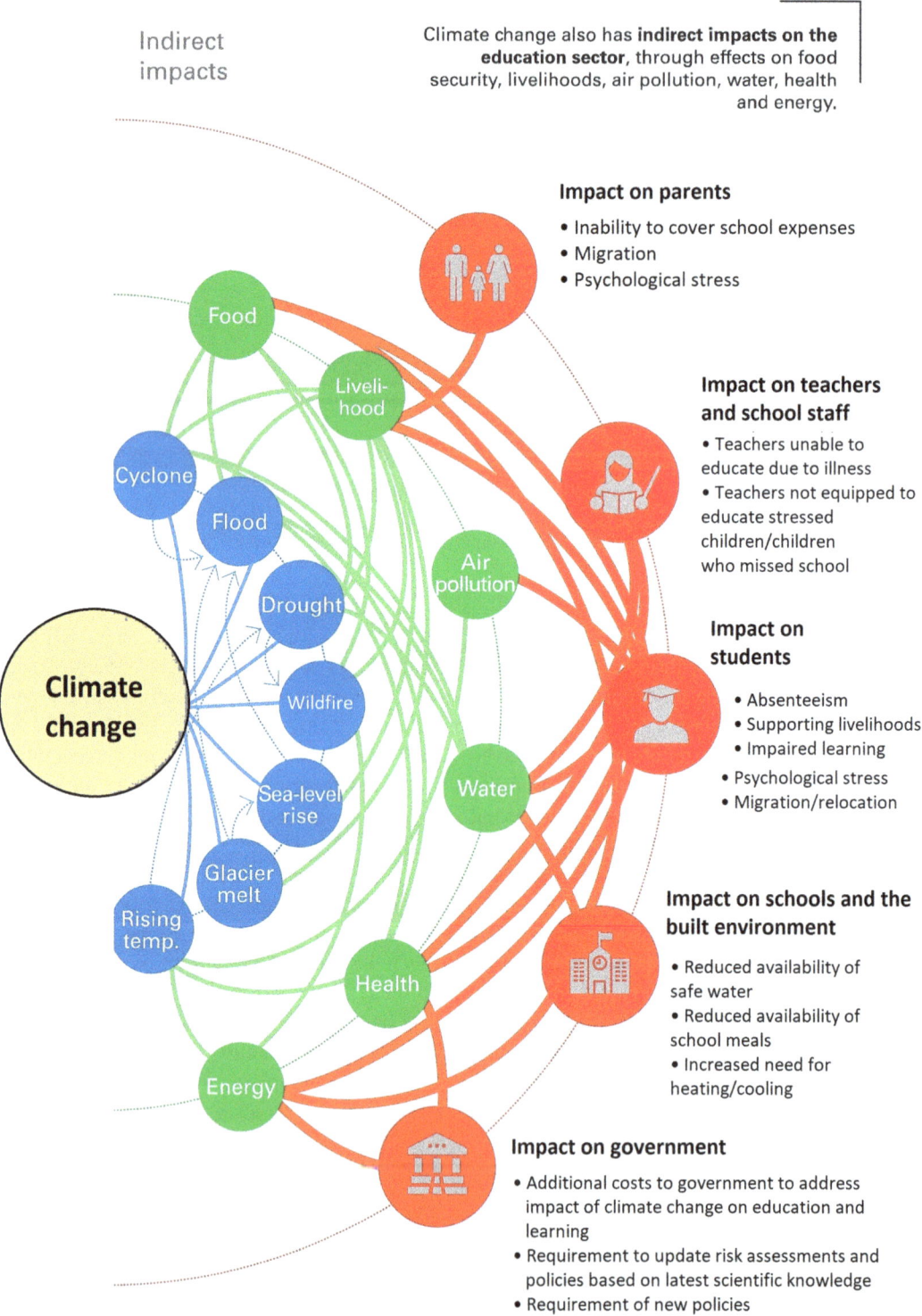

Fig. 8.8 Indirect impact of climate change on education. (Source: UNICEF (2019b))

and adequate food—thereby causing illness and death among all ages, but children and youth are worst affected, whose education and learning process is disrupted too. The World Health Organization estimates that children are to suffer more than 80% of the illnesses, injuries, and deaths attributable to climate change (McMichael et al. 2004). These can arise from both indirect long-term impacts (e.g., droughts, rising sea levels) and sudden-onset events (floods, wildfires), and many may have lifelong repercussions (Garcia and Sheehan 2016). Figure 8.9 shows the path through which climate change and related environmental, social, and economic changes affect children's well-being, health, and schooling. Threats to children's physical health have been well documented along with fatalities and injuries due to heat-related illnesses; exposure to environmental toxins through increased pollution; infectious, gastrointestinal, and parasitic diseases that are more prevalent in warmer temperatures (Sheffield and Landrigan 2010) and are also caused by polluted water and inadequate

sanitation; as well as malnutrition (USGCRP 2016).

Temperature increases, for example, especially in urban areas, can undermine air quality—increasing risks of respiratory diseases such as asthma and other illnesses to which children are very vulnerable (US EPA 2021), thereby reducing their ability to attend school. The impacts may, however, vary. For example, it was found that a high proportion of deaths occurred in Delhi (48%) during periods of high ambient temperatures among younger children (less than 15 years old), which was much higher than the proportion in São Paulo (10%), and enormously greater than the proportion in London (1%) (Hajat et al. 2005). This finding indicates that children in low-income countries may experience greater climate change-related health risks and hence more prone to learning losses than those in developed countries. This area, however, needs further investigation.

Children's education is also jeopardized when families can no longer afford to send their chil-

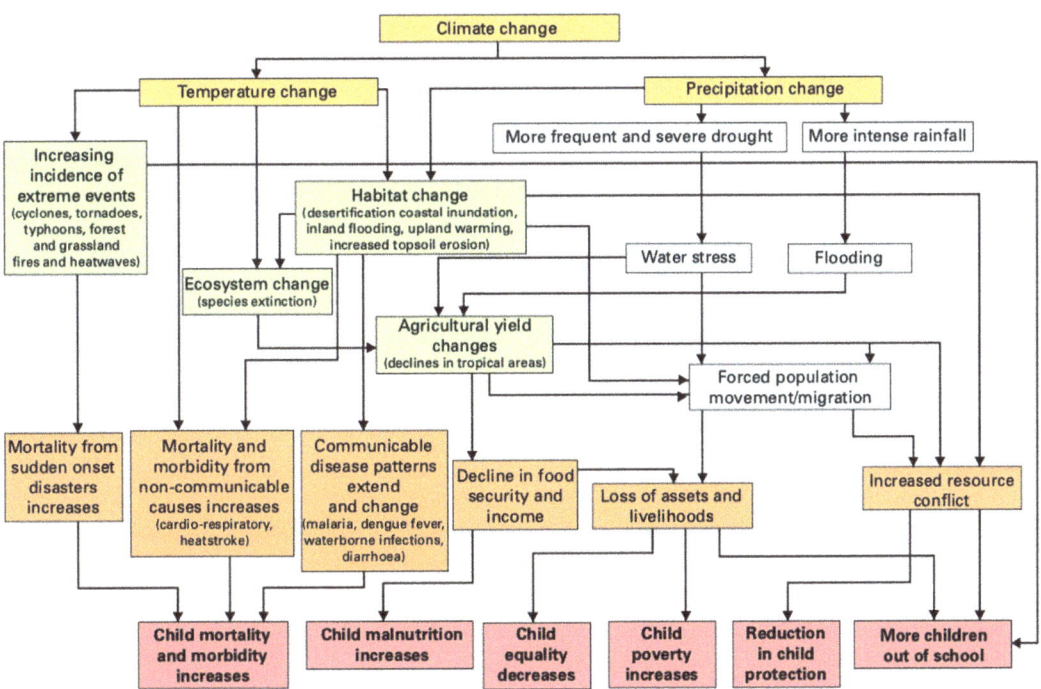

Fig. 8.9 The impact of climate change on children's health and their schooling. (Source: UNICEF (2012))

dren to school for loss of livelihoods or income reduction due to climate hazard (both rapid-onset like flood or cyclone or slow-onset hazards like droughts). Such phenomenon affects girls more than boys since their education tends to be less highly valued in many parts of the world (Gibbons 2014). For example, in droughts as well as floods in India, Indonesia, Mongolia, Nicaragua, Pakistan, and Uganda, significant declines in school attendance occurred, some of which lasted a decade (Plan International 2015). Furthermore, climate-induced displacements and forced migration lead to a range of symptoms of trauma and problems with adjustment (Pfefferbaum et al. 2016; Sanson et al. 2018a, b).

The psychological and mental health impacts of extreme climate events can be equally devastating (Burke et al. 2018; Clayton et al. 2017) with serious implications for educational attainments. Such impacts are in the form of significant increases in post-traumatic stress disorder (PTSD), depression and anxiety, sleep disturbances, cognitive deficits, and learning problems (Garcia and Sheehan 2016; Majeed and Lee 2017). For example, after the 2010 floods in Pakistan, 73% of 10- to 19-year-olds displayed high levels of PTSD—displaced girls were affected most seriously (Gibbons 2014). Extreme weather events have also caused distress, grief, and anger, loss of identity, feelings of helplessness and hopelessness, higher rates of suicide, and increased aggression and violence (Clayton et al. 2017), which seriously jeopardize education and learning.

The impact of sea level rise on education was noted in the coastal communities of Azerbaijan, where the rising level of the Caspian Sea resulted in a very significant drop in school attendance. Consultations with residents revealed that relocation of schools owing to sea level rise, deteriorating health among children, and loss of teachers due to financial problems contributed to the decline in attendance (Kudat et al. 1999).

While the direct impact of climate events in the form of damage to education infrastructure and mortality/injury has been relatively well-captured in disaster risk reduction, loss, and dam-

age literature (e.g., Gupta et al. 2004; Chinowsky et al. 2015; Mochizuki et al. 2015), the magnitude of indirect impacts is not so well covered. Nevertheless, emerging literature has increasingly been covering the indirect pathways through which climate can affect socioeconomic conditions (e.g., livelihoods and income) and ultimately education outcomes (Randell and Gray 2019).

8.4.3 Climate Change: Estimated Costs to the Education Sector

Global analyses of missed education show that limited educational opportunities for girls and barriers to completing 12 years of education cost countries between US$15 trillion and US$30 trillion in lost lifetime productivity and earnings (World Bank 2018). According to UNICEF (2019b), climate change is likely to increase these costs—by direct and indirect losses. Direct loss usually results from damages to schools and other education infrastructure as well as mortality of students, while the indirect losses are incurred through missed school. Together, the total cost of these impacts of climate change on education is likely to be in trillions of dollars.

8.5 Response to Climatic Impacts

Considerable work has been done across the world to address the impacts of climate change and climate-related risks on education and learning. It has included preparedness, emergency response, and medium- and long-term planning including adaptation to climate risks. An example of preparedness is Cambodia where the Ministry of Education, Youth and Sports decided to start and end the academic session 1 month later to prevent disruption in education—after recognizing that the flood season used to coincide with the beginning of the academic year (UNICEF 2019b).

8.5.1 Emergency Coping

Governments have explored the possibility of using alternative education modalities in climate emergencies to deal with severe disasters. For example, in Pakistan, over 12,000 children in the flood-stricken provinces in 2010 were given the opportunity to continue education in some 73 temporary learning centers initiated through UNICEF's support (UNICEF 2010). Likewise temporary learning spaces were set up in countries to ensure that education can continue even if school buildings are destroyed or access to schools is compromised—as in Bangladesh (Education Cluster Bangladesh and Nirapad 2014) or Cambodia (UNICEF 2019b).

Mongolia also piloted distance learning approaches with varying levels of success (UNICEF 2019b)—where mobile ger kindergartens were used tor nomadic families. Such mobile gers are more suitable and cost-effective for temporary use and can be used as temporary replacement for school buildings damaged by floods or heavy snowfall. Alternative education modalities, however, have been limited in the past, and students who missed classes due to extreme climate events often relied on remedial education or catch-up classes. Nonetheless, coronavirus pandemic has opened new avenues for remote learning. These need to be explored, tested, and where possible introduced as alternative education modalities during large-scale disasters resulting from extreme climate events—as happened during Pakistan's floods of 2010 that affected 11,000 schools and during cyclone Nargis in Myanmar, which affected 4000 schools (UNICEF 2012). In both instances, extremely large-scale disruptions were caused, forcing many children to lose months and months of education, contributing to big learning losses. These apart, even temporary disruptions to education can have long-term impacts on school attendance, with many students dropping out of school to never return. Therefore, focusing on education as a priority even during emergencies will mean that children are less likely to stop their education or drop out of school.

8.5.2 Medium- and Long-Term Actions

Regarding containment of climate change learning losses in the medium and long term, progress across the developing countries is fragmented. Much more work needs to be done to prepare educational stakeholders to manage the impacts of climate change and develop resilience against it. Ministries of education in developing countries still perceive their role in climate change to be limited to incorporation of climate change in curriculum or emergency response to climate-related events. This attitude was clearly reflected in CoP 26 in Glasgow in November 2021 during an Environment and Education Ministers meeting (Steer 2021). Delegates considered teaching of climate change issues and introduction of environmental education in the curriculum very important for lifestyle and behavioral change in managing climate change. However, the role of education in resilience or adaptation to climate impacts and climate justice was virtually nonexistent in the discussions and ministerial statements (Steer 2021).

While incorporating climate change science and management is an integral part of twenty-first century education, it is even more important for the ministries of education to play an active role in climate change adaptation and mitigation. Moreover, it is crucial for them to participate in disaster risk reduction efforts that encompass the comprehensive school safety framework. The concept of Comprehensive School Safety Framework (CSS) evolved through a movement of international humanitarian agencies, national governments, nongovernmental organization, researchers, practitioners, and advocates, who coalesced to identify strategies to address school safety in the presence of disaster risks (Varchetta 2019). The goal of CSS is to protect students and teachers from death, injury, and harm, plan for the continuity of education through not only climate but all expected hazards and threats, safeguard education sector investments, and strengthen risk reduction and resilience by educating people (UNISDR and GADRRRES 2017).

Save the Children's 2016 Comprehensive School Safety (CSS) Baseline Dataset identified the presence of CSS policies across 68 countries in three regions of the world: Africa, Asia Pacific, and Latin America and Caribbean. A master's thesis (Varchetta 2019) analyzed these results, which indicated that overall, countries studied have adopted about 48% of CSS policies. Asia Pacific countries were found to be ahead in adopting policies compared to the two other sampled geographic regions (Latin America including Caribbean and Africa). The analysis revealed that most countries have adopted disaster management policies that address the education sector. Further, they have also enacted policies for safer school construction, although fewer have allotted funding for multi-hazard risk assessment and retrofit of schools identified for reconstruction. Less than half use schools as temporary shelters. Some 25% include climate change and disaster risk reduction in their school curriculum, but far fewer train teachers in these subjects. The evidence of disaster impacts and advocacy have proved important facilitators in CSS policy development, while insufficient funding and poor technical capacity were the impeding factors. The overall results show policy gaps and practices that require attention.

While talking of Comprehensive School Safety, it is also important to mention that a publication—"Guidance Notes on Safer School Construction"—has been issued by the Global Facility for Disaster Reduction and Recovery (GFDRR 2009) which gives information on the needs and steps for establishing safer schools and basic design principles to safeguard from geo-hazards as well as climate-related extreme events: windstorms, floods, and wildfires. Similarly, a guidance book has been prepared by International Finance Corporation for disaster and emergency preparedness of school. A resource manual has also been provided by UNICEF (2012) on climate change adaptation and disaster risk reduction in education sector. As discussed earlier, however, "climate-proofing" the education sector goes far beyond simply improving the quality of school buildings and infrastructure to ensure resilience to floods, storms, droughts, and sea level rise. It involves targeting the entire education sector—which means building an education system that identifies immediate and long-term key climate risks, assesses vulnerabilities and exposure of children and education systems, and prioritizes and implements climate adaptation actions, in education as well as monitors and evaluates the effectiveness of these actions.

8.6 Conclusion

Both the mega risks coronavirus and climate change have had negative impacts on the global objective of providing universal education to all—as articulated in Sustainable Development Goal 4. Infectious diseases have disrupted the education systems in the world in the past, but one does not find any parallel to the magnitude and scale of the disruption caused by the ongoing pandemic. It has affected nearly 1.6 billion students in about 190 countries. The concern is that even after the advent of new year 2022, 21 months since the start of the pandemic, schools remain closed for millions of children and youth, while millions more are at risk of never returning to schools. In terms of monetizing the learning losses due to the pandemic alone, simulations show that the current generation of students may have lost an estimated 17 trillion US dollars in lifetime earnings at present value. This projected loss is equivalent to 14% of today's global GDP. The other mega risk—climate change—has affected 450 cities world over, each with a population of 1 million or more. They face recurring climate-related extreme events like cyclones, floods, etc., the deadliest and costliest of disasters. They directly affect education system through denied schooling time and lost instruction by destroying schools and relevant infrastructure and learning materials. The climate change also impact education indirectly by affecting the family livelihood and well-being through social and environmental channels like health, food, and shelter, again resulting in increased absenteeism, dropouts, and deteriorating quality of learning. Their financial impacts are also colossal and believed to be in trillions of dollars.

Besides education and learning, schools also provide critical services to students; hence, their closures whether due to pandemic or climate-related extreme event also deprive children of opportunities for their growth and development by denying them—school meals, health services including vaccination, and psychosocial support. The disadvantaged and underprivileged children and youth are the worst sufferers. Long school closures also result in decreased content coverage and test scores and erode students' academic confidence and perception of themselves. Progress in the gender equality is under threat too—for example, the pandemic has placed an estimated 10 million more girls at risk of early marriage in the next decade. Further, school closures due to any mega risk put children and youth into workforce causing an increase in school dropout rates. UNESCO predicts that 24 million students may drop out of the school system as a result of the pandemic alone. Additionally, both mega risks coronavirus pandemic and climate change made children and youth susceptible to lingering and compound psychosocial impacts such as depression, anxiety, sleep disorders, and behavioral problems. Children' safety has been jeopardized too with increased domestic violence due to parents' loss of job or income. An effective way of supporting children under stress and adversity is play, but school closures cut or limit access to playful spaces.

School closures increased exponentially educational institutions' reliance upon technology—digital solutions or broadcast media like radio and television. Higher learning institutions and universities also shifted to online, distance, or hybrid modes of instruction due to international travel restrictions. Globally, online platforms were the most common remote learning strategy, but the digital divide made it a remote learning paradox with 1.3 billion school-age children lacking Internet access at home. Electricity availability was another problem in such regions as sub-Saharan Africa, where only 47% of the population has electricity access, posing a barrier to learning even through broadcast media. At least 463 million children worldwide were not reached by digital and broadcast remote learning programs. The paradox is that countries with longer school closures had lower rates of Internet connection or digital access at home for school-age children.

The choice of remote learning technology in countries depended on their economic development or income group within which they fell. For example, radio-based instruction was more common in nations belonging to low-income or lower-middle-income group. In contrast, online platforms and TV-based instruction were more common in a large group of middle-income countries. Higher-income countries relied principally on online platforms. The deployment, application, and effectiveness of remote learning programs in cities and communities varied greatly. The key obstacles in a shift to distance learning during COVID-19, noted by various studies, included lack of technical resources, differential access to education, difficulties in adaptation, and resistance to change. These studies have also presented the beneficial side of COVID-19—mainly as a springboard for future promotion of alternatives such as digitalization, promotion of 5G technologies, and public-private educational collaborations. Undeniably, the abrupt or sudden changes disrupted the general "comfort zone" of teachers, but the situation is likely to improve over time, with experience and a harmonious mix of online and off-line programs. The digital divide, however, will need to be mitigated by developing cooperation between developed and developing nations.

The crises also had silver lining in facilitating educational institutions to connect students globally through such means as Zoom, Microsoft Teams, and interactive webinars. Moreover, the educational solutions and innovations not only attracted the attention of typical government-funded or nonprofit-backed social project, but investment from the private sector. The pandemic could pave the way for much larger-scale coalitions of companies and corporations and public-private partnerships to promote remote education. This may require more investments from public sector which is the main provider of education.

However, an analysis of massive stimulus packages deployed by countries in response to the crisis shows that very limited resources were allocated to the education and training sector. On average, countries allocated 3% to education, but in low- and lower-middle-income countries, that figure was less than 1%.

Overall, the principal lesson from the pandemic for both mega risks is to develop a more resilient education system. The one capable of dealing with all situations be it for emergency or for learning during school closures or opening schools safely after closure. It should also cater for accelerating recovery or building teachers and digital capacity in the medium term or for realizing a hybrid system of learning in the long term. Preparing for the new normal, it is important to make the education sector risk proof—by integrating health, climate, and education policies for promoting risk-resilient education system and ensuring support to students, teachers, and families under different risk scenarios. The pandemic has provided motivation, impetus, and opportunity to all those involved in education to speed up reform and prepare future generations to be both productive and resilient to face both mega risks through education and practice. The key to building forward better is building system-wide resilience so that lessons from the current pandemic response will help to build crisis response systems against all future disruptions be that from climate-related disasters, conflicts, and closures of public facilities and strikes, not just health-related disruptions.

References

ADB Asian Development Bank (2021) COVID-19 and education in asia and the pacific: guidance note, manila

Akresh R (2016) Climate change, conflict, and children. Futur Child 26:51–72. https://doi.org/10.1353/foc.2016.0003

Arndt C, Tarp F, Thurlow J (2015) The economic costs of climate change: a multi-sector impact assessment for Vietnam. Sustainability 7(4):4131–4145

Azevedo JP (2020) Learning poverty: measures and simulations. Policy research working papers. The World Bank. https://doi.org/10.1596/1813-9450-9446

Azevedo JP, Montoya S, Akmal M, Wong YN, Gregory L, Geven KM, Cloutier MH, Iqbal SA, Imhof AG, de Andrade Falcão N, Kouame CS (2021a) Learning poverty updates and revisions: what's new? Working paper. World Bank, Washington, DC. https://openknowledge.worldbank.org/handle/10986/36082

Azevedo JP, Hasan A, Goldemberg D, Geven K, Iqbal SA (2021b) Simulating the potential impacts of COVID-19 school closures on schooling and learning outcomes: a set of global estimates. World Bank Res Obs 36(1):1–40. https://doi.org/10.1093/wbro/lkab003

Benson C (1997) The economic impact of natural disasters in Fiji. Overseas Development Institute (ODI)

Bhamani S, Makhdoom AZ, Bharuchi V, Ali N, Kaleem S, Ahmed D (2020) Home learning in times of COVID: experiences of parents. J Educ Educ Dev 7(1):9–26. https://doi.org/10.22555/joeed.v7i1.3260

Bhatia A et al (2021) Violence against children during the COVID-19 pandemic. Bull World Health Organization 99(10):730–738. https://doi.org/10.2471/BLT.20.283051

Bhengu L (2021) Gauteng records more than 23 000 teen pregnancies in one year, some moms as young as 10. News. https://www.news24.com/news24/southafrica/news/gauteng-records-more-than-23-000-teen-pregnancies-in-one-year-some-moms-as-young-as-10-20210817

Bignardi G, Dalmaijer ES, Anwyl-Irvine AL, Smith TA, Siugzdaite R, Uh S, Astle DE (2021) Longitudinal increases in childhood depression symptoms during the COVID-19 lockdown. Arch Dis Child 106(8):791–797. https://doi.org/10.1136/archdischild-2020-320372

BizED (2020) International student mobility and the impact of pandemic. Retrieved from bized.aacsb.edu: https://bized.aacsb.edu/articles/2020/june/covid-19-and-the-future-of-international-student-mobility. Accessed 18 Jul 2020

Borokowski A, Bundy DAP, Burbano C, Hayashi S, Lloyd-Evans E, Neitzel J, Reuge N (2021) COVID-19: missing more than a classroom. The impact of school closures on children's nutrition. Innocenti Working Paper. UNICEF. https://www.unicef-irc.org/publications/1176-covid-19-missing-more-than-a-classroom-the-impact-of-school-closures-on-childrens-nutrition.html

Bourgault S, Peterman A, O'Donnell M (2021) Violence against women and children during COVID-19 – one year on and 100 papers: a fourth research round up. Center for Global Development, Washington, DC. https://www.cgdev.org/publication/violence-against-women-and-children-during-covid-19-one-year-and-100-papers-fourth. Accessed 12 Jul 2021

Brossard M, Carnelli M, Chaudron S, Di-Gioia R, Dreesen T, Kardefelt-Winther D, Little C, Yameogo J-L (2021) Digital learning for every child: closing the gaps for an inclusive and prosperous future. Task force 4 digital transformation. Think 20 Italy 202. https://www.t20italy.org/wp-content/uploads/2021/08/TF4-PB2-Brossard.pdf

Burke S, Sanson A, Van Hoorn J (2018) The psychological effects of climate change on children. Curr Psychiatry Rep 20:35. https://doi.org/10.1007/s11920-018-0896-9

Cano R (2020) Wildfires set school plans ablaze for more than 70,000 students, CAL MATTERS. https://calmatters.org/education/2020/08/wildfires-school-plans-students/

Cappa C, Jijon I (2021) COVID-19 and violence against children: a review of early studies. Child Abuse Negl 116(Pt 2):105053. https://doi.org/10.1016/j.chiabu.2021.105053

Chinowsky P, Schweikert A, Strzepek N, Strzepek K (2015) Road infrastructure and climate change in Vietnam. Sustainability 7(5):5452–5470

Chuang E, Pinchoff J, Psaki S (2018) How natural disasters undermine schooling. https://www.brookings.edu/blog/education-plus-development/2018/01/23/how-natural-disasters-undermine-schooling/anuary

Clayton S, Manning C, Krygsman K, Speiser M (2017) Mental health and our changing climate: impacts, implications, and guidance. American Psychological Association and ecoAmerica, Washington, DC. Retrieved from http://ecoamerica.org/wp-content/uploads/2017/03/ea-apa-psych-report-web.pdf

COVID-19 Global Education Recovery Tracker (2021) Education systems response to COVID-19, Brief September 10th 2021. https://www.covideducationrecovery.global/stories/2021-09-10-brief/

Crawford J, Butler-Henderson K, Rudolph J, Malkawi B, Glowatz M, Burton R et al (2020) COVID-19: 20 countries' higher education intra-period digital pedagogy responses. J Appl Learn Teach 3(1). https://doi.org/10.37074/jalt.2020.3.1.7

Dhawan S (2020) Online learning: a panacea in the time of COVID-19 crisis. J Educ Techn Syst 49(1):5–22. https://doi.org/10.1177/0047239520934018

Dorn E, Hancock B, Sarakatsannis J, Viruleg E (2021) COVID-19 and education: the lingering effects of unfinished learning. https://www.mckinsey.com/industries/education/our-insights/covid-19-and-education-the-lingering-effects-of-unfinished-learning

Edmonds EV, Schady N (2012) Poverty alleviation and child labor. Am Econ J Econ Pol 4(4):100–124. https://doi.org/10.1257/pol.4.4.100

Education Cluster Bangladesh and Nirapad (2014) Education in emergency: exploring options for continued education during disasters in Bangladesh. https://reliefweb.int/sites/reliefweb.int/files/resources/04.Study-on-Education-in-Emergency_15-June-2014.pdf

European Commission (2020) Educational inequalities in europe and physical school closures during covid-19. Science for Policy Brief. Fairness Policy Brief Series. European Commission. https://ec.europa.eu/jrc/sites/default/files/fairness_pb2020_wave04_covid_education_jrc_i1_19jun2020.pdf

Garcia D, Sheehan M (2016) Extreme weather-driven disasters and children's health. Int J Health Serv 46:79–105. https://doi.org/10.1177/0020731415625254

GFDRR Global Facility for Disaster Reduction and Recovery (2009) Guidance notes on safer school construction. http://gadrrres.net/uploads/files/resources/INEE_Guidance_Notes_Safer_School_Constr_EN.pdf

Gibbons E (2014) Climate change, children's rights, and the pursuit of intergenerational climate justice. Health Hum Rights 16:19–31. https://www.hhrjournal.org/2014/07/climate-change-childrens-rights-and-the-pursuit-of-intergenerational-climate-justice/

Gupta AD, Babel MS, Ngoc PH (2004) Flood damage assessment in the mekong delta, Vietnam. In Proceeding of the second Asia Pacific Association of Hydrology and Water Resources Conference, Singapore (Vol. 1, pp. 109–117)

Hajat S, Armstrong BG, Nelson G, Wilkinson P (2005) Mortality displacement of heat-related deaths: a comparison of Delhi, São Paulo, and London. Epidemiology 16:613–620. https://doi.org/10.1097/01.ede.0000164559.41092.2a

International Association of Universities (2020) The impact of COVID-19 on higher education around the world. Retrieved from iau-aiu.net: https://www.iau-aiu.net/IMG/pdf/iau_covid19_and_he_survey_report_final_may_2020.pdf

Islam N, Beer M, Slack F (2015) E-learning challenges faced by academics in higher education: a literature review. J Educ Training Stud 3(5). https://doi.org/10.11114/jets.v3i5.947

Khan MA (2021) COVID-19's impact on higher education: a rapid review of early reactive literature. Educ. Sci. 11:421. https://doi.org/10.3390/educsci11080421

Kousky C (2016) Impacts of natural disasters on children. Future Children:73–92

Kudat A, Musayev A, Ozbilgin B (1999) Social assessment of the Azerbaijan National Environmental Action Plan: A focus on community responses to Caspian Sea environmental disaster (Social Development Papers, No. 32). The World Bank, Washington, DC

Lee J (2020) Mental health effects of school closures during COVID-19. Lancet Child Adolesc Health 4(6):421. https://doi.org/10.1016/S2352-4642(20)30109-7

Majeed H, Lee J (2017) The impact of climate change on youth depression and mental health. Lancet Planet Health 1:e94–e95. https://doi.org/10.1016/S2542-5196(17)30045-1

Mavhunga C (2021) Zimbabwe reports major rise in teen pregnancies during pandemic. VOA. https://www.voanews.com/a/covid-19-pandemic_zimbabwe-reports-major-rise-teen-pregnancies-during-pandemic/6204648.html

McMichael AJ, Campbell-Lendrum D, Kovats S, Edwards S, Wilkinson P, Wilson T et al (2004) Global climate change. In: Ezzati M, Lopez A, Rodgers A, Murray C (eds) Comparative quantification of health risks: global and regional burden of disease due to selected major risk factors. World Health Organization, Geneva, pp 1543–1649

McMillen C, North C, Mosley M, Smith E (2002) Untangling the psychiatric comorbidity of posttraumatic stress disorder in a sample of flood survivors. Compr Psychiatry 43(6):478–485

Ministerio de Educación Nacional de Colombia (2021) Datos SIET. https://www.mineducacion.gov.co/portal/micrositios-superior/Educacion-para-el-Trabajo-y-el-Desarrollo-Humano-SIET/Sistema-de-Informacion-Para-el-Trabajo-y-Desarrollo-Humano-SIET/353023:Datos-SIET

Mochizuki J, Vitoontus S, Wickramarachchi B, Hochrainer-Stigler S, Williges K, Mechler R, Sovann R (2015) Operationalizing iterative risk management under limited information: fiscal and economic risks due to natural disasters in Cambodia. Int J Disaster Risk Sci 6(4):321–334

NSC Research Center (2020) COVID-19: transfer, mobility, and progress – first look fall 2020. https://nscresearchcenter.org/transfer-mobility-and-progress/covid19-transfermobilityprogress-firstlookfall2020/

Nyamweya N (2020) Displacement, girls' education and COVID-19. Global Partnership for Education. https://www.globalpartnership.org/blog/displacement-girls-education-and-covid-19

OCHA Office for the Coordination of Humanitarian Affairs (2017) South asia floods. Reliefweb. https://reliefweb.int/report/bangladesh/south-asia-floods-18000-schools-damaged-and-fears-thousands--children-may-never-go

Oulo B, Sidle AA, Kintzi K, Mwangi M, Akello I (2020) Understanding the barriers to girls' school return: girls' voices from the frontline of the COVID-19 pandemic in East Africa. Echidna Giving and Amplify Girls. https://static1.squarespace.com/static/5c86d4507fdcb8fc46e7d529/t/60d4fb74d835f0200cdea521/1624570756879/Understanding+the+Barriers+to+Girls%27+School+Return+–+Girls%27+Voices+from+the+frontline+of+t   he+COVID-19+Pandemic+in+East+Africa.++.pdf

Ozamiz-Etxebarria N, Idoiaga Mondragon N, Bueno-Notivol J, Pérez-Moreno M, Santabárbara J (2021) Prevalence of anxiety, depression, and stress among teachers during the COVID-19 pandemic: a rapid systematic review with meta-analysis. Brain Sci 11:1172. https://doi.org/10.3390/brainsci11091172

Pannen P, Riyanti RD, Ridwan (2021), Indonesia cyber education institute, quality of online education in Indonesia, conference paper. https://www.researchgate.net/publication/351548440_Indonesia_Cyber_Education_Institute_Assuring_Quality_of_Online_Education_in_Indonesia

Patel M (2008) Disaster prevention for schools, guidance for education sector decision makers. UNISDR, Geneva

Pfefferbaum B, Jacobs AK, Van Horn RL, Houston JB (2016) Effects of displacement in children exposed to disasters. Curr Psychiatry Rep 18:71. https://doi.org/10.1007/s11920-016-0714-1

Plan International (2015) We stand as one: children, young people, and climate change. https://www.plan.org.au/-/media/plan/documents/resources/we-stand-as-one-children-young-people-and-climate-change.pdf

Pokheral S, Chethhri R (2021) A literature review on impact of COVID-19 pandemic on teaching and learning. Higher Educ Future 8(1):133–141

Psacharopoulos G, Patrinos HA (2018) Returns to investment in education: a decennial review of the global literature. policy research working paper no. 8402. World Bank, Washington, DC. © World Bank. https://openknowledge.worldbank.org/handle/10986/29672 License: CC BY 3.0 IGO

Racine N, McArthur BA, Cooke JE, Eirich R, Zhu J, Madigan S (2021) Global prevalence of depressive and anxiety symptoms in children and adolescents during COVID-19: a meta-analysis. JAMA Pediatr 175(11):1142–1150. https://doi.org/10.1001/jamapediatrics.2021.2482

Randell H, Gray C (2019) Climate change and educational attainment in the global tropics. Proc Natl Acad Sci USA 116(18):8840–8845

Reimers FM, Marmolejo F (2022) Leading learning during a time of crisis. Higher education responses to the global pandemic of 2020. In: Reimers FM, Marmolejo FJ (eds) University and school collaborations during a pandemic. knowledge studies in higher education, vol 8. Springer, Cham. https://doi.org/10.1007/978-3-030-82159-3_1

Sanson AV, Burke SEL, Van Hoorn J (2018a) Climate change: implications for parents and parenting. Parenting 18:200–217. https://doi.org/10.1080/15295192.2018.1465307

Sanson AV, Wachs TD, Koller SH, Salmela-Aro K (2018b) Young people and climate change: the role of developmental science. In: Verma S, Peterson A (eds) Developmental science and sustainable development goals for children and youth, Social Indicators Research Series, 74. Springer, New York, pp 115–138. https://doi.org/10.1007/978-3-319-96592-5_1

Save the Children (2020) Tens of thousands forced out of school by floods in sub-saharan africa. https://www.savethechildren.net/news/tens-thousands-forced-out-school-floods-sub-saharan-africa

Save the Children (2021) The impact of COVID-19 on children in west and central africa: learning from 2020. https://resourcecentre.savethechildren.net/pdf/rapport_covid_anglais.pdf/

Schwartz HL, Diliberti MK, Berdie L, Grant D, Hunter GP, Setodji CM (2021) Urban and rural districts showed a strong divide during the COVID-19 pandemic. Rand Corporation

Sharma, M., P. Idele, A. Manzini, CP. Aladro, A. Ipince, G. Olsson, P. Banati, and D Anthony. 2021. "Life in lockdown. Child and adolescent mental health and well-being in the time of COVID-19." Florence: UNICEF Office of Research – Innocenti

Sheffield PE, Landrigan PJ (2010) Global climate change and children's health: threats and strategies for prevention. Environ Health Perspect 119:291–298. https://doi.org/10.1289/ehp.1002233

Solis SL, Liu CW, Popp JM (2020) Learning to cope through play: playful learning as an approach to support children's during times of heightened stress and adversity. LEGO Foundation. https://www.legofoundation.com/media/3298/learning-to-cope-through-play.pdf

Steer L (2021) COP 26, a turning point for education as a key solution for climate challenges, global center on adaptation. https://gca.org/cop26-a-turning-point-for-education-as-a-key-solution-for-climate-challenges/

Treve M (2021) What COVID- 19 has introduced into education: challenges facing higher education institutions (HEIs). Higher Educ Pedagogies 6(1):212–217. https://doi.org/10.1080/23752696.2021.1951616

UNESCO (2020) How many students are at risk of not returning to school. https://education4resilience.iiep.unesco.org/en/resources/2020/unesco-covid-19-education-response-how-many-students-are-risk-not-returning-school

UNESCO (2021a) School closure calendar, as of the end of October 2021

UNESCO (2021b) Adverse consequences of school closures. https://en.unesco.org/covid19/educationresponse/consequences

UNESCO (2021c) Uneven global education stimulus risks widening learning disparities. https://unesdoc.unesco.org/ark:/48223/pf0000379350

UNESCO (2021d) COVID-19: reopening and reimagining universities, survey on higher education through the UNESCO national commissions. Paris. https://unesdoc.unesco.org/ark:/48223/pf0000378174

UNESCO, UNICEF, World Bank, and OECD (2021) What's next? Lessons on education recovery: findings from a survey of ministries of education amid the COVID-19 pandemic. World Bank, Washington, DC. https://openknowledge.worldbank.org/handle/10986/36393

UNESCO-UNICEF-World Bank Joint Survey (2020). http://tcg.uis.unesco.org/survey-education-covid-school-closures

UNICEF (2010) In makeshift temporary classrooms, children in flooded Pakistan go to school. https://reliefweb.int/report/pakistan/makeshift-temporary-classrooms-children-flooded-pakistan-go-school

UNICEF (2012) Climate change adaptation and disaster risk reduction in education sector. Resource Manual, New York

UNICEF (2019a) The state of the world's children, 2019, children food and nutrition: growing well in a changing world. United Nations Children's Fund, New York. https://www.unicef.org/media/63016/file/SOWC-2019.pdf

UNICEF (2019b) It is getting hot: call for education systems to respond to the climate crisis. https://www.unicef.org/eap/reports/it-getting-hot

UNICEF (2021a) Reopening with resilience: lessons from remote learning during COVID-19, Florence, Italy

UNICEF (2021b) Education disrupted: the second year of the COVID-19 pandemic and school closures. United Nations Children's Fund, New York. https://data.unicef.org/resources/education-disrupted/

UNICEF and ILO (2021) Child labour rises to 160 million – first increase in two decades. https://www.unicef.org/press-releases/child-labour-rises-160-million-first-increase-two-decades, New York

UNICEF and International Telecommunication Union (2020) How many children and young people have internet access at home? Estimating digital connectivity during the COVID-19 pandemic. United Nations Children's Fund, New York. https://www.unicef.org/media/88381/file/How-many-children-and-young-people-have-internet-access-at-home-2020.pdf

UNICEF, FAO, IFAD, WFP, and WHO (2021) The state of food security and nutrition in the world 2021. https://data.unicef.org/resources/sofi-2021/

UNICEF United Nations Children's Fund (2020a) Covid-19: are children able to continue learning during school closures? A global analysis of the potential reach of remote learning policies using data from 100 countries. UNICEF, New York. https://data.unicef.org/resources/remote-learning-reachability-factsheet/

UNICEF United Nations Children's Fund (2020b) Policy brief: COVID-19: a reason to double down on investments in pre-primary education. https://www.unicef-irc.org/publications/1137-covid-19-a-reason-to-double-down-on-investments-in-pre-primary-education.html

UNISDR, & GADRRRES (2017) Comprehensive school safety: a global framework in support of the global alliance for disaster risk reduction and resilience in the education sector (No. STC00972; p. 8). United Nations Office for Disaster Risk Reduction, Global Alliance for Disaster Risk Reduction in the Education Sector, Save the Children

United Nations (2020) Policy brief: education during COVID- 19 and beyond, New York

US EPA, United States Environmental Protection Agency (2021) Air quality index and health. https://www.airnow.gov/air-quality-and-health/

USGCRP United States Global Change Research Program (2016) In: Crimmins A, Balbus J, Gamble JL, Beard CB, Bell JE, Dodgen D, Ziska L (eds) The impacts of climate change on human health in the United States: A scientific assessment. USGCRP United States Global Change Research Program, Washington DC. https://doi.org/10.7930/J0R49NQX

Varchetta A (2019) Evaluating comprehensive school safety through a global baseline survey of disaster risk reduction policies in the education sector WWU graduate school collection 878. https://cedar.wwu.edu/wwuet/878

Villegas CC, Peirolo S, Rocca M, Ipince A, Bakrania S (2021) Impacts of health-related school closures on child protection outcomes: a review of evidence from past pandemics and epidemics and lessons learned for COVID-19. Int J Educ Dev 84(July):102431. https://doi.org/10.1016/j.ijedudev.2021.102431

Wang Y, Avanesian G, Kamei A, Mishra S, Mizunoya S (2021) Which children have internet access at home? Insights from household survey data. Evidence for Action May 21. https://blogs.unicef.org/evidence-for-action/where-do-children-have-internet-access-at-home-insights-from-household-survey-data/

World Bank (2018) World development report, Washington DC

World Bank (2020) Pivoting to inclusion: leveraging lessons from the COVID-19 crisis for learners with disabilities. https://thedocs.worldbank.org/en/doc/147471595907235497–0090022020/original/IEIIssuesPaperDisabilityInclusive EducationFINALACCESSIBLE.pdf

World Bank (2021a) What is learning poverty? https://www.worldbank.org/en/topic/education/brief/what-is-learning-poverty

World Bank (2021b) Learners with disabilities and COVID-19 school closures: findings from a global survey conducted by the world bank's inclusive education initiative. World Bank, Washington, DC. https://openknowledge.worldbank.org/handle/10986/36326

World Bank (2021c) Policy actions for school reopening and recovery. https://www.worldbank.org/en/news/factsheet/2021/04/30/notes-on-school-reopening-and-learning-recovery

World Bank, UNESCO, and UNICEF (2021) The state of the global education crisis: a path to recovery. World Bank, UNESCO, and UNICEF, Washington DC, Paris, and New York

World Economic Forum (2020) Three ways the coronavirus pandemic could reshape education. Retrieved July 17, 2020, weforum.org: https://www.weforum.org/agenda/2020/03/3-ways-coronavirus-is-reshaping-education-and-what-changes-might-be-here-to-stay/

Yusuf M (2020) Teen pregnancies spike in kenya as schools remain shuttered. VOA. https://www.voanews.com/a/episode_teen-pregnancies-spike-kenya-schools-remain-shuttered-4344051/6111054.html

Zemrani B, Gehri M, Masserey E, Knob C, Pellaton R (2021) A hidden side of the COVID-19 pandemic in children: the double burden of undernutrition and overnutrition. Int J Equity Health 20(1):44. https://doi.org/10.1186/s12939-021-01390-w

Mega Risks, Social Protection, and Sustainability

9

9.1 Introduction

Social protection in this work has been defined as actions, programs, or policies that are being undertaken to help individuals and societies in managing the adverse social effects of mega risks and protect them from poverty and deprivation which are socially unacceptable within a given urban society. Both the mega risks have caused tremendous social and economic hardships to city dwellers especially vulnerable groups across the world. The pandemic's unprecedented magnitude in such a short time brought forth an intensive or widespread need for social protection responses (Hebbar and Phelps 2020). The huge increases in global poverty in 2020 because of COVID-19 exemplify the multidimensional impacts of a mega risk. As far as climate crisis is concerned, even before the pandemic, it had threatened to push over 130 million more people into poverty in the next decade alone. It is for the same reason that a recent brief entitled "Social Protection and Climate Change: Scaling up Ambition" (Costella et al. 2021) considers it essential to have risk management systems in place to save livelihoods, property, and lives.

This chapter has been divided into four sections. This introduction is followed by a section which discusses the short- and long-term social impacts of mega risks both COVID-19 and climate crisis on city residents. It covers financial insecurity; job and income shocks; housing-related problems, e.g., rents; utilities; health coverage; social isolation; and mental health. The section highlights how the two mega risks have exposed and exacerbated existing inequalities in cities and, in so doing, affected some groups of people more than others particularly the most disadvantaged and vulnerable workers and groups—poor and slum dwellers, immigrant workers, ethnic and racial minority, elderly, women, and children. The next section discusses the cities' response to social impacts, challenges faced, and lessons learned for sustainability and resilience. Besides reviewing short-term relief and recovery measures, the section examines effectiveness of existing or upgraded social assistance programs and measures undertaken by governments including social safety nets. Actions undertaken by civil society and community-based efforts have also been included in the discussion. The concluding section highlights the findings of the chapter.

9.2 Social Impacts of Mega Risks in Cities

The cities are bearing the brunt of both mega risks COVID-19 and climate crisis with a major impact on urban poor and vulnerable population. This section investigates the social problems and stresses faced by city residents due to the twin mega risk with the focus on the main effects and consequences that have been produced on the urban social system.

9.2.1 COVID-19 Social Impacts

The COVID-19 crisis has brought forward deep-seated inequalities and significant gaps in social protection coverage and inadequacies in them across cities and countries. The pandemic's socioeconomic impacts have made it difficult for policymakers to ignore several population groups—including children, older persons, and women and men working in diverse forms of employment and particularly those engaged in the informal economy—who were covered either inadequately or not at all by existing social protection measures.

The coronavirus crisis besides being a health emergency also brought an economic recession and social crisis. The social impacts of the COVID-19 pandemic on cities have varied widely depending on (i) the development status of the city, (ii) scale of the outbreak of the virus, (iii) policy responses, (iv) the preponderance of socially vulnerable population or deep-rooted inequalities in cities, and (v) how resiliently populations were able to adapt to the pandemic.

Although the pandemic had an impact on the cities universally, it had particularly harsh and inequitable impacts in cities of the developing world for several reasons. They are deficient in resources and house over 1 billion people in urban informal settlements or slums, who live in extremely overcrowded dwellings with inadequate infrastructure, and minimal services (Wilkinson 2020). Simultaneously, they house most of 2 billion informal workers of the world (ILO 2021a). The challenges of the lockdowns have been felt most severely by these informal workers, the poor, and the vulnerable, who are often without savings and safety nets. The actual vulnerability of the people to the pandemic and their resilience to the related social transformation in cities have depended on several variables including demographic characteristics such as age, gender, race, and socioeconomic status: the characteristic of city and place—including quality and size of dwellings and workplace, access to public services such as reliable energy and clean water, access to digital services, and pattern of urban governance. Deep-rooted inequalities in

cities have influenced how seamless and workable a transition to the new realities of life under the pandemic has been, and how resiliently populations have been able to adapt, in both the developed and developing countries (Boza-Kiss et al. 2021). Among the challenges faced due to the pandemic in both developing and developed countries included high levels of socioeconomic insecurity due to job loss or reduction in work hours, health, and associated harms, rising poverty and sharpening inequality, and a fragile social contract and predicaments faced by vulnerable groups.

9.2.1.1 Socioeconomic Insecurity

The crisis exposed the vulnerability of billions of not only poor but also those who appeared to be getting by relatively well but were not adequately protected from the socioeconomic shock of the pandemic (ILO 2021b). According to the International Labour Organization (ILO 2021a), losses in working hours in 2020 alone were approximately four times greater compared to the 2008/2009 financial crisis—the group that suffered worse included women, youth, and self-employed and low-skilled workers. During the pandemic, especially in its initial phase, strict lockdowns, economic downturn, and disrupted supply chains had serious impacts on employments and earning. Further, human rights violations in the form of police brutality and forced evictions (Kihato and Landau 2020) caused increased suffering. Informal livelihoods, which employed 2 billion informal workers worldwide (including 75% of all workers in Africa and India), were devastated. As a result, many cities saw enhanced poverty and food insecurity particularly among informal workers. A survey of informal workers in Durban revealed that four-fifths of informal workers experienced incidents of hunger among adult members of their household during lockdown. In households with children, 90% reported incidents of hunger among children (WIEGO and AeT 2021).

The COVID-19 pandemic had contrasting health and related social impacts across nations, cities, communities, and social groups. Although urban areas have been the centers of the pandemic, no decisive correlation between infection rates and city size or density was found. The key factors responsible for rates of infection and mortality have already been discussed in detail in Chaps. 2 and 7. Unequal access to basic services including health services, poverty and overcrowded living conditions, preexisting health conditions, and some forms of proximity jobs (UN Habitat 2021) and public policy played a major role. For example, pandemic's health, social, and economic burdens have been staggering in most developing countries of the Global South (Sverdlik and Walnycki 2021). However, cities in some developing countries like Vietnam (Lakner et al. 2021) and Senegal (Kirby 2021) and developed countries such as Korea and New Zealand successfully kept their caseloads low. The growth of more transmissible variants of the virus affected cities in both developed and developing countries; however, the impact has been devastating in the Global South. The situation was aggravated by overstrained health systems, shortage of vaccines, and mounting policy and governance challenges to contain the new strain of virus. A major concern is that the short-term health burdens of the pandemic are likely to have major long-term impacts on the economic and social well-being of people due to delayed recovery.

Rising Poverty The World Bank (Lakner et al. 2021) estimated that COVID-19 pushed 119–124 million people into poverty in 2020. The estimated number of COVID-19-induced poor in 2021 was set to rise to between 143 and 163 million (Lakner et al. 2021). The main causes pointed out for this growth in poverty were economic stagnation, job losses, and the decrease of about 40% remittances to low- and middle-income countries (United Nations 2020a). Large income and employment losses occurred in several cities and countries, at least for temporary period (Hill and Narayan 2020). The result of phone surveys by the World Bank indicated that an average of 62% of households across 27 developing countries suffered income loss between April and July 2020 (World Bank 2020). The poverty crisis, however, has aggravated enormously in 2022

when more than a quarter-billion (263 million) more people are likely to fall into poverty, according to Oxfam International (2022) due to impacts of COVID-19, rising inflation, and the war in Ukraine. According to the report, altogether as many as 860 million people could be living in extreme poverty (less than $1.90 a day) by the end of 2022.

Food Insecurity A closely related problem to poverty was food security and nutrition that resulted from income losses, and price hikes leading to hunger. Some 720–811 million people in the world faced hunger in 2020—an increase of as many as 161 million from 2019 (FAO, IFAD, UNIICEF, WFP and WHO 2021). Among the respondents surveyed by the World Bank (2020), 38% across 20 developing countries reported that an adult in their household had skipped a meal in the last month before the survey due to lack of money or resources (World Bank 2020). The problem of food security was acute with almost 135 million persons worldwide facing acute food shortages even before the coronavirus crisis (WFP 2020a). The number of these people has now gone up to 276 million people (WFP 2022), and the World Food Programme estimates it to rise to 323 million people by the end of 2022 driving at least 47 million people in 81 countries to the edge of famine (WFP 2022). The agency reported that the COVID-19 impact on livelihoods and on vulnerable people's access to food forced WFP to expand its program into cities at an unprecedented scale in most countries, where it operates (WFP 2020b). The pandemic disrupted food systems in many urban areas and impacted production and supply chains globally. In cities of developing countries, it devastated the local livelihoods of informal food vendors who were severely hindered during lockdowns. In the wake of increased food prices and supply shortages, the poor suffered more from insufficient food and inadequate nutrition (FAO 2020). The pandemic has clearly shown the importance of livelihood security and social protection measures in facing future shocks.

Eviction and Homelessness The pandemic also triggered disruptions in the security of housing tenure due to growing income instability that build up into mortgage and rental arrears. An Aspen Institute study in the USA, in August 2020, for example, estimated that between 30 and 40 million people were at risk of eviction at that time. An overwhelming majority (80%) of those facing evictions in cities across the country belonged to ethnic minorities—Black and Latino communities (Benfer et al. 2020). The developing country cities were in even worst situation—for example, in Nairobi, thousands of people were forced out of their homes with little warning, and even without compensation or alternative housing, and in Kariobangi informal settlement, for example, about 8000 residents were forced out of their homes, which were destroyed despite a court order in place prohibiting the eviction (UNHCR 2020). Similar situations were faced by residents of other East African cities—some 65,000 people were evicted in Somalia in the first half of 2020 including more than 33,000 in Mogadishu alone (NRC 2020). Even in developed country like the USA, landlords were able to evict tenants if they did not file a hardship form or if there were errors in their paperwork. The Princeton University's Eviction Lab's (2021) data collected from 27 cities in the USA indicated that almost a quarter of a million (247,463) evictions took place between the beginning of the pandemic and February 13, 2021. A major concern around the risk of evictions and foreclosures was an increase in homelessness and a further rise in transmission. At present, there are an estimated 330 million homeless urban households worldwide (UN Habitat 2021), a figure projected to grow by 440 million (a total of 1.6 billion people) by 2025 in the absence of strong measures to tackle the problem. Besides undertaking emergency measures, it underscores the need for the implementation of long-term social protection measures to reduce vulnerabilities around tenure security and accommodating homeless population.

Psychological and Mental Health Issues The pandemic and its multifaceted and repetitive virus variant shocks also had impact on psychological and mental well-being of people. The changes brought in daily lives, together with the feeling of loneliness, job losses, financial difficulty, and grief over the loss of loved ones, had the potential effect on the mental health of communities (Pietrabissa and Simpson 2020) particularly in cities. In the USA, a comparative study of National Health Interview Survey of 2019 and Household Pulse Survey of January 2021 showed that average share of adults reporting symptoms of anxiety disorder and/or depressive disorders rose from 11% between January and June 2019 to 41% in January 2021 (Kaiser Family Foundation 2021). Likewise, another Kaiser Family Foundation health tracking poll found specific negative impacts of the pandemic on citizens' mental health and well-being including difficulty sleeping (36%) or eating (32%), increases in alcohol consumption or substance use (12%), and worsening chronic conditions (12%). These visible consequences of the pandemic are likely to considerably outlive the pandemic because the crisis of this nature, even when its most severe impacts are short-lived, is likely to spark processes that enhance preexisting gaps in capacities and endowments within a society. The processes once they start are likely to continue during the recovery phase and may persist even afterward when economic activities rebound. Hence, it is important for the policymakers to act immediately to reverse the processes that pose longer-term risks to equity (Hill and Narayan 2020). According to Pierce et al. (2020), while the pandemic had greatest physical health risk to older people, the mental health of the younger was affected disproportionally. Moreover, the most vulnerable people like poor, racial, and ethnic minorities were most severely affected because they endured other hardships and discriminations as well as institutional barriers to upward mobility.

9.2.1.2 Sharpening Inequalities

The crisis, while increasing poverty, and food insecurity, has sharpened and reinforced the inequalities between and within cities and communities. Temporarily closing the economy sent shock waves through cities—exposing the fault lines created by race, gender, and age even in developed cities. Simultaneously, it exacerbated the multiple deprivations and divide created by the social inequalities between slum and informal vs non-slum dwellers in developing cities. All in all, preexisting inequalities worsened, made economic resiliency weak, and reinforced disadvantage. Several studies conducted in the developed countries (Adams-Prassl et al. 2020; Gonzalez et al. 2020; Martin et al. 2020; Witteveen 2020) have highlighted how the pandemic has exacerbated inequalities.

Racial Minorities In developed cities, the pandemic affected most the racial minorities and people at the bottom of the socioeconomic ladder. For example, in New York City, the mortality rate of the Black and Latino people was double that of Whites. They suffered more because of their bigger exposure to risks, economic problems, and lack of access to services (Wade 2020). Likewise, along economic lines, in areas with poverty levels of at least 30% in New York City, average death rates were 232 per 100,000 compared to 100 per 100,000 persons in areas where poverty level was less than 10% (Durkin et al. 2020). Findings of a study (Perry et al. 2021), in another area that examined employment and food, housing, and financial insecurity in a population-representative sample of Indiana residents in the USA comparing data before COVID-19 and during the initial stay-at-home orders, are also relevant in this regard. The research observed that the pandemic's socioeconomic shocks affected vulnerable groups disproportionately. According to the study, these findings were consistent with patterns of inequality found in climate-related disasters, such as Hurricane Katrina, the Chicago heat wave, and other shocks such as economic recession of 2008/2009. The study concluded that as with previous disasters, additional surges of the pandemic are likely to escalate short-term hardships that may translate into social devastation possibly translating into durable inequality.

Slum Dwellers and Urban Poor In developing cities, the pandemic hit slum dwellers and urban poor harder. A combination of factors such as crowded living conditions and stark reality of deficiencies in basic infrastructure and services, as well as precarious livelihoods, operated in tandem to make life difficult (Wasdani and Prasad 2020). Evidence showed that these factors were repleted in many cities of Asia, Africa, and South America. In India, for example, per capita floor area for the poorest 60% of city dwellers is 72 square feet which falls even below the recommended floor area for prisoners; the conditions were even worse in slums (Biswas 2020). Kihato and Landau (2020) have discussed similar issues in the context of some African and de Oliveira and de Aguiar (2020) and Castro (2020) for Brazilian cities. Among the latter, one study conducted in Belém metropolitan region of Brazil showed that 14.5% of the residents there were unable to control the viral contamination because of either poor housing or crowded living conditions with three or more people staying in the same room without access to basic sanitation. The study used a model in the metropolitan region that was run both—with and without social inequality scenario. It revealed that social inequality was a very strong factor in the propagation of COVID-19, which also affected the intensity of damage and had accelerated impact on the collapse of health infrastructure (Cardoso et al. 2020). Analysis of the pandemic situation in São Paulo, Brazil, showed similar results—where residents living in poorer areas were ten times more likely to die from the virus than the inhabitants of affluent areas (Mello 2020). In India, antibody tests conducted on a sample of thousands of people in three wards of Mumbai City in July 2020 also pointed out that on average, 57% of those residents living in slums had been infected compared to only 16% living in the non-slum areas (NITI Ayog et al. 2020).

Migrant Workers Migrant populations were also badly affected by the pandemic, particularly because they were housed in large numbers in substandard conditions. According to OECD (2018), an average of 17% of immigrants are living in overcrowded dwellings in the region, compared to only 8% of native residents. It is no surprise therefore that studies in several OECD countries indicated a higher infection risk among immigrants—at least twice as high as that for the native residents (OECD 2020a). In Singapore, overcrowded dormitories of immigrant workers and detention centers where they were held in the wake of lockdown caused a high prevalence of virus infections. Similar patterns were observed in other Southeast Asian and Gulf countries of the Middle East, where large migrant populations were concentrated in cramped and inadequate serviced conditions, ripe for virus attack. Qatar's industrial zone, for example, that housed many migrant workers, was a hotspot (Reuters 2020). The impacts of COVID-19 on the 13 million migrant workers from Bangladesh to other areas particularly the Middle East and their 30 million dependents studied by Karim et al. (2020) showed other impacts besides health—in shrinking remittance flows, depleted savings, and the emergence of socioeconomic crisis.

Gender   In terms of gender, the women faced greater risks of job and asset losses and suffered more due to food insecurity and domestic violence (Wenham et al. 2020). Gender gap was significant in business closure rates as women entrepreneurs were disproportionately affected by the economic contraction during the pandemic. Globally, a survey carried out in May 2020 involving 26,000 business owners and managers in 50 countries (using a Facebook business page) revealed that female-owned businesses were 5.9% points more likely to have closed their businesses than male-owned businesses. The primary reason for the gender gap in business closure was that most women-owned businesses depend on consumer-facing services including hospitality, retail, etc., where the demand shock was one of the highest (Gavas and Pleeck 2021). The pandemic also saw a highly inequitable declining paid work among women even in developed countries like the USA and UK (Hill and Narayan 2020), particularly because their majority was employed in informal jobs and sectors like tourism, leisure, and hospitality as well as increased caring duties that were hard hit.

Limited or nonexistent childcare also affected their jobs and business.

Women also shouldered the rising family care burdens. The care duties of women became especially challenging when they enhanced in tandem with severe shortfalls in water, sanitation, and clean energy—the deficiencies that are very common in informal settlements (Sverdlik and Walnycki 2021). Women were also disproportionately affected by insecurity of housing tenure—for example, 57% of women in the informal settlements, according to a survey organized by UN Habitat (2020a), in Yangon reported insecurity due to fear of eviction, compared to 49% of men. In another UN Habitat (2020b) survey of 16 informal settlements on the island of Viti Levu, Fiji, over three-fourths of all respondents who felt insecure about evictions were women. Domestic violence against women was also triggered and enhanced due to the pandemic enforced restrictions and lockdowns. Such violence happened in many parts of the world, but in Latin American cities, its magnitude was so high that a UN Habitat (2021) report refers to it as a "pandemic of violence," as it resulted in a flurry of femicides too (Prusa et al. 2020). In several cases, the need to escape from an abusive household resulted in homelessness. Hence, to deal with it expeditiously, refuges and shelters had to be established by city authorities to accommodate victims of domestic violence. For example, in Brussels, a hotel had to be requisitioned by city authorities to provide shelter to victims of domestic abuse during lockdown (Galindo 2020).

9.2.2   Climate Change Social Impacts

While the coronavirus pandemic was ravaging cities, extreme weather events also continued to strike countries and cities. Four hundred thirty-three extreme climate events resulted in over 17,000 casualties and affected some 139 million people from the beginning of the pandemic to August 2021 only. The two mega risks, in this scenario, simultaneously affected livelihoods, creating compound impacts, and reducing resilience to future shocks by making people poorer and more vulnerable. In Kenya, for instance, the combination of COVID-19 and floods in 1 year and the virus and droughts in the other aggravated food insecurity for the most vulnerable in both urban and rural areas. In Bangladesh, early action before Typhoon Amphan, and later ahead of massive flooding, was adjusted to reduce infection risk, and saved countless lives. Climate risks also overlapped with COVID-19 in the richest countries. For example, in Canada and the northwestern USA, an unprecedented heat wave killed hundreds of people. Options for vulnerable groups to seek shelter posed trade-offs with the risk of infection. Additionally, many risks associated with COVID-19 and climate-related disasters were amplified in urban settings—where poor informal workers were doubly hit by the rising cost of staples and limited wage-earning options due to the pandemic shutdowns.

9.2.2.1 Socioeconomic Insecurity

The climate change, on its own, is already having adverse and multidimensional impacts on human well-being and socioeconomic insecurity in cities globally. Rising temperature and increasing extreme weather events related to climate change have been contributing to asset loss/damage and enhanced poverty, hunger, and social inequities. They have also created nutritional and health problems and placed lives and livelihood at risk (Costella et al. 2021; Pozarny 2016) while sharpening inequalities. Some communities or vulnerable groups of people such as poor, racial, and ethnic minorities, women, aged, and people with preexisting illness or diseases and with disabilities are exposed to and affected by climate threats disproportionately. Differential vulnerability results from several factors including poverty or unequal resource access, cultural practices, governance shortcomings and knowledge nonavailability, etc.

Urban Poverty The poor and disadvantaged groups of people are more exposed and vulnerable to climate hazards often because they tend to live in slums—more often in hazard-prone zones along riverbanks, on hillsides and slopes prone to landslides, and along waterfronts in coastal areas susceptible to coastal flooding or sea surges or areas affected by seasonal storms, etc. This is especially true for developing country cities, where more than one billion people live in slums and informal settlements, often occupying sites highly vulnerable to climate change-related extreme events. The impacts of climate change itself exacerbate poverty as individuals lose their livelihoods and possessions particularly through climate-related sickness or loss of assets due to extreme weather events. A vicious cycle is thus triggered whereby marginalized groups bear the greatest burdens, making them more vulnerable to future change preventing them from escaping poverty.

The vulnerability of a city's residents to climate change depends not only on the nature and magnitude of disaster event but also on the socioeconomic characteristics of its population. For example, two or more cities that experience the same category of flood or cyclone or heat wave may experience varied mortality levels, economic losses, and social impacts depending upon relative wealth, infrastructure, and sociodemographic characteristics. The reasons why most of the poor inhabitants of informal settlements are so much at risk is because of low-quality housing with inadequate foundations, high levels of overcrowding, lack of infrastructure, etc. Many of these poor and low-income groups of people live in housing without even fans what to say of air-conditioning. Hence, during heat waves, the very young, elderly, and people in poor health are particularly at risk. In the heat wave of 2015 in India, the fifth deadliest in the world history, 2300 people were killed (Di Liberto 2015). In many Asian and African cities, slum dwellers are persistently confronted with floods—like the residents of slums near Lyari River in Karachi, Pakistan. In Accra, Ghana, more than ten floods occurred between 1995 and 2007 resulting in deaths, household displacements, and huge loss of properties, infrastructure, and capital (Aboagye 2012).

Food Insecurity Climate change has affected food security in all its dimensions—access, availability, utilization, and stability (FAO 2015). It is also causing destruction of crops, and livelihoods, while undermining people's ability to feed themselves (WFP 2022). With unabated emissions, "The Future We Don't Want" analysis suggests that 2.5 billion people, living in over 1600 cities, will experience declining agricultural outputs (C40 Cities undated). A snapshot of the scenario is available from a UN Habitat, WFP report hunger and malnutrition in sub-Saharan Africa, where hunger has long been associated with rural areas. The pandemic exacerbated climate change affects has brought forward the changing face of hunger, exposing vulnerabilities of the urban poor (UN Habitat 2022). With over 90% of the COVID-19 cases recorded in cities, an estimated 68.1 million women, men, and children among the urban population face the risk of acute food insecurity in sub-Saharan Africa.

Psychological and Mental Health Issues Climate change also has impact on psychological and mental well-being of people. The impact of climate change on mental health can be direct or indirect and can be both short term and long term (Cianconi et al. 2020). Direct impact can occur immediately as in case of heat waves. The indirect effect, on the other hand, can be both in the short term due to floods, tornadoes, and hurricanes and in the long-term because of prolonged droughts or rise in the sea levels resulting in forced migration, etc. The extreme weather events can cause psychiatric conditions such as PTSD and mood disorders including depression, anxiety, increased suicide rate, and substance use, as well as increased aggressive behavior. In addition, other extreme weather events can act through mechanisms like traumatic stress, leading to well-understood psychopathological patterns (Cianconi et al. 2020). The consequences of exposure to extreme or prolonged climate events can also be delayed leading to post-traumatic stress or disorders. These can also be transmitted, even to later generations (Cianconi et al. 2020). Compared to others, the vulnerable populations

such as racial and ethnic minorities, poor, the elderly, women, and children, as well as those with lack of social network, have a higher probability of developing psychiatric illnesses or psychopathologies due to climate change. Among such vulnerable groups, those with previous psychiatric diseases may worsen their mental condition.

9.2.2.2 Sharpening Inequalities

Climate change has sharpened or exacerbated existing inequalities (IPCC 2014, p. 796) because the disadvantaged households suffer disproportionately from both direct and indirect effects of climate hazards (Islam and Winkel 2017). For example, the destruction of assets by climate change-induced flooding is a direct effect, whereas the increasing climate hazards cause the insurance premiums to rise, making it difficult for the disadvantaged households to purchase the coverage. This is the indirect, market-mediated effect (Islam and Winkel 2017). Both direct and indirect causes enhance inequality of disadvantaged households, which aggravates in three major ways: (a) increased exposure to climate hazards, increased susceptibility to damage resulting from climate hazards, and (c) reduced ability to cope with and recover from the damage. For example, inequality in affordability forces the disadvantaged group of households to live in flood-prone areas, thereby resulting in their exposure to floods resulting from climate change. Likewise, of all the people living in a flood-prone area, the houses of disadvantaged groups are more susceptible to flooding because they are made of flimsy materials. Moreover, rich people often have their houses insured and therefore they get compensated for any damage to their houses caused by the flood. In contrast, the disadvantaged households, simply because they cannot afford insurance, bear the brunt of the entire loss and have no or little ability to recover and thus get locked into a vicious cycle of poverty and inequality.

Even if equal exposure and susceptibility to damage is assumed between advantaged and disadvantaged households in a city or community, a highly unlikely case, the rate of recovery can be an important determinant in future inequality. If both the advantaged and disadvantaged households recover at the same rate, then the inequality (measured as proportion) will remain constant. However, if the disadvantaged groups fail to recover at the same rate as the advantaged ones, a very likely scenario is the increase in inequality beyond any doubt. Indeed, numerous studies (Kraay and McKenzie 2014; Verner 2010; Carter et al. 2007; Ravallion and Jalan 2001) have shown that recoveries of the disadvantaged groups from adverse impacts of climate change are slower. In fact, the studies by Barbier (2010) and Barrett et al. (2011) illustrate that the disadvantaged households due to the deficiency of resources are often forced to cope with climate hazards in such detrimental ways that they put their future adaptive and growth capacity at risk.

9.2.2.3 Differential Impacts

Different population groups may be affected differently by the same extreme weather event such as a flood or cyclone depending on affecting factors such as demographic features like race and ethnic composition, gender, age, socioeconomic status, characteristic of a city or a place, as well as quality and characteristics of dwellings. The magnitude of distributional impact of climate change in cities also depends on how many vulnerabilities the resident individual, household, or community faces (Thomas et al. 2018)—the more the number of vulnerabilities, the less the ability to cope with it. Some groups like indigenous people, minorities, and women may be affected more for two reasons—firstly because they face more than one vulnerability and secondly because often explicitly or tacitly, they are excluded from decision-making processes and, in some cases, have limited access to insurance, information, and resources (UN Habitat 2011). Hence, they are not only less prepared for physical hazards but are also less able to adapt. These effects are manifested more in developing cities compared to developed ones. However, there are many cases where poor communities even in developed cities fared worse than the wealthier groups during the same disaster. For example, during Hurricane Katrina in 2005 in New Orleans,

USA, residents without cars and financial resources could not evacuate and were left behind. Moreover, residents of some of the hardest hit low-lying neighborhoods were the poorest, who experienced the worst devastation (UN Habitat 2006).

Racial, Ethnic, and Other Minorities Racial, ethnic, and other minorities are increasingly vulnerable to climate change in developing as well as developed countries. Discriminatory practices and attitudes lead minorities to segregate into groups and locate into the hazard-prone and risky neighborhoods. This together with lack of access to insurance and loans as security against climate change impacts makes them highly vulnerable. The caste-based discrimination, although banned in India, continues and during a climate extreme event or calamity becomes a greater menace as the ostracized people are denied access to facilities. For example, during Cyclone Fani in 2019, elements of discriminatory behavior were observed in caste-based entry to shelters, in unfair access to tube wells, and in various other forms (Patra and Patro 2021). Similarly, when Hurricane Katrina struck New Orleans, the African Americans living in low-lying areas were hard hit. They experienced the worst flooding and hence suffered more severe housing damage which, in turn, led to their delayed return to the city in aftermath of the hurricane (Fussell et al. 2010).

Often, provision of government assistance following disasters has also been less accessible to racial and ethnic minorities (UN Habitat 2011). For example, in the USA, the Federal Emergency Management Agency's assistance to Haitian residents of Florida following tropical storm damage based on single family household proved insufficient because several families tended to occupy a single household (Fothergill et al. 1999). Assessments of the relief and reconstruction efforts following disasters have further revealed discriminatory practices and human rights abuses against women, the poor, indigenous groups, and the disabled (Brooking institution 2009). For instance, outright exclusion of certain groups from disaster relief was noticed during the flooding of the Koshi River in 2008 in India and Nepal (Brookings Institution 2009). Discriminatory policies in housing and urban development have also resulted in the development of risky climate-related neighborhoods. For example, in the USA, communities of color are at higher risk of heat-related illness or death than their white neighbors (Witze 2021; Hoffman et al. 2020). During the extreme heat, while people of rich neighborhood benefit from green parks, spacious lawns, and sprawling trees, residents of poor areas swelt in vast expanses of asphalt (Witze 2021) and face the worst impact of high temperatures.

Gender In general, women, especially poor women, are more likely than men to suffer both physically and socioeconomically when climate-related disasters strike. A tropical cyclone in Bangladesh killed five times as many females as males in 1991 (UNDP 2007). Higher impacts on women have been observed in developed country cities as well, particularly in poor communities. For example, about 70% of fatalities in the French heat wave of 2003 were women, although this number may be artificially high since there were more women than men in older age groups (Toulemon and Barbieri 2008). Similarly, a sample of 141 countries studied found that in unequal gender societies, more women died from disaster events than men over the period 1981–2002 (Neumayer and Plümper, 2007). In contrast, where access to resources and the social status of women were nearly equal to men, the mortality difference between the sexes was much smaller or, in fact, negligible (Neumayer and Plümper, 2007).

Women in developing cities, particularly poor women, have unequal access to resources, credit, and insurance; hence, they are more exposed to direct harm from climate-related extreme events such as flooding or cyclones. In terms of disaster-related displacements, studies have found that 80% of those displaced by climate change are women (WEN 2010). Moreover, their sociocultural roles and caregiving responsibilities often prevent women from not only migrating but also seeking shelter before and after disaster events.

Further, in some cases, it is not considered important for women to learn skills that could help them in their survival during a disaster. Lower-economic status of women also makes them more vulnerable to disasters. In addition, when destruction or damage to homes during climate-related events occur, often it affects income of women more than men—simply because their income earning activities are based in their homes that get destroyed with the destruction of their dwellings. The economic vulnerability of women to climate change also increases in both developing and developed countries due to restrictions on their livelihoods. Moreover, they often earn lower wages than men especially in developing cities for the same or equivalent work. Even in developed countries, gender differences in employment opportunities and pay result in increased poverty rates among women (Ruth and Ibarrarán, 2009). Together these factors limit women's ability to recover in the aftermath of disasters too.

Women also face the brunt of the economic and health impacts of more prolonged droughts, reduced food production, and severe weather events disproportionately. The world's poorest cities and countries are the most vulnerable, where the majority of the people living on less than $1 per day are women (UNFPA 2009). After extreme climate events, women are the ones who besides health problems including trauma also face domestic violence. The medical needs of women, at times, are inadequately met in disaster relief programs especially those related to reproductive and psychological health. For example, in the aftermath of Hurricane Katrina, women were 2.7 times more likely than men to display clinical symptoms of post-traumatic stress disorder (Overstreet and Burch 2009). In many cases, the disease symptoms remained untreated after the event for years because of limited access to public assistance programs and lack of health insurance (Overstreet and Burch 2009). In addition, trauma programs even if they existed were often not tailored to the specific, and sometimes unique, needs of women (WEDO 2008).

Women also have limited capacity to secure post-disaster economic relief aid, due to lacunae either in formal assistance policies or cultural norms (Enarson 2000). In Bangladesh, for example, women had difficulty traditionally in receiving relief aid after disasters because of the problem of waiting in long line—especially with their children at home, who needed their care. Expanded recovery systems that provide door-to-door service are needed to address this issue (Enarson 2000). Further, women-headed households are sometimes ignored or fail to receive the assistance in post-disaster relief. This happens especially in the relief programs that are designed to count only men into the workforce or where male-headed households get the privilege for relief aid (Enarson and Phillips 2008). Thus, in the aftermath of Hurricane Andrew in Miami, relief checks were distributed to men as traditional heads of households, ignoring the reality that many families were then headed by women (Enarson 2000). In addition, if men leave their families, as often occurs following a natural disaster, women may become ineligible for public assistance or may go unrecognized by the system (UN Habitat 2011).

Disabled, Aged, and Chronically Sick The vulnerability to climate change could also be due to disability, age, chronic illness, etc. Disabled, older persons, and children often have physical limits to their mobility and coping capacity, which may prevent or create problems for them in evacuation or seeking shelter in emergency situations. Elderly and disabled preexisting illnesses create additional handicap. Also, their bodies adjust rather slowly to physical conditions; therefore, they may not perceive excessive heat quickly enough to prevent heat stroke. Evidence suggests that the aged persons or elderly are also affected disproportionately—with higher injury rates after climate disasters and high mortality in heat waves (Bartlett et al. 2009). A study of ten Australian and two New Zealand cities found that 1100 people aged over 65 die every year in these cities because of heat waves (McMichael et al. 2003).

No doubt adaptive measures are available to help the elderly and disabled, but these mechanisms are often available to rich only. Further, elderly,

disabled, and children are more likely than young adults to require assisted transportation out of a dangerous situation. Lack of transportation and distrust of volunteers may also hinder them from receiving financial assistance from public recovery and aid programs (Langer 2004). The vulnerability of the elderly and disabled, to a large extent, is dependent upon their economic status. Nevertheless, most elderly and disabled live disproportionately under poverty because they have neither a sustained source of income nor adequate resources or asset endowments (Ruth and Ibarrarán, 2009). Although rates and magnitude of poverty are greater in developing countries, family support is somewhat better. In contrast, older and disabled persons may be financially a little better in developed country cities but are more likely to live alone and are socially more isolated.

Children are also less equipped to handle disaster risks resulting from climate change. They are more susceptible to diseases like diarrhea and malaria. In addition, they are affected more by the displacement resulting from climate-related disasters. The time they spend in emergency or temporary shelters after disasters or evictions often has serious impact on their life. It may lead to an erosion of the social controls that normally regulate behavior within households and communities. With the increasing climate-related extreme events, the number of displaced persons is growing, and the accompanying dysfunctional environment is becoming a norm for many children, hence affecting more of them, who are forced to spend their early years in these shelters.

9.3 Social Protection Policy Response to COVID-19 and Climate Change

The mega risks both COVID-19 and climate change have no doubt generated serious social effects across the urban systems. Nevertheless, they have also opened a window of opportunity to reevaluate the efficacy of social protection programs to increase the resiliency of cities in facing the ongoing and future risks. It is important to mention that definition of social protection and contents of its programs have had somewhat varied connotation in existing literature (Highet 2021; Abdoul-Azize and El Gamil 2020). For example, the International Labour Organization (ILO) in 102 conventions mentions social protection as social security. Fiszbein et al. (2014), on the other hand, consider it as "an instrument for the goals of reducing poverty, reducing inequality, reducing risk and vulnerability." They included social insurance, labor market interventions, and social assistance programs as the three main components of social protection. However, the World Bank Atlas of Social Protection Indicator of Resilience and Equity (ASPIRE), the most acknowledged classification of social protection (also followed by this study), incorporates social insurance, labor market, social assistance, and private transfer programs as well in its ambit (Abdoul-Azize and El Gamil 2020). Irrespective of the mechanism or target of social protection programs, the basic intention as noted by Norton et al. (2001) is "to promote dynamic, cohesive, and stable societies through increased equity and security." The objective according to Koehler (2011) is also to cut or reduce harmful coping strategies like selling off assets or resources to meet immediate needs or pulling children out of school to make them work to enhance the family or household income that can cause serious disadvantage to the household in the long term.

When coronavirus pandemic hit the globe, many countries were facing significant challenges in making social security a reality for all. Less than half the global population (46.9%) in 2020 was effectively covered by at least one social protection benefit (excluding healthcare and sickness benefits)—Sustainable Development Goal (SDG) indicator 1.3.1 (ILO 2021b). This meant that over half the world (53.1%)—some 4.1 billion people—were left completely unprotected. Moreover, there were significant inequalities across the world and even between and within regions and cities. For example, coverage rates in Europe and Central Asia were exceptionally high (83.6%), while the coverage in Africa was dismal at 17.4% and quite unsatisfactory in Asia and the Pacific and Arab states of the Middle East (Fig. 9.1). Spending on social protection also varied significantly—countries on average spent 12.8% of their GDP on social protection (excluding health), with a big contrast between countries. High-income countries spent 16.4% and low-income countries only 1.1% of their GDP on social protection.

The situation regarding coverage of vulnerable population worldwide was also not very good—among children only one in four children (26.4%) received a social protection benefit; among women with newborns, only 45% received a cash maternity benefit; and among persons with severe disabilities, only one in three (33.5%) received a disability benefit. Condition with respect to unemployment benefits was even worse with only 18.6% of unemployed were effectively covered globally (ILO 2021b). Regarding elderly, while 77.5% of people above retirement age received some form of old-age pension, there were major disparities between regions, among cities and rural localities, and across gender (ILO 2021b).

9.3.1   COVID-19 and Social Protection Policy Response

9.3.1.1   Policy Measures Adopted by Governments

The gravity of socioeconomic problems posed by COVID-19 triggered an unparallel social protection policy response. Governments at both national and city/local levels undertook measures for social protection as a frontline response to protect people's health, job/business, and incomes. They also initiated actions to promote and ensure social stability. According to the World Bank (Gentilini et al. 2021), between March 20, 2020, and May 14, 2021, a total of 3333 social protection measures were planned or implemented in 222 countries or territories (Fig. 9.2)—an increase of nearly 148% since December 2020. While social assistance and insurance increased by about 120% and 110%, respectively, active labor market interventions surged by nearly 330% (Gentilini et al. 2021).

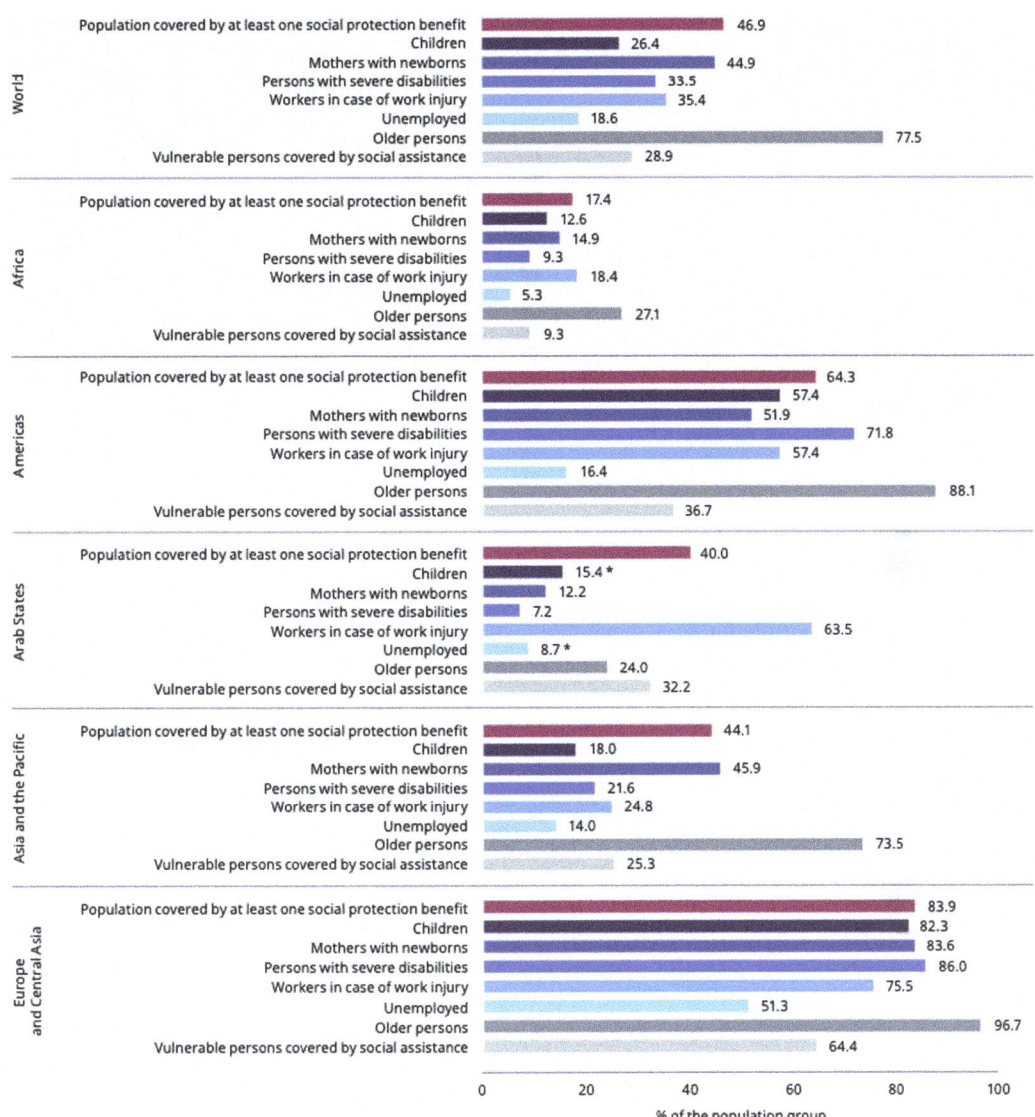

Fig. 9.1 Effective social protection (SDG indicator 1.3.1) coverage—global and regional estimates, by population group, 2020 or latest available year. (Source: ILO (2021b))

Many of these measures provided vital relief to individuals and families in need of support. The social protection measures were provided mostly in the form of social assistance representing 55% of global programs and were the most common form of support in all regions except for Eastern Europe, Central Asian states, and high-income countries. It accounted for over 70% of social protection responses in low-income countries including South Asia. Social insurance and active labor market programs ranged between 11% and

nearly 30% of the global portfolio (Gentilini et al. 2021).

Among various programs for social protection, cash transfer was the most widespread mechanism used in response to the pandemic in cities in terms of both area and population covered. According to the World Bank (Gentilini et al. 2021), the estimated number of planned individual social assistance beneficiaries in the world was 2.4 billion. However, the actual number of beneficiaries—or those participating in

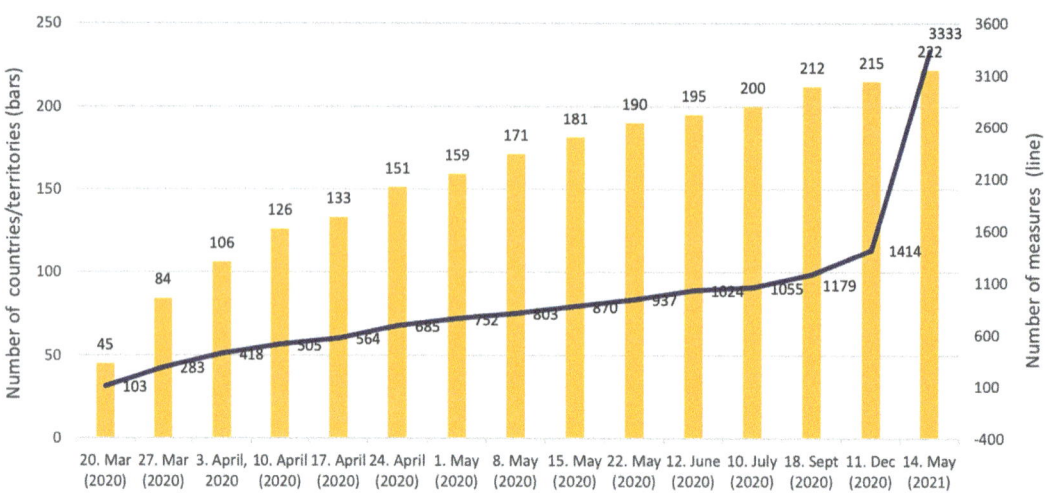

Fig. 9.2 Evolution of social protection measures and countries implementing them by numbers (March 2020–May 2021). (Source: Gentilini et al. (2021))

programs for which reporting on implementation progress is available—amounted to a little over 1.5 billion or one-fifth of the world's population (Gentilini et al. 2021). The number of planned and actual cash transfer beneficiaries was 1.8 and 1.5 billion, respectively. These numbers refer to people benefitting from either expanded coverage or more generous transfers from existing cash transfer schemes. Overall, almost 17% of the world's population was covered with at least one COVID-related cash transfer payment between 2020 and 2021(Gentilini et al. 2021). According to the United Nations (2020b) by September 2020, about 94 countries had already extended their social assistance program through new cash transfers including "Spain's *Ingreso Mínimo Vital*, Colombia's *Ingreso Solidario*, the US stimulus check, the Philippines' Social Amelioration Program, Japan's *tokubetsu teigaku kyūfukin* (special cash payment), Pakistan's *Ehsaas* Emergency Cash Program, or Hong Kong's Cash Payout Scheme" (United Nations 2020b). Several other cash transfer, rescue, and stimulus packages have already been discussed in Chap. 5 of this book.

The governments, where they considered necessary, extended coverage to groups that had remained unprotected till that time, enhanced level of benefit or instituted new benefits, modified delivery and administrative mechanisms, and

mobilized additional financial resources. According to the United Nations (2020b), 130 countries or territories, at minimum, increased budgets to supplement existing unemployment insurance, improved eligibility and timelines, and waived conditionalities to access social assistance programs. Brazil allocated US$1.7bn to its *Bolsa Família* program, decreasing the wait list to about a third of its previous size, for a total of over14.29 million beneficiary families, and passed a law allowing future regulations to raise the income thresholds for specifically vulnerable groups in its existing wage replacement program (*Benefício de Prestação Continuada*). Australia extended eligibility for its unemployment scheme, Job Seeker, to include casual, self-employed, and care workers affected by or caring for someone affected by COVID-19. The UK's Jobseeker's Allowance (JSA) and Denmark's *a-kasser* unemployment insurance funds eliminated certain conditionalities to receive unemployment benefits (United Nations 2020b).

In terms of social protection programs, the resource endowments made a big difference between developed and developing countries. Many low- and middle-income countries despite getting some international support had to struggle to launch social protection and stimulus response measures to contain the pandemic's adverse impacts in cities, which were worst affected. Moreover, the programs that were launched in low- and middle-income countries were no match to those introduced by high-income countries in terms of either coverage or financing (ILO 2021b). For example, in response to the COVID-19 pandemic by December 18, 2020, the USA among developed countries had deployed a total relief package of US $8 trillion compared to $2.3 trillion package by People's Republic of China and $412 billion by India (ADB 2021). Further, in terms of per capita investment, high-income countries on average invested an additional average of 695 US dollars per person in social protection between March and October 2020. In contrast, the average per capita invested in low-income countries was only 4 US dollars (Almenfi et al. 2020). It is also important to note that in most low-income countries, social protection as a response strategy is still in nascent stage (Hebbar and Phelps 2020). For example, in many African countries, investment in social protection has been too meager (Renzaho 2020). In contrast to the cash transfers and job retention programs of high-income countries, African countries have depended more on tax payment deferrals and reductions, loans, and moratorium on debt payments (Renzaho 2020).

Innovation in Targeting Cities A combination of census data and satellite imageries was utilized in Nigeria, using machine learning algorithms to identify the location and sizes of dwellings. Combined with a set of data from the National Social Safety Nets Coordinating Office (NASSCO), it helped in mapping of the poorest urban areas as shown in Fig. 9.3. Payments to three million target households in these areas were scheduled to begin in April 2021 (Gentilini et al. 2021). Innovation is also being deployed in payments. For example, across developing nations, at least 155 programs in 58 countries leveraged digital payments for the delivery of at least one of their new or expanded social assistance programs in response to COVID-19.

Health Protection According to ILO (2021b) almost two-thirds of the global population is covered through a health scheme. However, factors like out-of-pocket payments on utilized health services; physical distance; limitations in the range, quality, and acceptability of health services; long waiting times; and opportunity costs such as lost working time are acting as barriers to accessing health services. The pandemic has brought forward the urgent need not only to invest in healthcare services but also to improve coordination within the health system (ILO 2021b). The crisis, at its peak as discussed in Chap. 7 on health in this book, laid bare the challenges faced in deploying trained, supportive, and motivated healthcare workers even in developed countries' cities to ensure the delivery of ordinary what to say of quality healthcare service. It has also highlighted the need of coordinated approaches in meeting special and emerging needs due to increasing burden of long and chronic diseases, and population aging. In addition, the impact of the pandemic on vulnerable and elderly has clearly shown the need for coordination between health and social care (ILO 2021b).

Livelihood Support Several governments attempted to reduce the pandemic's impact through wage subsidy schemes, which were designed to assist workers in economic sectors badly affected by the crisis. Thus, Cambodia instituted a temporary program to support employees in the garment and tourism industries (ILO 2020). Sometimes, the level of this support was enormous like in the OECD, in which about 50 million jobs were being supported by May 2020 through different forms of job retention schemes such as dismissal bans, short-time work schemes, and temporary wage subsidies. This job support was ten times more than that deployed during the global financial crisis of 2008/2009. In several cases, preexisting short-time work (STW) schemes were used in response to the decline in demand due to the pandemic (OECD 2020b). As discussed in Chap. 5, some cities took specific measures to support local businesses and safeguard jobs. Milan, for example, provided productive project programs and mutual aid funds. Likewise, Barcelona, long renowned as one of Europe's creative capitals, provided subsidies,

Fig. 9.3 Relative wealth of urban areas in Nigeria. (Source: Gentilini et al. (2021))

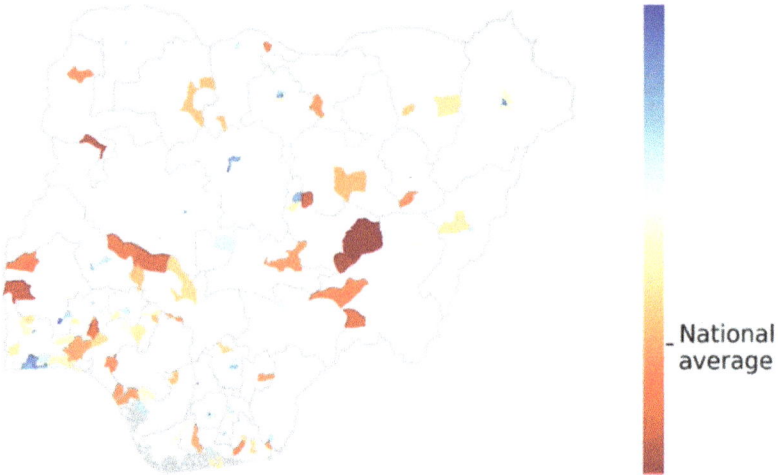

National average

tax exemptions, special investments, and advance payments to art companies and cultural programs and shows to mitigate the effects of the crisis (UN Habitat 2021). These temporary measures implemented by many governments and city authorities were crucial to support workers and employers to survive the economic shock of the pandemic.

Social Protection to Vulnerable Groups Assistance was also provided to vulnerable groups including both working and non-working population. Among working population, the most vulnerable were those retrenched during COVID-19 crisis and informal workers. In terms of population benefitting from social protection, a large segment of informal workers did not benefit from any social protection packages at least initially. Especially those on daily wages in insecure informal employment who faced disruptions in supply chains and ceased activities during the lockdowns, some 1.6 billion people suffered a 60% decline in the first month of the crisis (United Nations 2020a). Realizing this, some governments started to target informal workers and introduced programs such as "the Philippines' TUPAD conditional program, Argentina's *Ingreso Familiar de Emergencia*, Brazil's *Auxílio Emergencial*, the Indian state of Karnataka's relief package, or the relatively low-interest microcredits to informal workers provided by Mexico's federal government and Mexico City."

(United Nations 2020b). Altogether, some 21 countries had developed protections for informal workers by September 2020 (United Nations 2020b).

In the wake of lack of adequate social protection systems to compensate for loss of employment or to meet housing and other basic needs during the crisis, governments undertook an array of emergency measures. Besides combating the spread of the disease, these measures provided economic relief and to some extent covered the gaps in existing social provision. However, in many cases, the effectiveness of these measures as a crisis response instrument was limited in the absence of wage subsidies and unemployment schemes. In developing countries, even if existing social protection schemes had coverage, the focus of these were salaried workers, largely excluding self-employed and informal workers, despite the latter being in majority in many cases. Moreover, the emergency income support measures where provided were not adequate to meet the urgent needs of workers in the informal economy because of weak institutional and financial capacities.

Several social protection programs were instituted for groups of vulnerable people including destitute persons program, repurposing of urban spaces for homeless, infant and child grants, school meal programs, disability grants, eviction moratorium, establishing of handwashing stations in slums, etc. Some examples of these programs were as follows:

- Destitute persons program to deliver cash, food, and coupons to several categories of vulnerable people instituted in Botswana (Devereux 2021). Social Relief of Distress (SRD) grant for 6 months, and the distribution of food parcels by the Department of Social Development in South Africa (Devereux 2021).
- Nationwide moratoria on evictions, where the US Federal Government put a ban on eviction, which ended on August 26 after a 6–3 vote by the Supreme Court. As a result, an estimated 11 million renters fell behind in terms of rent and some 3.6 million households faced evictions (Colby and Smith 2021).
- Repurposing urban spaces to shelter the homeless—for example, from New York City to Jackson City, Michigan, and to Los Angeles in the USA, hotels were used to accommodate homeless population (DuBois et al. 2020).
- Financial and food support for the poor and jobless from government and community organizations. For example, in Los Angeles, the number of beneficiary households getting Supplemental Nutrition Assistance Program (SNAP) increased by 20%, from 686,378 in March to 822,356 in July 2020. Some of the people tracked managed to become food-secure after getting government food and financial help (Kayla 2020).
- Migrant communities' and workers' coverage was also initiated through programs, as in the case of California's Disaster Relief Fund in the USA. Under this program, undocumented adults were provided with a one-off cash transfer through community-based organizations under public-private partnership. The state was to provide $ 75 million and the remaining $ 50 million came from philanthropic partners (SPECTRUM News 1). Similarly, in Portugal and Italy, temporary mass regularizations made immigrants eligible for social protection under certain conditions (United Nations 2020b).

- Though limited, programs targeting women were also developed like Argentina's *Ingreso Familiar de Emergencia*, so did Brazil's *Auxílio Emergencial*, which contained women-specific provisions, and India's Pradhan Mantri Garib Kalyan Yojana provided cash transfers to women beneficiaries between April and June 2020.
- Programs were also initiated to temporarily cover disabled during the COVID-19 pandemic, such as Mauritius's home delivery of food, South Africa's increase of disability allowance, Kenya's additional temporary disability scheme, and Sierra Leone's distribution of rice and one-off cash payment. Among others were Botswana's disability grant (Devereux 2021) and Indian government's one-off pension of Rs 1000 paid in 3 months through the National Social Assistance Program (NSAP—Indira Gandhi Disability Pension), specially to people with disability (Romano 2020).

9.3.1.2 Challenges and Constraints

The implementation of social assistance measures in thousands looks impressive. However, it is not possible to ascertain if these measures did reach the poor or those with precarious employment or deserving and were also effective. In this connection, the United Nations (2020b) mentions several challenges faced—some of these relevant in the context of cities and urban communities were as follows:

- Duration of social protection coverage and adequacy of outlays/amounts allocated
- Adaption of social protection to the multifaceted realities of urban poverty
- Inclusion of migrant and undocumented urban workers in social protection
- Realization of the need of informal workers in social protection
- Promotion of gender-responsive social protection
- Acknowledgment of human rights principles in social protection

Duration and Adequacy In terms of duration of social protection measures, it was realized early on that short-term, contingent support measures were not enough for the sustenance of needy families and households in the ever-increasing pandemic. Most social assistance measures were designed with a maximum average duration of 3 months (Gentilini et al. 2021), and countries and cities had to renew schemes several times. Perhaps it would have been better to continue the relief measures until after recovery rather than extending these several times, which would have prevented people in poverty from getting temporarily unprotected. It would also be in confirmation with ILO's Recommendation No. 202 on Social Protection Floors (United Nations 2020b). Regarding adequacy, cash transfers were often inadequate particularly in the case of developing countries. For example, social grant of Rand 350 (US$18.44) per month in South Africa was less than a third of the country's own upper-bound poverty line (Government of South Africa, 2019). Similarly, the Government of Chile—with a 2019 national poverty line of Chilean Peso (CPL)164,605 (USD 212.18)—provided CPL 65,000 once and then CPL 100,000 three times for single households through its bureaucratic cash transfer program. This was below its own poverty line and created confusion as it was renewed at the last moment several times (United Nations 2020b). Pakistan's *Ehsaas* Emergency Cash program provided a one-time Pakistan Rupees 12,000 (about US$72) per household. This was again under national poverty line of PK Rs 3250.28 (GoP 2020).

Multifaceted Realities of Urban Poverty and Undocumented Workers Many schemes on social protection were not in line with the realities of people living in poverty or in precarious employment or without a fixed residential status. For example, need of official registrations and provision of address to authorities created big problems for migrants, undocumented workers, homeless, and those living in shelters and sleeping in their workplaces, or in informal dormitories. Some new social programs, like Canada's

Emergency Response Benefit, or the US CARES Act, required social identification numbers and therefore excluded informal and undocumented workers (United Nations 2020b). Spain's *Ingreso Mínimo Vital* also needed proof of effective residence through a municipal certificate, which was difficult for many people in poverty to produce. Likewise, special one-time cash payment program of Japan was paid only to officially registered heads of households (*setainushi*), making its access difficult to other household members. This created problem particularly for women already facing hardship, or victims of domestic violence without court orders to potential economic abuse (United Nations 2020b). In addition, complex and bureaucratic procedures as well as age, sex, and disability vulnerabilities created problems in procurement of social protection.

Informal Workers There were problems in providing relief measures to informal workers as well. For instance, in many countries especially in China, implementation of support to informal workers was generally left to subnational or city governments—although over the years, local/city governments have been known to have pushed low-wage workers, particularly street vendors, and/or dislodged them in the name of city beautification. This has happened in many other cities of the Global South even outside China. Moreover, providing street vendors with social security has no real impact on their livelihood if cities obfuscate their permit applications (United Nations 2020b), although cities in China did offer some compensation in the form of informal workers getting *di bao* allowance in their localities (between RMB 500 and 1000 (US$71–143) per month depending on the province of the country. In addition, they could apply for business loans at low interest rates, but these schemes were not designed for low-wage workers. Another problem in their case was that they became ineligible if their workplace or residence fell outside their household registration (*hukou*) locality. Further, despite efforts of certain municipalities to increase amounts of unemployment benefits, those who had not enrolled for at least a year (which many informal workers do not) were not entitled to full insurance amounts (United Nations 2020b).

Cash transfers to informal workers in developing countries had two significant problems: (a) disbursement of low amounts and (b) sometimes they were developed on inaccurate databases. In Egypt, for example, where an estimated two-fifths of the workforce or about 18.5 million people (half of these women) are engaged in the informal economy. An allowance of 500 Egyptian Pound or US$ 32.60 allocated per head expected to cover only 2.5 million beneficiaries officially recognized as "irregular workers" was grossly insufficient (Saafan 2020; Middle East Eye 2020). Likewise in India, where over 90% of the labor force is employed in the informal economy, about 66% did not receive any cash transfer under its Pradhan Mantri Garib Kalyan Yojana (PMGKY) package because of lack of registration in the social security system as the database on residence and income was not updated (Tiwari 2020). With millions of workers engaged in informal sector world over, it has become important to build social protection programs whether emergency or long term on "permanent, adapted, and integrated basis" by taking into consideration the needs of the different groups of workforce often clustered as informal workers (United Nations 2020b).

Gender-Related Issues The challenges in social protection related to gender also continued despite initiation of many specific programs to assist women like Brazil's *Auxílio Emergencial* initiative. For example, India's Pradhan Mantri Garib Kalyan Yojana (PMGKY), which provided Rs 500 (US$6.59) cash transfer to 204 million women beneficiaries, was a good effort. However, besides being inadequate in amount, it potentially excluded over half of women in poverty under $2.50/day PPP (Pande et al. 2020). Moreover, several women in poverty did not have ration cards that granted access to the food ration system, and despite the portability of ration cards, the problem was exacerbated due to the large number of internal migrant women workers who

moved because of COVID-19 (ICJ 2020). In Japan, under unemployment schemes, many women could not access support due to differences in employment type—as of March 2020, 54% of working women were nonpermanent employees in the country (United Nations 2020b).

Human Rights Regarding human rights principles in social protection, Article 9 of the International Covenant on Economic, Social and Cultural Rights puts a clear obligation on governments—to develop mechanisms for participation, transparency, and accountability when implementing the right to social security. Further, the UN High Commissioner for Human Rights has emphasized "genuine participation of affected groups and individuals in decision-making processes," to ensure compliance with human rights obligations. However, none of the national-level social protection measures reviewed by the United Nations (2020b) report included mechanisms to institutionalize the participation of poor people or their representatives in the development, application, or evaluation of recovery plans. However, it is encouraging to note that cities and local authorities together with community-based organizations have taken this burden to a large extent that governments should have taken up for participation.

9.3.1.3 Contribution of NGOs, Charitable Organizations, Communities, and Businesses

In addition, to national and local/city governments, communities, businesses, and philanthropic organization also came forward and helped in social protection. For example, a partnership between Cities Alliance and Slum Dwellers International supported informal communities in responding and recovering from COVID-19. The initiative spanned over 17 countries across Africa, Asia, and Latin America, and its activities were implemented mainly through existing community-based systems and networks (Cities Alliance 2021). Another good example, among others, included "the digital cash transfers from global citizens to slums of African cities and to households unable to pay rent in the US cities." This was a kind of community support "through a peer-to-peer model" (UN Habitat 2021). The pandemic also saw giving to human services increase by 9.7%, while donations to public-society benefit organizations enhanced by 15.7% (Ware 2021). Religious and faith-based organizations, which have played their helping role historically, were also active.

Most importantly, most of COVID-19 social protection measures were only temporary. There is a need to transform many of these into longer-term commitments to maintain social protection of the weakest and vulnerable. In cities, this is important and essential to build urban resilience to future pandemics and another looming mega risk—the climate change. Together, the twin mega risks underscore the critical importance of achieving universal social protection. It is essential that cities, governments, social partners, and other stakeholders resist the present pressures to reverse the step and fall back on a low-road social protection trajectory. They must pursue a high-road social protection strategy to not only contend with the pandemic but to secure a human-centered recovery (ILO 2021b) and appropriately encounter the challenges of climate change.

9.3.2 Climate Change Response

Cities are getting vulnerable to climate-related risks for two reasons. Firstly, even though before the pandemic, poverty had declined globally, the share of poor people living in urban areas was rising worldwide (Galvin and Maassen 2020). Secondly, cities have become increasingly exposed to flooding and extreme climate events because more and more people are occupying hazard-prone areas, which get hardest hit by these events. Hence, the need for social protection solutions is increasing in cities to address both poverty and climate-related risks and to provide relief during acute economic and livelihood disruptions caused by extreme climate events.

The extreme climate events not only have immediate- or short-term and visible impacts but devastating long-term impacts by affecting people's assets, livelihoods, health, and ability to save. This may be due to a lack of savings and limited access to finance and/or insurance. Their immediate reaction to protect their well-being and consumption after a shock is often to adapt negative "coping strategies" in the form of pulling children out of school to work and earn extra income for the family or taking high-interest loans or selling productive assets. Such short-term coping strategies for obvious reasons lead to the long-term detriment of families and households. Assistance from social safety nets can help in avoiding such negative coping strategies after an extreme climate event.

9.3.2.1 Social Protections Programs and Climate Change

Several social action programs have helped in building resilience to climate-related natural disasters. For example, in Bangladesh, the Chars Livelihood Program (CLP), a large social action package launched for promoting livelihood opportunities, also strengthened the resilience to climate-related extreme events. A combination of public works, asset transfers (cash and in-kind), livelihoods-related training, market development, and social development activities was undertaken in this program. There were four main climate resilience features of the program including (a) public works for reducing the risk of flooding, (b) innovative social safety net mechanisms to cushion the program's beneficiaries against disaster impacts, (c) post-disaster relief and recovery support to protect and restore the assets, and (d) income built up through the program and direct measurement of climate resilience outcomes in its monitoring and evaluation systems (World Bank 2013). The program helped protect 95% of recipients from losing their assets after 2012 floods (Vaziralli 2021). Similarly, Mexico's Temporary Employment Program (Programa de Empleo Temporal) contributed to the development of a highly collaborative and formalized institutional relationship between social protection, disaster management, and sectoral agencies. Along with creating a quick disaster response mechanism and contingency fund, it included a climate-sensitive and disaster targeting criteria into sectoral public work programs (World Bank 2013). Mexico also introduced the national cash transfer program "Prospera" meaning opportunity. The target population of the program was households with an estimated per capita income lower than the minimum necessary to purchase the basic food items or households with weak socioeconomic conditions to procure nutrition, education, and health benefits (Lárraga 2016). The implementation of the program therefore saved the households from adapting negative coping strategies like withdrawing their children from school after a climate shock.

Similarly, a Productive Safety Net Program (PSNP) in Ethiopia targeted a highly climate-vulnerable population. The PSNP entitled poor households to a secure, regular, predictable government transfer while protecting them from the impacts of climate-related disasters. It has enabled core beneficiaries to meet consumption needs, mitigate risks, and avoid selling productive assets during crises (World Bank 2013). The program has ensured that losses or reductions in consumption of the participants during drought were 25% less compared to those not participating in the program (Vaziralli 2021). The Horn of Africa Risk Transfer for Adaptation (HARITA) was another Productive Safety Net Program that promoted public works to increase resilience to climate-related shocks. The program protected its beneficiaries' assets and income from low-frequency but severe recurrent disaster impacts like droughts. The insurance scheme in the program made its first payouts to clients following a major drought in 2011. Its encouraging results have led to scaling up the program in Ethiopia and extending it to Senegal, under the direction of Oxfam America and the World Food Programme (World Bank 2013).

Two developing countries in Asia that offer good example of social protection against climate-related extreme events are Vietnam and the Philippines. The 2020 Vietnam's national strategy for disaster mitigation and management articulates that disaster mitigation and management actions must be compatible with poverty reduction and natural resource protection measures, so that development can be equitable and sustainable (Care 2015). After the Typhoon Ketsana, which caused 286 deaths and affected four million people in Vietnam, several cash transfer programs were executed by different organizations in response to the emergency. While actions were being taken under these, the Government of Vietnam initiated a national Community-Based Disaster Risk Management Program and integrated it into overall development plan. The program established a coordination mechanism to integrate climate-related risks across the strategies and targets of various ministries and civil society organizations that were responsible for delivering social protection. As a result, several programs in the country currently have dedicated flexible funds that can be repurposed to respond to extreme climate events and disasters (Vaziralli 2021).

In the Philippines, Typhoon Yolanda struck in 2013 killing 8000 people, severely affecting about 200 municipalities, damaging, or destroying a million houses and causing an estimated damage of US$13 billion (World Bank, 2014). The country already had a comparatively strong social protection system in existence—various components of which were mobilized to channel relief to affected people. For example, conditionalities in Pantawid Cash Transfer Program were dropped, and its coverage was extended by using an existing database of the poor and vulnerable that was available in the National Household Targeting System for Poverty Reduction (World Bank, 2014). Various rehabilitation programs developed included cash-for-work and cash-for-asset rebuilding. In addition, three flagship programs of the country—conditional cash transfer, community-driven development, and livelihood support—were converged to form an effective tool in making the households more resilient to disasters (World Bank, 2014). A system is now being developed that can be scaled up quickly in response to an extreme event—which involves building technical infrastructure, improving linkages between existing targeting systems and disaster response, and efficient data management and IT systems (Vaziralli 2021).

9.3.2.2 Coverage Gaps and Coping Approaches

Despite growing evidence on the economic benefits of social protection, spending and coverage in low-income countries are inadequate to combat the effects of climate change. No doubt, there is a significant variation, but most disaster-prone countries have large coverage gaps—leaving coverage of people most at risk, in their safety net programs (World Bank, 2018). Furthermore, the lack of coverage is even more evident among the poorest countries. For example, the South Asian and African regions, home to the world's largest share of poor, have safety net coverage below levels commensurate with their disaster risk. A comparison of India and Portugal provides some insights into it. The gross domestic product (GDP) per capita of India is ten times less than that of Portugal, but its social protection spending is 137 times less. According to the World Bank (2018), it may be beneficial for such countries to invest in Adaptive Social Protection (ASP) or safety nets and their use for building household resilience to shocks. A household's resilience to a shock is the product of its capacity to prepare for, cope with, and adapt to it (Bowen et al. 2020). ASP was conceptualized as a series of measures aimed to build resilience of the poorest and most vulnerable people to climate change. It combines elements of social protection, disaster risk reduction, and climate change (Arnall et al. 2010). The term "adaptive" has crystalized around two interrelated approaches over time. The first focuses on building household resilience before climate-related shock occurs and the second concentrates on increasing the capabilities of safety nets to respond to climate-related shock after they occur (World Bank, 2018).

The first approach based on boosting the role of social protection in building the resilience seeks to break the damaging cycle of poverty and vulnerability beforehand. In other words, it aims for more resilient households that will be better able to withstand the risk event if household members have more human capital, are able to access job opportunities, accumulate physical capital, and diversify their livelihoods. An example of this approach cited by the World Bank

(2018) is Brazil's Bolsa Família program which delayed the entry of children into the labor market. The Programa de Asignacion Familiar of Honduras provides another example where children in beneficiary families are less likely to work. Other productive inclusion interventions in this approach referred as "graduation models" by the World Bank (2018) support sustainable exits from poverty or resilience building. Several countries in West and East Africa are using this approach to boost household resilience in the face of repeated and chronic drought, along with other shocks (World Bank, 2018).

The second approach concentrates on enhancing the capability of safety nets to respond to climate-related disasters after they occur by making program design more flexible and scalable. It is also referred to as "vertical" (increasing grant amount to beneficiaries) and "horizontal" (horizontally expand and to reach more affected households) program expansion (Oxford Policy Management 2018). Increasing grant amounts or vertical expansion has been used as conduits to rapidly inject assistance to pre-targeted and enrolled poor households in affected areas—as also done in the case of the Philippines through additional grants from humanitarian actors such as World Food Programme and UNICEF following Typhoon Yolanda and Typhoon Ruby and in Fiji following Tropical Cyclone Winston in 2016 (Bowen et al. 2020). An example of horizontal expansion is the social safety net of Ethiopia's Productive Safety Net Program (PSNP), referred above that was implemented in response to the 2011 drought.

Operations of social safety nets can be improved further by additional investments to expand programs outreach and include additional households. Following this approach, several countries have invested to horizontally expand a safety net program. For example, Ethiopia with the Productive Safety Net Program and Kenya with the Hunger Safety Net Program—each is prepared to undertake horizontal expansions based on household needs generated by drought and related food insecurity in drought-prone areas (Bowen et al. 2020). The hallmark of an adaptive safety net is dynamic delivery systems, which allows the required flexibility and scalability to achieve horizontal and/or vertical expansion, depending on post-shock needs. The horizontal expansion becomes more effective when it is linked to information system related to vulnerabilities and risks. For example, early warning and related risk information (e.g., hazard mapping, meteorological monitoring, climate variance mapping, and geospatial data), along with information on household composition and characteristics, can provide vital information about the nature, location, and depth of a shock from a climate-related extreme event. As such, it can be of great help in not only expanding but also in undertaking appropriate response.

Adaptive Social Protection as a strategy has been further articulated in a World Bank report (Bowen et al. 2020) which provides insights into the means through which social protection systems can be made more capable of building household resilience using its four building blocks—programs, information, finance and institutional arrangements, and partnerships. The report mentions the elements of existing social protection systems that are the cornerstones for building household resilience to shocks. It also highlights the additional priorities and core investments that can be instrumental in enhancing the outcomes and making the social protection system more prepared for any future crisis.

In terms of social protection, despite the progress made so far, significant gaps in program coverage persist around the globe. These gaps are especially sharp in low-income countries, where before the pandemic only 18% of the poorest quintile were covered by SSN programs (World Bank, 2018). Even in lower-middle-income countries, less than 50% of the poor had access to SSN programs. Coverage is much better in upper-middle-income countries and high-income countries, although gaps exist even there. A major reason for low coverage and gap is low spending on social protection due to low GDP per capita or lack of fiscal space. However, constraints like lack of capacity, technology, and data as well as leakage of benefits, mistargeting, lack of transparency, or imperfect markets as well as huge informal workforce also contribute to the prob-

lem (Vaziralli 2021). Regarding the fiscal gap, help from outside has contributed positively. For example, the Caribbean Risk Insurance Facility (CCRIF SPC), a regional catastrophe fund for governments in the Caribbean and Central American region, has helped limit the financial impact of tropical cyclones among other disasters. The insurance facility provides liquidity to the member states for the purpose when a threshold is triggered. For example, following Hurricane Dorian, CCRIF paid about 11 million US dollars to Bahamas. CCRIF payments are made within 14 days but in this case, it was made even earlier. Pacific Catastrophe Risk Assessment and Financing Facility (PCRAFI) is another risk pool. It provides for disaster risk management and finance solutions to increase the resilience of Pacific Island states. Countries can insure themselves against tropical cyclones and some other hazards under this facility. No doubt these initiatives provide an important step forward in addressing the needs of cities in vulnerable countries and for providing financial backup. Nonetheless, overall direct access to international climate finance through national entities is still limited (Eckstein, Kunzel and Schafer 2021). Perhaps a Global Fund for Social Protection can help to ensure the availability of a protection floor to cover even those countries that do not have the financial means for the purpose. Beyond that, in crisis situations, the fund could also help those countries that are dependent on international support due to short-term financial bottlenecks.

9.3.2.3 Role of International Relief Organizations, Civil Society, and Private Sector Charities

International humanitarian organizations such as International Federation of Red Cross and Red Crescent (IFRC), NGOs, and civil society organizations have played crucial role in climate change disaster relief and raising awareness and conducting training in communities about the risks of extreme climate events. They have also been providing emergency care to people already registered in social protection beneficiary lists in cities and countries. However, the magnitude of climate change exacerbated risks, and the impacts they are having on all dimensions of people's lives—from their safety to their health, to their access to water or food, and to their livelihoods—demand more involvement and greater action by these nongovernment entities in terms of urgent humanitarian action. An analysis by IFRC has revealed that the number of people needing life-saving international humanitarian assistance could double by 2050, if climate change goes unchecked and the world fails to invest in bold adaptation and mitigation efforts (Grayson 2020). It is therefore imperative to utilize all available resources for mitigation, adaptation, and social protection related to climate change. The participation and advocacy of NGOs and civil society is also vital in social protection. It can help in greater investments in, and continual improvements of, government social protection systems around the world.

9.4   Conclusion

Nations and cities are at a crossroads concerning the trajectory of their social protection systems in the wake of two mega risks COVID-19 and climate change. The critical importance of investing in social protection has never been so glaring as it is now. No doubt, several countries, particularly those of the Global South, face significant fiscal constraints. However, countries and territories, irrespective of their level of development, face a crucial choice—whether to pursue a high-end strategy of investing to reinforce their social protection systems or a low-road strategy of minimalistic provision, succumbing to fiscal or political pressures. The gravity of socioeconomic problems posed by COVID-19 triggered an unparallel social protection policy response. Governments at both national and city/local levels undertook measures for social protection as a frontline response to protect people's health, job/business, and incomes. They also initiated actions to promote and ensure social stability. Countries and cities can use this policy window that has been opened to build on their crisis response measures and strengthen their social protection

systems. They can gradually fill the protection gaps and make sure that everyone is protected against mega risks and ordinary life-cycle threats. This would involve increased efforts to build universal, comprehensive, adequate, and sustainable social protection systems, including a solid social protection floor that guarantees at least a basic level of social security for all over the course of time.

Both the mega risks have caused tremendous social and economic hardships to city dwellers especially vulnerable groups across the world. The pandemic's unprecedented impact in such a short time brought forth an intensive or widespread need for social protection responses. The huge increase in global poverty in 2020 and beyond because of COVID-19 is just one example of the multidimensional impacts of a mega risk. As a frontline response to protect people's health, job/business, and income, governments at both national and city/local level had to initiate actions to promote and ensure social stability. Hence, between March 20, 2020, and May 14, 2021, over 3300 social protection measures were planned or implemented in 222 countries or territories—an increase of nearly 148% since December 2020. While social assistance and insurance increased by about 120% and 110%, respectively, active labor market interventions surged as much as 330%. Regarding climate crisis, even before the pandemic, it had threatened to push over 130 million more people into poverty in the next decade alone. Moreover, based on climate scenarios, an IFRC (2019) cost of doing nothing report projected that the potential increase in the number of people in the need of international humanitarian assistance would double by mid-century.

The most worrying aspect in facing the mega risks was the overlapping of extreme weather events with the COVID-19 crisis. From the start of the pandemic in March 2020 to August 15, 2021, over 430 extreme weather events affected about 139 million people and killed 17,000 across the world, and these figures do not include those who were affected by temperature extremes and drought events. It is important to note that there was no exceptional immunity, and the compound impacts of mega risks affected even the richest countries, like the USA and Canada, which besides the virus attack also suffered from the unprecedented summer heat wave of 2021. Although the mega risks have not made any exceptions, however, their social impacts on urban areas have varied widely depending on (i) the development status of the city, (ii) scale of the outbreak of the virus or intensity of weather event, (ii) policy responses, (iii) the preponderance of socially vulnerable population or deep-rooted inequalities in cities, and (iv) resiliency of the city and its population. Nevertheless, some communities or groups of people such as poor, slum dwellers, racial and ethnic minorities, women, and other vulnerable groups including aged and people with chronic illness and with disabilities were exposed disproportionately to and affected by both the pandemic and climate threats. Differential vulnerability results from these as well as several other factors including unequal resource access, cultural practices, governance shortcomings, and knowledge and social network availability or nonavailability.

In terms of social protection response measures and programs, the resource endowments made a big difference between low- and high-income countries. Many low- and middle-income countries despite getting some international support had to struggle to launch social protection and stimulus response measures to contain the pandemic's adverse impacts in cities, which were worst affected. Moreover, the programs that were launched in low- and middle-income countries were no match to those introduced in high-income countries whether in terms of coverage or financing. For example, in terms of relief package for the pandemic, the USA among developed countries by December 18, 2020, had allocated US $8 trillion compared to $2.3 trillion by People's Republic of China and $412 billion by India. Regarding per capita investment, high-income countries on average invested an additional average of 695 US dollars per person in social protection between March and October 2020. In contrast, the average per capita invested in low-income countries was only 4 US dollars. It is also important to note that in most low-income

countries, social protection as a response strategy is still in nascent stage. For example, in many African countries, investment in social protection has been "very meager and insufficient." Moreover, in contrast to the cash transfers and job retention programs of high-income countries, they were geared more to tax payment deferrals and reductions, loans, and moratorium on debt payments in low-income countries.

Furthermore, the lack of coverage is more evident in the poorest countries. For example, the South Asian and African regions, home to the world's largest share of poor, have safety net coverage well below levels commensurate with their disaster risk. It may be beneficial for such countries to invest in Adaptive Social Protection (ASP) or safety nets. ASP was conceptualized as a series of measures aiming to build resilience of the poorest and most vulnerable households to climate change. Adaptive Social Protection systems are particularly suitable because in the event of a disaster from a mega risk, social protection often needs expansion at a short notice to meet increased needs. The term "adaptive" has worked around two interrelated approaches. The first focuses on building household resilience before climate-related shock occurs and the second on increasing the capabilities of safety nets to respond to climate-related shock after they occur. An example of the first approach is Brazil's Bolsa Família program which delayed the entry of children into the labor market. The example of second case is the Philippines, where additional grants from humanitarian actors like the World Food Programme and UNICEF) following Typhoon Yolanda and Typhoon Ruby helped.

Finally, social protection's vital and indispensable role in ultimately building citizens' resilience to mega risks as well as other future shocks at both national and city level has become undeniable. One important consideration that needs full recognition in this respect is the differentiated needs of vulnerable groups of people that can be affected by crises or shocks due to mega risks. There is a need to prepare special programs focusing on socioeconomic needs of such groups including racial and ethnic minorities, poor and slum dwellers, disabled, and other vulnerable groups. Moreover, it is imperative to realize that social protection is a task that needs to be primarily financed from national resources and international co-financing of the systems can only be a temporary solution. However, a Global Fund for Social Protection is needed to (a) act as a protection floor for those countries that lack in financial means and (b) help countries that are dependent on international support due to short-term financial bottlenecks during a crisis. Contribution of informal social protection systems, which use community-based institutions such as savings groups or grain banks, in parallel to formal and state-run protection systems, particularly at city/local level also needs both recognition and promotion.

References

Abdoul-Azize HT, El Gamil R (2020) Social protection as a key tool in crisis management: learnt lessons from the COVID-19 pandemic [published online ahead of print, 2020 Sep 1]. Glob Soc Welf 8(1):107–116. https://doi.org/10.1007/s40609-020-00190-4

Aboagye D (2012) The political ecology of environmental hazards in Accra, Ghana. J Environ Earth Sci 2(10):157–172

Adams-Prassl A, Boneva T, Golin M, Rauh C (2020) Inequality in the impact of the coronavirus shock: evidence from real time surveys. J Public Econ 189:104245. Google Scholar

ADB (2021) One year of living with COVID-19: an assessment of how ADB members fought the pandemic in 2020, Manila

Almenfi M, Breton M, Dale P, Gentilini U, Pick A, Richardson D (2020) Where is the money coming from? Ten stylized facts on financing social protection responses to Covid-19. In: The World Bank/UNICEF/OECD (ed) Social Protection and Jobs, Policy and Technical Note No. 23. World Bank Group, Washington, DC

Arnall A, Katy Oswald, Mark Davies, Tom Mitchell, Cristina Coirolo (2010) Adaptive Social Protection: Mapping the Evidence and Policy Context in the Agriculture Sector in South Asia. https://doi.org/10.1111/j.2040-0209.2010.00345_2.x

Barbier EB (2010) Poverty, development and environment. Environ Dev Econ 15(6):635–660

Barrett CB, Travis AJ, Dasgupta P (2011) Biodiversity conservation and poverty traps. Proc Natl Acad Sci U S A 108(34):13907–13912

Bartlett S, Dodman D, Hardoy J, Satterthwaite D, Tacoli C (2009) Social aspects of climate change in urban areas in low- and middle-income nations. In: Paper prepared for the Fifth Urban Research Symposium, Cities and Climate Change: Responding to an Urgent Agenda, 28–30 June, Marseille, France

Benfer E, Bloom Robinson D, Butler S, Edmonds L, Gilman S, McKay K, Neumann Z, Owens L, Steinkamp N, Yentel D (2020) The COVID-19 eviction crisis: an estimated 30–40 million people in America are at risk, Aspen Institute. https://www.aspeninstitute.org/blog-posts/the-covid-19-eviction-crisis-an-estimated-30-40-million-people-in-america-are-at-risk/

Biswas PP (2020) Skewed urbanisation and the contagion. Econ Polit Wkly 55(16):13–15

Bowen T, del Ninno C, Andrews C, Coll-Black S, Gentilini U, Johnson K, Kawasoe Y, Kryeziu A, Maher B, Williams A (2020) Adaptive Social protection: building resilience to shocks. World Bank

Boza-Kiss B, Pachauri S, Zimm C (2021) Deprivations and inequities in cities viewed through a pandemic lens. Front Sustain Cities 3. https://doi.org/10.3389/frsc.2021.645914

Brookings Institution (2009) Protecting and promoting rights in natural disasters in south Asia: prevention and response, report on the project on internal displacement, Chennai, India

C40 Cities (undated) Food security. https://www.c40.org/what-we-do/scaling-up-climate-action/adaptation-water/the-future-we-dont-want/food-security/

Cardoso E, Silva Da Silva M, De Albuquerque F, Júnior F, Venâncio de Carvalho S, de Carvalho A, Vijaykumar N, Francês C (2020) Characterizing the impact of social inequality on COVID-19 propagation in developing countries. IEEE Access 8:172563–172580

Care (2015) Rethinking resilience: social protection in the context of climate change in Viet Nam. http://careclimatechange.org/wp-content/uploads/2015/09/Learning-Series-2-Rethinking-Resilience-EN-2015_09_21.pdf

Carter M, Little PD, Mogues T, Negatu W (2007) Poverty traps and natural disasters in Ethiopia and Honduras. World Dev 35(5):835–856

Castro AM (2020) ORBIA and UNICEF donate handwashing stations in Ecuador to prevent the spread of COVID-19. https://www.unicef.org/lac/en/stories/orbia-and-unicef-donate-handwashing-stations-ecuador-prevent-spread-covid-19

Cianconi P, Betrò S, Janiri L (2020) The impact of climate change on mental health: a systematic descriptive review. Front Psych 11:74. https://doi.org/10.3389/fpsyt.2020.00074

Cities Alliance (2021) COVID – 19 fostering community resilience amid the pandemic. https://www.citiesalliance.org/newsroom/news/results/covid-19-response-informal-settlements-fostering-community-resilience

Colby C, Smith D (2021) The federal eviction moratorium is gone, what renters should know now. https://www.cnet.com/personal-finance/the-federal-eviction-moratorium-is-gone-what-renters-should-know-now/

Costella C, McCord A, Van Aalst M, Holmes R et al (2021) Social protection and climate change: scaling up ambition. SPACE, UKAID, German Cooperation, GTZ and Australian Aid

de Oliveira LA, de Aguiar AR (2020) Neighborhood effects and urban inequalities: the impact of covid-19 on the periphery of Salvador, Brazil. City Soc 32(1). https://doi.org/10.1111/ciso.12266

Devereux S (2021) Social protection responses to COVID-19 in Africa. Global Soc Policy 21(3):421–447. https://doi.org/10.1177/14680181211021260

Di Liberto T (2015) India heatwave kills thousands. Climate.gov, https://www.climate.gov/news-features/event-tracker/india-heat-wave-kills-thousands

DuBois N, Williams A, Batko S (2020) Temporary hotel placements kept people safe during COVID-19 but will not curb growing homelessness crisis. https://www.urban.org/urban-wire/temporary-hotel-placements-kept-people-safe-during-covid-19-pandemic-will-not-curb-growing-homelessness-crisis

Durkin E, Groneworld A, Bocanegra M (2020) Neighborhood death tolls show virus disparities—western New York cleared to begin reopening—Deeper economic decline projected, Politico. https://www.politico.com/newsletters/new-york-playbook/2020/05/19/neighborhood-death-tolls-show-virus-disparities-western-new-york-cleared-to-begin-reopening-deeper-economic-decline-projected-489266

Eckstein D, Kunzel V, Schafer (2021) Global climate risk index, germanwatch. http://germanwatch.org/sites/default/files/Global%20Climate%20Risk%20Index%202021_2.pdf

Enarson E (2000) Gender and natural disasters, ILO, In: Focus programme on crisis response and reconstruction, Working Paper 1, pp 4–29

Enarson E, Phillips B (2008) Invitation to a new feminist disaster sociology: integrating feminist theory and methods. In: Phillips B, Morrow BH (eds) Women and disasters: from theory to practice. International Research Committee on Disaster, Xlibris, pp 41–74

EvictionLab (2021) Eviction tracking. https://evictionlab.org/eviction-tracking/

FAO (2015) Climate change and food security: risks and responses. https://www.fao.org/3/i5188e/I5188E.pdf

FAO (2020) Urban food systems and COVID-19: the role of cities and local governments in responding to the emergency. http://www.fao.org/3/ca8600en/CA8600EN.pdf

FAO, IFAD, UNICEF, WFP and WHO (2021) The state of food security and nutrition in the world, key messages. https://www.fao.org/3/cb4474en/online/cb4474en.html#chapter-Key_message

Fiszbein A, Kanbur R, Yemtsov R (2014) Social protection, and poverty reduction: global patterns and some targets. World Dev 61:167–177. https://doi.org/10.1016/j.worlddev.2014.04.010

Fothergill A, Maestas EGM, Darlington JD (1999) Race, ethnicity and disasters in the United States: a review of the literature. Disasters 23(2):156–174

Fussell E, Sastry N, Vanlandingham M (2010) Race, socioeconomic status, and return migration to New Orleans after Hurricane Katrina. Popul Environ 31(1–3):20–42. https://doi.org/10.1007/s11111-009-0092-2

Galindo G (2020) Brussels hotel to shelter domestic violence victims amid coronavirus quarantine. The Brussels Times. https://www.brusselstimes.com/brussels/105172/brussels-hotel-to-shelter-domestic-violence-victims-amid-coronavirus-quarantine/

Galvin M, Maassen A (2020) 5 big ideas to address the climate crisis and inequality in cities. https://www.wri.org/insights/5-big-ideas-address-climate-crisis-and-inequality-cities

Gavas M, Pleeck S (2021) Global trends in 2021, how COVID-19 is Transforming International Development. Center for Global Development, https://www.cgdev.org/publication/global-trends-2021-how-covid-transforming-international-development

Gentilini U, Almenfi MB, Dale P, Palacios RJ, Natarajan H, Galicia Rabadan GA, Okamura Y, Blomquist JD, Abels M, Demarco GC, Santos IV (2021) Social protection and jobs responses to COVID-19: a real-time review of country measures. World Bank, Version 15, May 14, Washington, DC. © World Bank. https://openknowledge.worldbank.org/handle/10986/33635 License: CC BY 3.0 IGO

Gonzalez D, Karpman M, Kenney GM, Zuckerman S (2020) Hispanic adults in families with noncitizens disproportionately feel the economic fallout from COVID-19. https://www.urban.org/sites/default/files/publication/102170/hispanic-adults-in-families-with-noncitizens-disproportionately-feel-the-economic-fallout-from-covid-19_2.pdf

Government of Pakistan (2020) Ehsaas emergency cash: a digital solution to protect the vulnerable in Pakistan during the COVID-19 crisis, p 4

Government of South Africa (2019) National Poverty Lines, Statistical Release, P0310.1. https://www.statssa.gov.za/publications/P03101/P031012019.pdf

Grayson CL (2020) A climate for change: towards a humanitarian climate and environment charter. https://rcrcconference.org/blog-stories/a-climate-for-change-towards-a-humanitarian-climate-and-environment-charter/

Hebbar M, Phelps L (2020) Rapid literature review: social protection, https://socialprotection.org/discover/publications/rapid-literature-review-social-protection. Accessed 6 Sept 2021

Highet M (2021) Rapid scoping review on the topic of ensuring social protection and basic services to inform the united nations framework for the immediate socioeconomic response to COVID-19. Int J Health Serv 51(4):462–473. https://doi.org/10.1177/00207314211024896

Hill RV, Narayan A (2020) COVID-19, and inequality: a review of the evidence on likely impact and policy option, Prevention Web, UNDRR. https://www.preventionweb.net/publication/covid-19-and-inequality-review-evidence-likely-impact-and-policy-options

Hoffman JW, Shandas V, Pendleton N (2020) The effects of historical housing policies on resident exposure to intra-urban heat: a study of 108 US urban areas. Climate 8(1):12

ICJ, Independent Commission of Jurists (2020) India on the brink of hunger crisis during COVID-19 pandemic, briefing paper

IFRC (2019) The Cost of Doing Nothing, Geneva. https://www.ifrc.org/sites/default/files/2021-07/2019-IFRC-CODN-EN.pdf

ILO (2020) Temporary wage subsidies: country examples. https://www.ilo.org/wcmsp5/groups/public/%2D%2D-ed_protect/%2D%2D-protrav/%2D%2D-travail/documents/publication/wcms_745667.pdf

ILO International Labour Organization (2021a) ILO monitor: COVID-19 and the world of work. Seventh edition. updated estimates and analysis. https://bit.ly/34KwWuV

ILO International Labour Organization (2021b) World social protection report 2020–2022, Geneva

IPCC (2014) Climate Change 2014, Synthesis Report. https://www.ipcc.ch/site/assets/uploads/2018/05/SYR_AR5_FINAL_full_wcover.pdf

Islam N, Winkel J (2017) Climate change and social inequity, UNDESA Working Paper No. 152, ST/ESA/2017/DWP/152

Kaiser Family Foundation KFF (2021) The implication of COVID – 19 for mental health and substance use. https://www.kff.org/coronavirus-covid-19/issue-brief/the-implications-of-covid-19-for-mental-health-and-substance-use/

Karim MR, Islam MT, Talukder B (2020) COVID-19's impacts on migrant workers from Bangladesh: in search of policy intervention. World Dev 136:105123. https://doi.org/10.1016/j.worlddev.2020.105123

Kayla DLH (2020) Los Angeles County: Americans are not getting enough to eat during the COVID-19 pandemic. https://www.usnews.com/news/cities/articles/2020-12-23/americans-arent-getting-enough-to-eat-during-the-coronavirus-pandemic-heres-whats-happening-in-los-angeles-county

Kihato CW, Landau LB (2020) Coercion or the social contract? COVID 19 and spatial (in)justice in African cities. City Soc 32(1):1–11

Kirby J (2021) How Senegal stretched its health care system to stop Covid-19. Vox. https://bit.ly/3gaexGF

Koehler G (2011) Transformative social protection: reflections on South Asian policy experiences. IDS Bull 42(6):96–103

Kraay A, McKenzie D (2014) Do poverty traps exist? Assessing the evidence. J Econ Perspect 28(3):127–148

Lakner CN, Yonzan D, Mahler G, Aguilar AC, Wu H (2021) Updated estimates of the impact of COVID-19 on global poverty: looking back at 2020 and the outlook for 2021. World Bank. https://bit.ly/3peextg

Langer N (2004) Natural disasters that reveal cracks in our social foundation. Educ Gerontol 30(4):275–285

Lárraga LGD (2016) How does prospera work? Best practices in the implementation of conditional cash transfer programs in Latin America and the Caribbean,

Inter-American development bank, social protection and health division, Technical Note IDB-TN-971

Martin A, Markhvida M, Hallegatte S, Walsh B (2020) Socio-economic impacts of COVID-19 on household consumption and poverty. Econ Disaster Clim Change 4(3):453–479. https://doi.org/10.1007/s41885-020-00070-3

McMichael A, Woodruff R, Whetton P, Hennessy K, Nicholls N, Hales S, Woodward A, Kjellstrom T (2003) Human health and climate change in Oceania: a risk assessment 2002. Commonwealth Department of Health and Ageing, Canberra, Australia

Mello D (2020) Risco de morrer por coronavírus varia até 10 vezes entre bairros de SP', Agência Brasil. https://agenciabrasil.ebc.com.br/saude/noticia/2020-05/risco-de-morrer-por-coronavirus-varia-ate-10-vezes-entre-bairros-de-sp

Middle East Eye (2020) Why Egypt's coronavirus response failed

Neumayer E, Plümper T (2007) The Gendered Nature of Natural Disasters: The Impact of Catastrophic Events on the Gender Gap in Life Expectancy, 1981–2002. Annals of the Association of American Geographers 97(3):551–566. https://doi.org/10.1111/j.1467-8306.2007.00563.x

NITI Ayog, BMC and TIFR (2020) Technical details: SARS-CoV2 serological survey in Mumbai by NITI-BMC-TIFR, Technical details: SARS-CoV2 Serological Survey in Mumbai by NITI-BMC-TIFR

Norton A, Conway T, Foster M (2001) Social protection concepts and approaches: implications for policy and practice in international development. Overseas Development Institute

NRC (2020) An unnecessary burden: forced evictions and COVID-19. https://www.nrc.no/globalassets/pdf/briefing-notes/an-unnecessary-burden-forced-evictions-and-covid-19-in-east-africa/an-unnecessary-burden_nrc-briefng-note_july-2020.pdf

OECD (2018) Settling in 2018: indicators of immigrant integration. OECD Publishing, Paris/European Union, Brussels

OECD (2020a) What is the impact of the COVID-19 pandemic on immigrants and their children? https://www.oecd.org/coronavirus/policy-responses/what-is-the-impact-of-the-COVID-19%20-19-pandemic-on-immigrants-and-their-children-e7cbb7de/)

OECD (2020b) Job retention schemes during the COVID-19 lockdown and beyond. http://www.oecd.org/coronavirus/policy-responses/job-retention-schemes-during-the-COVID-19-19-lockdown-and-beyond-0853ba1d/

Overstreet S, Burch B (2009) Mental health status of women and children following Hurricane Katrina. In: Willinger B (ed) Hurricane Katrina and the women of New Orleans. Newcomb College Center for Research on Women, New Orleans

Oxfam International (2022) First crisis, then catastrophe. https://www.oxfam.org/en/research/first-crisis-then-catastrophe

Oxford Policy Management (2018) Shock- Responsive Social Protections Systems Research, Synthesis Report, Oxford, UK. https://www.opml.co.uk/files/Publications/a0408-shock-responsive-social-protection-systems/srsp-synthesis-report.pdf?noredirect=1

Pande R, Schaner S, Moore CT, Stacy E (2020) Reaching India's poorest women with covid-19 relief. Yale Economic Growth Centre. https://egc.yale.edu/reaching-indias-poorest-women-covid-19-relief

Patra D, Patro E (2021) Discrimination during pandemic, cyclones put lives of Dalits at risk, down to earth. https://www.downtoearth.org.in/blog/governance/discrimination-during-pandemic-cyclones-puts-lives-of-dalits-at-risk-78580

Perry BL, Aronson B, Pescosolido BA (2021) Pandemic precarity: COVID-19 is exposing and exacerbating inequalities in the American heartland. Proc Nat Acad Sci USA 118(8):e2020685118. https://doi.org/10.1073/pnas.2020685118

Pierce M, Hope H, Prof Ford T, Prof Hatch S, Prof Hotopf M, Prof John A et al (2020) Mental health before and during the COVID 19 pandemic: a longitudinal probability sample survey of the UK population. Lancet Psychiatry 7(10):883–892

Pietrabissa G, Simpson SG (2020) Psychological consequences of social isolation during COVID-19 outbreak. Front Psychol 11:2201. https://doi.org/10.3389/fpsyg.2020.02201

Pozarny P (2016) Climate change and social development: topic guide. GSDRC, University of Birmingham, Birmingham

Prusa, A, García Nice B, and Soledad O (2020) 'Pandemic of Violence: Protecting women during COVID-19', Weekly Asado, 15 May. https://www.wilsoncenter.org/blog-post/pandemic-violence-protecting-women-during-covid-19

Ravallion M, Jalan J (2001) Household income dynamics in rural China. World Bank policy research working paper. World Bank, Washington, DC

Renzaho A (2020) The need for the right socio-economic and cultural fit in the COVID-19 response in sub-Saharan Africa: examining demographic, economic political, health, and socio-cultural differentials in COVID-19 morbidity and mortality. Int J Environ Res Public Health 17(10):3445

Reuters (2020) Coronavirus hits migrant workers in Qatar. https://www.reuters.com/article/uk-health-coronavirus-qatar/coronavirus-hits-migrant-workers-in-qatar-idUKKBN2162D4

Romano C (2020) Inclusion of persons with disabilities in social protection for COVID-19 recovery and beyond. https://socialprotection.org/discover/blog/inclusion--persons-disabilities-social-protection-covid-19-recovery-and-beyond

Ruth M, Ibarrarán ME (2009) Distributional Impacts of Climate Change and Disasters. Books, Edward Elgar Publishing, number 13215. https://www.elgaronline.com/view/edcoll/9781848440371/9781848440371.xml

Saafan (2020) 2.5 million beneficiaries of sisi's grant for irregular workers. State information service SPECTRUM News1, 2020). Governor Newsum announces initiative to help workers, immigrants impacted by COVID-19. https://spectrumnews1.com/ca/la-west/health/2020/04/15/gov%2D%2Dnewsom-announces-new-initiative-to-help-workers%2D%2Dimmigrants-impacted-by-covid-19

Sverdlik A, Walnycki A (2021) Better cities after COVID-19: transformative urban recovery in the global south, Issue Paper, IIED, ISBN 978-1-78431-897-0Williams

The World Bank (2014) World Development Report, Risk and Opportunity: Managing Risk for Development. http://hdl.handle.net/10986/16092

The World Bank (2018) The State of Social Safety Nets. https://openknowledge.worldbank.org/bitstream/handle/10986/29115/9781464812545.pdf?sequence=5&isAllowed=y

Thomas K, Hardy RD, Lazrus H, Mendez M, Orlove B, Rivera-Collazo I, Roberts JT, Rockman M, Warner BP, Winthrop R (2018) Explaining differential vulnerability to climate change: a social science review. WIREs Climate Change 10(2):e565. https://doi.org/10.1002/wcc.565

Tiwari S (2020) No documents, no benefits: how India's invisible workforce is left to fend for itself, India Spend. https://www.indiaspend.com/no-documents-no-benefits-how-indias-invisible-workforce-is-left-to-fend-for-itself/

Toulemon L, Barbieri M (2008) The mortality impact of the August 2003 heat wave in France: investigating the "harvesting" effect and other long-term consequences. Popul Stud 62(1):39–53

UN Habitat (2006) The state of the world's cities 2006/2007: the millennium development goals and urban sustainability. Earthscan, London

UN Habitat (2011) Global report on human settlements: cities and climate change. Earthscan, London

UN Habitat (2020a) Rapid assessment of informal settlements in Yangon: covif-19 pandemics and its impacts on residents of informal settlements. UN Habitat, Yangon

UN Habitat (2020b) Rapid assessment of COVID-19 in informal settlements in Fiji: insights on socio-economic impacts on residents in 16 communities across Viti Levu. UN-Habitat, Fiji

UN Habitat (2021) Cities and pandemics: towards a more just, green and healthy future. https://unhabitat.org/sites/default/files/2021/03/cities_and_pandemics-towards_a_more_just_green_and_healthy_future_un-habitat_2021.pdf, Nairobi

UN Habitat (2022). https://unhabitat.org/food-insecurity-a-real-concern-among-the-urban-poor-in-sub-saharan-africa-following-pandemic-new

UNDP United Nations Development Programme (2007) Human development report 2007/2008. Palgrave Macmillan, New York, p 77

UNFPA, United Nations Population Fund (2009) Facing a changing world: women, population and climate. http://www.unfpa.org/publications/state-world-population-2009

UNHCR (2020) COVID-19 crisis: Kenya urged to stop all evictions and protect housing rights defenders. https://www.ohchr.org/en/NewsEvents/Pages/DisplayNews.aspx?NewsID=25901&LangID=E#:~:text=GENEVA%20(22%20May%202020)%20%E2%80%93,on%20the%20situation%20of%20human

United Nations (2020a) The sustainable development goals report 2020. https://unstats.un.org/sdgs/report/2020/

United Nations (2020b) Special Rapporteur on extreme poverty and human rights, looking back to look ahead: A rights-based approach to social protection in the post-COVID-19 economic recovery. https://www.ohchr.org/sites/default/files/Documents/Issues/Poverty/covid19.pdf

Vaziralli S (2021) Building climate responsive social protection. International Growth Center. https://www.theigc.org/blog/building-climate-responsive-social-protection/

Verner D (ed) (2010) Reducing poverty, protecting livelihoods, and building assets in a changing climate: social implications of climate change for Latin America and the Caribbean. World Bank, Washington DC

Wade L (2020) An unequal blow. Science 368(6492):700–703

Ware A (2021) Giving 2021: pandemic lessons and the future of religious giving. https://blog.philanthropy.iupui.edu/2021/07/06/giving-2021-pandemic-lessons-and-the-future-of-religious-giving/

Wasdani KP, Prasad A (2020) The impossibility of social distancing among the urban poor: the case of an Indian slum in the times of COVID-19. Local Environ 25(5):414–418

WEDO – Women's Environment and Development Organization (2008) 'Gender, climate change and human security, lessons from Bangladesh, Ghana and Senegal', Paper prepared for Hellenic Foundation for European and Foreign Policy (ELIAMEP). WEDO, New York

WEN – Women's Environmental Network (2010) Gender and the climate change agenda. The impacts of climate change on women and public policy. http://www.wen.org.uk/wp-content/uploads/Gender-and-the-climate-change-agenda-21.pdf

Wenham C, Smith J, Morgan R (2020) COVID-19: the gendered impacts of the outbreak. Lancet 395(10227):846–848

WFP World Food Program (2020a) COVID-19 will double number of people facing food crises unless swift action is taken. https://www.wfp.org/news/covid-19-will-double-number-people-facing-food-crises-unless-swift-action-taken

WFP World Food Program (2020b) COVID-19 level 3 emergency: external situation report 17. https://www.ilo.org/global/about-the-ilo/newsroom/news/WCMS_755875/lang%2D%2Den/index.htm

WFP, World Food Program (2022) Unprecedented needs threaten a hunger catastrophe. https://docs.wfp.org/api/documents/WFP-0000138231/download/

WIEGO and AeT (2021) COVID-19 crisis and the informal economy: informal workers in Durban, South Africa. https://bit.ly/2STJh6u

Wilkinson A (2020) Local response in health emergencies: key considerations for addressing the COVID-19 pandemic in informal urban settlements. Environ Urban 32(2):503–522

Witteveen D (2020) Sociodemographic inequality in exposure to COVID-19-induced economic hardship in the United Kingdom. Res Soc Stratif Mobil 69:100551

Witze A (2021) Racism is magnifying the deadly impact of rising city heat. Nature 595(7867):349–351. https://www.nature.com/articles/d41586-021-01881-4

World Bank (2013) Building resilience to disasters and climate change through social protection. World Bank, Washington, DC

Part V

Mega Risks, City Governance, and Vision for Cities of Tomorrow

An effort has been made in this part to create a vision for a better urban future in the wake of challenges created by the twin mega risks. In terms of "governance," the part highlights how integrated governance and preparedness planning at the local level in some cities facilitated timely actions to prevent the diffusion of the virus and reduced the socioeconomic impacts of the pandemic. It builds a case for strengthening local governments and promoting stakeholders' involvement in urban governance in the wake of mega risks. It also focuses on how appropriate governance can make cities more resilient to future shocks. The need for integrated urban governance has been stressed. In addition, emphasis has been placed on visioning, pre-risk event planning, preparedness, and early warning as well as synchronization of stakeholders' efforts. The necessity for both horizontal and vertical coordination between the tiers of government has been highlighted as the basic ingredient of judicious and appropriate governance response to urban mega risks.

The part also highlights the role of long-term visioning and appropriate planning in preventing and mitigating the impacts of mega risks and adapting to them. It summarizes the main ideas on how to shape the future of cities in the light of lessons learned from the pandemic and climate crisis. The most important lessons from mega risks are that ongoing economic recovery and future development should divulge from "business-as-usual" environmentally destructive investment patterns and activities. It demands careful selection of development options based on trade-offs between economic, social, and environmental dimensions. Basically, the selected economic growth or development option should not cause adverse social impacts or have environmental footprints. It also highlights the monitoring of both short- and long-term recovery and development measures to keep development on the right track. Regrettably, all the tracking initiatives put in place so far to analyze the greening of recovery measures depict disappointing results. They highlight that a very large proportion of recovery measures are neither green nor geared to achieving sustainable development. This shows that both the mega risks have failed to change the direction of development trajectory to the right level, although science has informed that both the mega risks have emerged from ignoring the green measures. For example, the uncontrolled expansion of agriculture and other commercial enterprises as well as human settlements into forests and wilderness has resulted in habitat loss—contributing to risky overlaps between human and animal species, thus allowing easy transfer of viruses like coronavirus from the latter to the former. Similarly, comparable pursuit in the industrial sector resulted in excessive emissions, pollution, and climate change. The biggest challenge for the cities in coming years is therefore to decouple the resource use from the development process. The issue becomes particularly pertinent in the likely scenario of continuous growth of cities and towns in the world.

A big lesson from the mega risks is to shift from "reactive approach to emergencies" to "proactive preparedness, planning, and decision-making." In order to be truly resilient, the cities need to be able to not just bounce back from a catastrophic event but also to move forward with resilience. The part has made a case that the best option to tackle these and other related issues is to follow the framework of Sustainable Development Goals of the Agenda for Sustainable Development 2030. It is a good omen that cities and mayors are increasingly taking the initiative and responsibility for local-level monitoring and local-level plan development to implement both Agenda 2030 and the Paris Agreement on climate change.

The cities had already embarked on the path of promoting technologies even before the pandemic and their growing use had already led to the idea of smart cities. The use of digital technology, however, enhanced exponentially during the pandemic for remote functioning, putting early warning systems in place and in designing protective mechanisms. Ideas on making cities resilient through city layout and URBAN design also emerged more strongly and were practically applied in the wake of mega risks. One idea entitled "15-minute city" became especially popular and has been put to test in several cities. The part, while outlining the city design theories from garden city movement to 15-minute city, discusses the later concept in detail in relation to both mega risks: COVID-19 and climate change.

10.1 Introduction

Urban resilience, preparedness, and response to mega risks whether coronavirus pandemic or climate crisis has hinged largely upon cities' governance systems and prevalent sociopolitical milieu. The twin mega risks have illustrated that while the national and state governments have a crucial role to play in responding to such risks, it is the cities and urban systems where the actual battle is being fought. The pandemic has precipitated a historic resurgence of state intervention and state power across public health and varied social and economic support. Simultaneously, increased permissiveness associated with crisis has enhanced and promoted trends in urban innovation. These are evident in expanding repertoire of governance mechanisms across public space, as well as encouragement of new mobility patterns, use of technology in administration, and human service delivery (McGuirk et al. 2020).

This chapter has been divided into four parts. After this introductory part, the next part focuses on the urban governance in relation to the pandemic. It has four sections. The first section examines the institutional mechanisms and governance approaches and practices followed during the pandemic highlighting the municipal governments' role in the COVID-19 crisis management and recovery. Since the challenges posed by the pandemic were mammoth, the city governments alone did not have either the capacity or resources to manage these and had to depend on other tiers of government for assistance in this endeavor. The next section as such examines the coordination aspect of governance. It deals with the actions undertaken by city governments in collaboration with national and subnational governments probing into the coordination mechanisms. The purpose is to examine coordination frameworks used and how they could be strengthened while taking care of or ensuring the boundaries and autonomies of the partners. The third section concentrates on innovation in governance mechanisms adopted to deal with the pandemic. The fourth section highlights the limitations of governments in general and city/municipal governments, particularly in dealing with the crisis. It also discusses the lessons derived from the pandemic for improving the governance system in cities and making them more resilient to face future shocks including the climate crisis. The next part of the chapter deals with the governance system and policies of cities in relation to climate crisis. It has three sections. The first section traces the evolution of urban governance in relation to climate change and examines the modes of climate governance and institutional mechanisms adopted for the purpose as well as enablers and barriers to climate action. The second section on multilevel climate governance basically deals with coordination frameworks used for the purpose. The third section while discussing the achievements of urban climate governance highlights the constraints and shortcomings faced. The findings of the chapter are summarized in the concluding part.

10.2 Urban Governance and Coronavirus Pandemic

Initially in terms of governance, when the crisis hit the cities, the national governments were the first to take the control. They raised the alarm, undertook lockdown measures, coordinated health responses, implemented border control, and undertook the main economic measures for cushioning the impact. However, in this context, the denialism of leaders like Trump in the USA, Bolsonaro of Brazil, and Orbán in Hungary raised the "infodemic" to pandemic levels (Losada and Abdullah 2020) and caused serious negative impacts. To counter this, many cities established their own online platforms and provided their citizens with reliable information. For example, São Paulo in Brazil launched an online channel to provide citizens with accurate information on the virus and to discredit fake news (São Paulo undated). In Rome, the website RomaAiutaRoma was launched by authorities to serve as a one-stop information source for residents to access updates, news on local initiatives, advice on family well-being, and other useful content (UN Habitat 2021).

While the national governments were taking the lead, the cities, as the hotspot of the pandemic, did not remain aloof in terms of governance. Within a short time, they became active and moved to the forefront of the response at least in terms of meeting some of the basic needs of their citizens despite their limited budget. Besides providing essential services and adapting public space to enable social distancing, they offered care to the most vulnerable; supported businesses, professionals, and workers affected by the crisis; and made efforts to strengthen their healthcare systems (Losada and Abdullah 2020). Additionally, the crisis required many local and city governments, to take or lead initiatives in areas, which were not necessarily in the scope of their normal responsibilities, either at the request of the central government or to respond to emergencies that arose (UN Habitat 2021). A UN Habitat (2021) report divided cities' actions in

response to the pandemic into three categories: (a) implementation of health-related measures aimed at controlling the spread of the virus including physical distancing, mandatory mask wearing, school closures, restrictions on movement, curfews, and other actions associated with "lockdown"; (b) support-oriented initiatives to cushion urban populations from the socioeconomic effects of the pandemic (these included the provision of food, water, and immediate needs to economic relief—including tax breaks and housing subsidies, as well as social support—such as counseling and initiatives to address gender-based violence); and (c) forward-looking strategic actions and investments for stimulating the economy and increasing the resilience of cities to future shocks including investments in inclusive transport, schools, clean energy, safer public spaces, and other measures—to support both the immediate response and also deliver sizeable benefits to cities in the long term.

10.2.1 Governance Institutions and Practices

This subsection has been divided into two parts; the first one concentrates mainly on the institutional aspects of pandemic governance with the focus on cities including the use of existing institutional mechanisms as well as evolution of new institutional mechanisms. It discusses the approaches followed in terms of centralization or decentralization and coordination processes and procedures adopted. The mechanisms followed for vertical coordination between cities and other tiers of government and horizontal coordination between various stakeholders and among cities and their regions have been highlighted. The second part focuses mainly on the innovation in governance practices in terms of both hard (digital technology) and soft tools (partnerships, community engagement, adjustment of regulations, etc.) that became necessary to contain the multidimensional impact of evolving coronavirus crisis.

10.2.2 Institutional Mechanisms

In response to the pandemic, cities used both existing governance mechanisms and new mechanisms to address the challenge. Irrespective of the mechanisms adopted, strong institutions with multilevel coordination and cooperation and previous experiences led to successful response to the pandemic in several cities of Canada, China, Japan, the Republic of Korea, Singapore, Taiwan, and Vietnam. Most cities in these countries had learned from previous public health crisis—like SARS outbreak in Asia. Many African cities learned from Ebola crisis of 2014–2016 in Africa, and still other cities in the USA and Europe learned from heat wave episodes in their respective countries. As the coronavirus crisis unfolded, at least in the initial stages, cities with preexisting institutional structures and mechanisms and practices for addressing emergencies fared better than those that did not have these. In contrast, absence of proactive planning and emergency plans in many developing countries such as Bangladesh, Brazil, India, and Indonesia made it difficult for cities to effectively respond to the crisis. Cities in these countries did not assess the risk or analyzed the situation properly, failed to enhance the capacity of the healthcare system and take required measures in a timely manner (Shammi et al. 2020; Tanveer et al. 2020), and suffered more.

10.2.2.1   Existing Institutional Mechanisms

A UN Habitat (2021) report classified the preexisting governance mechanisms into three types: (a) existing institutionalized bodies for multilevel governance; (b) existing local administrative structures and networks; and (c) existing governance practices, instruments, and infrastructure.

Public Health Agency of Canada, a fully institutionalized body created during 2003 SARS epidemic, provides an example of the first type. The agency activated its Health Portfolio Operations Centre as focal point for response activities and emergency operations. Kenya's Council of Governors constituting leaders of the 47 counties provides another example. It created a county government emergency fund and facilitated engagement with the national government on responding to the pandemic. The example of second type that utilized existing local administrative structures and networks that preceded the pandemic were those of Singapore, based on SAR's experience; Congo, based on Ebola experience for raising awareness and communicating health information and for undertaking policy measures; and Chicago, based on its experience of 1995 severe heat wave, which helped the city in swift repurposing of some buildings and infrastructure to serve as emergency hospitals and homeless shelters during COVID-19 (Kling 2020).

The third type, in terms of utilization of existing mechanisms, utilized governance practices and principles rather than institutions, for example, practices such as public participation and principles such as social cohesion, communal solidarity, transparency, and trust in public institutions that had yielded multiple positive results in the past. Thus, in cities of Uruguay, preexisting governance modes of collective decision-making ensured widespread support for containment measures. The business associations and unions in cities there made voluntary agreements to abide by restrictive measures and supported campaigns of the government even in the absence of mandatory regulations. In some cases, governance instruments on disaster prevention and management or health emergencies were also effectively used. Several cities in Colombia, for example, used a 2012 law under which a disaster management system was created to establish local disaster management strategies in response to the crisis (Government of Colombia 2020). Similarly empowered by a 2019 Regulation No. 30895, Peru strengthened its Ministry of Health's capacity to respond to the crisis—making it one of the first country in Latin America to impose restrictions and approve a national plan for addressing the pandemic, even before the country registered its first case of COVID-19 (Republic of Peru 2020).

10.2.2.2 New Institutional Mechanisms

In terms of new mechanisms, ad hoc task forces and special bodies were created to address the pandemic in cities and ensure coordination at different levels. They took varied forms—some with broader scope included nongovernmental actors and stakeholders. Their breadth helped in increasing compliance and governess effectiveness—while also contributing to enhanced trust in institutions, accountability, transparency, and community engagement. Chile's Social Committee for COVID-19, comprising municipal association's representatives, government authorities, academics, and health workers and professionals (OECD 2020a), provided a good example. Task forces were also formed at subnational/city levels in other countries, spearheaded by various levels of government such as Toronto Office of Recovery and Rebuild in Canada (Toronto undated) and Chicago's COVID-19 Recovery Taskforce in the USA. The latter included a broad range of membership from industry, regional governments, and community-based organizations to focus on a wide range of areas—from business and policy development to mental health and social change (Chicago undated). Cities also saw the formation of special bodies, for example, in Senegal, local authorities partnered with the Senegalese national fund to form a task force called "Force COVID-19." Overseen by the president of the Association of Mayors of Senegal, it raised funds, received medical equipment from donors, and disbursed these to health facilities in a coordinated manner (UCLG Africa 2020). In Turkey, decentralized urban response and coordination was undertaken through "pandemic boards" established by governorates (UN Habitat 2021). Vertical coordination, besides institutional means, was promoted through agreements, as in Germany, where tougher measures were agreed between 11 cities and the national government (DW 2020). The Government of Georgia established the National Intersectoral Coordination Council, which became the main decision-making body regarding the rules and restrictions related to COVID-19 within the country (Bieliei et al. 2020). In Armenia, a Crisis Management Center, under the authority of the Deputy Prime Minister, was established to ensure centralized management of the crisis (Council of Europe 2021).

Overall preexisting governance mechanisms were more successful in responding to COVID-19 in cities compared to the ad hoc arrangements. Among existing governance mechanisms, institutionalized ones had relative advantages (UN Habitat 2021). In many cases, however, existing structures were not fully utilized in the pandemic, and sometimes even parallel new structures were created which resulted in duplication of efforts and confusion in implementing a coordinated response strategy (UN Habitat 2021).

10.2.3 Approaches and Practices

This subsection discusses the various approaches that were adopted to centralize or decentralize governance functions. It also describes methods followed to promote coordination between various levels of government.

10.2.3.1 Centralized Versus Decentralized Approach

Both centralized and decentralized approaches existed in the unitary and federal types of government when the pandemic struck, and power delegation or centralization was exercised in both types of government. The delegation in federal state was exercised through transfer of power from national to subnational (regional/metropolitan) as was done in Germany (Han et al. 2020) and Russian Federation (UN Habitat 2021). It was also exercised through intracity transfer of power from the city council to mayor—the city council of Toronto in Canada did this by giving extended power to the mayor to declare public health emergencies and reallocate city resources (DeClerq 2020). Powers were also delegated in unitary states like Italy, which granted extended power to regions—to apply restrictions higher than those imposed at the national level (Government of Italy 2020). Similarly in the UK, a legislation of July 2020 gave local authorities (including London borough councils) new control powers to exercise restrictions on access to indoor, outdoor, and event spaces. In some unitary states, local/city governments were also given more fiscal space (UN Habitat 2021).

Conversely, recentralization in federal and quasi-federal states was achieved through transfers of power from the subnational to the national and from the city to the subnational level. For example, in India, the national government used the Disaster Management Act to impose a nationwide lockdown, overriding the authority of states (James 2020). Likewise, cities in the USA saw their powers preempted by subnational authorities—for example, the governor of Georgia issued an executive order that effectively annulled any prior local "stay-at-home" mandates, reopening public spaces such as beaches that had been closed by local authorities; similar actions were taken in Florida, Mississippi, and Arkansas (Foster 2020). Unitary states also saw a similar transfer of power. For example, Colombia issued a decree which gave power to the president to overrule the provisions of governors and mayors (Republic of Colombia 2020).

A comparative case study of two cities—Shanghai in China under centralized system and Los Angeles in a decentralized system by Weng et al. (2020)—provides useful and interesting insights. The study demonstrated that besides centralization versus decentralization, several other factors such as immediacy versus thoroughness, transparency versus secrecy and security, and state-driven solutions versus coproduction (state in cooperation with other stakeholders) have played important role in early control of the disease in the two cities. The findings of the study showed that although the two cities differed in the roles of the governments due to variation in the institutions, existing values, political milieu, and administrative capacity, some common attributes of good managerial response enabled both cities to manage the situation better in early stages (Weng et al. 2020).

No doubt, national policies helped Shanghai, but the city's own efforts also deserved recognition for having effective planning and policy execution of its "four early steps policy"—early

detection by finding any patients as soon as possible; early reporting of cases by local officials; early isolation of patients to contain the possible transmission; and early treatment of patients when found (Weng et al. 2020). The administration in Shanghai took a high government-centric approach and intruded considerably into the privacy of individuals. In contrast to such a strong-handed approach due to the restrictions of power, local government in Los Angeles relied more on voluntary compliance, coproduction, public education, and the sense of civic duty in fighting the pandemic.

All in all, the pandemic demonstrated the importance of cities and regional governments in dealing with external shocks. The factors that helped included their closer proximity to affected populations and greater potential to deliver fast and make response flexible to emerging situations as they evolved. In this context, there is a need to enhance resources of local governments and promote cooperation and coordination between various tiers of government whether vertical—top down from national and subnational to local level or horizontal intra—metropolitan cooperation between municipalities. In fact, cities and regions need the autonomy to develop policies that make sense in the context of their own territories, issues, and populations as well as resources to operationalize governance. Nevertheless, while regional, metropolitan, and local governments are best informed of local circumstances and well positioned to implement measures at the local level, national governments are often best placed to oversee the design and implementation of coherent and equitable action plans across the country. A coordinated approach is therefore imperative to combine both these positive aspects which can lead to the achievement of better results.

10.2.3.2 Coordination and Cooperation Approaches

Coordination and cooperation in governance were crucial in the management of cities in the wake of mega risks both between various tiers of government and between the government and other stakeholders. This section discusses first the coordination approaches and strategies for the intergovernmental coordination in designing and implementing measures to combat the pandemic. It then describes the aspects of cooperation between the governments and nongovernmental stakeholder including private sector, scientific community, and civil society.

10.2.4   Vertical Coordination Between Cities and Upper Tiers of Government

Vertical coordination involved cooperation between cities and upper tiers of government, i.e., at subnational or national levels. In terms of scale of collaboration, cities engaged with higher levels of government individually as well as collectively. In the former case, a specific city coordinated its actions directly and independently with another level of government. For example, Mexico City worked in close cooperation with the state government to coordinate the capacity of hospitals in the region (Saliba 2020). In the latter case, coordination was fostered jointly by several cities (through an alliance of cities/local authorities) with an upper level of government. Thus, the Federation of Municipalities and Provinces besides playing an important role in the management of the pandemic has been coordinating with the national government to draft agreements for the post-pandemic recovery. In terms of cooperation in response to the pandemic, city/local authorities played key role in three areas—implementation of health and related lockdown measures; easing associated socioeconomic problems; and promoting forward-looking strategies. Vertical coordination in these endeavors involved a wide range of programs—from tackling health emergency to economic response and social protection.

10.2.4.1   Cooperation in Implementing Healthcare Measures

Cooperation between cities, subnational, and national governments became very important in the implementation of healthcare measures related to masks, social distancing, school and restaurant closures, lockdowns, testing and tracing, etc. Many countries adopted territorial-based approaches for cities or localities, rather than applying a national approach to avoid or limit the economic impact. This was the approach, for example, adapted in Aberdeen (Scotland), Auckland (New Zealand), Barcelona (Spain), Melbourne (Australia), certain provinces in India, and some districts of Germany (OECD 2020a). These differential territorial approaches enabled to avoid the huge costs of a national confinement while providing more targeted responses to the problems where they occurred. In federal countries like the USA, policies differed depending on approaches adopted at the state or subnational level. However, effective coordination between local authorities, subnational governments, and health agencies as well as the central government proved essential to manage local outbreaks even there.

Most of the measures adopted for vertical coordination between cities and above level of governments related to health even in developing countries. For example, in Douala, Cameroon, city officials cooperated with the Ministry of Public Health for the distribution of PPE (masks) and in the adoption of other safety measures (UCLG Africa 2020); in Mombasa, Kenya, in constructing emergency health facilities and distributing food (The Council of Governors 2020); in Kisumu, Kenya, in enforcing curfew (Kiruga 2020); and in Kiambu, Kenya, to promote social distancing (County Government of Kiambu 2020). In cities of South Africa, measures were enforced by metropolitan police, national police services, and national defense force jointly (Dullah Omar Institute 2020).

City's role in implementing the "track, isolate, test, and treat" strategy also played significant role. For example, in Italy, testing all 3300 residents of the town of Vò Euganeo facilitated taking containment and other measures that eventually stopped all new infections. The approach applied locally used extensive testing and proactive tracing and relied more on home diagnosis and care, as well as monitoring medical personnel and other vulnerable workers. The experience of Vò Euganeo's city was replicated successfully in other localities of Veneto region of Italy (RFI 2020) to control the impact of the pandemic. In Japan, local governments had the responsibility for implementing the testing strategy through local institutions and local outpatient and testing centers (OECD 2020a). In Korea, local governments had specific Subnational Centers for countermeasures against the pandemic and helped in coordinating local measures with central authorities (OECD 2020a). In the UK, the government provided a funding package of GB pound 300 million for local authorities to develop tailored outbreak control plans, in cooperation with local National Health Service and other stakeholders. Plans focused on identifying and containing potential outbreaks in workplaces, housing complexes, care homes, and schools (OECD 2020a). In many cases, local/city and subnational governments also assisted the central governments in implementing reopening plans as in the USA, the UK, and France. In the USA, several state governors conceded to allow local officials to determine the best and safest path to transition from a strict "stay-at-home" mode to a gradual reopening of their economies (Benton 2020).

10.2.4.2   Cooperation in Economic Policies/Measures

Cities collaboration with other level of governments on economic front has not been as common as in the field of health. Nonetheless, there are two examples of this type of collaboration between national and local government from Iceland and Finland. In the former case, the national government and local municipalities started a special investment program within the economic response package to the COVID-19 crisis with the focus on transport, public works, and tourism (Government of Iceland 2020). In the latter case, Finish municipalities were given responsibilities for delivering lump sum aid to the self-employed to make sure that the companies continued operation profitably after the pandemic-driven economic crisis. The central government compensated cities/municipalities for their support to the self-employed.

10.2.4.3   Cooperation in Social Policies/Measures

Utilization of existing local capacities at city and local level also became important for upper-level governments not only in meeting pandemic-driven social challenges but also in extending/maintaining or scaling up much needed social assistance. New Orleans (2020) City and the Federal Emergency Management Agency of the USA, for example, partnered to provide food to vulnerable families. In the Dominican Republic, the national government defined a municipal-level subsidy in the form of an economic benefit to informal workers to allow them to stay at home during the crisis. This benefit, created by the Ministry of Finance, granted a transfer of RD$5000 (equivalent to US$93) for all beneficiary households and an additional RD$2000 (US$37) for households headed by aged—older than 60 years (UNDP 2020a). The social support actions were also extended to cooperation between levels of administration on burial arrangements. For example, in Quito city of Ecuador, the mayor coordinated with national-level bodies such as the Criminalistics and Forensic Medicine Unit of the National Police and the Civil Registry to deal promptly with the large number of casualties (EMGIRS 2020).

10.2.5   Intergovernmental Horizontal Coordination

Besides vertical coordination, the frequency of horizontal coordination between local governments, cities, municipalities, and small towns grew exponentially. This emerged from realization by the city administrations that the virus does not respect the territorial jurisdictions, has spillover effects, and demands a multilevel approach in implementing mitigation measures. Hence, partnerships were established between municipalities within metropolitan region; between different cities; and between cities and regions to combat the pandemic. This type of cooperation flourished particularly in implementing health regulations and mobility or public transport measures. In the USA, for example, coordination between the governors of New York, New Jersey, Connecticut, and Pennsylvania in the formulation of health policies led to the creation of a common set of guidelines on social distancing and putting limitations on recreational activities (Edwin 2020). In another case in the USA, three coastal counties of South Carolina (Charleston, Georgetown, and Horry), with tourism as a major driver of their economy (Wilson 2020), forged an alliance to restrict access to the resort locations and transient housing used by vacationers. In Europe, local governments in Serbia created emergency task forces to enforce social distancing and other containment measures in cooperation with one another (UNDP 2020b). Cooperation between cities and their surrounding region was also forged to facilitate the management of food supply and flow of goods across them as well as in providing basic services like water, sanitation, and waste management during the pandemic. In addition, existing city networks like ICLEI and C40 Cities were used for promoting horizontal coordination between cities for exchange of experience and case studies.

10.2.5.1 Coordination Between Government and Non-state Actors

The cooperation and support of the non-state actor was crucial in the COVID-19 crisis. Without their cooperation, it may not have been possible for the cities' governments to accomplish what they did in combating the pandemic, or at least not with the speed and efficiency that were required for it (Losada and Abdullah 2020). These accomplishments ranged from establishment of strong networks to provide support to the vulnerable community (Parnell and Claassen 2020) to innovative solutions that covered fields as diverse as health, biomedicine, mobility, urban planning, and sociology (Ng 2020). For example, as discussed in Chap. 7 on health, meeting the urgent needs of shortages in healthcare equipment like ventilators and PPEs due to rising needs and disruptions in supply chains during the pandemic would not have been possible without private sector initiatives. Their quick designing

for production of PPE was initiated through 3D printers (Abdullah and Reynés 2020).

Partnerships were also developed by city authorities with the private sector and R&D institutions to provide essential goods and services to citizens as well as offer economic reprieve. For example, in Mexico City, authorities in close collaboration with commercial bodies such as Santander México and BBVA México built a mobile application to help residents stay informed, assess symptoms, and provide reliable real-time information to the city government (BBVA 2020). The involvement of non-state actors, in some cases like Montréal, was to look at the post-pandemic future whereby city authorities got engaged with universities and businesses to conduct research on the future opportunities related to work, study, and physical recreation (OECD 2020b). Similarly, the Australia/New Zealand Smart Cities Council forged link with private sector by drawing together consulting, data analytic, and tech firms to form *Bouncelab*—a think tank to advance strategies and projects for promoting "digitally enabled and data-driven" urban recovery post-COVID (McGuirk et al. 2020). Likewise, *City Possible*—a global network promoted by Mastercard—arranged meetings of global municipal decision-makers to exchange experience on strategies related to addressing COVID-19 in their communities (WEF 2020). The company also extended access to transport and delivery services for vulnerable communities by forming a partnership with Lyft. The big challenge in this collaboration is to avoid a wholesale merging of corporate and broader urban governance agenda while taking advantage of the capacity gains of private sector resources (McGuirk et al. 2020).

Cities also explored and maintained partnerships with civic society organizations, NGOs, philanthropies or charitable organizations, and other community actors for mobilizing resources of these important stakeholders in urban governance during the pandemic. These actors contributed to city response strategies in filling gaps by providing targeted assistance. They contributed through direct cash transfers, provision of food and other commodities, as well as production and distribution of health kits. For example, São Paulo's *Cidade Solidária*, a partnership between the city and civic society organizations, coordinated donations and services of volunteers to tackle the pandemic's social and economic effects (UN Habitat 2021). In Québec, an established network of municipal workers, community activists, and academics RÉMIRI held regular meetings for knowledge sharing and response to the pandemic in their city (Council of Europe undated).

10.2.5.2 Community Engagement by Governments

Community engagement by governments took many forms and included different actors and approaches, as was done in the previous outbreak of infectious diseases (Al Siyabi et al. 2021; WHO 2002). One approach taken in Ho Chi Minh City (Vietnam) during the COVID-19 crisis was to promote this engagement through community-based organizations. It helped city and local governments significantly in information dissemination, provision of economic and social support to vulnerable groups, sanitizing public spaces, as well as in implementing social distancing and "stay-at-home" measures (Thoi 2020; Santos et al. 2020). According to Al Siyabi et al. (2021), Oman benefitted considerably from three types of community-based approaches that had been tried there previously. During the pandemic, they facilitated the involvement and collaboration with various segments of the community and promoted the bottom-up approach. They also allowed people to identify their needs and take appropriate actions to fulfill them through financial and in-kind contributions—thus ensuring their ownership of the initiatives and establishing reliable monitoring and evaluation system. Most importantly, they were instrumental in diffusing vital information successfully through different community networks, planned media handling, and mass communication methods.

A rapid review of community engagement in infectious disease prevention (Bhattacharyya et al. 2020) identified 37 initiatives, in which community engagement was used at different stages of risk reduction from planning and gaining community entry to strengthening confidence, risk communication, and surveillance and tracing etc. The review emphasized that localities and countries need to assess existing community engagement structures and to use them to support COVID-19 control measures.

Trust in urban institutions and the city health authorities served as a key pillar for community engagement in the management of the pandemic. In most cases, it depended on communication strategies—which varied considerably from city to city. In the Buenos Aires Metropolitan Area, for example, an innovative method was developed by municipalities (Foglia and Rofman 2020), through an agile and engaged communication network. It was established and used to announce regulations, raise awareness, and reach at-risk groups such as children, slum dwellers, and persons with disabilities with targeted information drives throughout the crisis (ACIJ undated). While communicating messages, cities realized that it was important to access all segment of city population irrespective of their residential status (local or immigrant), culture, language, or economic situation. Many cities, for example, to extend their outreach to specific communities tailored specific campaigns, often in partnership with NGOs or local groups. For example, in Singapore the immigrant workers were supported through a nongovernmental organization IRR (ItsRainingRaincoats). Cities also adopted innovative strategies in community engagements through the process of discussion and dialogues. This proved crucial, in some Australian cities to bolster public confidence in city governments (UN Habitat 2021).

10.2.6 Innovations in Governance Mechanisms

In addition to institutional mechanisms, gravity and complexity of challenges posed by the pandemic stimulated cities to make innovations in urban governance processes, procedures, and practices. Hard as well as soft tools were utilized and developed for the purpose. They ranged from increasing the use of digital technologies to adjustment in governance mechanisms and regulations through partnerships with non-state actors and enhancing community engagements. Amendments in regulation and procedures were made to contain the multidimensional impact of the evolving crisis.

10.2.6.1   Digital Innovations and Urban Governance

Digital technologies were introduced to meet the mounting urban governance challenges and better service delivery even before the COVID-19 pandemic in terms of smart city solutions (see Chap. 11). The crisis accelerated the trend and brought new needs along with increased demands on digital government services. Many cities and local governments, recognizing the importance and opportunities offered by these technologies, made quick efforts to acquire and integrate them into their governance practices. They used them for tracking and surveillance; information sharing through digital platforms; provision of social services (telehealth, remote education); day-to-day administration such as remote meetings, online surveys, and other e-government practices as well as to encourage public participation; and data collection and policy design.

One area where the digital tools played a very important role initially was tracking and stopping the spread of the coronavirus. In London, cameras, sensors, and artificial intelligence algorithms, normally intended to control traffic, were used to measure distance between pedestrians to control social distancing (UNDESA 2020). The second area was information dissemination to share pandemic-related information with citizens—to promote compliance with containment measures, reduce fear and anxiety, enhance trust in city authorities, and negate misinformation. The third area included support in the provision of social arena to provide education, health, and social services as well as delivery of food and other essential items to most needy. Examples of use of digital tools in the mentioned three areas have been given in Chaps. 3, 4, and 6 of this book. The fourth area, where these tools contributed effectively, was participatory governance, whereby

opinions and feedbacks were sought from the citizens and stakeholders on strategies for recovery. Some Australian cities provide good examples of this. Melbourne (2020), for example, opened an online platform for citizens to share their priorities and perspectives on the future of the city, in terms of long-term recovery plan. Likewise, Sydney (2020) sought inputs from local businesses, property owners, organizations, residents, workers, students, and other groups through an online survey on City's Recovery Strategy. Digital tools were also deployed by city and local governments using online streaming platforms to continue their business under the watch of the citizens/public. These tools allowed cities and localities to move their meetings to a virtual setting at a time when in-person meetings were not possible. For example, Albemarle County in the USA did not hold a single virtual meeting in 2019, but in 2020, it held 255 meetings via the Internet as the pandemic moved the public's business online (Wrable 2021). In certain countries like the Netherlands and Spain, representatives at local level were allowed to deliberate and even take legally binding decisions through online sessions (OECD 2020b). Moreover, besides existing online administrative services, new services such as issue of e-permits during curfew were added to promote the convenience of citizens.

Finally, the pandemic boosted development of a smart city framework, which has been discussed in some details in Chap. 11. Further, as mentioned in Chap. 4, services of robots and drones were also utilized—in Chinese cities they were used for providing security and sanitation thus reducing staff exposure to risk. Patrol robots using facial recognition and thermal cameras were also deployed at airports and public places to scan crowds and identify potentially infected people. Sterilization robots equipped with ultraviolet lights proved helpful in disinfecting hospitals and contaminated areas. Other robots monitored vital parameters from medical devices or allowed patients to communicate remotely with the doctors and nurses. Governments also used drones with similar technologies to monitor streets, deliver medical supplies, or disinfect public spaces.

In terms of time, the digital technologies helped solve short-, medium-, as well as long-term problems. They helped resolve health problems in the short term, mitigate socioeconomic impacts in the mid-term, and reinvent existing policies and tools in the long term. A national-level survey by the United Nations capturing the scope and quality of online services, status of telecommunication infrastructure, and existing human capacity showed Denmark, the Republic of Korea, and Estonia in the lead, followed by Finland, Australia, Sweden, the UK, New Zealand, the USA, the Netherlands, Singapore, Iceland, Norway, and Japan. Among the least developed countries, Bhutan, Bangladesh, and Cambodia were in the lead in digital government development. Mauritius, the Seychelles, and South Africa were leading the e-government ranking in Africa. Overall, 65 percent of member states were at the high or very high e-government development index level (UNDESA 2020).

10.2.6.2 Use of Soft Tools: Adjustment in Governance Mechanisms

Cities and local governments developed new means primarily to avoid public hardships, confusion, and incoherence on measures adopted to face the crisis and to promote transparency and citizens' trust in institutions. For example, innovations in regulations and administrative procedures were tried to contain the multidimensional impact of the evolving crisis. Some of these were "forced experiments" in governance (Larcom et al. 2017) necessitated by the pandemic. In certain cases, they caused some disruptions in the institutional forms, relations, and practices (McGuirk et al. 2020). However, in general they were considered to accelerate longer term trends toward "innovation," a trend that had already started shaping or reshaping the ecosystems of urban governance even before the pandemic (Phelps and Miao 2019; Timeus and Gascó 2018)—like creation of flexibility in regulation and procedures.

Some governments recognizing its need embarked on relaxation in administrative procedures to enable cities to respond appropriately to the pandemic. China, for example, put emergency provisions in place to construct temporary health facilities, with the usual bidding and procurement requirements relaxed or suspended to prevent delays. Similarly, in Italy, procedures were simplified in 14 regions to avoid bureaucratic requirements for smaller businesses (OECD 2020c). Likewise, municipalities in Iceland and Slovenia were allowed to reorient their budget priorities to accommodate the needs of the pandemic (OECD 2020a). Mexico City adjusted public procurement process by providing more flexible rules to speed up contracting (Mexico City 2020). No doubt this kind of relaxation was the need of the time. However, in the long term, it may lead to the erosion of accountability and increase the possibility of mismanagement of resources. Therefore, it will be crucial to keep oversight and transparency in place for ensuring that governance structures are not corroded by corruption.

10.2.7 Governance Challenges and Constraints

Among multiple governance challenges, the most important faced by cities during the pandemic related to resources both financial and technological. The findings of a survey of 57 cities and regions in 35 countries worldwide also confirmed this as shown in Fig. 10.1 (UCLG, Metropolis and LSE Cities 2020). Most cities and regions focused their limited resources on the immediate management and response to COVID-19. This was no doubt important, but the critical challenge for cities and local governments was to consider how the short-term actions on which limited resources were being spent aligned with medium- and long-term measures. The cities and regions, which were taking this into consideration, gained advantage toward building back better. Further, a crucial need in this effort was to synergize all actions, as far as possible, at both national levels with the global Sustainable Development Agenda 2030 as well as the Paris Climate Agreement.

The findings of the above survey (UCLG, Metropolis and LSE Cities 2020) revealed that

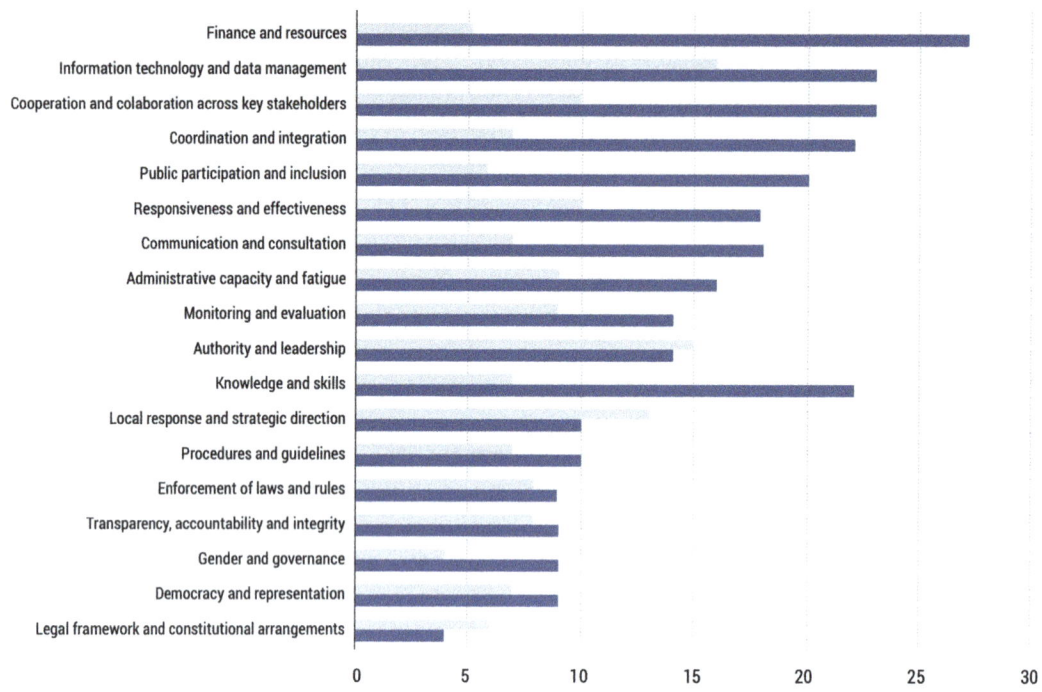

Fig. 10.1 Knowledge gaps and innovative practices by emergency governance domain in cities/regions. (Source: UCLG, Metropolis and LSE Cities (2020))

beside financial resources, access to information technology and data management along with cooperation and collaboration between key stakeholders constituted other important gaps (Fig. 10.1).

10.2.8 Governance Lessons from the Pandemic

Urban shocks and crises, while not discounting their human toll and economic and social impacts, carry some silver lining—as they can be turned into opportunities for making cities safer, healthier, environmentally sustainable, resilient, and inclusive particularly through improvement in urban governance. Reflecting on the problems and challenges as well as governance-related response of cities to combat the pandemic carried important guidelines not only for future management of cities but also for increasing their resilience to future shocks. The crucial takeaway is the need to strengthen city/local governments and improve coordination between tiers of govern-

ments. No doubt, in response to the pandemic, different levels of governments took initiatives, but their actions were not always appropriately coordinated, and their priorities differed. For example, in Australia, the federal government concentrated more on reducing the economic impacts or designing economic stimulus packages to revive the economy. The state governments, on the other hand, were focusing on reducing the pressure on hospitals and on healthcare restrictive measures. These varying priorities resulted in confusion and undermined the effectiveness of city-level response. Strong city governments whether with federal status like Berlin, Buenos Aires, Vienna, and Zurich or metropolitan governments like London and Barcelona fared better in managing the pandemic and comparatively excelled in designing and implementing recovery strategies as they were better equipped in terms of resources and tools (Losada and Abdullah 2020).

In many cases, however, city-level governance particularly in developing countries proved weak. For example, in countries like Pakistan, the local governments have limited powers, are not even recognized in the constitution, have little fiscal clout, and are at the mercy of higher tiers of government for financial resources. It is important to reconcile the administrative and fiscal authority of city/local government with the delivery of functions expected from them. Moreover, performance-based incentive structures are needed to make city governments financially independent. Important lessons for improving urban governance particularly in strengthening administrative and financial autonomy of local/city governments brought forward by the pandemic were as follows:

• Increase empowerment of local governments and enhance the city mandates, roles, and responsibilities for their effective functioning.
• Develop mechanisms for coordination of different actors and sectors in management of crises and emergencies to avoid confusions/conflicts and ensure effective and efficient use of limited resources.
• Use mechanism to enhance community engagement and trust in government as a basic tenet in urban governance ecosystem.
• Keep urban governance open to "Transformative Innovations" both hard and soft—that helped reshape the ecosystems of urban governance during the pandemic "state of exception"—to make urban governance progressive, smart, and more inclusive.
• Strengthen urban networks and other mechanisms—to allow cities to learn from each other in implementing effective emergency responses, recovery, as well as short-, medium-, and long-term actions for enhancing resilience.
• A long-term approach needs to be factored into urban governance even when responding to the immediate response to a crisis, as much as possible.
• A key signal especially important in a post-COVID fiscal context is to enhance the fiscal capacity of the city governments.

10.3 Climate Change and Urban Governance

Like coronavirus mega risk, cities are also at the front line of climate crisis both as the driver of climate change and as its victim. This section deals with both local and multilevel dimensions of urban climate governance highlighting the innovative mechanisms adopted as well as the problems and prospects. Ever increasing urban energy needs together with growing waste is generating a big carbon footprint. Although cities occupy only 3% area of the world (Schriber 2005), they account for two-thirds of global greenhouse gas emissions. Therefore, slowing down, if not reversing, climate change depends a lot on the cities' climate action. Governance is important in handling climate action at urban scale for several reasons. First and foremost, it guides in developing a common vision among key stakeholders. Second, it ensures their engagement in reaching a common set of goals to handle. Third, it helps in pooling scarce resources at local level for combating a common issue. Fourth, it allows to alleviate key areas of potential conflict between climate change response measures and policies/programs for other core urban governance needs. Fifth, city size or urban scale is just right to contribute to policy innovation and carries potential gains from learning by doing—without imposing large costs on an entire country or the world with untried options (Oates 2002). Moreover, the scale also allows cities and local governments greater flexibility to adapt to new situations quickly and to modify existing climate policies or to define new ones relatively faster (Galarraga et al. 2011). Last but not the least, cities and local authorities are often more effective in reducing GHG emissions from some sources over which they have direct control like land use planning, building codes, waste management, and organizing infrastructure and public transport.

10.3.1 Historical Perspective

Cities' fight against climate change, which was initiated some 30 years ago, has evolved and broadened in terms of governance scope, focus, stakeholders' involvement and goal and target setting. From a limited focus on energy efficiency to mitigation, the urban climate action has now also incorporated adaptation within its ambit. Moreover, the approach to treat climate change as an environmental issue is being transformed into mainstreaming it into development process. At the same time, stakeholder involvement has been expanded to include private sector and civil society in the process. Further, goals and targets are being set more realistically and are being aligned with national and global goals and targets. This is largely a result of a growing number of urban local governments joining city climate networks at national (from the American Cities Climate Challenge and the Swedish *Klimatkommunerna* network to Climate Action Network Australia and Climate Action Network Tanzania) and international level (the Global Covenant of Mayors [GcoM]; C40 Cities Climate Leadership Group; International Council for Local Environmental Initiatives [ICLEI]; and United Cities and Local Governments [UCLG]).

There has been some variation in type of climate actions undertaken by cities in time and space. For example, the initial focus of climate actions in developed countries was improvement of energy efficiency and mitigation or the reduction of GHG emissions with varied targets. Some cities adopted Kyoto targets to compensate for their national governments not ratifying Kyoto Protocol as in the USA, others because their national governments were not likely to achieve their Kyoto pledges like Italy (OECD 2008). Developed country cities also started their emission inventories to explore the potential for GHG emission reductions and to monitor their performance. In this effort, to date, most cities have had consensus around a few common frameworks for tracking carbon emission. One popular system is *Global Protocol for Community-Scale Greenhouse Gas Emission Inventories (GPC)*, that provides a framework for accounting and reporting city-wide greenhouse gas emissions (GPC undated). World Resources Institute, C40 Cities and ICLEI have partnered to develop this protocol. Other frameworks have also been developed. Despite variation in categorizing different emission sources, these frameworks are fundamentally alike in their function of summing up estimates of emissions (Squires 2021). However, it is still important for cities to connect their inventory approaches or protocols to existing Intergovernmental Panel on Climate Change (IPCC) guidance and United Nations Framework Convention on Climate Change (UNFCCC) national reporting systems to support international monitoring, review, and verification process under the Convention.

The inventory of GHG emissions enabled cities to integrate climate change strategies and action plans for specific sectors as the next step in mitigation policy at local levels. In the city mitigation policy, four areas targeted included—(a) improvement of energy efficiency in buildings particularly public buildings and promotion of renewable energy solutions - use of only renewable energy in Public buildings e.g. Heidelberg, Germany (b) Greening of transport sector particularly improving public fleet operation - using low-carbon fuels e.g. Seattle and some other cities of the US; public transport systems and promoting alternate forms of transport, such as walking and cycling e.g. Quito, Ecuador (c) waste—focusing on waste prevention e.g. San Francisco and Oakland zero waste programs; reuse such as waste to energy and recycling e.g. Seoul, Korea (d) urban and land use planning actions, ranging from standards for new buildings to development of eco-cities like Tianjin in China and planning for new eco-neighborhoods London, Stockholm (Renault Group 2021). For other examples, see Chap. 6 on Energy.

Initially, while cities accepted scientific uncertainty in mitigation option and promoted "no-regret" actions, they could not do so for adaptation as they needed sound scientific basis in developing an adaptation strategy (Lindseth 2005). Hence, many cities somewhat delayed actions on adaptations until they faced extreme weather events. Although improvements in global climate models and projections have made them more robust, predicting local impacts on cities and their neighborhoods continues to be a challenge (Chu et al. 2019). Adaptation plans initiated originally (within these limitation) ranged from preliminary studies on how to prepare an adaptation plan (Cape Town), and impact assessments (London), to the early stages of mere sectoral action plans (health sector in the Toronto/Niagara region and for Stockholm). The first countries to take adaptation into consideration at the local level were the USA (see Chap. 2 for adaptation plan of New York), Canada, and the UK. The emergence of transnational networks provided a boost to adaptation. For example, in 2005, the Climate Alliance of European Cities started the AMICA project to campaign for combining mitigation and adaptation and for developing an integrated methodology to address both. In 2006, the ICLEI added the topic of adaptation to its strategic plan (OECD 2008). Since then, numerous new networks like 100 Resilient Cities and C40 Cities Climate Leadership Group have surfaced and promoted adaptation action in cities (Chu et al. 2019). They have contributed through dissemination of best practices, supporting pilot projects, recognizing adaptation priorities in municipal budgets, and producing downscaled climate projections (Chu et al. 2019).

The main factor that made cities to complement mitigation approach with adaptation (OECD 2008) and/or fast track adaptation approach was the change in risk landscape due to enhanced extreme climate events particularly in developing countries. Adaptation measures against these events and related hazards such as temperature change, extreme precipitation, and sea level rise were increasingly used as pretexts for adaptation by cities to reduce economic losses and protect human lives, health, and well-being. A relatively recent study (Aylett 2014) indicates that the adaptation planning has established itself quickly in urban policy and governance space previously dominated by mitigation only. Conducted across 350 cities of the world, primarily member of ICLEI, to assess progress in their climate governance and challenges faced, it shows that the climate action is currently getting more balanced—with 73% of respondent cities

reporting that their focus now is on both adaptation and mitigation. About a quarter (24%) reported that they focus solely on mitigation, and only 3% pointed out that they focus on adaptation alone. A final synthesis of responses pointed to the fact that while most cities focused on both adaptation and mitigation, they conducted their planning in these areas in an integrated fashion.

Another study (Dulal 2018) focusing on Asia, primarily developing country cities, showed that 180 urban adaptation activities were carried out between 2004 and 2014 in 74 Asian cities. Out of these, a majority (103) were on groundwork, while only 77 were on actions related to adaptation. This depicts that the developing country cities are still at an early stage of climate change adaptation planning and governance. Countries where cities implemented adaptation activities in greater number were the ones where central governments provided both policy directives and support to local governments. In contrast to their counterparts in developed countries, the adaptation activities in developing cities emerged first mostly as a reactive action to combating climate-related disasters. Some good examples of these were Greater Dhaka Integrated Flood Protection Project in Bangladesh and flood control measures in Ibadan undertaken by the state government (Egbinola et al. 2017). Likewise, the city of Ahmedabad in India was one of the first cities to address the issue of extreme heat by preparing a Heat Wave Action Plan. The importance of mitigation, however, is also being realized and incorporated into the scope of urban climate actions in cities of developing countries. Nevertheless, the efficacy of their actions is affected by such constraints as lack of financial resources, technology, and institutional capacity.

10.3.2 Enablers and Barrier to Climate Action

The main driving force behind city climate action in cities of developing countries is central and subnational government mandates and interest. However, a multitude of other factors such as partnerships with transnational bodies or organizations, motivation and political will, civil society initiatives, stakeholders' alliance, and facilitation by existing city networks have also played important role as depicted in the following examples.

- Partnership with transnational organization or funding through external aid: Many actions were steered through the financial or technical assistance of a transnational organization or developed country. For example, Ho Chi Minh City Climate Change Adaptation project was initiated in 2011 with the assistance from Rotterdam. It has led to the development and implementation of the Climate Adaptation Strategy which divides the city into implementation zones, with ongoing pilot projects in each (C40 Cities undated-a). Likewise, Jakarta's mitigation of land subsidence was driven by international assistance and sharing of lessons learned from other delta cities. It was based on acknowledgment of the need for an integral water management approach, in which sanitation, wastewater treatment, surface water quality, alternate access to clean drinking water, and flood protection were all addressed in an interconnected way (C40 Cities undated-a). Similarly, climate action in Surat, India, was initiated by the Asian Cities Climate Change Network.
- Motivation of city leadership/political willingness: Action pioneered by local leaders/elected representatives—for example, Tri Rismaharini, a former mayor of Surabaya in Indonesia was quite active in this regard. She built parks to act as carbon sinks in the city and promoted solar power—particularly in the operation of traffic lights. Similarly, in Baoding, Hebei Province of China, the mayor facilitated a low-carbon city development initiative hoping that his pioneering efforts will set a role model for low-carbon city development in the country (Qi et al. 2008).
- Civil society initiative and litigation: Lobbying and promotion of litigation by citizens and led by civil society organizations have also promoted climate actions. Climate litigation has become very important in recent years with

the spike in court cases. A recent UNEP (2021) report noted that climate cases nearly doubled over 2017–2020 period and have compelled increasingly governments and corporations to implement their commitments on climate actions. It was observed that although climate litigation continues to be concentrated in the developed world, the trend is also on the rise in the Global South—the report listed recent cases from Colombia, India, Pakistan, Peru, the Philippines, and South Africa.
- Partnerships and informal collaboration between city administrations/private sector/civil society stakeholders/NGOs: These types of partnerships or collaborations were forged to tackle the issue of differential vulnerabilities within the city—often, successful implementation of one action led to a sustained partnership with the city administration. For example, under Climate Adaptation Road Map for 2030, Jakarta plans to relocate close to 400,000 illegal squatters from riverbanks and nearby reservoirs through "a humanized and participative process." Under the plan, government has already shifted thousands of squatters to newly constructed subsidized high-rise, low-cost housing. While providing these communities with basic amenities (electricity, water), considerations have included their job security through an economic empowerment Scheme (C40 Cities undated-b). Another example is from Ahmedabad municipality in India, which collaborated with the NGO MHT in dealing with extreme heat during summer. Some philanthropic institutions such as Clinton Foundation and Bloomberg have also helped. For example, the Clinton Foundation has aided through partnering with cities to develop climate change adaptation and mitigation strategies in both developed and developing countries by technology provision. For example, it has helped some large cities in the world from Seoul to Johannesburg and Houston in retrofitting their municipal, commercial, and residential buildings (OECD 2008). Moreover, the foundation has assisted Lagos, Delhi, Mexico City, and London in handling waste, which release methane—A 22 times more

powerful GHG than carbon dioxide. It has also assisted in popularizing new energy-saving LED technology to replace conventional streetlights and traffic lights that use energy 24 h a day (OECD 2008). In São Paulo, C40 Cities Clinton Climate Initiative is helping promote public-private partnership for developing sustainable urban mobility that will reduce traffic and related GHG emissions. Another important philanthropy Bloomberg has also supported city mayors and other partners at local level and collaborated with them to combat climate change.

- Support by city networks: There are various networks like C40, 100 Resilient Cities (100RC), and International Urban Cooperation (IUC) program funded by the European Union that facilitate knowledge and experience sharing across the world. A Regional Learning Network has also been set up in Latin America including six large cities—Bogotá, Buenos Aires, Lima, Mexico City, São Paulo, and Santiago (UN-ECLAC 2014).

The results of a study by Aylett (2014) reveal some more enablers of climate action—for example, two-fifth of the respondent 350 cities covered by the study reported that their climate action was motivated by a comprehension of local climate-related risks. In addition, the study confirmed that city membership of transnational municipal network was a major catalyst in climate action. However, the study observed that networks were biased toward cities in developed or wealthy countries and there was a need for these networks to diversify themselves by involving cities from developing or low-income countries. This is particularly important because most current megacities are in developing countries and furthermore new megacities in the future will also emerge there (see Chap. 2). Three other motives which incentivized cities to take climate action according to the study (Aylett 2014) were cities' desire to demonstrate leadership globally, nationally, or regionally; to promote sustainable urban development; and to improve the quality of life of their citizens more generally.

There are barriers too that have affected world cities on not taking climate actions. In addition, issues like the uncertainty of the climate impacts, the short-term perception surrounding investment returns, and the lack of political will in leadership as well as coordination between institutions have prevailed as pervasive problems. Based on a literature review, Salon et al. (2014) came up with several other factors that acted as barrier to city climate action including lack of financial and technical resources, existence of a conceptual disconnect between the global scale of the climate challenge and the local scale action, a severe lack of detailed data on local energy use, and institutional barriers such as lack of jurisdictional authority to mandate energy efficiency. The study also pointed out that local actions have been most successful when climate action or policies have concurrently targeted both climate and local development goals and when climate policy champions have existed within the local government.

10.3.3 Vulnerability and Capacity Variation

Cities response to climate change has varied across the world cities depending on their vulnerability to climate hazards, their capacity and commitment to handle climate-related issues, support from the governments at national and provincial/regional levels, and as mentioned before on cities' participation in national and international city networks that are spearheading climate change. Cities vulnerability to climate change has also varied depending on type of climate hazards and their exposure to it. For example, cities located in coastal areas are more vulnerable to flooding while others to rising sea levels; still others located in deserts are more prone to heat waves and so on. Their capacities and commitment to combat climate hazards also vary depending on their development status (whether located in developed, developing, or least developed country) and the financial and technical resources they possess. The cities in some developing countries have so meagre resources that they can only have commitments to solve day-to-day problems on a firefighting basis, and as such they rely more on external funding for undertaking climate action. In contrast, others with much larger resources in developed countries besides combating climate hazards have firm commitments to shift to climate-friendly urban development patterns—with the realization that it offers enormous potential and opportunity for cost-effective action. Although urban and local climate change policy initiatives date back to the 1990s (Alber et al. 1996; Collier and Löfstedt 1997), these initiatives in the past were largely decoupled from national policy frameworks. Among other things, this resulted in restricting the resources available to cities to support experimentation with innovative urban practices on combating climate change. It also prevented or limited their ability to identify and diffuse good urban and territorial development policy practices to address climate change (Corfee-Morlot et al. 2009). The trend is changing, but more needs to be done in this area for filling in the "policy gaps" between levels of government through strong vertical and horizontal coordination and cooperation.

10.3.4 Modes of Urban Climate Governance

Four modes of urban governance (Kern and Alber 2008) have been identified for implementing climate change policies in cities as follows:

- Self-governance: the municipality/city as consumer—Whereby a municipality or a city controls or limits its own carbon footprint through management of municipal or city operations by enhancing energy efficiency of public buildings, by greening its own fleet of buses and vehicles, as well as by promoting green procurements.
- Governance by provision: the municipality/city as provider—Whereby a municipality or a city provides or develops green infrastructure and supplies or delivers sustainable utilities and

services. For example, appropriate waste management or providing mass transit for transportation or developing cycling and walking infrastructure, etc. Likewise developing green housing and managing natural resources in an environmentally sustainable manner.

- Governance by regulation: the municipality/city as regulator—Whereby a municipality or a city enacts regulations to curb GHG emissions in those areas which fall under its jurisdictions by using instruments such as land use plans, building codes, and regulating urban transport and transit systems, waste generation and energy use, etc.
- Governance by facilitation: the municipality/city as facilitator—Whereby a municipality or a city coordinates and cooperates with private and community actors by promoting public-private partnerships and voluntary agreements, etc.

The first "self-governance or the municipality as consumer" refers to municipalities or cities' government managing their own activities in a climate-friendly manner. For mitigation, it means promoting the energy efficiency of municipal buildings, renewable energy solutions, and the greening of public transport vehicles and promoting alternate forms of transport, such as walking and cycling. For adaptation, it means that the planning and the management of public buildings need to be adapted to climate change, for example, sites facing risks from flooding are to be avoided and sufficient cooling in cases of heat waves are guaranteed. These are very common local action in mitigation and adaptation, leading to financial as well as other benefits. A good example of this is the city of Los Angeles (LA), which has become the top solar city in the USA—with enough energy from the sun to power 82,500 homes and save more than 187,000 metric tons of greenhouse gas emissions (Los Angeles 2019). Encouraged by the progress, the city has committed to net zero carbon emissions by 2050, the most ambitious effort to achieve the goals of the Paris Climate Agreement. To help electrify its own fleet and those of cities across the country, the mayor of LA has launched the Climate

Mayors Electric Vehicle (EV) Purchasing Collaborative—a new online portal that lowers the cost of electric vehicles and charging infrastructure by enabling cities to bid jointly for these ventures in large numbers. The new platform has enabled 20 member cities of Climate Mayors' network and 2 counties to purchase 391 electric vehicles—a figure that is expected to grow as the program continues to engage other cities across the USA. Further, the mayor of LA and 11 other international mayors have signed the Fossil Fuel Free Streets Declaration, pledging to purchase only zero-emission buses beginning in 2025 to ensure that a major area of their city is zero emission by 2030 (Los Angeles 2019).

In developing cities, Bogotá in Colombia is trying to slash GHG emissions by half by 2030, officials are expanding bike lanes and pedestrian paths, using more electric buses, and extending the reach of electric cable cars—some partly driven by renewable solar power—that serve poor areas in the city's south (Moloney 2021). Similarly, Johannesburg in South Africa has taken measures including retrofitting of council buildings, energy savings in water pump installations, and methane gas recovery (Rosenzweig et al. 2011). Mexico City has also established environmental certification systems for buildings and provided funding for new housing that will integrate sustainability criteria. Energy efficiency programs have also been introduced to reduce emissions, including improvement of efficiency in lighting of buildings and street lighting, as well as promoting solar energy in government buildings. The city has taken additional action to reduce emissions from septic systems by constructing sewerage and water treatment services in areas of low methane gas. In the transportation sector, actions include an obligatory school transportation system, which will reduce carbon dioxide emissions by 470,958 tons per year by ensuring that students take public instead of private transportation to school. The study of 350 cities mentioned before (Aylett 2014) found that most common areas where emission reductions have been made are municipal government buildings (89%) and vehicle fleets (72%) and waste reduction (55%).

The second mode or governance by provision is important in mitigation because it enables local governments to create infrastructure that generates fewer GHG emissions from transportation (including mass transit networks), waste, energy provision, etc. Regarding adaptation, it demands establishment of warning systems and emergency planning by taking climate change and extreme weather events into consideration within the service provider areas to avoid disruptions (Kern and Alber 2008). Cities in many countries, in several sectors such as energy, transport, waste, etc., either provide services directly or are majority shareholder in the local utility companies and have an advantageous position to steer local action on climate change both in mitigation and adaptation. Several examples on this include two from China—Meishan City (innovative solution for cleaner energy infrastructure) and Hohhot City (wind power)—and one from Korea, Seoul (waste), which have already been given in Chap. 6 on Energy in this book.

The third mode governance by regulation is an area, where city governments have a direct control through urban planning mechanisms in both mitigation and adaptation at city/local level. This mode allows city/local governments to meet climate policy goals by directing, prohibiting, or adding costs (using economic instruments) to activities such as land use development, vehicle use, building energy efficiency, generation and use of energy, and waste production and disposal. Although implementation of climate policy goals by city authorities in most countries is still voluntary, the cities have been using their regulatory authority to promote climate action since long. A case in point is Barcelona's Solar Thermal Ordinance of 2000 requiring the installation of solar thermal collectors for hot water supply. Similar ordinances were subsequently adopted in 40 municipalities in Catalonia and in almost 30 municipalities in other parts of Spain (Ekelund and Sigurdson 2007). This is an example that demonstrates clearly that (a) governing by regulation can be very effective at local level and (b) that such regulations can be used as a model for similar initiatives in other cities. Municipalities in the USA initiated climate change policy somewhat later than their European counterparts, but pioneering cities did catch up fast. The city of Boulder in Colorado, for example, attracted considerable media attention in 2006 when voters approved a citywide carbon tax to fund Boulder's GHG emissions reduction strategy, laid down in its Climate Action Plan (Betsill and Bulkeley 2006). Among other actions under regulatory measures, in Melbourne, a mandatory energy performance requirement has been introduced under the municipality's planning powers for office developments greater than 2500 square meters (Hoornweg et al. 2011).

Among developing cities, in New Delhi, solar water heating systems have been made mandatory in government offices, hospitals, educational institutions, and the hospitality sector, while the use of incandescent bulbs has been banned in all new and existing government establishments (Bulkeley et al. 2009). In São Paulo since 2007, mandatory regulations demand buildings with more than three bathrooms, whether they are homes, apartments, trade, services, or industrial buildings, to have solar water heating systems (Bulkeley et al. 2009). Mexico City has devised a new "Clean Building Label" for all new construction required as part of its Plan Verde (Bulkeley et al. 2009). These examples show that, even in the absence of direct municipal competencies for establishing building standards, local governments may act. They may be motivated to go beyond standard practice and use a range of other tools at their disposal to augment the energy performance and reduce GHG emissions from the built environment.

Governing through facilitation, motivation, or incentivization is another important governance mode where local governments have a lot of clout particularly because of their proximity and familiarity with the local stakeholders like private sector, businesses, and civil society. It allows local governments to persuade and influence private actors and civil society as well as individual residents to do their share in mitigation and adaptation. Numerous mechanisms have been used for this purpose including information campaigns, public-private partnerships, voluntary agreements, as well as use of economic instruments.

While awareness raising and promotion campaigns are increasingly being undertaken in most cities for climate action, the establishment of new partnerships for the transition of existing infrastructure and services is picking up slowly. Such arrangements particularly public-private partnerships appear to be better suited to the needs of large and competitive cities of developed countries, such as London, Munich, and New York, which are highly committed to climate change policy and pursue ambitious GHG reduction goals and where private sector is also motivated by economic incentives and prospects (Kern and Alber 2008). For example, London Climate Change Agency (LCCA), owned and controlled by the London Development Agency, was established with the direct support of private companies. Its main mission is to deliver projects that reduce London's GHG emissions and implement the Climate Change Action Plan and the Energy Strategy. In Munich, the city has not only created an energy commission but also established various other fora that directly involve private stakeholders (Kern and Alber 2008). Similarly, the New York City is involving the stakeholders like private sector, local institutions, and community through its One City: Built to Last Program, which aims to reduce emission of GHGs by 3.4 million metric tons a year up to 2025—the equivalent to taking 715,000 vehicles off the road (Cadham 2016). Some developing country cities like São Paulo have also encouraged public-private partnership for climate action (United Nations – ECLAC 2014). Grants and subsidies have also been offered, and economic instruments such as tax credits and other economic and financial incentives have been used to promote the adoption of climate-friendly technologies and mechanisms by stakeholders.

Overall review of urban climate action reveals that cities/municipalities in most developed countries with human and financial resources and access to technology have promoted various modes of climate governance to address climate change. In contrast, many cities in developing countries have not. The cities with good practices on climate governance are, therefore, like a few islands surrounded by a sea of "business as usual." However, the growing trend of development of policy frameworks at the national and international level after the Paris Agreement appears to have made some change. This is likely to enhance further after the 26th COP of United Nations Framework Convention on Climate Change held in Glasgow which has provided some additional incentives to cities for moving beyond business-as-usual practices in undertaking proactive climate action.

10.3.5 Institutional Mechanism for Urban Climate Governance

In terms of institutional development, cities or local governments started formalizing climate action within their jurisdiction, by setting up urban climate units (sometimes in mayor's office as in Bogotá). Through these offices, they initiated programs as well as drafted regulations, policies, and codes (Chu et al. 2019; Roberts 2008). The unit (with a few or one dedicated personnel) has been overseeing relevant climate change policy issues or serving as a focal point for a climate policy steering group. Sometimes it has been functioning as a climate protection coordination office, with appropriate competencies for mainstreaming climate change policy in the work of relevant sectoral departments. This is sometimes combined with various task forces to coordinate the activities around specific issues and across all relevant policy areas within the city government.

More often, the responsibility for climate change policy development has been assigned to the environmental policy unit or agency within city administration, as in most German cities (Kern et al. 2005). In developing countries, such tasks are handled by one personnel or officer responsible for environmental affairs. Often it has given rise to coordination and integration problems in cases where the environmental units/agencies or the responsible personnel do not have the capability to implement the program or project or have limited capacity—particularly so because climate change policy is an issue that affects a variety of departments within local gov-

ernment like those dealing with finances, procurement, urban planning, economic development, education, etc. The dilemma is comparable to the one faced by the environment ministries in the national governments.

According to the urban climate survey of 350 cities (Aylett 2014) mentioned above, in 40% of reporting cities, the unit/agency or employees' team mainly responsible for the climate change planning (including both adaptation and mitigation) was quite small consisting of 1–5 employees. Less than a quarter city (23%) had only a single staff member (a sustainability or environmental coordinator) for the entire local government. A smaller proportion (15%) of respondent cities reported that responsibility for the climate-planning portfolio had yet to be clearly assigned. Eight percent reported having a large team (of six or more full-time employees), and only 4% had two distinct teams for adaptation and mitigation. Among regions, Africa was an exception where a significant percentage of cities reported that the climate planning there was led by a consultant (Aylett 2014).

Where they existed, most dedicated climate change mitigation teams were in either the bureau/department responsible for environmental issues (42%), sustainability (17%), or planning (12%). In the rare cases where they existed, dedicated adaptation teams were in the bureau/department responsible for planning (33%), environmental issues (25%), or the mayor's office (17%). These figures changed in cities where there was only one staff member focusing on the climate change planning. There, 28% were based in the bureau/department responsible for environmental issues, 24% in planning, and 21% in the mayor's office (Aylett 2014).

10.3.6 Multilevel Urban Climate Governance

The debate on the multilevel role of institutions and policies in climate governance from international to local level is not new. According to conventional collective action theory, climate mitigation requires a global treaty (Lindholm

1959). Because even though individual cities and countries may develop climate policies and make the necessary transformations toward a low-carbon economy, the impact on climate change will not be of much value unless a global approach is taken that meaningfully includes the biggest greenhouse gas emitters. This means that no governments will act without global regulation or externally enforced requirements (Lindholm 1959). An alternative view of scholars argues that cities may provide the best or most suitable entry point for tackling climate problems (Betsill and Bulkeley 2007) because the urban scale is better suited to experiment and develop new solutions (Lindholm 1959; Bulkeley and Betsill 2005; Biesbroek and Lesnikowski 2018). These scholars claim that cities have become de facto leaders in climate policy development and implementation and are less susceptible to national budgets, large and complex administrations, political coalitions, and competing interests. Thus, many cities in the USA have arguably committed to GHG reduction targets despite the federal government negative attitude during Trump presidency. Likewise, cities such as Copenhagen or Leipzig framed climate mitigation and adaptation as strategic concerns backed by resource allocation even prior to national and European political commitments (Jensen et al. 2020). Other scholars argue that in fact climate governance offers potential benefits across multiple scales, providing incentives for action even in the absence of global coordination.

Besides vertical coordination, climate governance also has a horizontal coordination aspect, which besides involving several local governments like municipalities within a metropolitan area also involves a host of other actors from individuals and local communities to civil society and private sectors including corporations. Therefore, the debate on what level can best address climate change needs to focus more on how to ensure an appropriate distribution of decision-making competencies among various tiers of government vertically and horizontally and how to ensure coordination across actors. In this context, climate governance at urban scale is equally important for both city and national gov-

ernments. For example, national governments need city/local governments to play an important role in implementation if country's emission reduction and adaptation targets and goals are to be met in the context of international governance/obligations. Similarly, in both mitigation and adaptation policy areas, city/local governments often find themselves at constraint if emission sources or some aspects of land use planning fall outside their boundary to a neighboring area/city jurisdiction. In such cases, national approaches and regulations are essential to trigger action, giving a mandate or at least a framework to cities/local authorities to coordinate.

10.3.6.1 Models of Vertical Coordination on Climate Change

Institutional models that influence or guide policy action on climate change across levels of government vary, and there is no one-size-fits-all framework at the national level to guide or incentivize local action or promote national-local coordination. One set of national-local governance actions on climate change has been driven from the top by national or provincial/regional authorities. This top-down approach mandates or specifies a set of policy-related actions at local/urban scale. The second set of policies moves from the bottom to the top, where local policy innovations provide model for provincial/regional or national actions. The third set of policy approach is a hybrid of the two.

Although in the top-down model national policies targeting local authorities have taken different forms, ranging from regulation to enabling, enabling policies or frameworks are more common where mandated approaches play a role. However, their general framework leaves wide latitude to city/urban authorities to tailor policies on climate change in their local contexts. Norway provides the earliest and a good example of a national enabling policy framework on climate change at urban scale. The White Paper on the Kyoto Protocol issued in 1998 by the Government of Norway formed the basis to send a circular to cities/municipalities requesting them to develop local climate plans aiming at reducing carbon emissions and increasing sequestration through forestry projects. These plans were to be developed in partnership with the national and regional governments—Norwegian krone seven million (US$ one million) were allocated to stimulate action. In addition to financial help, a web-based information source and emissions calculation tool were also put in place by the national government (Aall et al. 2007). Top-down approaches were also promoted in China, France, Portugal, and the UK (Corfee-Morlot et al. 2009).

The "bottom-up" approach encouraged cities/local governments or permitted them to go beyond national requirements or incentives—allowing them to act independently in addressing climate change as an active part of national policy or even in the absence national policy. In this model, learning and experience acquired through successful city or local programs were not only replicated but helped steer government policy-making at regional or national levels. The City of Portland in the Oregon State of the USA is an example of the case (Corfee-Morlot et al. 2009). As early as 1994, the city aggressively developed a green building sector, and the innovation had a direct impact on state-wide policy. The program was initiated by a voluntary citizen group, after exploring the potential for a local green building technical assistance scheme (City of Portland Bureau of Planning and Sustainability 2009), but was implemented by city authorities to address climate change. It focused on policy development, demonstration projects, technical assistance, education, and financial incentives (City of Portland Bureau of Planning and Sustainability 2009). It was funded through local residential and commercial solid waste fees and grants. In 2007, Portland had the highest number of LEED-certified buildings (Leadership in Energy and Environmental Design—the most widely used green building rating system in the world) in the USA. Moreover, the city was already attracting firms and qualified workers from around the country (Allen and Potiowsky 2008). The Portland experience provided insights into regulating building energy performance and enabled greening of buildings across the State of Oregon. In Spain, the state Government of Catalonia

played an active role in supporting cities in its region, notably in the Barcelona metropolitan area, to address climate change in the context of sustainable urban development (Laigle 2009). Other leading cities that promoted green buildings included Bogotá, Boston, Cape Town, London, Melbourne, and Oslo (Bell 2018). Several examples of local/urban leadership in addressing climate change through enhancing urban energy efficiency and management have been given in Chap. 6 on Energy.

The hybrid approach—where national and/or regional governments work closely with local authorities to encourage experimentation and innovation at the local level to respond to climate change—stimulates two-way learning. The successful lessons from these once identified can also be replicated/expanded for practice elsewhere. An example of this approach is the Swedish KLIMP funding program, which is a voluntary framework to incentivize action through funding or supporting activities. Under this program, local communities compete for central funding to undertake mitigative or adaptive action. The key disbursement criteria are based on performance or implementation evaluation (Corfee-Morlot et al. 2009). Another example from Japan provides an even more decentralized approach, where enabling legislation explicitly recognizes and summons local governments to act. In this context, innovative ideas have been initiated at the local level (such as the use of energy labeling) and tested locally before replicating or diffusing to other areas. Kyoto Protocol implementation in Japan, for example, required development of a labeling system informing consumers about the environmental impact of their air-conditioning appliances and televisions.

In another type of hybrid model, the private sector is a central actor, whereby it increasingly plays active role in to steering local action through international or national carbon markets. These markets were created in part under the Kyoto Protocol as a tool for national governments to achieve aggressive mitigation targets. The model has been used to promote public-private partnership for multilevel action on climate change. São Paulo in Brazil presents a good example of this hybrid model, where local and national governments worked together with the private sector to facilitate action at local scale under Joint Implementation and Clean Development Mechanism (CDM). Brazil has no national target for emission reductions as it is a Non-Annex I Party to the UNFCCC, but the country is active in promotion and development of CDM projects and helped São Paulo accordingly. The city completed an inventory of its emissions, with the help of ICLEI in the early part of this century. The inventory showed that about half of GHG emissions (48.6%) came from land transportation and about another one-fourth (23.5%) from the landfills. Based on this, the city developed several initiatives focusing on the use of CDM to mitigate the emissions. A landfill project was implemented jointly with a local private company Biogás Energia Ambiental, in cooperation with German Bank KGW and the private firm Van der Wiel and Arcadis (Netherlands) (UNFCCC 2005). The project generated energy from the landfill methane emissions that reduced GHG emissions in its early phase by as much as 11% in the city (Cunha and Rei 2006). Revenues from the sale of carbon credits, amounting to 34 million real (US$ 16 million), were invested in social projects in the landfill area and on climate change mitigation (Puppim de Oliveira 2009).

Japan presents another example where although environmental and energy policies are national government's responsibility, it can delegate authority to governors and mayors to implement national laws. It encourages regional and local governments to develop their own policies and measures whenever the central government does not act in the climate change arena (Sugiyama and Takeuchi 2008). In fact, the Kyoto Protocol Target Achievement Plan, introduced in 2005 and revised in 2008, aimed to stimulate municipal and regional initiatives in the energy efficiency, transport, and regional planning sectors (Government of Japan 2008). It encouraged the use of Joint Implementation and Clean Development Mechanism (CDM) to achieve low-carbon objectives at urban scale. Therefore,

throughout Japan, local and regional governments implement and monitor their own climate action plans (Sugiyama and Takeuchi 2008). Tokyo, for example, introduced a mandatory cap and trade system in Japan as part of its climate change strategy (City of Tokyo 2008).

10.3.6.2   Horizontal Coordination on Climate Change

As stated earlier, the dynamics and consequences of climate change cut across and spill over jurisdictional boundaries of cities and therefore require horizontal coordination. This type of coordination refers to interactions between various city governments—such as one local authority and another, which may also result in the development of national network of cities. It also covers interactions between various municipalities within a metropolitan government—which is particularly important in addressing jurisdictional boundaries to build metropolitan governance systems to better address climate change. For example, metropolitan Mexico City extends over the territories of municipalities of two states as well as the federal district to include as many as 58 municipalities: City of Buenos Aires has 32 municipalities; City of Abidjan encompasses 196 local governments; Metro Manila has 10 cities and 7 municipalities; Tokyo Metropolitan Region has 365 municipalities; and Minneapolis-Saint Paul in the USA has 188 cities and townships (McCarney et al. 2011). Another aspect of horizontal coordination involves partnerships between local government and private sector, scientific community, and civil society actors for climate action.

Despite constraints of jurisdictional boundaries and lack of or weak institutional capacity and resources, however, city and metropolitan governments in cooperation and coordination with other actors have been able to establish several good practices in response to climate change. Mexico City is one example which has developed important efforts to curb its greenhouse gas emissions by developing collaboration between policy networks, political leaders, and research groups. Policymaking in the city has been constrained by two sets of institutional factors: the problem of fragmentation in local governance and a lack of institutional capacity. Despite this, in 2008, Mexico City was the first city in Latin America to implement a Climate Action Program (C40 Cities 2015), which enabled the reduction of 1.4 million metric tons of CO_2 equivalent to 4% of the city's emissions. Components of the program involved several actors and numerous actions. Among actions included were a zero-emission transport corridor, a public bicycle system, replacement of minibuses and taxis with lower-emitting vehicles, a sustainable housing program, regulations to encourage the use of solar collectors in commercial and services sectors, green roofs, restoration of ecosystems outside the city, and development and use of an environmental management system. The city implemented the Second Climate Action Program (PACCM) 2014–2020 continuing actions to increase resilience and combat climate change.

Another example of good practice is Quito city in Ecuador, where many of the issues raised by residents on health, environmental quality, security, and safety are addressed in the Quito's Climate Change Strategy (QCCS). The strategy was prepared by the city in collaboration with key stakeholders, including the academia and civil society. The municipality is currently implementing a series of adaptation and mitigation measures under the strategy in key sectors (Zambrano-Barragán et al. 2011). This shows that through such horizontal coordination with stakeholders, climate action can become a process by which planners and public officials can find creative ways to respond to general needs and concerns of the city residents.

Horizontal coordination through transnational city networks has also been very useful. Besides enabling the exchange of information and experience and expertise, these networks have been a source of funding while also promoting political stature of city officials and administrations undertaking climate action. Moreover, with the passage of time, these networks have been able to mobilize private actors (Bulkeley et al. 2009).

For example, the C40 Cities Climate Leadership Group while bringing global cities together to address climate change (Hodson and Marvin 2009) has entered in a partnership with philanthropies like the Clinton Climate Initiative (CCI). It has also expanded its collaboration with corporations like Microsoft to produce software for the accounting of GHG emissions at the city scale. Moreover, under CCI's Energy Efficiency Building Retrofit Program, it brings together cities, building owners, banks, and energy service companies to reduce GHG emissions from large corporate buildings. These networks have also expanded in time with the involvement of cities from developing countries or Global South (Bulkeley 2010). In this connection, networks like C40, CCI, and the Resilience Network are now explicitly targeting cities in middle- and low-income nations.

Most importantly, these transnational urban networks are also taking a more overtly political stance toward the climate change, seeking to position cities as critical sites for addressing the issue. In some cases, they have taken stance that opposes even the national governments (as in the USA and Australia) and in so doing have advanced claims for the strategic importance of urban governance (Hodson and Marvin 2009). In this connection, the increasing political clout of the transnational networks has raised some concerns and arguments that these subnational or local actors are disproportionally increasing their influence in relation to, if not compared with, nation-states. These arguments need negation as in reality, the networks are neither seeking to eclipse nor reduce the role the nation-states are playing in climate change negotiation through UNFCCC regime nor competing with them. It would, therefore, be more productive if the debate between the "transnationalists" and "multilateralists" in climate change governance research (Betsill et al. 2015) focuses less on "competition" and more on coordination between the two processes as well as on linkages between them in terms of climate governance for mutual benefit.

10.3.7 Achievements of Climate Governance

Considerable progress has been made in urban climate governance during the last 25 years in terms of establishment of institutions, initiation of policies, programs, and projects. However, how much difference it has made in real terms on the ground is still not very clear. In the absence of in-depth assessment, it is not possible to present in definite terms the overall achievements made by the initiatives undertaken across world cities so far—particularly in terms of mitigation (reducing GHG emissions) and adaptation (enhancing urban resilience to climate change-related risks). A survey of 350 cities (Aylett 2014) made some exploration toward the impact of mitigation efforts toward reduction in GHG emissions, but no attempt was made in this survey to assess the impact of adaptation measures on cities resilience.

In the survey of 350 cities (Aylett 2014) across the world, 70% of respondent cities reported that their mitigation efforts have produced positive results toward measurable emission reductions. The three most common areas where cities' governments reduced emissions included local government buildings (89% respondents), local government vehicle fleets (72% respondents), and waste reduction (55% respondents). Emission reductions from the private sector efforts were less common—24% cities reported reductions from commercial green building programs, 17% reported reductions from local industry, and 4% reported reductions in the freight and shipping sector. About a quarter of the surveyed cities (23%) reported emission reductions by addressing the procurement (upstream emissions associated with municipal purchases).

In terms of world regions, cities of the European Union and Australia reported that measurable emissions reductions were between 72% (EU) and 82% (Australia). In African and Asian cities, the two outliers, the percentages of reduction fell to 27% and 43%, respectively. In sectoral terms, Latin American cities did not report any reductions of emissions in local government vehicle fleet or residential energy use.

They were however leaders (along with Canadian cities) in reductions through increased use of public transportation. Although lagging in public transportation, Asian cities reported that they had cut their GHG emissions by waste reduction measures. African cities displayed the most uneven response. They were leading in the emission reduction by landfill gas capture (along with Canadian cities) and by reducing residential energy use but reported no reductions in half of other sectors included in the survey.

At country level in the USA, some 600 local/city governments have developed action plans for greenhouse gas inventory and reduction targets, but are these efforts working? A Brookings Institution study (Victor and Muro 2020) of 100 cities on emission reductions in the USA noted that as of 2017, only 45 of the 100 largest cities had any serious climate pledge at all, and many of those pledges were more aspirational than realistic. Further, about two-thirds of cities with climate pledges were lagging in their targeted emissions cuts, while 13 others did not appear to have available emissions tracking in place. Some of the largest achievements have emanated from the electric power sector. For cities such as Los Angeles, the closure of coal plants (that sent their power to the city long distances via wire) has offered big reductions. Half of the top six cities having emissions reductions were in California. Other top performers included cities in the Democratic-leaning arc that swings up the West Coast of the USA, across much of the North, and down half of the Eastern Seaboard. But even more interesting are the areas that are shifting politically and where topics such as climate change are centrist: Greensboro and Durham in North Carolina; Cincinnati in Ohio; and Richmond in Virginia and many others. Geographically, local climate pledges are relatively evenly distributed—a good sign that action on climate change is spreading throughout the urban heartland.

In terms of adaptation, even the anecdotal evidence of investments provides its apparent cost benefits. For example, putting early warning systems alone can make a big difference. According to the Global Commission on Adaptation (GCA 2019), "Just 24 hours warning of a coming storm or heat wave can cut the ensuing damage by 30 percent and spending $800 million on such systems in developing countries would avoid losses of $3–16 billion per year" (GCA 2019). Further, every 1 dollar invested in early warning systems could save loss of 9.7 dollars caused by disaster, and similarly every dollar invested in making new resilient infrastructure could save 5–8 dollars (GCA 2019). However, no systematic study is available to show aggregated resilience to climate change-related adaptation.

10.3.8 Governance Challenges and Constraints

A global urban climate-related survey (Aylett 2014) of cities and the *Urban Climate Change Research Network's Second Assessment Report on Climate Change in Cities* (Romero-Lankao et al. 2018) have brought forward some major governance challenges cities are facing in tackling climate change. Among these, the first challenge is to have an empowered local governance in political, fiscal, and human resources terms. The second associated problem relates to deep-seated vulnerabilities related to poverty and inequality particularly in cities of developing countries. The third challenge is data and measurement shortcomings at spatial scale, while the fourth is fragmented urban jurisdictional boundaries that hamper addressing climate change issues in metropolitan areas. The fifth is building effective partnerships for more inclusive governance. Then there is a myriad of other challenges such as lack of legal frameworks and mandates and realization of co-benefits between climate change work and other key local development priorities and aspirations. These mainly arise from lack of mainstreaming more effectively climate change issue in planning and urban management practices.

In terms of political clout, local governments in some countries are very weak to take effective actions. Regarding financing, a survey of 350 cities across the globe (Aylett 2014), bulk of respon-

dent cities (81%) indicated that local/city funding was the main source for staff and operational costs. About one-fourth (23%) respondents identified support from national or state/regional governments as the second important source of funding. Revenue from fees or service charges derived from areas under the climate unit's jurisdiction, or under the jurisdiction of its parent department (i.e., fees charged for waste collection and recycling, etc.), was the next important source of funding. The survey results also showed that while a diversity of funding sources is available for supporting specific projects and programs, funding for staff is much more limited. In terms of human resources, a comparative study of Johannesburg and Cape Town on climate change policy and action (Holgate 2007) showed that Cape Town municipality with comparatively stronger human resource base made better progress compared to Johannesburg. The latter city had only one official responsible for addressing not only climate change but also all-encompassing environmental challenges.

Among other problems were (a) poverty and inequality; (b) lack of scientific information and standardized data at disaggregated scale in undertaking adaptation actions; (c) problems between the relevant decision-making agencies and levels of government and between government and communities because of information source, its transmission, and use (Romero-Lankao et al. 2018); (d) uncertainty in information due to contradiction between science and policy that sometimes lead to political exploitation and become a reason for not taking any action; and (e) fragmentation in governance due to administrative boundaries or jurisdictional coordination (both vertical and horizontal) or lack of effective coordination between various actors.

The Climate Secretariat in the municipality of Aarhus in Denmark provides an example of how to overcome the last constraint. With a goal of carbon neutrality by 2030, it promoted networking both within the municipal government and between the government and private actors for coordinated action. The Secretariat attracted private companies by focusing on the municipal marketing and use of development potential in promoting climate change action (Romero-Lankao et al. 2018). It managed to bring 32 businesses together in a formal network involving housing, clean technology, Aarhus University, and the engineering sector where linked activities brought benefits to all parties (Cashmore and Wejs 2014). This innovative networking approach also helped catalyze urban government and bypassed some intraorganizational constraints. Ultimately, despite fragmentation and conflicting ambitions and values, the Climate Secretariat leveraged around 50 million DKK (approximately US$9.2 million) in climate change investments in 2012 (Cashmore and Wejs 2014). This best practice has become an example of building effective partnership for inclusive climate governance. There are other examples available on best practices for promoting mitigation and adaptation compiled by C40 Cities in their best practice guide on their website (https://www.c40.org/researches/unlocking-climate-action-in-megacities).

10.4 Conclusion

The significant and pivotal role of cities as well-governed, well-managed, and well-financed global entities was emerging fast in recent years, but the two mega risks coronavirus pandemic and the climate crisis have added fuel to the fire. Discussion in this chapter, while bringing forth myriad of new governance challenges cities are facing due to twin mega risks, has also shown starkly the need for an urban agenda to meet these challenges effectively. No doubt, individual cities are facing their own challenges in responding to mega risks and developing their own solutions that are most suited to their own needs. However, their globalized impact, more than anything else, call for an effective and long-term solutions that are based on an empowered and integrated urban governance approach. The approach that acknowledges the roles and responsibilities of all actors both state (from local to national) and non-state as well as transnational actors.

The debate on the role of international governance in local and vice versa in terms of effectiveness of policy development and implementation to address mega risks, be they pandemics or climate change, is not new. According to conventional collective action theory, the solution lies in an international treaty—because even though individual cities and countries may develop policies for mitigation, they may not be effective without global action. For example, in case of climate change even though some cities and countries may make the necessary transformations toward a low-carbon economy, the overall impact on global climate change will be very little in the absence of action from the biggest greenhouse gas emitters. This means that regulation or externally enforced requirements (from the top) are essential in ameliorating the climate change mitigation. An alternative view proposes that cities may provide the best or most suitable entry point for tackling climate problems. The argument advanced is that the urban scale is better suited to experiment and develop new solutions because it is less susceptible to national budgets, large and complex administrations, political coalitions, and competing interests. Both these points of views appear to be valid. While global regulation is important for making meaningful change, cities and regions also need to undertake policies that make sense for their own territory, issues, and populations as well as resources to operationalize governance. In the same token, the national governments are often best placed to oversee the design and implementation of coherent and equitable action plans across the country. Hence, none of these approaches are mutually exclusive, and a coordinated approach is therefore needed that could combine all these positive aspects at various levels and lead to achieve the best results in the fight against mega risks.

The case studies in this chapter on the response to mega risks have provided many lessons for urban governance. An important one has shown clearly that there is no single governance approach (top down or bottom up or hybrid) that can be prescribed as a solution. Nevertheless, a coordinated multilevel governance and an integrated approach involving relevant actors both state and non-state including health professionals, scientists, NGOs, communities, and private sector are key to addressing the mega risks and implementing a needed and meaningful mitigation or recovery strategy in the context of any mega risk be it a pandemic or climate change. For example, the partnerships established between municipalities within metropolitan region; between different cities; and between cities and regions helped combat the pandemic more effectively by jointly implementing health regulations and mobility or public transport measures. Thus, in the USA, coordination between the governors of New York, New Jersey, Connecticut, and Pennsylvania led to the formulation of joint health policies and the creation of a common set of guidelines for social distancing and recreation (in contiguous urban areas) that were also subsequently adopted by other states. Similarly, metropolitan governments in coordination with municipal governments and cooperation with other non-state actors have been able to establish several good practices in response to climate change. Mexico City, for example, developed important mitigation efforts to curb its greenhouse gas emissions by developing collaboration between policy networks, political leaders, and research groups. Similarly, creation of Greater London Authority for horizontal coordination led to the development of Climate Change Adaptation Strategy of London including mechanisms to tackle heat waves, floods, and droughts.

Cooperation and engagement of non-state actors have also been extremely important without which it may not have been possible for the cities to achieve what they did in combating the mega risks, or at least not with the required speed and efficiency. In the pandemic, these achievements ranged from enhanced use of digital technologies in detecting and isolating coronavirus cases and hotspots to conduct of urban administration in lockdowns. Partnerships with non-state actors also helped in the establishment of strong networks for providing support to the vulnerable community through innovative solutions that covered fields as diverse as health, biomedicine, education, and commerce. For example, to meet

the urgent needs of shortages in healthcare equipment and PPEs due to rising needs and disruptions in supply chains during the pandemic, quick designing for their production was initiated through 3D printers by the private sector. In tackling climate change, cooperation with non-state actor has guided in developing a common vision among key stakeholders; ensured their engagement in reaching a common set of goals to handle; helped in pooling scarce resources at local level for combating a common issue; and allowed to alleviate key areas of potential conflict between climate change response measures and policies/programs for other core urban governance needs.

Horizontal cooperation through intercity networks has also helped considerably in tackling mega risks. A global survey of cities revealed that professional contacts within government agencies and the networks that facilitate them have been critically important in urban climate action. Both national and international networks have acted as catalyst in diffusing knowledge and ideas on mega risks although less so on pandemic than climate change. These networks have facilitated the exchange of experience, transfer and transmission of best practice undertaking innovative solutions, strengthening capacities, developing standardized methodologies, and integrating cities' outlook at national and international levels. International networks such as ICLEI CCP, Global Covenant of Mayors, C40, United Cities and Local Governments (UCLG), and Climate Alliance—while increasing their importance as driver for municipal action on the mega risks like climate change—have also considerably increased their overall influence. Their growing political clout has raised some concerns and arguments that these subnational or local actors are disproportionally increasing their influence in relation to, if not compared with, nation-states. These arguments need negation as in reality, the networks are neither seeking to eclipse the role the nation-states are playing in climate change negotiation through UNFCCC regime nor competing with them. It would, therefore, be more productive if the debate between the "transnationalists" and "multilateralists" in climate change governance focuses less on "competition" and more on coordination between the two processes and understanding linkages between them in terms of climate governance for mutual benefit.

Finally, cities have been hotspots of both mega risks and have key role to play in their management. The most important message for national governments from these is to reconcile the administrative and fiscal authority of cities and local governments commensurate with the delivery of functions expected from them. Perhaps, performance-based incentive structures would be more appropriate to make city and local governments autonomous. Key lessons on administrative and financial autonomy, as well as other aspects in urban governance, brought forward by mega risks can be summarized as follows:

- Increase empowerment of local governments and enhance the city mandates, roles, and responsibilities.
- Integrate urban governance and coordination of different actors and sectors in management of crises and emergencies to avoid confusions/conflicts and ensure effective and efficient use of limited resources.
- Use mechanism to enhance community and stakeholders' engagement to build trust as a basic tenet in urban governance ecosystem.
- Keep urban governance open to "Transformative Innovations" both hard and soft—that helped reshape the ecosystems of urban governance during the Pandemic "state of exception"—to make urban governance progressive, smart, and more inclusive.
- Strengthen urban networks and other mechanisms—to allow cities to learn from each other in implementing effective emergency responses, recovery, as well as short-, medium-, and long-term actions for enhancing resilience.
- A long-term approach needs to be factored into urban governance even when responding to the immediate response to a crisis, as much as possible.

References

Aall C, Groven K, Lindseth G (2007) The scope of action for local climate policy: the case of Norway. Glob Environ Polit 7(2):83–101

Abdullah H, Reynés J (2020) Barcelona's coronavirus makers, coproducing local solutions to a global pandemic. In: Losada AF, Abdullah H (eds) Cities on the frontline, CIDOB, Barcelona Center for International Affairs

ACIJ (undated) Información legal. https://acij.org.ar/covid19yderechos/

Al Siyabi Huda, Al Mukhaini S, Kanaan M, Al Hatmi S, Al Anqoudi Z, Al Kalbani A, Al Bahri Z, Wannous C, Al Awaidy ST (2021) Community Participation Approaches for Effective National COVID-19 Pandemic Preparedness and Response: An Experience from Oman. Frontiers in Public Health 8. https://www.researchgate.net/publication/348795486_Community_Participation_Approaches_for_Effective_National_COVID-19_Pandemic_Preparedness_and_Response_An_Experience_From_Oman

Alber G et al (1996) Municipal climate protection strategies and their national framework conditions: analysis and recommendations. Klima-Bündnis, Ökoinstitut Südtirol and Österreichisches Ökologie-Institut

Allen JH, Potiowsky T (2008) Portland's green building cluster: economic trends and impacts. Econ Dev Q 22:303–315

Aylett A (2014) Progress and challenges in the urban governance of climate change: results of a global survey. MIT, Cambridge, MA

BBVA (2020, May 7) BBVA y Santander entregan una 'app' a la Ciudad de México para hacer frente al COVID-19'. https://www.bbva.com/es/mx/bbva-mexico-y-santander-entregan-una-app-a-la-cdmx-para-hacer-frente-al-covid-19/

Bell L (2018) Explore the 7 most sustainable cities across the globe, rate it green. https://www.rateitgreen.com/green-building-articles/the-7-most-sustainable-cities-in-the-world/102

Benton JE (2020) Challenges to federalism and intergovernmental relations and takeaways amid the COVID-19 experience. Am Rev Public Adm 50(6–7):536–542. https://doi.org/10.1177/0275074020941698

Betsill M, Bulkeley H (2006) Cities and the multilevel governance of global climate change. Glob Gov 12:141–159

Betsill M, Bulkeley H (2007) Looking Back and thinking ahead: a decade of cities and climate change research. Local Environ 12:447–456

Betsill M, Dubash ND, Paterson M, van Asselt H, Vihma A, Winkler H (2015) Building Productive Links between the UNFCCC and the Broader Global Climate Governance Landscape. Global Environmental Politics 15(2):1–10. https://doi.org/10.1162/GLEP_a_00294

Bhattacharyya S, Lopes CA, Nyamupachitu-Mago E, Ndejjo R, de Claro V, Diallo AA, Tchetchia A, Gilmore B (2020) Community engagement for COVID-19 infection prevention and control: a rapid review of the evidence. https://collections.unu.edu/eserv/UNU:7694/Lopes_et_al_COVID19.pdf

Bieliei S, Grigoryan H, Ichkiti G, Kandelaki S, Kulesa A (2020, December) Direction: an efficient state—polish experiences of decentralisation and modernisation, lessons learned for Armenia and Georgia. CASE. https://case-research.eu/files/?id_plik=6605

Biesbroek R, Lesnikowski A (2018) Adaptation. In: Forster J, Jordan A, Huitema D, Van Asselt H (eds) Governing climate change: polycentricity in action? Cambridge University Press, Cambridge, pp 303–319

Bulkeley H (2010) Cities and the governing of climate change. Annu Rev Environ Resour 35(1):229–253

Bulkeley H, Betsill M (2005) Rethinking sustainable cities: multilevel governance and the 'urban' politics of climate change. Environ Polit 14:42–63

Bulkeley H, Schroeder H, Janda KB, Zhao J, Armstrong A et al (2009) Cities and climate change: the role of institutions, governance and urban planning. Presented at 5th Urban Research Symposium 2009: Cities and Climate Change: Responding to Urgent Agenda, Marseille

C40 Cities (2015) Cities 100: Mexico City - comprehensive program increases resilience. https://www.c40.org/case-studies/cities100-mexico-city-comprehensive-program-increases-resilience/

C40 Cities (undated-a) Climate change adaptation in delta cities, good practice guide. http://c40-productionimages.s3.amazonaws.com/good_practice_briefings/images/5_C40_GPG_CDC.original.pdf?1456788885

C40 Cities (undated-b) Unlocking climate action in megacities. https://www.c40.org/researches/unlocking-climate-action-in-megacities

Cadham C (2016) NYC's public-private partnerships to fight climate change. Columbia Climate School. https://news.climate.columbia.edu/2016/03/30/nycs-public-private-partnerships-to-fight-climate-change/

Cashmore M, Wejs A (2014) Constructing legitimacy for climate change planning: a study of local government in Denmark. Glob Environ Chang 24:203–212

Chicago (undated) Recovery Task Force. https://www.chicago.gov/city/en/sites/covid-19/home/covid-19-recovery-taskforce.html

Chu E, Brown A, Michael K, Jillian D, Lwasa S, Mahendra A (2019) Unlocking the potential for climate transformative adaptation in cities, Background Paper. World Resources Institute

City of Portland Bureau of Planning and Sustainability (2009) City of Portland proposed high performance green building policy [Homepage of City of Portland] (Quoted in Corfee-Morlot et al. 2009)

City of Tokyo (2008) Introduction of Tokyo's mandatory cap & trade system. Tokyo Metropolitan Government

Collier U, Löfstedt R (1997) Think globally, act locally. Glob Environ Change 7(1):25–40

Corfee-Morlot J, Kamal-Chaoui L, Donovan MG, Cochran I, Robert A, Teasdale PJ (2009) Cities, climate change and multilateral governance. OECD Publication, Paris

Council of Europe (2021, February 10) Covid-19 and amalgamated municipalities in Armenia. https://www.coe.int/en/web/yerevan/news/-/asset_publisher/UATN4Wl8F3Wu/content/covid-19-and-amalgamated-municipalities-in-armenia?101_INSTANCE_UATN4Wl8F3Wu_viewMode=view/

Council of Europe (undated) Intercultural cities: COVID-19 special page. https://www.coe.int/en/web/interculturalcities/covid-19-special-page#{%2262433518%22:[6]}

County Government of Kiambu (2020, March 23) COVID-19 preparedness. https://kiambu.go.ke/2020/03/covid-19-preparedness/

Cunha KB, Rei F (2006) Sub-national climate-friendly governance initiatives in developing world: a case study of state of São Paulo – Brazil. Brazil Institute for Energy and Environment, Sao Paulo

DeClerq K (2020, April 30) Toronto city council approves extending mayor's emergency powers. CTV News. https://toronto.ctvnews.ca/toronto-city-council-approves-extending-mayor-s-emergency-powers-1.4918622

Dulal HB (2018) Cities in Asia: how are they adapting to climate change. J Environ Stud Sci 9(6–7). https://doi.org/10.1007/s13412-018-0534-1

Dullah Omar Institute (2020) Khosa v Minister of Defense: municipalities warned on enforcing the lockdown. Local Gov Bull 15(2)

DW (2020, October 9) Coronavirus: Merkel and mayors agree on tougher restrictions. https://www.dw.com/en/coronavirus-merkel-and-mayors-agree-on-tougher-restrictions/a-55215422

Edwin B (2020) Challenges to federalism and intergovernmental relations and takeaways amid the COVID-19 experience. Am Soc Public Adm 50(6–7):536–542

Egbinola CN, Olaniran HD, Amanambu AC (2017) Flood management in cities of developing countries, the example of Ibadan, Nigeria. J Flood Risk Manage 10:546–554

Ekelund N, Sigurdson B (2007) Städers Klimatarbete. Internationella Exempel, Stockholm stad

EMGIRS (2020, May 9) Municipio del Distrito Metropolitano de Quito activa protocolo para el levantamiento de cadáveres en la Capital. https://emgirs.gob.ec/index.php/noticiasep/541-municipio-del-distrito-metropolitano-de-quito-activa-protocolo-para-el-levantamiento-de-cadaveres-en-la-capital

Foglia C, Rofman A (2020) Local participatory governance in greater Buenos Aires: a current x-ray of the 24 municipalities. Ibero-Am J Munic Stud 21:113–145

Foster S (2020) As COVID-19 proliferates mayors take response lead, sometimes in conflicts with their governors. Georgetown Law. https://www.law.georgetown.edu/salpal/as-covid-19-proliferates-mayors-take-response-lead-sometimes-in-conflicts-with-their-governors/

Galarraga I, Gonzalez-Eguino M, Markandya A (2011) The role of regional governments in climate change policy. Environ Policy Gov 21:164–182. https://doi.org/10.1002/eet.572, ISSN: 1756-9338

Global Commission on Adaptation (GCA) (2019) Adapt now: a global call for leadership on climate resilience, 2019. Global Center on Adaptation and World Resources Institute. https://cdn.gca.org/assets/2019-09/GlobalCommission_Report_FINAL.pdf

Government of Colombia (2020, April 20) Informe Operación COVID-19. http://portal.gestiondelriesgo.gov.co/Paginas/Slide_home/Informe-Operacion-COVID-19.aspx

Government of Iceland (2020, March 23) Measures in response to COVID-19. https://www.government.is/default.aspx?pageid=5781e635-46bb-4c79-8218-03d44073071e

Government of Italy (2020, March 25) Decreto Legge 25 marzo 2020, n.19. Misure urgenti per fronteggiare l'emergenza epidemiologica da COVID-19. http://www.protezionecivile.gov.it/amministrazione-trasparente/provvedimenti/%2D%2D/content-view/view/1237817

Government of Japan (2008) Action plan for achieving a low-carbon society. www.kantei.go.jp/foreign/policy/ondanka/final080729.pdf

GPC, Greenhouse Gas Protocol for Cities (undated) GHG Protocol for Cities. https://ghgprotocol.org/greenhouse-gas-protocol-accounting-reporting-standard-cities

Han E, Tan M, Turk E, Sridhar D, Leung G, Shibuya K, Asgari N, Oh J, García-Basteiro A, Hanefeld J, Cook A, Hsu L, Teo Y, Heymann D, Clark H, McKee M, Legido-Quigley H (2020) Lessons learnt from easing COVID-19 restrictions: an analysis of countries and regions in Asia Pacific and Europe. Lancet 396(10261):1525–1534

Hodson M, Marvin S (2009) 'Urban ecological security': a new urban paradigm? Int J Urban Reg Res 33:193–215

Holgate C (2007) Factors and actors in climate change mitigation: a tale of two South African cities. Local Environ 12:471–484

Hoornweg D, Freire M, Lee MJ, Bhada-Tata P, Yuen B (eds) (2011) Cities and climate change. The World Bank, Washington, DC

James K (2020, April 3) Covid-19 and the need for clear centre state roles. Vidhi Centre for Legal Policy. https://vidhilegalpolicy.in/blog/covid-19-and-the-need-for-clear-centre-state-roles/

Jensen A, Nielsen HO, Russel D (2020) Climate policy in a fragmented world – transformative governance interactions at multiple levels. Sustainability 12:10017. https://doi.org/10.3390/su122310017

Kern K, Albar G (2008) Governing climate change in cities: modes of urban climate governance in multi-level systems, chapter 8 in OECD. Climate Change and Competitive Cities, Paris

Kern K, Niederhafner S, Rechlin S, Wagner J (2005) Kommunaler Klimaschutz in Deutschland - Handlungsoptionen, Entwicklung und Perspektiven. Wissenschaftszentrum Berlin, DP SP IV 2005-101. https://www.ssoar.info/ssoar/bitstream/handle/document/19672/ssoar-2005-kern_et_al-kommunaler_klimaschutz_in_deutschland_-.pdf?sequence=1

Kiruga M (2020, April 17) Kenya's fight against coronavirus difficult with its two-tiered governance system. The Africa Report. https://www.theafricareport.com/26354/kenyas-fight-against-coronavirus-difficult-with-its-two-tiered-governance-system/

Kling S (2020) Shut down or adapt? Pandemic-era infrastructure in a divided Chicago. In: de Losada AF, Abdullah H (eds) Cities on the frontline: managing the coronavirus crisis. CIDOB, Barcelona

Laigle L (2009) Vers des villes durables: Les trajectoires de quatre agglomérations européennes, PUCA (ed.), Ministère de l'Écologie, du Développement et de l'Aménagement Durables, Paris

Larcom S, Rauch F, Willems T (2017) The benefits of forced experimentation: striking evidence from the London underground network. Q J Econ 132(4):2019–2055. https://doi.org/10.1093/qje/qjx020

Lindholm CE (1959) The science of "muddling through". Public Adm Rev 19:79–88

Lindseth G (2005) Local level adaptation to climate change: discursive strategies in the Norwegian context. J Environ Policy Plan 7(1):61–83

Los Angeles (2019) Green new deal. https://www.lamayor.org/sustainability

Losada AF, Abdullah H (2020) Cities on the frontline. CIDOB, Barcelona Centre for International Affairs

McCarney P, Blanco H, Carmin J, Colley M (2011) Cities and climate change. In: Rosenzweig C, Solecki WD, Hammer SA, Mehrotra S (eds) Climate change and cities: first assessment report of the urban climate change research network. Cambridge University Press, Cambridge, pp 249–269

McGuirk P, Dowling R, Maalsen S, Baker T (2020) Urban governance innovation and COVID-19. Geographical Research 59(2). https://doi.org/10.1111/1745-5871.12456

Melbourne (2020) COVID-19 reactivation and recovery plan. City of Melbourne, Melbourne

Mexico City (2020, March 30) Gaceta Oficial de la Ciudad de México. https://data.consejeria.cdmx.gob.mx/portal_old/uploads/gacetas/ddcac298af1eb-9c3e3235ac7890a9a32.pdf

Moloney A (2021) Bogota crowdsources a green transport future to cut emissions. Reuters. https://www.reuters.com/article/us-colombia-climate-changetransportatio/bogota-crowdsources-a-green-transport-future-to-cut-emissions-idUSKCN2D7203

New Orleans (2020, June 30) City of New Orleans launches unprecedented COVID-19 meal assistance program in partnership with FEMA and local restaurants; residents encouraged to apply. https://nola.gov/mayor/news/june-2020/city-of-new-orleans-launches-unprecedented-covid-19-meal-assistance-program-in-partnership-with-fema/

Ng MK (2020) Honk Kong's head start in tackling the new coronavirus. In: Losada AF, Abdullah H (eds) Cities on the frontline. CIDOB, Barcelona Centre for International Affairs

Oates WE (2002) A reconsideration of environmental federalism. In: Recent advances in environmental economics. Edward Elgar, ISBN: 1858986117 9781858986111

OECD (2008, October 9–10) Competitive cities and climate change. OECD Conference Proceedings, Milan

OECD (2020a, November 10) The territorial impact of COVID-19: managing the crisis across levels of government. http://www.oecd.org/coronavirus/policy-responses/the-territorial-impact-of- covid-19-managing-the-crisis-across-levels-of-government-d3e314e1/

OECD (2020b, July 23) Cities policy responses. http://www.oecd.org/coronavirus/policy-responses/cities-policy-responses-fd1053ff/

OECD (2020c, April 22) Italian regional SME policy responses. https://www.helvetas.org/en/switzerland/how-you-can-help/follow-us/blog/inclusive-systems/Local-governments-during-COVID-19

Parnell S, Claassen CM (2020) COVID-19 in Cape Town, initial state and civil society responses. In: Losada AF, Abdullah H (eds) Cities on the frontline. CIDOB, Barcelona Centre for International Affairs

Phelps NA, Miao JT (2019) Varieties of urban entrepreneurialism. Dialogues Hum Geogr 10:304–321. https://doi.org/10.1177/2043820619890438

Puppim de Oliveira JA (2009) The implementation of climate change related policies at the subnational level: an analysis of three countries. Habitat Int 33(3):253–259

Qi Y, Ma L, Zhang H, Li H (2008) Translating a global issue into local priority: China's local government response to climate change. J Environ Dev 17:379–400

Renault Group (2021) Six eco-neighborhoods to discover in Europe. https://www.renaultgroup.com/en/news-on-air/news/6-eco-neighbourhoods-to-discover-in-europe/

Republic of Colombia (2020) Decreto 418 de 2020. https://dapre.presidencia.gov.co/normativa/normativa/DECRETO%20418%20DEL%2018%20DE%20MARZO%20DE%202020.pdf

Republic of Peru (2020, January 31) Resolución ministerial. https://cdn.www.gob.pe/uploads/document/file/505245/resolucion-ministerial-039-2020-MINSA.PDF

RFI (2020) Italy's Veneto region to launch population-wide testing for Covid-19. http://www.rfi.fr/en/europe/20200325-italy-s-veneto-region-to-launch-population-wide-testing-for-covid-19

Roberts D (2008) Thinking globally, acting locally—institutionalizing climate change at the local government level in Durban, South Africa. Environ Urban 20(2):521–537

Romero-Lankao P, Burch S, Hughes S, Auty K, Aylett A, Krellenberg K, Nakano R, Simon D, Ziervogel G (2018) Governance and policy. In: Rosenzweig C, Solecki W, Romero-Lankao P, Mehrotra S, Dhakal S, Ali Ibrahim S (eds) Climate Change and Cities: Second Assessment Report of the Urban Climate Change Research Network. Cambridge University Press. New York 585–606. https://uccrn.ei.columbia.edu/sites/default/files/content/pubs/ARC3.2-PDF-Chapter-16-Governance-and-Policy-wecompress.com_.pdf

Rosenzweig C, Solecki W, Hammer S, Mehrotra S (eds) (2011) Climate change and cities: first assessment report of the urban climate change research network. Cambridge University Press, New York

Saliba F (2020) Claudia Sheinbaum, maire de Mexico: «La mégapole de demain sera plus participative». https://www.lemonde.fr/smart-cities/article/2020/06/21/claudia-sheinbaum-maire-de-mexico-la-megalopole-de-demain-sera-plus-participative_6043603_4811534.html

Salon D, Murphy S, Sciara G-C (2014) Local climate action: motives, enabling factors and barriers. Carbon Manage 5(1):67–79. https://doi.org/10.4155/cmt.13.81

Santos A, Sousa N, Kremers H, Bucho JL (2020) Building resilient urban communities: the case study of setubal municipality. Portugal Geosci (Switzerland) 10(6):1–13

São Paulo (undated) SP contro o novo coronavirus: Saiba como se proteger. https://www.saopaulo. sp.gov.br/coronavirus

Schriber M (2005) Cities cover more of earth than realized, live science. https://www.livescience.com/6893--cities-cover-earth-realized.html

Shammi M, Bodrud-Doza M, Towfiqul Islam ARM, Rahman MM (2020) COVID-19 pandemic, socioeconomic crisis, and human stress in resource-limited settings: a case from Bangladesh. Heliyon 6(5, e04063)

Squires C (2021) Are cities underestimating carbon pollution. Bloomberg City Lab. https://www.bloomberg.com/news/articles/2021-03-26/how-to-tell-if-cities-are-underestimating-emissions

Sugiyama N, Takeuchi T (2008) Local policies for climate change in Japan. J Environ Dev 17(4):424–441

Sydney (2020) Survey: Sydney's recovery plan. https://www.cityofsydney.nsw.gov.au/vision-setting/survey-sydneys-recovery-plan

Tanveer F, Khalil AT, Ali M et al (2020) Ethics, pandemic, and environment; looking at the future of low middle income countries. Int J Equity Health 19:182, Benton. https://doi.org/10.1186/s12939-020-01296-z

The Council of Governors (2020, June 1) Mombasa governorate demonstrates great leadership in the COVID-19 response. https://www.cog.go.ke/component/k2/item/207-mombasa-county-demonstrates-great-leadership-in-the-covid-19-response

Thoi PT (2020) Ho Chi Minh City- the front line against COVID-19 in Vietnam. City Soc 2020:32. https://doi.org/10.1111/ciso.12284

Timeus K, Gascó M (2018) Increasing innovation capacity in city governments: do innovation labs make a difference? J Urban Aff 40(7):992–1008

Toronto (undated) COVID-19: impacts & opportunities report from Toronto's Office of Recovery & Rebuild. https://www.toronto.ca/home/covid-19/covid-19-reopening-recovery-rebuild/covid-19-about-reopening-recovery-rebuild/

UCLG Africa (2020, April 23) COVID-19: African local and regional governments on the front line. https://www.uclga.org/news/covid-19-african-local-and-regional-governments-on-the-front-line/

UCLG United Cities and Local Governments, Metropolis and LSE Cities (2020, July) The COVID-19 response: Governance challenges and innovations by cities and regions. LSE Analytics Note 2. https://www.lse.ac.uk/Cities/publications/Policy-Briefs-and-Analytics-Notes/Analytics-Note-02-The-COVID-19-Response-Governance-Challenges-and-Innovations-by-Cities-and-Regions

UN Habitat (2021) Cities and pandemics: towards a more just, green and healthy future. HS/058/20E, Nairobi

UNDESA (2020) Policy Brief # 61: COVID – 19: embracing digital government during the pandemic and beyond. https://www.un.org/development/desa/dpad/publication/un-desa-policy-brief-61-covid-19-embracing-digital-government-during-the-pandemic-and-beyond/

UNDP (2020a, July 23) Impacto económico y social del COVID-19 y opciones de política en la república dominicana. https://www.latinamerica.undp.org/content/rblac/es/home/library/crisis_prevention_and_recovery/impacto-economico-y-social-del-covid-19-y-opciones-de-politica-e.html

UNDP (2020b, October) COVID 19 socio economic response plan. COVID-19 Socio-Economic Impact Assessment. https://serbia.un.org/sites/default/files/2020-09/seia_report%20%281%29.pdf

UNEP (2021) Global climate litigation report: 2020 status review. https://www.unep.org/resources/report/global-climate-litigation-report-2020-status-review

UNFCCC (2005) Project design document form (CDM PDD): Bandeirantes landfill gas to energy project (BLFGE). UNFCCC

United Nations -ECLAC (2014) Adaptation to climate change in Latin America. Santiago

Victor DG, Muro M (2020) Cities are pledging to confront climate change but are their actions working. Brooking Institution. https://www.brookings.edu/blog/the-avenue/2020/10/22/cities-are-pledging-to-confront-climate-change-but-are-their-actions-working/

Weng S, Ni AY, Ho AT-K, Zhong R-X (2020) Responding to the Coronavirus pandemic: a tale of two cities. Am Rev Public Adm 50(6–7):497–504. https://doi.org/10.1177/0275074020941687

Wilson, K (2020, April 13) SC Gov. McMaster's plan to "revitalize" economy. The Georgetown Times/South Stand News. https://www.southstrandnews.com/news/sc-gov-mcmasters-plan-to-revitalize-economy/article_6eb5a530-7de2-11ea-94b4-77c5f353db2e.html

World Economic Forum (2020) Cities are especially vulnerable to COVID-19. These organizations are leading the urban response. https://www.weforum.org/agenda/2020/05/cities-urbanization-urban-living-coronavirus-covid19-pandemic-design/. Accessed 10 Aug 2021

World Health Organization (WHO) (2002) Community participation in local health and sustainable development: approaches and techniques. WHO Regional Office for Europe, Copenhagen

Wrable A (2021) As local government meetings moved online, public participation jumped. The Daily Progress. https://dailyprogress.com/news/local/govt-and-politics/as-local-government-meetings-moved-online-public-participation-jumped/article_f7e98970-86b0-11eb-b1b9-c36c20bb6b3c.html

Zambrano-Barragán C, Zevallos O, Villacís M, Enríquez D (2011) Quito's climate change strategy: a response to climate change in the Metropolitan District of Quito, Ecuador. https://www.researchgate.net/publication/236607813

Vision of a Sustainable, Smart, and Resilient City

11

11.1 Introduction

Pandemics, natural calamities, and economic downturns have impacted cities in the past no doubt, but they are dwarfed when the intensity of their impacts is compared to the two mega risks facing the world today—COVID-19 and climate change. The twin mega risks have caused multiple problems and concerns to humanity. The pandemic alone disrupted economic activities in human settlements causing the worst recession in the world since the Great Depression. Losses in working hours in 2020 were approximately four times greater than the 2008–2009 financial crisis (ILO 2021). There was also a 40% drop in foreign direct investment in 2020 compared to 2019. Although the official development assistance in 2020 enhanced to a historical high—$161 billion—it still fell short of the target of 0.7 percent of donors' gross national income (UNDESA 2021a). The impacts of the debt distress along with dearth of technological capacity, and lack of access to vaccines, have been limiting the fiscal and policy space of developing countries and their cities in pandemic recovery. They have also constrained critical investments in climate action and the achievements of SDGs and are threatening to prolong their recovery efforts.

Cities as epicenters of the pandemic and victims of climate change have faced the brunt of the health, economic, and social challenges. Their high level of global and local connectivity together with overcrowding of populations in their slums and informal settlements has made them fertile ground for the spread of the virus and high intensity of climate change-related calamities. The health issues of the pandemic simultaneously expanded into an economic and social crisis that disrupted the city economy and its social fabric. It created serious problems of business disruption, joblessness, reduced urban finance, interruption in provision of public services, infrastructure, and transport. It also had most serious impact on urban equity by disproportionally affecting the most vulnerable in society. The year 2020, which stands out for the beginning of the pandemic, also saw a big rise in climate-related disasters, which were responsible for the 389 recorded events that killed 15,080 people, affected 98.4 million, and caused a minimum economic loss of US$ 171.3 billion (CRED and UNDRR 2021). It is important to note, however, that twin mega risks while bringing devastation have also opened many avenues. They have opened opportunities for changing current human unsustainable relationships with the environment, as well as improving human-to-human social relationships. This makes it imperative (a) to reflect on the impact and response strategies to cope with the mega risks as carried out in the previous chapters and (b) to focus on the opportunities offered in terms of visioning for the future of cities.

The lessons from mega risks should help the world rise to current and future challenges through visionary thinking. This chapter outlines the vision for a sustainable, resilient, and livable city in the wake of mega risks—which have unleashed forces that have the potential to lead to relatively long-lasting transformation of cities at least in the short term. As the world enters the third year of the pandemic, it is very clear that the COVID-19 crisis is of monumental proportion, which has consumed over 5 million lives, and the economic and social toll has been unprecedented. Some progress has been made toward recovery, no doubt (Baller 2021), but overall recovery tracking has shown that efforts have been uneven, inequitable, and still not geared to achieving sustainable development. This is not a good omen—the most serious concern is that the current crises are threatening decades of development gains made in the past. Further, they are seriously endangering and delaying the urgently needed transition to greener, more equal and inclusive economies—as envisaged in the Sustainable Development Agenda 2030 and New Urban Agenda for cities. The two international agendas recognize urbanization not merely as a demographic phenomenon but rather as a transformative process capable of galvanizing momentum for sustainable and resilient development in nations and cities, a development which is economically vibrant and green, socially inclusive, environmentally safe, and resilient to risks.

The chapter has been divided into five parts. After the introduction, the next part deals with the forces unleashed by the mega risks that not only have led to short-term transformations of cities but also have the potential for bringing long-term changes at least in some form. The third part deals with the vision of sustainable development in building forward. It has four sections—the first section is on the vision for a resilient green urban economy; the next is on the vision of a socially inclusive city with judicial governance, equality, enhanced participation, improved quality of life, and enhanced social capital; the third section discusses the environmental sustainability aspect; and the fourth section deals with the vision of a technologically smart city which includes anticipatory and adaptive innovations and creativity. The fourth part of the chapter is on the vision for creating a resilient structure of the city through urban design—a design for its neighborhoods, public spaces, buildings, and infrastructure with due consideration to its functional coherence, as well as health

and well-being of its residents. The findings of the chapter have been summarized in the concluding section.

11.2 Mega Risks and Urban Transformation Trends

The experience of previous pandemics and climate-related disasters shows that while playing havoc, they also brought notable positive economic, social, and urban design changes. Moreover, none of them diminished the role of large cities nor detracted the growth of urbanization in the long term for at least two reasons—(a) the "innovation, creativity, and economic growth," which demand the clustering of talent and economic assets, face-to-face interaction, buzz, diversity, and the critical mass, can only be provided by cities (Storper and Venables 2004), and (b) cities, as dynamic and adaptive entities, have shown a high degree of resilience capacity (Sassen and Kourtit 2021). Hence, it appears highly unlikely that either COVID-19 or climate change despite their major devastation will derail the long-standing process of urbanization and the economic role of cities. Nevertheless, it does not negate the possibility of transformation in function or form of cities, which is already on the horizon and needs to be guided in the right direction. Mega risks, particularly the pandemic, for example, have transformed drastically several if not every aspect of urban functioning and living—from work to shopping and socialization and mobility and transport to trade and supply chain—and at all scales from individual buildings to neighborhoods, cities, and their region. Some aspects of these transformations have been outlined below along with forces that were responsible to bring about these changes.

- Transformation of urban functions in the wake of lockdowns and social distancing which promoted remote work as new job style; retail delivery instead of contact shopping at stores or malls; remote education as an alternative to classroom teaching; teleconferencing instead of in-person meetings; telemedicine as an alternative to hospital visits; provision of administrative services online instead of office visits; and entertainment through life streaming via digital media instead of visiting cinemas and theatres or participating in concerts, etc. These changes have made long-term transformations possible—ranging from a dramatic and permanent switch to all-online or adaptation of a hybrid in certain lines of work to a rebound to the preexisting status quo in others. The most likely future scenario appears to be a "new normal" consisting of work or function that has a greater mix of remote-from-home than prior to the forced experiment (Florida et al. 2021). However even a small shift in remote work or functions such as remote shopping could have important repercussions for mobility, real estate and infrastructure, and services as well as office design, commercial developments, and commuting patterns. The long-term impacts will largely depend upon the future behavior of the city administration, institutions, or commercial firms as well as workers and individuals in relation to the new circumstances (Florida et al. 2021).

- Transformation in the manufacturing and business processes through the use of technology, automation, robotics, and Internet of Things (IoT), as well as the use of big data and cloud computing. It was not just a boost or fast-track in technology deployments but also a test or trial of the ability of entrepreneurs to get their workers and clients or customers adopt to new forms of business engagement, transections, and interaction. This transformation appears to have prepared consumers as well as businesses, companies, and organizations alike to accept and adopt to digital change that is most likely to continue in future at a rapid pace.
- Transformation of urban transport—the enhancement of non-motorized transport was triggered during the pandemic by both public regulations and individual responses. In Quito, Ecuador, for example, around 70 kilometers of new bike lanes were constructed, and the number of bike trips increased by over 700

percent during just May 2020 (UN Habitat 2021). The pattern was not unique to Quito; several other cities across both the developed and developing world adopted it with the twin objectives of meeting the need of the pandemic and to help mitigate climate change. This may continue in the future.

- Transformation for the repurposing of public spaces—public spaces were quickly adapted to support emergency services through the setup of temporary hospitals, warehouses, and other facilities that helped improve cities and their neighborhood response capacities. Similarly, many activities were moved from indoor to outdoor locations. For example, businesses such as restaurants, cafes, theatres, cinemas, and gyms started operating in the open spaces or in streets outdoors. This became the new normal to support livelihoods and business activities and important social and cultural services to the city residents. The use of green spaces as main recreational areas increased enormously. According to one study (Venter et al. 2020), "outdoor recreational activities" increased by 291 percent during lockdown. New concepts, such as Gastro Safe Zone (Harrouk 2020) in Europe and food streets and other form of outdoor eating in the USA, also provided safety to customers while offering at least some income to restaurant owners and food vendors. This collateral use of extended spaces despite some concerns (MacCarthy 2021) may become an integral feature of future urban landscapes in the world (which was already a norm in some cities of developing countries). Moreover, stores, restaurants, theatres, museums, and other gathering places may need redesigning to promote social distancing as a safety measure for the future. There is a considerably more potential for the use of streets and open spaces if Wi-Fi hotspots are provided outdoors. Among others, it can enable children remote access to education and adults a means of remote work from home.
- Transformation created through movement to second-tier, secondary, and smaller cities, as some people started moving out of large cities

at least initially in the pandemic. *The New York Times* highlighted surges in exodus of households from New York City to suburbs in March and April 2020 (Hughes 2020). In addition, Harris Poll reported that the pandemic prompted nearly 40% of urban dwellers to consider moving to less crowded places. In case remote work remains the norm postpandemic, many of these outmigrants may not return (Hart 2020). Moreover, it may encourage medium and small cities to invest in services like broadband, healthcare, cultural, and entertainment services to attract more high-tech remote workers. However, recent studies show that the urban exodus was overblown, and data as of April 2021 suggested that some population had even returned to large cities and real estate business improved—New York City, for example, saw a strong rebound in residential sales and rentals in early 2021 (Florida et al. 2021).

- Transformation in the urban form and shape through urban planning and design is also on the horizon. While an emerging strand of research is exploring to better proof world cities from the ill effects of future contagions (Lennon 2021; Martínez and Short 2021), there are ongoing debates on the new city design. These have involved organization of public spaces, land use, mobility, connectivity, etc. and led to the initiation of such ideas as 15-minute city and its application to cities like Paris. Such ideas are likely to prompt architects, designers, and planners to innovate and to respond to the threat of future pandemics and climate risks. However, a major concern in this kind of transformation of urban form and function is whether the urbanization emerging from these ideas will be inclusive, public-oriented, and friendly. There are already negative messages from responses to climate change in places such as Miami. The construction of super-luxury towers in that city has attempted to isolate the rich from the growing impacts of flooding while diverting future floodwaters onto the areas of less advantaged (Stewart and García-Navarro 2019). It is therefore critically important to

carefully examine the impacts of new ideas on designing for resilience more carefully. It has raised the necessity for a rigorous debate on such schemes before their adaptation as a practice or using them in the formulation of new regulations (Florida et al. 2021).

- Another transformation was the revival of cottage industry triggered by the urgent demand of medical supplies—like ventilators and basic but essential items like facemasks. The homemade facemask industry emerged—as an urgent need to protect the masses (Livingston et al. 2020) and to address the critical shortages (Rubio-Romero et al. 2020) and as part of a post-lockdown exit strategy (Allison et al. 2021). This may possibly change the landscape of global production at least in the short term, probably leading to an attenuation of consumerist tendencies (Ibn Mohammed et al. 2021). Moving forward, there is a likelihood of its impact on R&D, whereby companies may take short-term views as a repercussion of recession. They may change long- and medium-term R&D in favor of short-term product development for profit that was observed in the past in the case of automotive and aerospace sectors during previous recessions (Ibn Mohammed et al. 2021).

- Transformation in industrial economic model is also in the air for decoupling of economic growth from resource consumption, waste management, and wealth creation. This is leading to heightened interests in circular economy. Ibn Mohammed et al. (2021) have identified the areas in which governments are increasingly recognizing the need for policies on circular economy.

- Finally, in terms of governance transformation, while the pandemic has shown the virtues of strong central government that can take quick control of the situation, it has also shown the virtue of pandemic-resilient urban government—"well-informed and effective with the credibility, capacity and good information at the local scale" (Nathan 2021). Hopefully, the transformation in giving more power to local/urban government toward reopening after lockdowns may evolve into giving them more power to manage their affairs.

Even with the transformation taking place or on the horizon, there are certain aspects of urban living that are highly unlikely to change drastically—especially in global cities. For example, New York, London, Hong Kong, and Singapore are likely to continue as the world's great financial centers; the Silicon Valley cities like San Francisco may continue to remain the hub of high technology; and Los Angeles, New York, and Paris may stay as centers for entertainment and film, while Paris, Milan, London, and New York may continue their high rank as fashion capitals. Similarly, all of these and many others like Tokyo, Shanghai, Seoul, Toronto, and Berlin are likely to maintain their status as global cities. Nevertheless, some changes may take place in the morphologies and urban form of even these cities—for example, their downtowns may lose out somewhat to their suburbs and nearby small towns, but in general, winner-take-all geography of cities is likely to persist (Florida et al. 2021). Nevertheless, as they recover and rebuild, from the visionary perspective, there is a window of opportunity even for these cities to reconsider both their functions and forms in relation to sustainability, human prosperity, and health of the planet.

11.3 Building Forward and Sustainable Development

In terms of development, the far-reaching impacts of the two mega risks are a reminder of the fact that policymakers whether urban or national can no longer work in isolation. They cannot afford to consider "economic" (economy and trade) and "non-economic" shocks (health and climate emergencies) separately, firstly, because non-economic shocks also trigger economic shocks and, secondly, because both can reverse hard-won gains on the front as is apparent in the increased poverty and enhanced disparity in cities during the pandemic. The mega risks particu-

larly COVID-19 crisis starkly exposed the weaknesses and inadequacies in systems and institutions for economic development at both national and urban levels. Urban economies in this regard particularly matter because cities not only account for 80 percent of global GDP (World Bank 2020a), but their economic impacts spill far beyond their respective boundaries. Hence, the United Nations (2017) is right in predicting that battle for sustainable development will be won or lost in cities. Besides economy, cities will also be central in promoting two other dimensions of sustainable development: social (to promote inclusive growth) and environmental (to confront the threats of any future disease outbreak and climate change).

It is interesting to note that "sustainable development" as a term was first used in the context of "maximum sustainable yield" for exploitation of fisheries (Russell 1931). Its wider use started in the late 1970s after a book edited by Pirages (1977) discussing growth used the term in its title. The term's familiarity enhanced after its use in the *World Conservation Strategy* (IUCN, UNEP, and WWF 1980). However, its popular usage started soon after the publication of the report of the World Commission on Environment and Development (WCED 1987), entitled *Our Common Future*. The report defined sustainable development as a "development that meets the needs of the present without compromising the ability of future generations to meet their own needs." While it became the most frequently quoted definition of the term (IISD undated), it also opened a debate in the scholarly literature on the definition and concept of sustainability (Redclift, 1987, 1994, 2005; Ríos-Osorio et al. 2005; Lipschutz 2012; Ruggerio 2021) which is still ongoing. Meanwhile, the idea of "sustainable development" entered the global governance and was institutionalized by the United Nations during and aftermath of the Earth Summit held in Rio de Janeiro in 1992. Soon after the Summit, in December 1992, the United Nations established the Commission on Sustainable Development to monitor the progress on the implementation of Agenda 21 and other outcomes of the Earth Summit. However, various critiques in the early academic literature of the economic status quo from social and environmental perspective led to emergence of three pillars or dimension of sustainable development—economic, social, and environmental or ecological—reconciling trade-offs in economic growth as a solution to social and ecological problems (Purvis et al. 2019).

Whether called pillars or dimensions, their entry into the dominant development paradigm of both cities and nations led to the creation of connections between economic growth, social equity, and the environmental protection as important elements of sustainability. According to Baumol and Oates (1993) initially, the interrelationships between them were often overlooked and left to market dynamics, ambition, industriousness, and productivity in promoting growth. Despite some improvements, this is still a major challenge in the twenty-first century to sustainability as often developmental trajectories followed have not seen the desired changes. Even during COVID-19, much effort is going into predicting whether the economic recovery will be "V-shaped," "U-shaped," "L-shaped," or "K shaped." Although, rather than targeting the shape of the curve, it is more important to aim for a recovery and future development that is not only based on the GDP growth alone but the one that targets a holistic economic, social, and environmental reset (World Economic Forum 2020). The critical importance of sustainable development has never been as glaring as it is now. The need for the cities and countries is to change unsustainable development trajectories and mainstream non-financial economic performance measures—to track social and environmental gains or loss more fully into economic analyses and track them on a regular and sustained basis. While proceeding further in this chapter, it is important to emphasize that visions for building forward better whether by improving sustainability of cities through selection of right trade-offs or by making them smart by adopting appropriate technologies or making them resilient to mega risks by city, neighborhood, or building design are not isolated ideas. They need to be implemented within one overarching framework of sustainable development.

11.3.1 Vision of a Sustainable Urban Economy

The economic consequences of the pandemic were significant and widespread in cities and nations, affecting all areas of the economy, including capital flows, business operations, employment, and jobs (Djankov and Panizza 2020). They call not only to look at short-term measures to shore up livelihoods, employment, and economic growth but also to reflect on the driving forces and time span of development. This is an essential first step in "Building Forward Better"—in the context of sustainable and resilient economic recovery from COVID-19. For example, in economic terms, one major effect on urban production during the crisis resulted from vulnerabilities in the supply chain. As a result, many scholars started advocating a shift toward self-sufficiency and reduced reliance on global supply chains (Ibn Mohammed et al. 2021, Free and Hecimovic 2021; Kang et al. 2020; Klerkx and Begemann 2020; Lenzen et al. 2020; Moraci et al. 2020). No doubt, it is important to ensure that local supply chains do genuinely improve by enhancing resilience and reducing environmental impacts, through increased resource efficiency and augmented circularity of supply chains. Nevertheless, while the voice of self-reliance is getting louder, and interdependence is being portrayed as a vulnerability, it cannot be abandoned altogether. Interdependence, in the long term, is also an important source of mutual gain. Perhaps local circumstances can serve as an important guide in the selection of policy options on how much self-reliance is needed on case-by-case basis.

Reducing adverse social impacts and environmental footprints (Moglia et al. 2021; Le Quéré et al. 2020; Rume and Didar-Ul Islam 2020; ITC News 2020) will also be crucial in promoting economic growth. It is absolutely essential to take stock of the past mistakes that inflicted damages, draw lessons, and use corrective measures (OECD 2020a) in building forward better in the post-pandemic recovery. The business-as-usual recovery in GDP growth and firm profitability alone will not work. A paradigm shift is needed at both city and national level—using a compass composed of green and inclusive targets widely adopted and fully operationalized for promoting sustainable development. Sustainable Development Agenda 2030 appears to offer the best guide for the purpose. The agenda is especially relevant and attractive because (a) it was created through a process of participation which enabled its widespread ownership; (b) its universality and applicability to both developed and developing countries and at all spatial scales—from cities to subnational, national, and international levels; and (c) its commitment to tackle all important global challenges to sustainability (poverty and inequality, consumption, and production) as well as climate change, (d) partnerships for development, and (e) close interlinkages between its 17 goals.

11.3.1.1 Sustainable Urban Economy and SDG Linkages

Planning and development of sustainable urban economy is linked to many SDGs. For example, it is directly linked to goal 8 on work and economic growth, goal 12 on sustainable consumption and production, goal 7 on energy, and goal 13 on climate change. In addition, providing urban and regional infrastructure, as a component of goal 9, contributes to enhancing markets and value chains. Goal 11, "make cities inclusive, safe, resilient and sustainable" (United Nations 2020), provides one of the best examples of these linkages. The relationship of goal 11 on other goals (targets and indicators) has been presented in Fig. 11.1. An example of strong impact of goal 11 on goal 13 climate change can be gauged from the fact that in generating economic growth cities produce over 60% of the global GHGs. Therefore, any progress in achieving climate transition will be linked to local actions on curbing GHG emissions at city level.

It is encouraging to note that cities are formulating their sustainable development agenda along SDGs. Singapore, Seoul, Mannheim, and Quito are some notable examples. Among these, Seoul Sustainable Development Agenda has 17 Sustainable Development Goals and 96 indicators with a clear link between the agenda and the

goals. The City of Mannheim is another example of effective local integration of the goals and the agenda. The City of Quito has aligned its existing strategies and Vision 2040 toward implementing the agenda by localizing the Sustainable Development Goals. It has made significant progress which includes the adoption of a resilience strategy and disaster risk management plan under the agenda. Cities such as Jakarta, Madrid, and Durban are other cities that have incorporated urban planning approaches into their local development plans and connecting these to the 2030 Agenda and the New Urban Agenda.

At national level, the governments were making some progress in implementing SDGs, but unusual situation created by the pandemic has raised difficulties. Nevertheless, the new economic recovery packages developed to deal with the recession have provided opportunities for the implementation of SDGs. An analysis of national response to economic recession shows some variation in scope. For example, while the European Union recovery plans are focused more on the green transition, China's stimulus spending was more immediately targeted to building out the public health system. The US stimulus packages and investment plans, on the other hand, concentrate more on infrastructure. However, whatever long-term objective be in dealing with the economic shock, it is important that nations and cities respond aggressively to minimize the reversal of development gains and opt for a strong and swift response rather than end up with "too little, too late." In this connection, the *Economic and Social Survey of Asia and the Pacific* (UNESCAP 2021) found that "on average, a financial crisis lowers GDP per capita by less than 1 percent in countries that respond aggressively to shocks and have low pre-existing vulnerabilities compared with more than 3 per cent in other countries. An epidemic sets back educational outcomes by half a year in the former countries compared with a year and a half in the latter. A natural disaster sets back environmental performance by less than a year in the former countries compared with more than 6 years in the latter" (UNESCAP 2021).

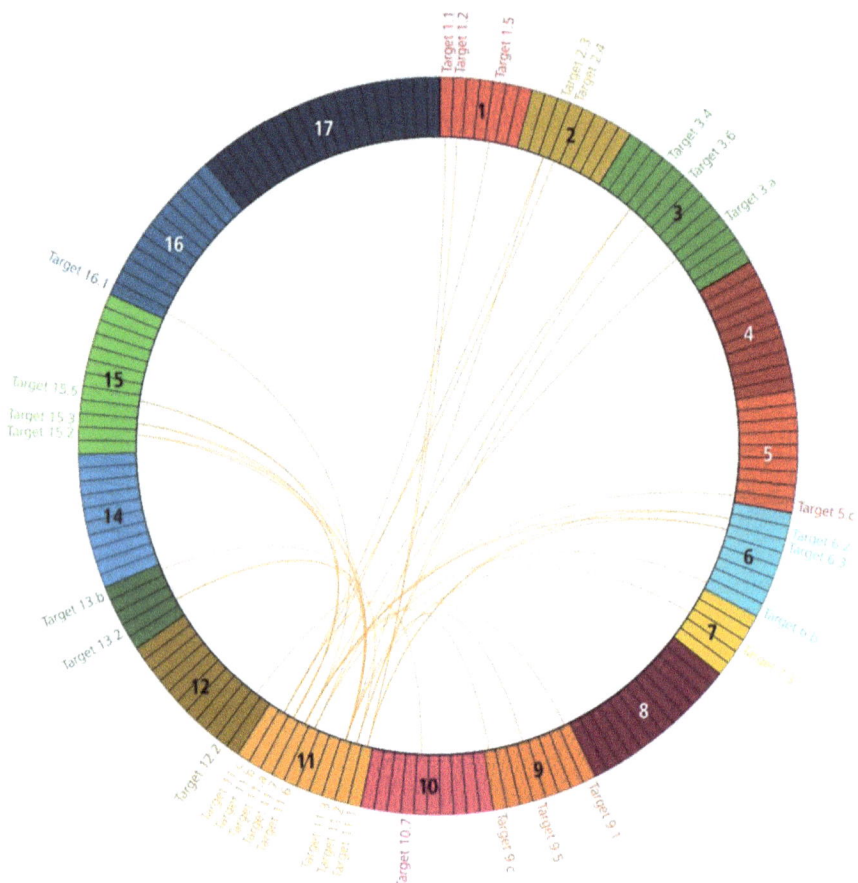

Fig. 11.1 Relationship of SDG 11 with other sustainable development goals (targets and indicators). (Source: United Nations (2018))

11.3.1.2 Monitoring and Tracking Economic Development

Monitoring of recovery and development measures is also essential. Several tracking initiatives have been developed by private, intergovernmental, and nongovernmental and academic institutions to assess the greening of recovery measures. Invariably all tracking initiatives present disappointing results. They show that a very large proportion of recovery measures are not geared to achieving sustainable development. For example, *Greenness of Stimulus Index* designed by Vivid Economics (2021), a strategic economics consultancy firm, evaluates the impacts of governmental rescue and recovery packages on climate and nature in G20 and ten other emerging economies. The findings of the index released in July 2021

showed that only US\$ 4.8 trillion out of a total US\$ 17.2 trillion public stimulus money spent indicate a positive environmental effect. The Energy Policy Tracker (2021) launched by a consortium of NGOs and universities gather publicly available information on approved policies concerning energy production and consumption for 31 major economies and 8 multilateral development banks. The August 2021 findings of the tracker showed that 4% of energy stimulus funding is targeted toward fossil fuels totaling US \$336 billion. The *Sustainable Recovery Tracker* developed by the International Energy Agency (IEA 2021) monitors the impact of total COVID-19-related government spending on clean energy measures across more than 50 countries globally. The last assessment, released in July 2021, revealed that around US$ 380 billion of total pandemic-related fiscal spending has been assigned to clean energy measures—only 2% of the total. *OECD Green Recovery Database* assessment released in July 2021 showed that although overall green measures increased in both number and budgetary spending, they account for only a small share of total spending on recovery—around 21% or about one-fifth of the total economic recovery spending is currently allocated to environmentally positive measures in OECD, EU, and key partner countries.

A comparatively comprehensive development tracker is *Global Recovery Observatory* initiative. It is led by Oxford University and has the support of the Green Fiscal Policy Network, UNEP, and the IMF. The observatory seeks to comprehensively track sustainability of all COVID-19-related spending, not just those with environmental implications, across around 50 leading countries and an additional 39 emerging and developing economies. Measures are assessed based on not only environmental impact (covering greenhouse gas emissions, air pollution, natural capital) but also social impact (wealth inequality, quality of life) as well as economic impact (multiplier, speed of implementation). The methodology involves mapping the measures first into 40 exhaustive and mutually exclusive archetypes, as well as 158 sub- archetypes (O'callaghan and Murdock 2021). The observatory is a live database of all COVID-19-related government spending in the 50 largest economies, with over 3500 policies as of February 2021. The findings of the observatory showed that 1 year after the pandemic the world had not moved on track for a green recovery (UNEP 2021a). The spending announced in 2020, according to the observatory, paints a disappointing picture of overall efforts to build forward with green priorities. Only 18.0% of recovery spending in 2020 and 2.5% of total spending had positive green characteristics (Global Recovery Observatory 2021). Most of the $14.6 trillion COVID-induced spending (rescue and recovery) in 2020 (UNEP 2021a) related to "rescue"-type measures, intended to save lives and protect livelihoods. The main features of green spending (UNEP 2021a) were as follows:

- *$341bn or 18.0%* of the total rescue and recovery spending was green—mostly accounted for by a small group of high-income countries.
- *$66.1bn was invested in low-carbon energy*—primarily Spanish and German subsidies for renewable energy projects and hydrogen and infrastructure investments accounted for this.
- *$86.1bn for green transport* was announced mainly for electric vehicle transfers and subsidies, investments in public transport, and cycling and walking infrastructure.
- *$35.2bn was pronounced for green building upgrades* to enhance energy efficiency, primarily by retrofits—mainly in France and the UK.
- *$56.3bn was declared for natural capital or nature-based solutions*—ecosystem regeneration initiatives and reforestation. 49% was allocated for public parks and counter pollution measures (mainly in the USA and China), improving the quality of life, and to deal with environmental concerns.
- *$28.9bn was announced for green research and development (R&D)*—including renewable energy technologies and technologies for decarbonizing sectors such as aviation, plastics, and agriculture and carbon sequestration.

Overall, the 2020 green investment covered a broad range of priorities but was skewed toward green energy and green transport. It is good though that some importance was given to nature-based solutions and green research and development (R&D). Without progress in the latter (green R&D), meeting the Paris Agreement targets would demand far-reaching pricing and lifestyle changes. On a per capita basis, in 2020, the total spending in advanced economies was 17 times greater than in emerging market and development economies. The latter will require substantial concessional finance from international partners—without which, debt constraints will restrict recovery and economic health, widening the already stark inequality between nations. The world, according to the Global Recovery Observatory (2021), has missed the initial opportunity to invest in green recoveries for securing growth and setting a new course for prosperity. This is a wake-up call for correction of course—governments can still use green recovery spending to bring stronger economic growth and social benefits. For this purpose, the Global Recovery Observatory data offers to policymakers a rich new set of points and insights. According to UNDP administrator—"expanding access to such resources will help to increase the transparency, accountability, and effectiveness of the investments being made now and their impact on our sustainable future (UNEP 2021a).

An intergovernmental process is also in place to review long-term measures and efforts toward implementation of sustainable development. This review is carried out in the United Nations system through High-Level Political Forum (HLPF), which was established in 2013. The forum replaced the Commission on Sustainable Development. It meets annually under the auspices of the Economic and Social Council for 8 days, including a 3-day ministerial segment. It also meets every 4 years at the level of Heads of State and Government under the auspices of the General Assembly for 2 days. The HLPF is now the main United Nations platform on sustainable development that reviews the implementation of Agenda for Sustainable Development 2030 and its 17 Sustainable Development Goals (SDGs) at the global level (UNDESA 2021b).

11.3.1.3 Sustainable Urban Development: Local Approaches

The debate on urban sustainability over time has crystallized into two types of perspectives—one viewed by practitioners and the other by scholars in scientific literature. The first one or the practitioners' view considers the world as mechanistic, linear, and deterministic—allowing for steady-state and bounce-back interpretations of sustainability and resilience. This view commonly held among city officials, is based on their need to operationalize, and show concrete results in time

and space (Meerow and Stults 2016). The other view, however, sees urban world as dynamic, complex, adaptive, and multifaceted. The complexity in this view also implies a recognition of sustainability and resilience as emergent and contested concepts that are continually reframed. "In practice, this reframing sometimes follows the need to respond to operational realities, real or imagined" (Romero-Lankao et al. 2016).

Whatever the case in promoting sustainable development, the big challenge for cities is to check the environmental and social impacts of economic growth. Many cities in low-income countries with low quality of life have small ecological footprint because of low production and consumption. In contrast, cities in high-income countries have superior quality of life but often with large ecological footprints. The challenge therefore is to combine a high quality of life with a small ecological footprint or to delink high quality of life from excessive consumption and unsustainable production—the issue becomes particularly important in the likely scenario of continuous growth of cities and towns in the world. The way a city or an urban system is planned, managed, and governed has important implications for sustainability. Examples of innovation and good practice exist from cities in high- as well as middle- and low-income nations. Singapore, a high-income nation, is one of the most densely populated and high-tech cities but is also one of the most forward-looking for sustainability in Asia. The Sustainable Development Blueprint of the city-state is based on the Sustainable Development Goals 2030. Its development targets include improving energy efficiency by 35 percent, ensuring that 80 percent of its buildings are certified green, and having 80 percent of households within a 10-minute walk to a train station (The climate reality Project 2017). The city's drastic changes in transportation for improving sustainability among others include limiting car ownership, building effective public transit system, and promoting bike lanes. San Francisco in the USA has taken a lead by cutting its water consumption drastically to an average of about 49 gallons of water per day against the national average of 80–100 gallons per day. This, together with innovations in technologies to improve energy efficiency in building and transportation system as well as other advances in recycling and composting, has contributed to making the city a leader in sustainability (The Climate Reality Project 2017). Additional examples from cities are available in C40 Cities Good Practice Guide at https://www.c40cff.org/knowledge-library/c40-cities-good-practice-guide-city-climate-funds.

Examples of innovation and good practice are also available from cities of middle- and low-income nations. Many come from city governments in nations where they were given more power and resources under the national decentralization programs like Brazil. Porto Alegre in the country is one such example. The city has grown very rapidly in recent decades, from under half a million inhabitants in 1950 to over 3.5 million current residents. Despite this explosive growth (which is usually blamed for most problems in cities of Global South), the efforts of local government have succeeded in maintaining a high-quality living environment and innovative environmental policies. The average life expectancy and indicators of environmental quality in the city are comparable to other cities in Western Europe and North America (Satterthwaite 2010). Besides governments, many innovative and good practices for slum upgrading and poverty alleviation have also been initiated by local civil society groups or grassroots organizations in cooperation with local nongovernmental organizations (NGOs), which at a later stage also resulted in the involvement of local or city governments. To start with, these good practices were seen as one-off projects for upgrading or community development in "targeted" neighborhoods for water, sanitation, drainage, and garbage collection. Mostly undertaken in informal settlements or slums, often they also had programs to improve education and healthcare and alleviate poverty. There are several examples of these in UN-Habitat Best Practice database available at https://unhabitat.org/knowledge/best-practices.

These initiatives, which started as projects, ultimately led to two big achievements. The first one was the formation of organizations or federations of slum and shack dwellers at city and national levels—all of which have saving groups and interact with and support each other. They have formed their own umbrella organization as well—like Slum/Shack Dwellers International (SDI) at global level. SDI is a network of community-based organizations of the urban poor in 33 countries in Africa, Asia, and Latin America (SDI undated). Its Know Your City website combines hard data and rich stories from urban poor communities in cities across the Global South. SDI and its affiliated federations use their data and collective capacity to co-produce solutions for slum upgrading (SDI undated) or in providing/improving services. They also assist dwellers of informal settlements in their negotiation on getting land tenure. Hundreds of thousand households have received land possession/land tenure through their efforts. Likewise, several million low-income people were provided better access to services including water and sanitation due to support provided by these federations (Satterthwaite 2010). SDI has a financing mechanism that provides loans to communities for people-centered upgrading projects such as public toilets, water kiosks, incremental housing improvements, and securing tenure. The organization has also assisted many communities in building partnerships with local governments on service provision and settlement upgrading. An extremely important component of the SDI approach has been peer-to-peer learning through targeted exchanges between slum dwellers in different countries. These exchanges were focused on diverse issues including enumerations and mapping of slums, project implementation, and financial management.

The second important achievement of these initiatives was attracting the attention and recognition of local and national governments. For example, in Thailand its recognition by the central government led to the establishment of the Community Organizations Development Institute (CODI) that channels infrastructure subsidies and housing loans directly to saving groups formed by low-income inhabitants in informal settlements. By 2018, CODI had supported 1051 citywide slum upgrading projects (covering

105,000 families). It also helped community welfare funds in 5949 wards (out of a total of 7825 wards in the country) with 5.3 million members. The funds utilized amounted to US$ 420 million, of which 64 percent came from people's own contributions (Boonyabancha and Kerr 2018). Overall success and achievements of CODI show that substantial and large-scale changes can be made in the lives of the poor by supporting a community-driven process, a process that opens space for negotiation and collaboration between government and other partners on community development and improvement of housing and related environment (Boonyabancha and Kerr 2018). This example shows that small initiatives can be scaled up with competent governance and political will—particularly willingness of local governments to work with those most at risk in improving living conditions and attaining sustainability.

11.3.2 Vision of a Socially Resilient Inclusive City

There are three arenas that demand actions on the social front. However, these areas of action are not mutually exclusive and can be undertaken simultaneously. The first and foremost is to control the loss of development gains made over the years particularly in reducing poverty, hunger, inequality, and improvements in the distribution of social facilities. The second is to articulate vision for structural transformation in future development path to promote and develop healthy and happy cities for people and to attain the highest possible level of urban quality of life. In this connection, it is important to make use of the opportunities provided by the pandemic which made changes that were not possible during normal times. The pandemic made people not only to see things differently but also to act differently. This made them ready for changes that faced political barriers in pre-COVID-19 time as given in Chaps. 2, 4, and 6. Policymakers need to make use of these possibilities to facilitate social progress especially in achieving the SDGs on poverty and inequality (SDGs 1, 2, and 10). The third

area is to build stronger social protection measures based on efforts already initiated. For example, the gravity of socioeconomic problems posed by COVID-19 triggered an unparalleled social protection policy response. Governments at both national and city/local levels undertook measures for social protection as a frontline response to protect people's health, job/business, and incomes. They also initiated actions to promote and ensure social stability, as mentioned in Chap. 9 of this book. Countries and cities need to use this policy window that has been opened to build on their crisis response measures for upgrading and strengthening their social protection systems. They can gradually fill the protection gaps and make sure that everyone is protected against mega risks and ordinary life cycle threats.

11.3.2.1 Controlling Loss of Development Gains

The pandemic and its accompanying economic recession has resulted in endangering the development gains made over the years. It has reversed years of progress made toward ending poverty and contributed to unprecedented rise in extreme poverty for the first time since 1998, from 8.4 percent in 2019 to 9.5 percent in 2020, nullifying the progress made since 2016. The World Bank (Lakner et al. 2021) estimated that COVID-19 pushed 119–124 million people into poverty in 2020, a substantial increase from earlier estimates, and in 2021, the estimated number of COVID-19-induced poor was estimated to have risen further to between 143 and 163 million in 2021(Lakner et al. 2021). For Asian and the Pacific region alone, estimates show that 89 million people may have been pushed back into extreme poverty ($1.90 per day threshold), *erasing years of progress in poverty reduction* (UNESCAP 2021). Current projections (Fig. 11.2) depict that the global poverty rate is expected to be 7 percent in 2030 (around 600 million people)—which means that the target of eradicating poverty under SDG1 is likely to be missed (UNDESA 2021a).

The surge in poverty and other social problems was accompanied by a widening of inequality gaps in cities. Besides poor and slum dwellers

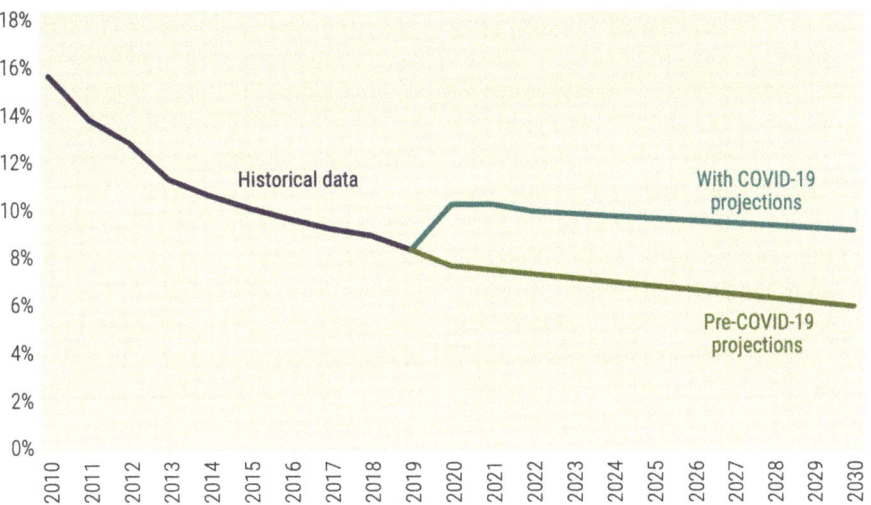

Fig. 11.2 Global poverty trends and projections 2010–2019 (pre-pandemic and with pandemic). (Source: UNDESA (2021d))

and informal worker, the most vulnerable groups which were disproportionately affected included aged, women, racial and ethnic minorities, refugees and migrants, children and youth, persons with disabilities, self-employed and low-skilled workers, and other groups due to their specific health and socioeconomic circumstances. One major indicator of vast inequities is in vaccination and vaccine distribution: As of June 2021, around 68 vaccines were administered for every 100 people in Europe and Northern America compared with fewer than 2 in sub-Saharan Africa. Health and care workers have been designated to be top priority in vaccination, but September 2021 data from 119 countries suggest that two in five health and care workers (HCW) were vaccinated on average. Among these, the differences across regions and economic groups were stark. For example, less than 1 in 10 such workers were fully vaccinated in Africa, while 4 in 5 were vaccinated in 22 mostly high-income countries (WHO 2021). Among variation in vaccination rates among cities, India is a good example where South Delhi is on top compared to other urban areas and even within the capital district. It is one of the richest and the least populated of the city's 11 districts. By mid-June 2021 in South Delhi, the first shot had been given to 43% of its 1.1 million residents, while some lagging districts had given first shot to only 3% of their population by that time (Arya 2021).

While talking of inequalities, it is important to note that although the mega risks particularly the pandemic had an impact on the cities and human settlements universally, it had extremely harsh and inequitable impacts in urban areas of the developing world for several reasons, firstly, because they are deficient in resources, and then they house over 1 billion people in urban informal settlements or slums, who live in extremely overcrowded dwellings with inadequate infrastructure and minimal services (Wilkinson 2020). Many of these are informal workers, the poor, and the vulnerable, often without savings and safety nets. They fell in poverty because their jobs and income were badly affected by the coronavirus crisis. Large income and employment losses occurred in several cities and countries, at least for temporary period (Hill and Narayan 2020). The result of phone surveys by the World Bank indicated that an average of 62% of households across 27 developing countries suffered income loss between April and July 2020 (World Bank 2020b).

Along with poverty and hunger, myriads of other social issues have emerged that are threatening cities and countries such as health, education, water, sanitation, etc., where previous development gains are in jeopardy. The year 2020 witnessed the first ever decline in Human Development Index since it was introduced in 1990, some three decades ago (Gavas and Pleeck 2021). In terms of health, which has a central place in SDG 3 "Ensure healthy lives and promote well-being for all at all ages," beyond vaccination and the disease itself, some 90 percent countries, during the pandemic, reported one or more disruptions to essential health services. For example, in 2020, 23 million children around the world missed out on lifesaving vaccines—3.7 million more than in 2019. Available public data from a few countries indicated shortening in life expectancy too (UNDESA 2021a). SDG 4, which envisages ensuring inclusive and equitable quality education and promoting lifelong learning opportunities for all, also suffered seriously. Children and youth learning and well-being were also gravely affected by school closures. The pandemic caused an additional 101 million children in primary and lower secondary school to fall below the minimum reading proficiency threshold during the first year—2020—wiping out two decades of gains in education (UNDESA 2021a). Simulations by the World Bank showed that in monetary terms the current generation of students may have lost an estimated 17 trillion US dollars in lifetime earnings at present value. This projected loss is equivalent to 14 percent of today's global GDP.

Handling these issues during the pandemic became more difficult in cities or their parts which had inadequate social and physical infrastructure, poor or inadequate housing, and basic services (Bhardwaj et al. 2020) or with limited public spaces (UN Habitat 2020). Cities with weak local governments and high levels of crime and violence (UNODC 2020) also faced serious problems in responding to COVID-19. In the wake of mounting vulnerability and increasing needs, both national and local governments faced a dilemma of how to spend the limited resources on citizen's well-being. Their response in the initial emergency phase concentrated on preventing disease transmission and caring for the affected. This was soon accompanied by taking steps for the mitigation of impact on vulnerable groups.

This continued in the recovery phase by providing stimulus packages. Simultaneously, the focus is on visioning and future planning.

11.3.2.2   Development Agenda on the Social Front

The most critical issue on the current development agenda on social front is tackling poverty and inequality. The urgency of poverty reduction and inequality demands multi-prong strategies and policy approaches at both national and local levels. At the national level, achieving rapid economic growth has been advocated to significantly reduce income poverty (ADB 2019; Iradian 2005). However, the problem with this approach is that it is difficult to ascertain when growth will or will not trickle down to benefit the poor. It is for the same reason that the importance of reduction of inequality in reaching the poverty targets has been emphasized by the United Nations (UNDESA 2021d). Several studies have also shown human development approach (Alkire and Santos 2010) and growth as well as increased delivery of social services in contributing substantially to poverty reduction (Barrientos 2010). These strategies and approaches need to be combined along with others focused on developing the lagging areas, increasing investments to generate jobs, promoting small- and medium-sized enterprises, and developing social protection interventions.

At the city or community level, along with national actions, local development approaches will be needed, whereby the poor people or community and local stakeholders authentically participate in poverty alleviation efforts. They will need to organize a group within them to lead and identify the issues to tackle. This will also demand that they should have confidence as well as raised level of awareness about their situation and strategy for implementation. These local actions have broadly taken two forms in the past—the first used participatory human development approach which led to community-based and community-owned initiatives and good practices for slum upgrading and poverty alleviation as discussed in the previous section in this chapter. They worked primarily toward alleviating deprivation but often included income improvement element through microfinancing. The second—local economic development (LED)—is another approach that gained widespread acceptance by national and local governments, as a locality-based response to economic challenges (Mbeba 2014). The main purpose of LED according to the World Bank is to reduce the level of poverty by building up the economic capacity of a local area through utilization of local resources to improve quality of life for all (Masuku et al. 2016).

A classic example of community-based participatory human development initiative is OPP (https://en.wikipedia.org/wiki/Orangi) or the Orangi Pilot Project which was initiated in Orangi Town of Pakistan—a cluster of 113 low-income settlements with a population of 1.5 million in 1980. After the successful implementation of its five basic programs of low-cost sanitation, housing, health, education, and credit for micro enterprise, in 1988 OPP was upgraded into three autonomous institutions, which are assisting in the improvement of basic services in other areas. The approach followed was to encourage and strengthen community initiatives with social and technical guidance and evolve partnerships with the government for development based on local resources. A similar success in such an initiative in Bangkok, Thailand, led to government ownership and its upgrading into a nationwide urban development program known as Baan Mankong Program (Mitchell 2021). Under this program, government funds are provided directly to the urban communities through Community Organizations Development Institute (CODI). The objective is to alleviate deprivation and upgrade housing and associated environment through provision of basic services. Members of the communities work closely with local governments, professionals, and universities in this endeavor. This has helped communities in gaining multiple opportunities to access land and housing, improve drainage systems, install communal septic tanks for sanitation, and get better connections for water and electricity supply as well as gray water treatment units. The program's outreach has now been extended to 405 cities, in 76 (out of 77) provinces of Thailand. The communities in these cities have already implemented more than a thousand housing projects to provide decent housing with security of tenure to over 100,000 urban poor families (CODI 2020).

Regarding local economic development (LED), the World Bank states, "The purpose of local economic development (LED) is to build up the economic capacity of a local area to improve its economic future and the quality of life for all. It is a process by which public, business, and non-governmental sector partners work collectively to create better conditions for economic growth and employment generation" (Swenson et al. 2006). After its initial success in many developed countries (Rogerson 1999), the potential for LED is acknowledged well in the developing countries especially South Africa (1998), as a critical sphere for policy development. How far LED has helped in poverty alleviation in cities and municipalities of South Africa has been debated (Meyer 2014; Mbeba 2014; Rogerson 1999). Nevertheless, its usefulness has led to the development of two primers or guides by international agencies like the World Bank (Swenson et al. 2006) and UN Habitat (2013). These two also provide some best practices in LED. A qualitative case study of ILO decent work project in Ghana (Azunu and Mensah 2019) also highlights important insights and empirical assessment related to impact of development partners' role in conceptualizing, implementing, and monitoring local economic development interventions in sub-Saharan Africa. The findings of the case study show that the project has led to the creation of jobs and enabled participants/stakeholders to improve their businesses as well as their economic situation. The communities involved also benefitted in meeting their health needs and adopting strategies to pull themselves out of poverty.

It is important to emphasize here that for alleviation of poverty in cities of developing nations, the national government support would be crucial—putting the responsibility on local governments alone with limited resources will be flawed. The task requires massive effort through skill development programs and support to the informal sector by providing microcredits, workspace to vendors and other backup policy, and legislative and institutional measures especially by supporting social solidarity economy (SSE). SSE, as defined by the United Nations, "consist of enterprises and organizations that have explicit economic and social and often environmental objectives; that involve varying degrees and forms of cooperative, associative and solidarity relations between workers, producers, and consumers; and practice workplace democracy and self-management" (UNDESA 2021c). Cooperatives are the core members of SSE. For example, waste pickers and renewable energy cooperatives are playing key roles in SSE in cities of developing countries. In Brazil, they are active at the municipal level, where they have played a critical role in the enactment of federal legislations which acknowledges and accepts the role of waste pickers in collecting and recycling materials. In another case, a cooperative of waste pickers in Pune, India, has been distributing gloves and masks to informal waste pickers (Kaza 2020).

To tackle inequalities in the short term, and protect the groups of vulnerable people, priority actions would be needed again, by both national and local governments. Additionally, many short-term response measures undertaken during the initial outbreak will need to be continued to give enough time for recovery in the wake of repeated waves of the virus and its variants. At the same time, shortsighted response based on quick fixes would need to be avoided as these could worsen and entrench the impacts of the pandemic. In the long term, eradicating the inequalities and development deficits in cities and bolstering the resilience of vulnerable groups to future shocks will need consistent and sustained efforts by all levels of governments. The first step in this direction needed is the comprehension of inequalities through data collection, mapping, and analysis. It could then lead to adoption of measures for bridging or reducing the appropriate deficits such as in housing and basic services, etc. It may also involve slum upgrading, giving equitable access to facilities, as well as provision of equal opportunities. Involvement of representatives of mar-ginalized communities in planning and design of measures in this connection could ensure incorporation of their needs in the solutions and encourage participation of stakeholders in implementation.

11.3.2.3   Social Sustainability and SDGs

The social problems/issues that are threatening social sustainability are being raised today universally across all the world cities, and not limited to one specific city or nation. Globally, this "universality" was recognized, endorsed, and promised in the Sustainable Development Agenda (SDA) 2030 under various Sustainable Development Goals (SDGs)—2 (hunger), 3 (health), 4 (education), 6 (clean water and sanitation), and 8 (decent work and economic growth). These interlinked issues call for social protection, provision of sustainable livelihoods and basic income, support to informal economies, as well as provision of basic services including health and education for all. Moreover, SDG 11 calls for making cities inclusive, safe, resilient, and sustainable. Reducing inequality in opportunities and outcomes features throughout the Sustainable Development Agenda 2030—with 60 targets across the SDGs, excluding those in SDG 10, that are directly or indirectly linked to reducing inequality. There is a need now to tackle these deep-rooted social problems in cities through structural transformations and development of solutions guided by the SDGs.

The first step required in this endeavor of making cities inclusive, and safe, is to ensure equitable access to COVID-19 vaccines and treatments. Other actions include significantly strengthening of social protection systems (World Bank 2013) and public services, including health systems, education, water, sanitation, and other basic services. The crucial need is to increase investments in the care economy—to ensure full coverage so that no one is left behind. Investment in health and social protection systems would create an important line of defense in restraining the impact of the pandemic and shocks resulting from climate change that could aggravate poverty and inequalities.

11.3.3 Vision of an Environmentally Sustainable City

Whether cities or nations, environmental sustainability is extremely important as survival and well-being of humans depend on natural environment. Hence, it becomes imperative to create and maintain the conditions under which humans and nature can exist in productive harmony to support present and future generations.

11.3.3.1 Environmental Sustainability and SDGs

Sustainable Development Agenda 2030 carries several SDGs on environmental sustainability, and to tackle mega risks, the ones relevant specifically to the two pandemics or mega risks are SDGs 3, 12, 13, and 15. Goal 3 on health and goal 15 on biodiversity, forest, and desertification are more relevant to the coronavirus or other health-related pandemics, while goal 13 addresses climate change specifically. Goal 12 "Ensure Sustainable Production and Consumption Patterns" is highly relevant to both and one of the most versatile in its design. It covers a wide range of topics—one especially relevant to mega risks is efficient use of natural resources. Among other topics in the goal are minimizing the loss and waste of food, ecological management of chemicals throughout their life cycle, solid waste management, sustainable public purchases, sustainable tourism, environmental education, as well as the elimination of fiscal incentives for fossil fuels to help in climate abatement. Governments have committed to implement SDGs and need to accelerate progress on them including those on the mega risks in building forward better.

Among the two mega risks, while progress is being achieved in the pandemic control particularly through expansion of vaccination, the progress on climate front is hazy. Among the important steps needed to handle mega risks, the fundamental steps needed are to (a) decouple development process from resource use, (b) seek nature-based solutions and development pathways, (c) promote low-carbon growth, and (d) promote circular models of production and consumption.

11.3.3.2 Decoupling Resource Use from Development Process

The most important structural transformation required to achieve environmental sustainability and tackle mega risks is to decouple the development process from resource use. The relentless pursuit of higher levels of wealth has led to unbridled expansion of agriculture and other commercial enterprises as well as cities and human settlements into forests and wilderness resulting in habitat loss. This is contributing to risky overlaps between human and animal species, allowing easy transfer of viruses from the latter to the former—the consequence is recurrence of zoonotic epidemics and pandemics. Another serious concern emanating from it is biodiversity loss and degradation. The cities and nations therefore need to make a forceful move toward putting an end to further loss of forests and wilderness and to restore some of those already lost. It is also important to share the earth more equitably with nonhuman species in the interest of the survival of the human species itself (UNDESA 2021d).

Similar relentless pursuit in the industrial sector is contributing to rampant expansion of industrial sector with excessive material and energy use or consumption on the one hand and growing pollution and climate change on the other. At the beginning of the Industrial Revolution in 1750, the concentration of carbon dioxide in the atmosphere was 280 ppm, which by 2005 increased to 380 ppm, an increase of over one-third (Climate Policy Watcher 2022). Bulk of this increase occurred since 1959, after the word energy use rose exponentially causing increased GHG emissions and climate change. To sum up, therefore, one can only blame human actions on not decoupling resource use from development process, not only for triggering the two mega risks but also for enhancing the risk landscape of the future.

11.3.3.3 Promoting Nature-Based Solutions

Whether it is planting urban forests or restoring them, building green infrastructure, protecting mangroves, or practicing agroecology in urban agriculture, deployment of practical and implementable nature-based solutions is important to protect and enhance nature. The advantage in their use is that while contributing to the achievement of the Sustainable Development Goals, they also create, sustain, or enhance decent employment and generate income. Therefore, they have become particularly important in the wake of cities recovering from the recession created by the pandemic along with greening of economy in building forward better. A considerable body of experience of putting nature-based solutions into practice from local to national and regional levels around the globe is now available (Wickenberg et al. 2021; WWF and ILO 2020; European Commission 2020). Some of the most relevant urban examples that had positive impacts on climate mitigation and have delivered health benefits as well have been outlined in the WWF and ILO publication (2020) "Nature Hires."

A good example of nature-based solution is afforestation. According to the US Forest Service, over 130 million acres of forests are located within or near cities (USFS 2020). The economic benefits of these forests in the country are apparent in energy-saving, runoff reductions, and public health. In monetary terms according to McPherson et al. (2005), every dollar invested in urban forests generates benefits ranging from US$1.37 to US$3.09. The resourcefulness of nature-based solutions has also been recognized in other parts of the world. In Europe, it has led to the development of Emscher Landscape Park (ELP) in Germany consisting of an urban forest—one of the largest ecosystem restoration projects in the continent. The park involving 20 German cities is also a part of a river revitalization program comprising more than 100 complementary green projects (European Commission 2020).

In their application to climate change, nature-based solutions have shown substantial contribution in mitigation and/or adaptation outcomes that are economically, socially, and ecologically beneficial to both local stakeholders and subnational and national authorities. It is for the same reason that the nature-based solutions are increasingly being used/promoted in addressing the climate emergency by public and private institutions

as well as the international community. This has helped in building resilience of cities and communities in several ways by (a) reducing exposure to inland and coastal flooding; (b) building adaptive capacity and empowering local communities in managing their local environment; and (c) increasing resilience to climatic shocks (Seddon et al. 2019; Lavorel et al. 2019). It is important to note that most nature-based solutions that are used to address climate change mitigation also promote climate adaptation simultaneously (WWF and ILO 2020). For example, Reducing Emissions from Deforestation and Forest Degradation (REDD+) programs which seek to reduce emissions from forest loss or deforestation and forest degradation also reduce local communities' vulnerabilities to climate change (UN REDD 2014) protecting them from extreme climate events such as heat waves and floods.

11.3.3.4   Climate Change and Green Recovery

Initially the COVID-19 and its related economic slowdown led to reduced pollution in cities during the lockdown. Air pollution was down by 60% globally (Bauwens et al. 2020); a 5.6 percent reduction also occurred in fossil fuel related carbon dioxide. However, the trend did not last long, and atmospheric concentrations of the major greenhouse gases, carbon dioxide, methane, and nitrous oxide continued to increase in 2020 and 2021. Regrettably, the momentum created by the short-term decrease in global greenhouse gas emission was not pursued when lockdowns were relaxed and restrictions on movement were eased, so emissions enhanced—situation is likely to deteriorate if business-as-usual scenario is not changed. For example, if G20 countries continue old path and simply recover and maintain the existing "brown" economy, this would lead to permanent environmental damage (Barbier 2020); as a "business-as-usual" approach, it could result in a 3 °C increase in global temperatures by the end of the century (Hepburn et al. 2020).

A recent IPCC (2022) Working Group II report has raised the alarm that the window for climate action is narrowing as climate-resilient development has already become challenging at current warming levels. It will become more limited if global warming exceeds 1.5 °C (2.7 °F), and in some regions, it will be impossible if it exceeds 2 °C (3.6 °F). This key finding highlights the urgency for climate action, focusing on equity and justice. It demands adequate funding, technology transfer, political commitment, and partnership for effective climate change adaptation and mitigation to reduce emissions (IPCC 2022). Another report by World Meteorological Organization (WMO 2021) has also raised serious concerns mentioning that the previous 7 years are on track to be the seven warmest on record. WMO has predicted with a 40 percent chance that average temperatures of the globe and its several subregions will temporarily rise above 1.5 °C in at least one of the next 5 years. In fact, according to the report, this has already started happening raising red flag for Paris Climate Agreement. Besides warming, a major indicator of climate change is the sea level rise. The global mean sea level rose between 1993 and 2002 at the rate of 2.1 mm per year, but between 2013 and 2021, the rate enhanced to 4.4 mm per year—an increase by a factor of 2 between the periods (Fig. 11.3). This was mostly due to the accelerated melting of ice mass from glaciers and ice sheets (WMO 2021).

However, an encouraging trend appeared in the new pledges made by the governments of the world in preparation for and at the 26th Conference of Parties (CoP) to the United Nations Framework Convention on Climate Change (UNFCCC) held in Glasgow in November 2021. The UNEP (2021b), "Emission Gap Report," however, shows that these efforts are still not strong enough to reach the targets set by the Paris Agreement. According to the report, the new national climate pledges together with other mitigation measures could put the world on track for a global temperature rise of 2.7 °C by the end of the century. This, however, is well above the goals of the Paris Agreement (goal 2 °C; aspirational goal 1. 5 °C) and will bring catastrophic changes in the earth's climate. The UNEP (2021b) report says that to keep global warming below 1.5 °C this century, the aspirational goal of the Paris Agreement, the world needs to halve annual greenhouse gas emissions in the next 8 years. However, even if implemented effectively, net-zero emission pledges could only limit warming to 2.2 °C, closer to the well below 2 °C goal of the Paris Agreement. Hence, the Secretary General of the United Nations while addressing the 26th CoP of UNFCCC rightly warned that, "We are still knocking on the door of climate catastrophe. It is time to go into emergency mode or our chance of reaching net-zero emissions by 2050 will itself be zero." A glimpse of the catastrophe mentioned by the Secretary General was witnessed in the USA in 2021. The country experienced 20 major climate-related disaster events (damage cost of each exceeding one billion dollar). The total damage cost of all these events was US$145 billion along with 688 lives. Therefore, as the nations and cities move from immediate crisis to "building forward better," it is imperative to promote a "new green deal" while following SDGs in their letter and spirit. Cities are particularly important in this endeavor because they are responsible for over 60% of global GHG emissions.

11.3.3.5 Low-Carbon/GHG Economy and Climate Change

While the alarm is being raised, opportunities for low-carbon economy are also being proposed in new green deal. For example, a recent IFC (2021) report "Ctrl-Alt-Del: A Green Reboot for Emerging Markets" states that if cities in 21 emerging markets analyzed by the organization give priority to climate-smart growth in their recovery plans, they can get up to $7 trillion in investments which could generate 144 million new jobs by 2030. The investment also has the potential to curb up to 1.5 billion tons of GHG emissions, benefitting people and ecosystems worldwide (IFC 2021). Cities, which have been hotspot of pandemics, will be the prime candidate to benefit from this investment. According to IFC (2021), there are three areas where city-focused decisions and investments will have significant environmental/climate-friendly and job creation impacts—energy-efficient building

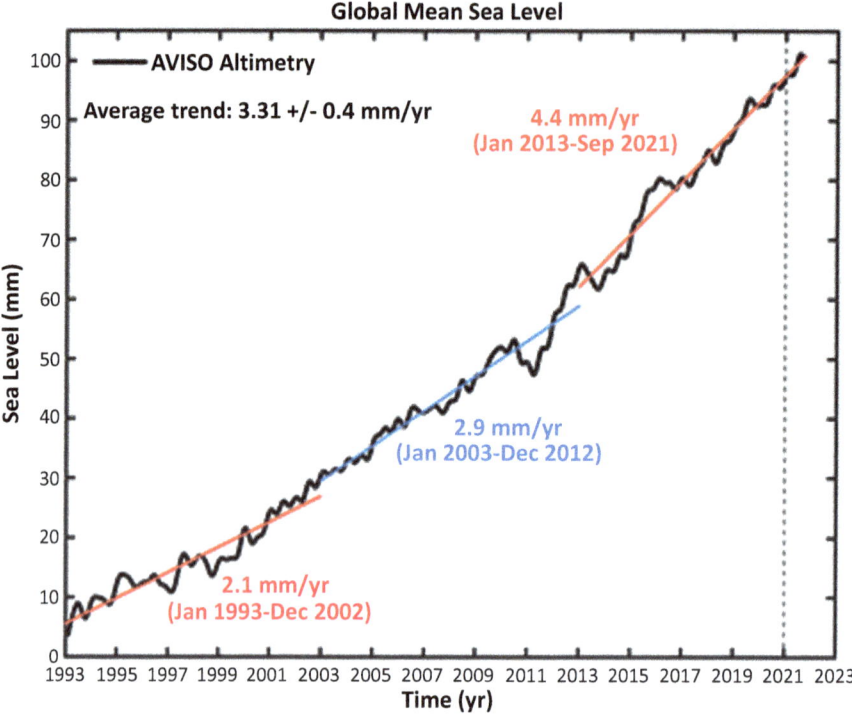

Fig. 11.3 Global mean sea level rise from January 1993 to September 2021. (Source: WMO (2021))

retrofits, municipal waste and water management, and urban transportation. Building retrofits present a $1.1 trillion investment opportunity with the potential to create nearly 25 million new jobs; low-carbon municipal waste and water present a $2 trillion investment opportunity and potential to create more than 23 million jobs, while green urban transport is a $2.7 trillion investment opportunity and has the potential to create more than 53 million jobs in emerging markets (IFC 2021).

IFC is assisting cities to structure green interventions by bringing together stakeholders and creating the enabling environment needed to promote private sector finance. IFC's tools like the new Advanced Practices for Environmental Excellence in Cities (APEX) app has been designed to help cities select precise projects suitable for green financing, develop strategies, and track performance (Da Silva and Robins 2021). Another report by the Coalition for Urban Transitions, *Seizing the Urban Opportunity*, also

highlights the potential of implementing a pack of currently available technologies and practices across six emerging countries including Brazil, China, India, Indonesia, Mexico, and South Africa. The package according to the report could collectively cut annual emissions from key urban sectors by 87 to 96% by 2050 beyond countries' initial commitments under the Paris Agreement. Furthermore, it could provide economic returns with a net present value of more than $12 trillion by 2050 by direct energy and material cost savings. In addition, it would potentially support 31 million new jobs in 2030 (Godfrey et al. 2021).

The available potential for promoting low-carbon/GHG economy, however, needs to be utilized by nations and cities effectively. Moreover, as pointed out by the IPCC (2022) Sixth Assessment report, equity and justice are needed in climate action. This demands assistance to developing countries and their cities through partnerships, funding, and technology transfer in the new green deal. Lack of resources and technology in developing countries and their cities is a major constraint that is undermining the pandemic green recovery and climate change mitigation and adaptation. Transparency Initiative data shows that the overall proportion of climate-focused projects within Official Development Assistance (ODA) went down and projects focusing on climate mitigation and/or adaptation decreased between 2019 and 2020 (UNDESA 2021e). Climate finance commitments has also become challenging for developed countries as well in the wake of economic pressures on many developed countries for meeting and promoting green recovery-related activities—like cutting down fossil fuel subsidies.

11.3.4 Vision of a Smart City

The biggest challenge for stakeholder in cities is how to manage urban dynamics while coping and adjusting simultaneously to current and future threats such as the COVID-19 pandemic and climate change with an intelligent transformation. Digital and ICT-based technologies have presented big opportunities for this transformation during both mega risks. They have shown enormous potential in dealing with both mega risks by engaging people, answering their needs, changing their behavior, and making cities move forward in new ways. The enhanced use of these technologies in cities during the pandemic has given a big boost to the smart city concept.

11.3.4.1 Smart City in Historical Perspective

Historically, the credit for promoting smart city concept goes to Professor Mitchell of MIT, who initiated the "smart city movement." His book *City of Bits*, published in 1995 by MIT press, discusses vast implications for planners and designers from post-Internet technology. Mitchell predicted today's smart roads, smart vehicles, and smart sneakers and forecasted "a matrix of digital telecommunication systems and reorganized circulation and transportation patterns." His biggest contribution was the establishment of MIT's *Smart Cities Group* in 2003 (Jackson

2019a). The smart city concept is still evolving. According to OECD, initially, "the 'smart city' concept referred to initiatives that use digital and ICT-based innovations to improve the efficiency of urban services and generate new economic opportunities in cities" (OECD 2020b). While the concept is still under debate (Borsekova et al. 2018), OECD currently defines it as "initiatives or approaches that effectively leverage digitalization to boost citizen well-being and deliver more efficient, sustainable and inclusive urban services and environments as part of a collaborative, multi-stakeholder process" (OECD 2020b). There is a vast literature available now on smart city tools and applications as well as best practices (C40 Cities 2022; Sanseverino et al. 2017; Nagargoje et al. 2016; Hassankhani et al. 2021; Barufi and Kourtit 2015). However, it is important to note that whereas these tools have the potential to change cities for the better, they also come with potentially hidden costs. Defining scalable, efficient, and realistically achievable smart city policies, therefore, requires a clear understanding of the strengths, weaknesses, opportunities, and threats facing smart cities (OECD 2020b). Among countries, apart from developed ones in the West, China has been active in building smart cities. Government figures show that work is ongoing on development of more than 500 smart cities across China, equipped with sensors, cameras, and other gadgets that can provide data on everything from traffic and pollution to public health and security (Mak 2020).

11.3.4.2 Smart Technologies: Innovations and Applications

The smart technology was in use even before the pandemic—mainly to enhance the potential of city-level governance systems (Kummitha 2020) and manage smart utilities by increasing efficiency in service delivery together with conservation of resources. For example, smart devices attached to rubbish bins electronically connected to a system alerted authorities when they got full. Likewise, smart devices reported outages and monitored energy usage and loss. Smart thermostats and customer portals were put in place so that service providers could give the residents of a smart city the best service for their money. Communication networks were used to support numerous uses in smart city—like metering, lighting control, and even pipe corrosion monitoring and leakage from water or sanitary pipes. Service provision outside homes also became the norm—for instance, at the airport or stadiums or civic centers—sensors guided people to vacant parking spots or charging stations for electric vehicles. With the utilities becoming smarter, cities also started benefitting in other ways, such as use of renewable energy (Tschirret 2019) and promotion of sustainable consumption and production. Global IT companies notably Siemens, IBM, Cisco, and Philips are luring governments by showing them the benefits of installing Internet-enabled telecoms and public lighting systems. They are still aggressively contesting for government development budgets worth trillions (Jackson 2019a).

Smart Cities Group of MIT, a group of younger professors, have shown startling urban design innovations in the twenty-first century, using electronically mediated interplays of light, energy, and data. At the Venice Biennale of Architecture in 2006, Carlo Ratti's SENSEable Cities Lab, for example, presented dynamic mapping of pedestrians carrying mobile phones in the streets of Rome, with crescendos of streaming data for a major soccer match and a Madonna concert at the stadium (Jackson 2019a). In addition, working with the City of Cambridge, Massachusetts, Ratti's team gathered weather and air quality data for different precincts, from sensors fixed to garbage trucks. Many more examples of smart applications for designing and studying the functions and environment in cities by various labs (Future Cities Laboratory, Sustainable Design Lab, SENSEable Cities Lab) and commercial ventures are available in the book *Data Cities* by Davina Jackson (2019b). These examples depict the big potential of new technology in planning and management of cities through simulations.

Referring to simulation, Microsoft made its Azure Digital Twins available in 2020 for the purpose. Digital Twins is like a blank canvas which can model any entity (Ravi 2021). For example, Energy Grid Ontology for Digital Twins is to help providers accelerate development of solutions for energy use cases—monitoring grid assets, outages and impact analysis, simulation, and predictive maintenance. It facilitates digital transformation and modernization of the energy grid and is used to enhance energy efficiency, contributing to climate change mitigation. Digital Twins can also aid urban authorities in quickly assessing the emergencies and their impacts on city operation through simulations well ahead of time. This may help local governments issue early warnings and operate better through measures that are likely to enhance the safety by taking critical decisions such as relaxing or imposing lockdowns. It is important to note that initially, Digital Twins were used to improve disaster management, which shows their importance in management of climate-related extreme events. However, during the pandemic, Digital Twins have been primarily used for better analysis and management of COVID-19 pandemic (Gade and Aithal 2021).

11.3.4.3 Mega Risks and Smart Technologies

Both mega risks of coronavirus pandemic and climate change have shown that beyond utilities, smart technology can be used to reduce the risks and vulnerability of populations and can be used as a planning tool to increase preparedness and recovery capacities in the face of crises. They have also indicated that smart technologies can be implemented in a much wider socioeconomic and political context in terms of their scope. Realizing this, cities and national governments and societies turned to digital technologies to respond to the pandemic in the short term, to resolve socioeconomic repercussions in the midterm, and to reinvent existing policies and tools in the long term (UNDESA 2020). Cities used technologies for public health response extensively during the pandemic. These technologies enabled them to identify infected individuals and predict diffusion dynamics, helped minimize human-to-human contact, and assisted in the enforcement and tracking of social distancing and quarantine rules (Sharifi and Khavarian-Garmsir 2020). Overall, they enabled in designing effective response and recovery measures.

The use of technology in public health response during the pandemic has been reviewed by Budd et al. (2020), while Kummitha (2020) and Vargo et al. (2020) have reviewed the use of smart technologies in health as well as other spheres in cities. Chapter 4 of this book also provides details on the use of technologies in response to mega risks. During the pandemic, digital technologies allowed cities to continue their functions through remote operations, at first during the lockdown and then to save people from getting infected. They were also used for communication between citizens, governments, and other stakeholders through digital portal and platforms, etc. (see Chaps. 3 and 4), especially for building public trust. In the case of climate change, ICT or information and communication technologies have been used for monitoring, preparedness, early warning, and management of extreme climate events and crises created by them. They have also been used in climate mitigation and adaptation. In view of their benefits, Sustainable Development Agenda 2030 in its SDGs has also stressed on leveraging technology to improve citizens' well-being and enhancing urban digital infrastructure.

Using technology in cities, especially in mega risks, has two major advantages among others—acquisition of data and information for use by planners and managers of the city and assistance in the provision of services. The data provided by the technology is usually in the form of interactive maps, satellite images (see Chap. 3), or such data type as location-based applications on mobile phones, tracking devices, sensor-based information, social media information, etc. (see Chap. 4)—all of which can help in increasing urban resilience. The details of services provided in various fields—remote work, education, healthcare, etc.—have been discussed in Chap. 4. These ventures besides increasing the resilience and crisis response/recovery capacity also enhanced the profitabil-

ity of investments. However, scholars have questioned whether the use of technology has improved citizen's quality of life or well-being as well—especially when the well-being is seen not only in material but also in terms of social health and happiness. Sassen and Kourtit (2021) have presented many arguments and counterarguments for the same. They gave a few practical examples from present cities while quoting that "in many cities the sky-rocketing expectations about smart city effects have not materialized." For example, Chicago for one refers to itself as a smart city, but the question is "does it excel in urban safety or income inequality?" Athens, another smart city- has it been 'able to solve its congestion and air pollution problems?' Beijing, yet another smart city which can control the movements of its citizens, but what happens 'when it comes to coping effectively with the smog during the summer?' Apparently, a significant gap exists between the myth of the smart city and its actual performance as a healthy or happy city. According to Sassen and Kourtit (2021), this is more striking, when one takes into consideration the range of new digital opportunities for the enhancement of urban livability and health conditions.

It is important to note that smart cities are neither a utopian dream nor a goal in themselves. They are realistic and meant to serve the citizens' well-being—particularly where involvement of various important stakeholders and responsible citizens is essential to promote sustainable development practices to enhance the quality of life (Meijer 2016; Sassen and Kourtit 2021). In addition to governments, in smart cities the knowledge sector and industry and civic organizations are considered as core players in sustainable governance and policy. Their role in building public-private partnerships for effective sharing of technologies, expertise, and tools can hardly be denied. Further, in envisioning a smart city, the citizen's role is of paramount importance as a key actor in a smart societal transformation toward a resilient society and economy in terms of awareness, behavior, practices, values, preferences, and civic engagement.

11.3.4.4   Challenges and Issues

The technology deployment in smart cities has certainly helped improve the well-being of their residents and maintain their functionality during crises. Nevertheless, it also generated many technical and non-technical challenges and issues. Among the technical constraints were digital divide, lack of access to the Internet and devices, etc. These were major issues primarily in cities of developing countries but were also present in developed nations. The main non-technical issues included privacy, security, confidentiality and trust, social inclusiveness, political bias, and dissemination of misinformation (see Chap. 4). Hence, a combination of technology-driven and human-driven approaches has been advocated to alley such concerns in the short term. Beyond the crisis, in the long term, cities, particularly in developing countries, will need to accelerate the implementation of innovative digital technologies to enhance the future resilience of the health, the economy, and the public services delivery (UNDESA 2020). It may be added here that some scholars are now looking "beyond smart cities." Tim Campbell (2012), for example, in his book *Beyond Smart Cities*, suggested that "clouds of trust"—networks of key actors in any community—provide the best platform for urban communities to continue to develop. He also suggested four essential factors for a learning city: gathering knowledge, fostering a milieu of trust, building institutional processes to support learning, and recording knowledge for future generations.

11.4   Vision of a Resilient City Design

The emergence of the two mega risks COVID-19 pandemic and climate change has brought increasing uncertainties to cities. While devastating cities, they have brought fresh ideas on the future form, shape, and functions of cities. Presenting new images of cities and visioning on city form, design, and architecture, they have covered issues such as the future of neighborhoods, public space, key infrastructure, mobility and accessibility, density, cyberspace, local businesses, adequate and affordable housing, etc. This section in the light of lessons from mega risks especially COVID-19, and previous pandemics and diseases, discusses the emerging design of cities, neighborhoods, and buildings. The idea is to identify ways and means that can assist cities in enhancing their resilience to future pandemics and other risks like climate change.

11.4.1   Health Risks and City Design

The physical layout of the city (including such features as density, streets design, public transport, public spaces, parks, and green areas and building design) in a manner that promotes health and well-being of its citizens has remained the focus of urban studies throughout ages.

11.4.1.1   Historical Perspective

Historically, urban planning and design theories, concepts, regulations, and practices emerged primarily in response to public health crises, including pandemics, pollution with rapid industrialization, congestion with urbanization, and loss of green space in cities (Banai 2020). For example, the radical improvements in the renaissance cities through removal of sordid and cramped living quarters and establishment of large public spaces were inspired by the fourteenth-century bubonic plague (Lubell 2020). Similarly, eighteenth-century yellow fever and nineteenth-century cholera and smallpox outbreaks brought innovations in broad boulevards, urban sewer systems, and the expansion of suburbs (Lubell 2020). The identification of contagious bacterium tubercle bacillus by physician Robert Koch in 1882 resulted in instigating the *sanatorium movement* in Europe and the USA. The movement was initiated to house, treat, and isolate patients. In the absence of drugs to control the disease, sanatoriums offered environmental treatment, where strict hygiene was practiced with ample exposure to sunlight, air, and nature. Likewise, the epidemic breakouts of tuberculosis, typhoid, polio, and Spanish flu incidences in the twentieth century prompted enhanced urban planning efforts and stimulated architectural modernism.

11.4.1.2 Sanitary Movement

Risks such as disease outbreaks have also transformed built environment of cities and initiated sanitary movement. For example, following cholera outbreaks in the nineteenth century brought renovation of Paris by undertaking extensive public works directed by the emperor Napoleon III and executed by Haussmann (Cuttle 2018), between 1853 and 1870. It included the demolition of medieval neighborhoods that were considered overcrowded and unhealthy. Simultaneously, it led to the building of wide avenues; new parks and squares; and the construction of new sewers, fountains, and aqueducts. Like Paris, London also saw changes in physical infrastructure especially development of its sewer system.

11.4.1.3 Garden City Movement

The emerging worst conditions in its overcrowded blackened cities in Britain after the industrial revolution led to human ingenuity and creation of visionary thinking. They led a pioneer British town planner, Ebenezer Howard, to come forward with his "Garden City" concept, which became famous as a garden city movement. Howard presented this concept—development of smokeless green cities—in his book, *Garden Cities of To-morrow*, first published in 1898. He designed and created the first garden cities, at Letchworth and Welwyn in Hertfordshire, England, by 1920. The garden city was to address the health, safety, and general welfare of industrial city residents exposed to all the negative consequences of industrialization, crowding, congestion, pollution, sprawl, and lack of parks and open green space. The small, low-density garden city land use in balance with the natural environment was a more attractive model of urbanism than the compact, high-density metropolitan cities like New York or London in Howard's time. Even today, in the face of pandemic or epidemic, it appears to be an attractive more resilient urban model according to the health guidelines, which call for social distancing and low population density of the public realm

(Banai 2020). However, in the wake of current speed of urban growth and limited space, can we promote garden cities everywhere today? It is an open question. Further, Howard's ideas of locating new towns in the countryside and zoning cities into different sections by type of land uses have also given "dead city syndrome" to some scholars. For example, American writer Jane Jacobs (1961) in her book, *The Death and Life of Great American Cities*, criticized Howard's ideas. She appreciated the vibrancy of 24-hour street life in New York, compared to the dead downtowns and precincts of less dense cities. Jacob had no formal training as a planner, but her book introduced groundbreaking ideas about how cities function, evolve, and fail. The impact of her activism and writing has led to a "planning blueprint" for generations of architects, planners, politicians, and activists to practice (CLC undated).

11.4.1.4 Towers in the Park

As opposed to Howard, the French modernist architect-urbanist Le Corbusier gave a different vision of cities in his tower-in-the-park skyscraper concept for a "Contemporary City." He considered this a better solution to the rapid urbanization of population and congestion spurred by industrialization. His idea accommodated 100 times more population in the contemporary city as compared to the garden city proposed by Howard for 32,000 people (LeGates and Stout 1998). However, Plan Voisin, a later version of the Contemporary City proposed for central Paris in 1925 (consisting of 18 identical skyscrapers, which were spread out evenly over an open plan of roads and parks), was rejected by the City of Paris, because it was considered too radical. However, Corbusier's concept of a dense, modern urbanism with the iconic skyscraper endures globally in cities, famously in the iconic New York City (Lang 1994).

11.4.1.5 Architectural Modernism

The architectural modernism in urban design during the twentieth century was triggered primarily by infectious diseases of the time. It promoted new and innovative technologies and materials for construction, particularly in the use of glass, steel, and reinforced concrete. It also promoted principles that affirmed purity of form, strict geometries, and rejected ornamentation. The architectural features such as terraces, balconies, flat roofs, and precisely designed interiors and furniture served the pursuit of a therapeutic lifestyle, even as they expressed the design theories of functionality and rationality linked to modernism. The clean, smooth surfaces of this era offered an anesthetic to disease and trauma. Architects of the period from Adolf Loos to Alvar Aalto designed these curative environments as cleansed physically and symbolically from both disease and pollution (Chang 2020). Famous Swiss born French architect Le Corbusier of the time envisioned a spartan city where every home is whitewashed and "there are no more dirty, dark corners. Everything is shown as it is. In term of *inner* cleanness." His ultramodernist Villa Savoye had this aesthetic. Painted clinical white, its living quarters were suspended on columns above the germ-ridden earth below (Chang 2020). Terraces, balconies, and flat roofs were common elements in modern architecture, even in climates less suited to these features. Beyond their aesthetic appeal, these features embodied modernist preoccupations with the healing effects of light, air, and nature (Chang 2020). Le Corbusier in his book *City of Tomorrow* stated, "Hygiene and moral health depend on the lay-out of cities. Without hygiene and moral health, the social cell becomes atrophied." The COVID-19 pandemic has made the truth of this statement agonizingly clear.

11.4.1.6   Polycentric City

There are many theories on how to design the city better. However, the gap between disciplinary dreams in theory and practice or real outcomes translated as urban design product has been growing over the last few decades, and it is also being debated and led to the emergence of new ideas. One such idea entitled "15-minute city" was floated in 2016 by Carlos Moreno, a professor at Paris-Sorbonne University (Moreno et al. 2021). Moreno's model, however, got more pop-

ularity during the pandemic especially after its advocacy by mayor Hidalgo of Paris and her decision to implement it in Paris. The model is based on the idea of a polycentric city with self-contained sustainable neighborhoods. The idea of polycentric cities has also been supported by others. For example, according to Zeljic (2020), the era of a traditional metropolis of 10 or 15 million or more with a central business core surrounded by rings of decreasing density may no longer be suitable for sustainable urbanization anymore. This needs to be replaced by a polycentric model, with several self-contained key districts—each functioning like a district or a "city within a city." Zeljic takes the concept to an extreme when he talks of the district as a city within a city and particularly when he says, "each district offering something different."

11.4.1.7   15-Minute City: An Emerging Concept

As opposed to Zeljic's polycentric city, Moreno's "15-minute city" is composed of lived-in, people-friendly, "complete," and connected neighborhoods. Each of which must cater for six basic social functions (Fig. 11.3): living, working, supplying, caring, learning, and enjoying (Reimer 2020). The neighborhood is so designed that each person living there can meet most, if not all, of his/her needs within a short walk or bike ride from their home. It means decentralizing city life and services and reconnecting people with their local areas. According to C40 network of cities as cities work toward COVID-19 recovery, the 15-minute city is more relevant than ever as an organizing principle for urban development (C40 Cities 2020). The concept, while helping cities to revive urban life with safety during the pandemic, also offers opportunities for developing sustainable cities in the long term—the vision of which may be developed by local governments and shared with their constituents. According to the city network C40 Cities (2020), more specifically, the concept "will help to reduce unnecessary travel across cities, provide more public space, inject life into local high streets, strengthen a sense of community, promote health and well-being, boost resilience to health and climate shocks, and improve cities' sustainability and livability" (C40 Cities 2020).

Physical and digital connectivity is at the heart of any 15-minute city strategy (C40 Cities 2020) and is considered critical to counter sprawling, car-dependent urban areas. It is also a complementary approach to transit-oriented urban development, a development that promotes denser mixed land use around public transport services, allowing a large-scale shift away from private vehicles. Fast, frequent, and reliable public transport connecting neighborhoods with each other and with centers of work are of central importance in the concept to enable car-free access to work and socialization in 15-minute city.

From the humanistic angle, Moreno envisions that if people can meet their needs near their homes, they will be happier and more willing to protect their communities and develop better relationships. This is a major shift from the land use planning of the past 100 years which separated houses from industry, entertainment, business, and retail centers. This kind of space categorization or land use zoning made sense at the start of the industrial revolution and in the years following it. Pollution from factories at that time posed serious health risk which made Howards suggest this kind of zoning. Later, suburbs dominated by cars further resulted in the split, leading to years of less-than-flattering big-box strip malls, huge industrial areas, and isolated offices (Reimer 2020).

Apart from the need of the pandemic, desire for promoting people-oriented or people-centered cities is resulting in a surge of interest in the "15-minute city"—with a popular vision of urban living that has already taken many names and shapes around the world (C40 Cities 2020). A leading example of this is Paris, where Professor Moreno and Mayor Hidalgo of Paris are working together. As per their insight, each of Paris' arrondissements (administrative district) will have all amenities accessible by foot or bike (Fig. 11.4). The implementation of the 15-minute city initiative in Paris has not only considered ease of travel, walkability, and public services but an all-encompassing approach. It has brought a greener aspect and integrated workplaces, cultural activities, and the more ephemeral nature of social connections. The aim according to commissioner of 15-minute city in Paris is to create "a city of proximities"—not only between structures but also people. The original ideas given by Moreno in terms of features of the 15-minute city were later consolidated into six main functions in a modified framework with four pillars in a paper published in 2021 (Table 11.1) by Moreno and others (Moreno et al. 2021).

The program's success in Paris is indicated by the Mayor Hidalgo's triumph in winning a second term in office (Reid 2020). The popularity of the concept is also indicated by its expansion in other cities and the attention and interest of international organizations it has attracted, which among others include the C40 Cities Network, the World Health Organization (WHO), the UN-Habitat, and the OECD (Moreno et al. 2021).

11.4.1.8   Diffusion of the Concept

With Paris leading, other cities promoting the original or similar initiatives include Bogota (Program Barrios Vitales), Portland (Complete Neighborhoods) and Melbourne (20 Minute Neighborhoods). Melbourne's long term plan for 2017–2050 is being guided by the principle of "20-minute neighborhoods", instead of 15 minute. The principle however is the same—all about "living locally" and creating inclusive, vibrant, and healthy communities. In Melbourne, this is defined as giving people the ability to meet most of their daily needs (Fig. 11.5) within a 20-minute return walk from home (800 m walkable catchment), with safe cycling and transport options. Hallmarks of a 20-minute neighborhood as given by the Victoria State Government (2022) include the following

- Be safe, accessible, and well connected for pedestrians and cyclists to optimize active transport.
- Offer high-quality public realm and open spaces.
- Provide services and destinations that support local living.

Fig. 11.4 Features of 15-minute neighborhood in Paris, France. (From the top center features include Education, Work, Knowledge Exchange, Shopping, Recreation, Community Engagement, Health, Public Transport, Exercise, and Nutrition. Source: Reimer (2020))

- Facilitate access to quality public transport that connects people to jobs and higher-order services.
- Deliver housing/population at densities that make local services and transport viable.
- Facilitate thriving local economies.

Many other cities (Barcelona, Edenborough, Madrid, and Milan) in Europe; (Houston, Ottawa, and Seattle) in North America; and (Guangzhou, Shanghai, and Chengdu) in China are following the same approach (C40 Cities 2020). C40 Cities guides posted on website of the network (C40 Cities 2020) explain how cities can improve services in all neighborhoods, starting with the most underserved. It gives updates on planning and zoning rules that are critical to meeting these objectives in both short (during the pandemic) and longer term. The guides also explain how equity and inclusivity can be made central in the 15-minute city and how people-centered streets and mobility can be developed. However, many

Table 11.1 Consolidated functions and main pillars/dimensions of 15-minute city

Original features (Paris)	Consolidated urban functions	Pillars/dimensions
Community engagement	Living	Density
Work	Working	Proximity
Shopping	Commerce	Diversity
Health	Healthcare	Digitalization
Education	Education	
Recreation	Entertainment	
Exercise		
Nutrition		
Knowledge exchange		
Public transport		

Source: Derived from Reimer (2020) and Moreno et al. (2021)

ideas and principles underpinning the 15-minute city are not new and can be traced back in the idea of American urban planner Perry (1929), who formulated the idea of the livable "neighborhood unit." Moreover, several cities already have areas that align with the principles of 15-minute city, even if by accident rather than by design. However, credit must be given to Moreno and Hidalgo who integrated old and new ideas into the concept with vigor that has gained momentum in the pandemic. "More cities are now embracing this model to support a deeper, stronger recovery from COVID-19 and to help foster the more local, healthy, and sustainable way of life that many of their citizens sought" (C40 Cities 2020).

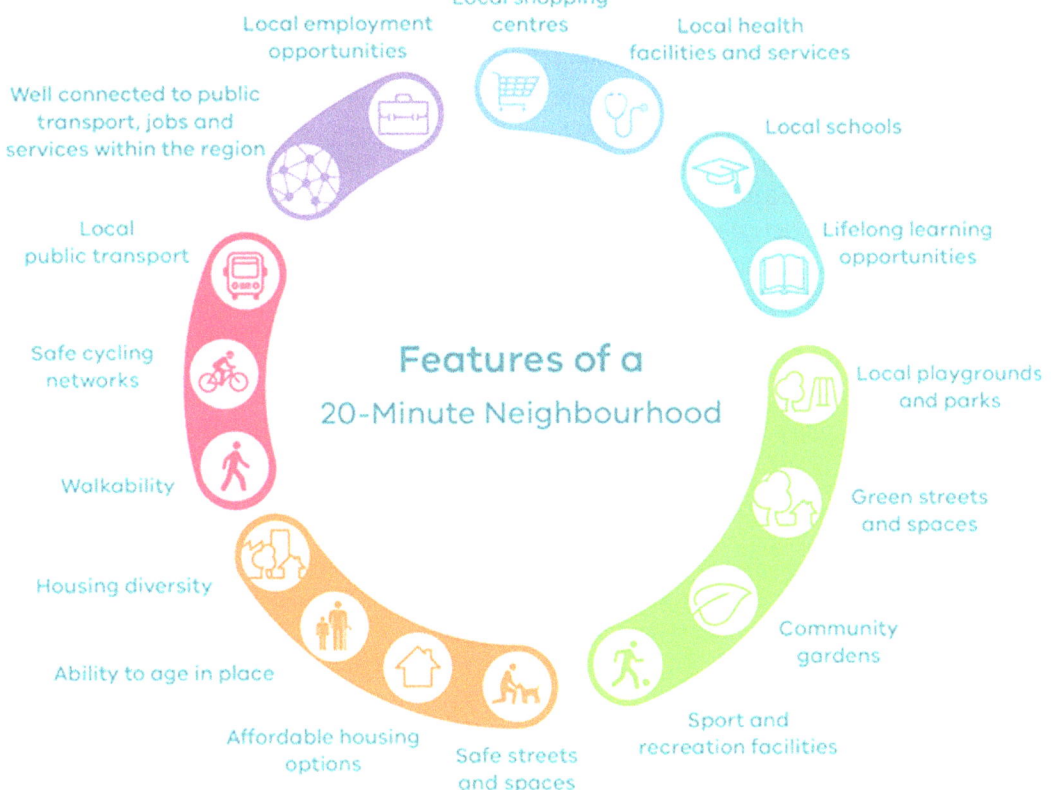

Fig. 11.5 Melbourne: features of 20-minute planned neighborhood. (Source: Victoria State Government, Australia (2022))

It will not be out of place to mention here that while the idea of 15-minute city has been applauded a lot, some concerns have also been expressed on it. One of it is on the potential quality of public services provided. For example, marginalized neighborhoods may end up getting "terrible" doctors and schools—which could bring about further discrimination and inequality and "territorial stigmatization" (Yeung 2021). Others consider it as an opportunity for a paradigm shift to change the disadvantaged Parisian neighborhoods that have long faced economic and social blocks to promote their progress.

Concern has also been expressed by a UK-based think tank that "the end of big cities" could also reduce creativity (Yeung 2021). Cities allow people to mix, to be together, and to share ideas that usually happen in the city center. The question raised is whether it will be possible to recreate the ingenuity and innovation if they no longer exist (Yeung 2021). Carlos Moreno believes that creativity and well-being will flourish in the 15-minute city because it is being developed for its inhabitants and will be led by them. He points to the role of participatory budgeting, which since 2014 has allowed residents to vote on 5% of the municipal spending in Paris—totaling half a billion euros.

With the growing popularity of the concept, there is a need to test and showcase the concept and its elements in cities of the Global South and those that may be financially constrained to undertake the extensive urban regeneration exercise that this kind of planning model demands (Moreno et al. 2021). It is rather interesting to note that many old cities of the Global South evolved in the form of neighborhoods. For example, the city in which the author grew up, the old walled city (Peshawar) had several mohalla (word for neighborhood in local language). These mohallas or neighborhoods had some basic services and amenities like schools, shops, and occasionally a Hakim (Ayurvedic healer) as well as a midwife to deliver child, etc. However, under the global influence in recent past, new developments in these cities followed the global trends of land use segregation and abandoned traditional urban development. The old neighborhoods in

these cities still maintain very close social relationships. Revival of the old model through 15-minute concept could be rewarding in such cities especially when combined with the LED or local economic growth and participatory social development approaches for slum upgrading mentioned in earlier part of this chapter.

11.4.2 Climate Change and Urban Design

Cities layout and urban design can also help both in climate change mitigation and adaptation. One scientific study (Chandler 2018) noted that street and public space layout, building design, as well as construction materials used all contribute to generate heat and cause heat island effects in cities. The study carried out by MIT and the National Center for Health Research over 50 cities observed that cities like Chicago and New York built with precise grid-like structure were far hotter than cities with a more chaotic arrangement such as Boston and London. The heat island effect is caused primarily by urban building materials, such as concrete and asphalt, that absorb heat during the day and radiate it back at night, much more than areas covered with porous material and vegetation. The effect can be serious, adding as much as 10 degrees Fahrenheit to nighttime temperatures in places like Phoenix, Arizona—where it can significantly increase both health problems and energy use during hot weather. The study also found that in the state of Florida alone, urban heat island effects cause an estimated $400 million in excess costs for air conditioning. Consideration of climate change mitigation and adaptation in city layout and building designs has therefore assumed great importance both for the sake of protecting human health and saving enormous costs.

Currently, commonly pursued efforts at ameliorating climate change, whether in cities or nations, have followed the nonintegrated mitigation and adaptation approach. Even majority of mitigation funds have been directed primarily to energy projects that produced no secondary benefits for local populations—in the form of heat management or enhanced protection from extreme climate-related damages to private property and public infrastructure (Raven et al. 2018). Forward-looking cities, however, have started to utilize the available potential in the built environment and natural systems including design of neighborhoods and buildings, as well as green infrastructure to mitigate climate change on the one hand and to "future proof" their built environment from climate extremes on the other—thus, integrating mitigation and adaptation strategies (Raven et al. 2018). For example, investing in pedestrian and cycling tracks, especially when integrated with parks and other green spaces in cities, has reduced carbon emissions, enhanced carbon sequestration, and also had cooling effect on cities through shading and evapotranspiration. Figure 11.6 shows the strategies that are being used by urban planners and designers to facilitate integrated mitigation and adaptation in cities. These include reducing waste heat-resistant construction materials and reflective surface coatings as well as increasing vegetative cover.

Several studies on climate and urban design show that densely occupied cities are better from climate standpoint. For example, a study of urban spatial structure and the occurrence of heat wave days across more than 50 large US cities (Stone et al. 2010) indicated that the annual frequency of extreme heat events has increased slowly in compact cities compared to the sprawling cities. Similarly, the US Department of Transportation, in a 2010 report to Congress, pointed out that land use strategies relying on compact, walkable, transit-oriented development could reduce US GHG emissions by 28–84 million metric tons carbon dioxide equivalent by the year 2030. Benefits would grow over time to possibly double that amount annually in 2050 (Raven et al. 2018). Compact cities are also more efficient in their use of energy (and therefore generate less waste heat) (Resch et al. 2016). This is especially true for cities with mass transit systems. Moreover, policies for increasing urban population density that also manage land use and transport networks have proved more effective in reducing urban GHG emissions (Dulal et al. 2011).

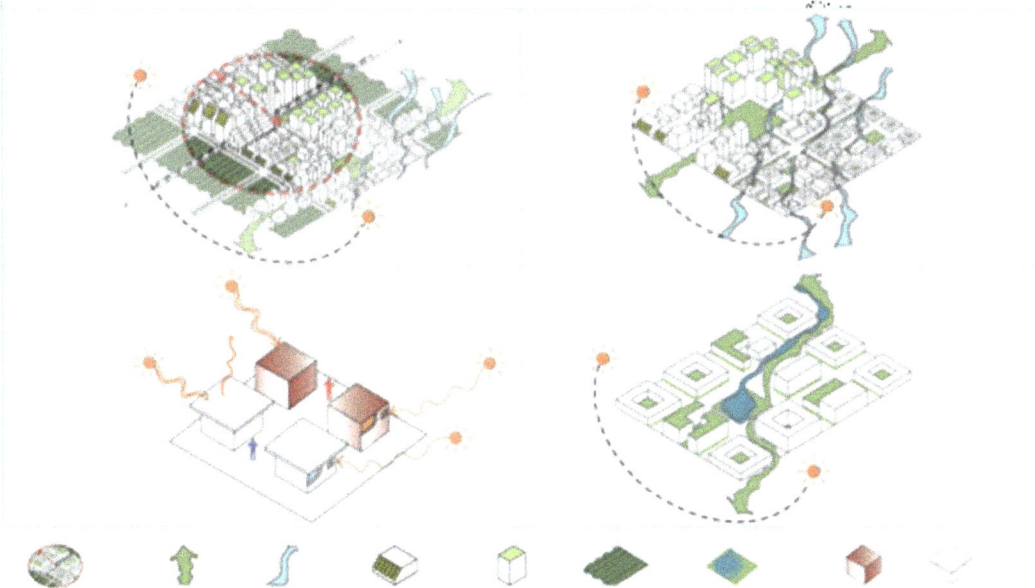

Fig. 11.6 Strategies used by urban planners and designers to integrate mitigation and adaptation. (Note: Top left, improvement in efficiency of urban system; top right, improvement in form and layout of neighborhoods. Bottom left, heat-resistant construction material; bottom right, increase in vegetative cover. Source: Raven (2011), Raven et al. (2018))

These ideas and principles in terms of land use, density, and designing neighborhood and buildings in various studies related to climate and urban design support the viewpoints presented in 15-minute city or similar models. As discussed earlier, in the pursuit of a 15-minute city model, the adoption of mixed-use neighborhoods predominates. It is adopted to ensure that an optimal density and proximity of essential amenities are achieved while also providing for development of walkable streets and bicycle lanes. This shows that there is no need to reinvent the wheel. Rather, it will be more important to help cities incorporate or adapt design techniques that would reduce temperatures within the existing city/neighborhood development models, techniques such as heat-resistant construction materials, reflective surface coatings, cool and green roofs, solar panels, cool pavements, and increasing vegetative cover. These and other similar measures are necessary to support integrated mitigation and adaptation approach to climate change in future city, neighborhood, and building designs.

Many local governments have already embarked on that road like New York City (NYC). Cool Neighborhoods program of NYC is a strategy developed by the Mayor's Office of Resiliency to coordinate multiple extreme heat mitigation and adaptation projects. The Mayor's Office of Resiliency uses the results of an intensive heat vulnerability mapping collaboration between the NYC Department of Health and Mental Hygiene and Columbia University to direct cool design interventions and tailor heat resilience social programs. In terms of building infrastructure, the city uses its cool roof programs to heat-vulnerable neighborhoods. As of April 2019, NYC has installed more than ten million square feet of reflective, cool roofs. The city estimates that cool roofs can lower building AC costs by 10–30 percent and reduce indoor air temperatures by up to 30 percent during the summer (ULI 2019). The city also leverages its green infrastructure programs, enacting policy to require green roofs in buildings. Moreover, the city has targeted tree planting along streets, in parks, and in forests; many of these trees are being planted in areas which have comparatively little vegeta-

tion cover and the highest levels of heat vulnerability (ULI 2019).

11.4.3 Mega Risks: Public Spaces and Mobility Design

During the pandemic, while city and neighborhood designs were being debated and considered for application, COVID-19 had already started reshaping the built environment in cities—causing transformation in the use of public spaces, parks and green areas, transport infrastructure including street design and use, and triggered several changes in building design, which also have implications for climate change.

11.4.3.1 Public Spaces

Public spaces in cities like streets, sidewalks, parking spaces, town squares, and stadiums as well as parks and green areas have always constituted important part of urban design. Their critical importance is apparent during the mega risks, when their repurposing contributed enormously to cities' socioeconomic resilience as well as the well-being of communities and individual citizens. For example, during the pandemic, scores of cities around the world revised traffic patterns and made more room for bikes, pedestrians, and selected commercial activities including food sale and dining. In cities of the Global South such as Bangkok, Kuala Lumpur, Jakarta, Manila, Delhi, and Karachi let alone in a pandemic, even in normal times, the role of accessible public space is vital for tens of thousands of informal vendors. Shutting down these areas, even for short duration, had devastating effects on urban informal vendors who supported themselves and their families on a day-to-day basis. Therefore, many vendors had no option but to adapt quickly to impose new restrictions. For example, in Kisumu, Kenya, the closure of popular Kibuye Market led retailers to build a makeshift alternative to keep their business going (UN Habitat 2021). A novel example of repurposing public spaces is provided by the city of Vilnius, the Lithuanian Capital in Europe. The city transformed itself into an *open-air café*, where hundreds of café and restaurants were set up in its plazas, squares, and streets—serving customers from a safe distance. Similar restaurants served food in other European and American cities. Moreover, a drive-in movie theater started operating at the Vilnius city's idle airport (Serhan 2020). Parks and beaches also featured social distancing circles to prevent overcrowding.

Talking of parks and beaches, in fact, access to green spaces has also proved important for city residents during the pandemic. The benefits from these, however, are only possible if they are distributed equally across neighborhoods. Their inequitable distribution in certain areas like slums and informal settlements in the Global South leaves many households from lower socioeconomic backgrounds worse off and intensifies their existing patterns of exclusion. A recent study of 610 cities across 95 countries found that only 47 percent of the population studied lived within 400 meters walking distance of open public spaces (UN Habitat 2021). The city of Vancouver in Canada has, therefore, developed a citywide master plan for parks and recreational areas to rectify such inequalities. Whether pandemic or no pandemic, it is important to note that parks, plazas, promenades, community centers, or streets will maintain their importance for meeting or congregating of citizens. While designing to keep them safe, therefore, it would be necessary to install early detection and warning systems for future emergencies.

As the pandemic prolonged, the urban designs have gone beyond the ad hoc measures of bollards-and-traffic-cones approach for widening sidewalks and carving out space for pedestrians. They have adopted now modular concepts for outdoor retail, public cleansing stations, and pop-up services like haircuts and mobile libraries. Some are no larger than a parking space; others can be scaled up (Wittenberg 2020). For example, City of Baltimore in the USA offered a set of solutions to public space challenges during the pandemic. The city released a document "Design for Distancing Ideas Guidebook"—prepared jointly by the Johns Hopkins Bloomberg School of Public Health, the Baltimore Development Corporation, and the city's nonprofit Neighborhood Design Center. It provides ten plans for creating temporary, low-cost spaces that permit physically distant social interaction in urban public spaces such as streets, alleys, vacant land, and parking lots. The *selected concepts were drawn from a pool of 162 submissions* from architecture and design firms. Although the plans were conceived around the needs of Baltimore's neighborhoods, according to Wittenberg (2020), they can be adapted to cities anywhere.

Adaptability of building designs and plans is extremely important in the face of mega risks—whether the pandemic or climate emergency. Coronavirus and previous pandemics and epidemics have demonstrated that if buildings and spaces within them are purposefully designed, they greatly assist in the prevention, containment, and treatment of infectious diseases. During the COVID-19 crisis, many places of work like manufacturing facilities and offices, as well as hospitals and care homes, showed an increased rate of infection among their occupants or users because of inadequate layouts or ineffective ventilation systems. Additionally, the outbreaks of the pandemic waves have been associated often with the physical form of buildings. There is a need, therefore, to reassess building design to avoid overcrowding. UN-Habitat (2021) has provided some guidelines on (a) designing buildings conducive to infection control taking into consideration social distancing and (b) precautions that need to be considered in their use.

Provision of good ventilation systems is also essential for ensuring ample, clean airflow to decrease the presence of contagion. Advanced ventilation was the most important tool for making buildings, especially hospitals, healthy during the pandemic. Numerous technology-based mechanisms (Lubell 2020) were used such as the following:

- Negative air pressure (which keeps pathogens from spreading to other parts of a building or a hospital).
- Displacement ventilation (in which cooler air enters from below and lifts contaminants).
- Clean air ventilation (which brings in fresh air, rather than recirculating existing air).
- Various filtration and humidity systems.

One essential aspect that needs special consideration in building design and associated space use is *automation*—this is needed to mitigate impact of the contagion. COVID-19 has already enhanced development of all types of touchless technology—automatic doors, voice-activated elevators, cellphone-controlled hotel room entry, hands-free light switches and temperature controls, automated luggage bag tags, and advanced airport check-in and security (Giacobbe 2020). Further, there is a need to introduce far more opportunities for handwashing and sanitizing as well as RIFD technology to make purchases, in addition to the metal detectors that have become commonplace in public venues. They may also install temperature screening or even some form of UV disinfecting devices if needed (Giacobbe 2020).

The need for quick creation of emergency facilities like hospitals, rescue centers or quarantine services, testing sites, and temporary lodging has also emerged during the pandemic. Among the response in the construction field, a good example was *modular construction*—based on developing prefabricated standardized components in a factory and then taking them on-site for quick assembly. This fast, flexible, and less expensive alternative to traditional building construction (see Chap. 7 on Health) has provided some good examples. In Wuhan, China, for instance, during the early days of the pandemic, construction of two large hospitals in 2 weeks was a great accomplishment—in normal times, it takes 2 years to build a hospital of the type. The flexible components of a modular building like movable walls have an added advantage of customization. Moreover, they can help buildings adapt to new needs—like shifting to a new place or enlarging health facility spaces for treatment

and quarantine or creating new spaces to accommodate ICU beds. Tents and containers have also been used for quick generation of health facilities, intensive care units, and drive-through testing sites. Several examples on this are available in an article by Lubell (2020). Their potential use could continue during any future outbreak of diseases as well as during climate-related disasters.

Adaptive reuse of buildings for meeting the emergency need has also been a norm during the pandemic. For example, New York's Javits Center became a 2900-bed hospital, the New Orleans Convention Center became a 3000-bed facility, and Chicago's McCormick Place became a 3000-bed complex. Many sports centers were also turned into medical facilities like London's ExCel center, Seattle's CenturyLink Field, and New York's Billie Jean King National Tennis Center (Lubell 2020). In São Paulo, Brazil, a stadium was converted into an open-air hospital, and in Vienna, a large exhibition hall was transformed into a temporary hospital. In Santiago, Chile, the Espacio Riesco Convention Center was reconfigured as an emergency hospital (UN Habitat 2021). These examples show the need for cities to identify and enlist multipurpose and flexible buildings that can be used, if needed, in future crises resulting from mega risks—whether infectious diseases or climate change-related extreme events.

11.4.3.2 Infrastructure and Transport

During the pandemic, many citizens started walking and cycling more frequently. To encourage and maintain this shift in behavior, city authorities moved to convert road spaces into pedestrian walkways and cycle lanes (Table 11.2).

The trend may lead to at least some infrastructure changes. For example, in Berlin, outbreak of the pandemic led to setting up of two temporary bike lanes (District Office Berlin 2020), which was followed by the pursuance of long-term policies on biking (Connolly 2020). The city of Hamburg plans to be car free by 2034 (Intertraffic 2021). In New York City (NYC), an upsurge in bicycle usage by over 67% in March led to calls for permanent transformation to make NYC a "bicycle city" (Hu 2020). In Beijing, bicycle

Table 11.2 Emergence of cycling infrastructure in cities

City	Cycling infrastructure	Date implemented
Berlin	Bicycle lane	March 2020
Edinburgh	Bicycle and walking zones	August 2020
Oakland	Bicycle lane	April 2020
Philadelphia	Bicycle lane	March 2020
Denver	Bicycle lane	April 2020
Minneapolis	Bicycle lane	April 2020
Bogota	Bicycle lane	March 2020
Vancouver	Bicycle lane	March 2020
Calgary	Bicycle lane	March 2020

Source: Adapted from Moreno et al. (2021)
Note: In addition to the cities mentioned, Kampala, Lima, London, Milan, Nairobi, and a host of other cities have also extended their walking and cycling infrastructure

usage increased by over 150% as people tried to avoid public transport and comply with the restrictions on private car usage (Morris 2020). Copenhagen declared bicycle shops as essential service providers, and they were never closed there, even when other nonessential service providers were forced to shut down during the lockdowns (Ibold et al. 2020).

Most present-day cities have inherited streets designed essentially for moving cars, and while there is a growing number of bike lanes and off-street trails now, they are still relatively few and sparse. However, the pandemic has propelled an image of a "bike-friendly city," and hopefully this will become a reality. It is important to realize that as car space dominates streetscape, it is not providing equitable access to the neighborhoods. Professor Moreno highlights it through the example of Paris as he states, "In Paris, 66% of the public space is streets for cars. But individual cars move only 17% of the population" (Reimer 2020). Each of the cars carries only two people, which according to Moreno is totally opposite to the concept of hyper-proximity (Reimer 2020). If traffic congestion, delays, and pollution are added, the situation looks grim. In the USA, for example, it has been estimated that on average, traffic delays lead to a consumption of US$ 1010 per year per person, which cumulatively translates to over US$ 166 billion per year nationally. Likewise, an average national tally of 54 hours a year is wasted by every average US driver on the road due to traffic (Liu 2019). This has created the need to shift priorities to promote walking and bicycling within neighborhoods while also promoting safe mass public transport or transit-oriented transport development between various neighborhoods in a city and other parts of it. The drastic reduction in the use of public transport at least initially due to "enochlophobia," or fear of crowds has, however, raised the potential need to enhance safety aspects in public transport and mass transit (Kakderi et al. 2021).

A review of this section shows that a lot of changes have taken place in city development and urban design due to mega risks. Although many of these were reactive, they have, nevertheless, presented an opportunity for proactive response to promote healthy cities and resilient urban spaces. It is therefore important to continue this initial work in building resilient and robust cities, neighborhoods, and urban spaces with a meaningful and sustained engagement across communities, planners, and designer as well as decision-makers.

11.5 Conclusion

The mega risks both coronavirus and climate change have revolutionized the urban life and changed the urban landscape beyond any imagination. However, this is not the first time that cities have faced these shocks. It is the great quality in "adaptability and flexibility of cities" that has made them survive all kinds of shocks—from natural calamities and diseases and pandemics to wars, invasions, and economic recessions.

Therefore, it appears highly unlikely that either COVID-19 or climate change despite their major devastations will derail the long-standing process of urbanization and the economic role of cities.

However, it does not negate the possibility of transformation in cities function or form, which is already happening—but needs to be guided in the right direction. There are two main lessons from the mega risks for future development of cities and nations. The first one is to learn from the driving forces that caused them and make sustained efforts to eradicate them. The second is to always be ready and prepared for such kind of risks to do a better job in handling them. Regarding the driving forces, it is the relentless pursuit of higher levels of economic growth with complete disregard of environment that has contributed to both mega risks. For example, unbridled expansion of agriculture, other commercial enterprises, and human settlements into forests and wilderness has resulted in habitat loss. This contributed to risky overlaps between human and animal species, allowing easy transfer of viruses from the latter to the former—the consequence is recurrence of zoonotic epidemics and pandemics like coronavirus. Similarly, comparable pursuit in the industrial sector contributed to rampant expansion of industrial sector with excessive material and energy consumption on the one hand and growing pollution and climate change on the other. Hence, the most important structural transformation required in attaining environmental sustainability and resilience against mega risks is to decouple development process from resource use.

Both mega risks the COVID-19 and climate change struck cities together with vengeance in 2020. Their far-reaching impacts in the form of economic recession and enhanced poverty and inequality are a reminder of the fact that policymakers whether urban or national can no longer work in isolation. They cannot ignore the connections between economic growth, social equity, health, and the environmental sustainability, all of which constitute important elements of sustainability. They should not only look at short-term measures to shore up livelihoods, employment, and economic growth but also reflect on the driving forces and social and environmental impacts of development process. This is an essential first step in "Building Forward Better"—in the context of sustainable and resilient economic recovery and promotion of future sustainable development.

Monitoring of development process is also essential—several tracking initiatives have been developed to assess the sustainability of recovery measures. However, invariably all tracking initiatives used so far present disappointing results. They show that a very large proportion of recovery measures are not geared to achieving sustainable development. The most comprehensive development tracker developed by the Global Recovery Observatory shows that only 18.0% of recovery spending in 2020 and 2.5% of total spending had positive green characteristics. In contrast, most of the $14.6 trillion COVID-induced recovery spending related to non-green investments. Whatever green investment was made in 2020 covered a broad range of priorities but were skewed more toward green energy and green transport. Some importance was given to nature-based solutions and green R&D. This is a good omen as without progress in the R&D, meeting the Paris Agreement targets may jeopardize. The world, according to the Global Recovery Observatory (2021), has missed the initial opportunity to invest in green recoveries and setting a new course for economic growth and prosperity. This is a warning bell and a wake-up call to governments and other stakeholders for correction of course in spending on recovery and future development.

Urban economies are vital in the course correction not only because cities account for 80 percent of global GDP but also because their economic impacts spill far beyond their respective boundaries. Hence, the United Nations is right in predicting that the battle for sustainable development will be won or lost in cities (United Nations 2017). Cities will also be central in promoting inclusive growth. This is especially true for the Global South where one billion slum dwellers need urgent attention. All-out efforts are needed to promote and expand exemplary successful participatory approaches of slum upgrading there. A combined local and national government support can do wonders in this regard as demonstrated in Thailand.

A major challenge for stakeholders in cities is how to manage urban dynamics and simultaneously cope and adjust to current and future threats such as pandemics and climate change with an intelligent transformation. They have started using digital and ICT-based technologies, in this regard, which has shown great promise. These technologies have shown enormous potential in dealing with both mega risks. During COVID-19, they enabled city authorities in detecting, isolating, and monitoring the spread of the virus as well as in engaging people, answering their needs, changing their behavior, and making cities move forward in new ways. They also enabled to provide telemedicine, efficient medical infrastructure, drive-through testing facilities, and dedicated command and control centers to effectively tackle any pandemic situation. Their capability is already well known in climate mitigation through smart technologies for energy production, transmission, and management as well as improving their efficiency in buildings, transport, and undertaking economic activities. In climate adaptation, they have helped through development of early warning systems and in designing protective mechanisms. Additionally, they have assisted in handling crisis situations during climate-related extreme events.

The application of these technologies in cities has given rise to the concept of smart city, the history of which goes back to 2003 in the establishment of the Smart City Group in MIT. Since then, applications and deployment of smart technologies for studying, designing, and managing the functions and environment in cities by various laboratories and commercial ventures have enhanced exponentially. No doubt the technology deployment in smart cities has certainly helped improve the well-being of their residents and maintained their functionality during crises. Nevertheless, it has also generated many technical and non-technical challenges and issues. Among the technical constraints are digital divide, lack of access to the Internet and devices, etc. These major issues not only were present primarily in cities of developing countries but also were present in developed nations during the pandemic. The main non-technical issues included privacy, security especially from cyber-attacks, confidentiality and trust, social inclusiveness, political bias, and dissemination of misinformation. Hence, a combination of technology-driven and human-driven approaches has been advocated to alley such concerns at least in the short term. However, effective solutions must be put in place in the long term. Utilities of technologies in improving citizen's quality of life or well-being have also been questioned—especially when seen in terms of social health and happiness rather than material progress.

Mega risks have also generated fresh ideas on form, shape, and structure of cities. In fact, epidemics and pandemics throughout history—from bubonic plague to breakouts of cholera and tuberculosis and flu pandemic—have presented new images of cities. They have brought to the fore basic issues on city design and architecture. For example, the radical improvements in the renaissance cities through removal of sordid and cramped living quarters and establishment of large public spaces were inspired by the fourteenth-century bubonic plague. Breakouts of cholera resulted in radical changes to improvements in Paris and London. Tuberculosis bacteria brought sanatorium movement, while breakouts of tuberculosis, typhoid, polio, and Spanish flu incidences in the twentieth century stimulated architectural modernism. Together with insanitary conditions created by industrial revolution, they also inspired visionary thinking that produced garden city movement and Le Corbusier tower-in-the-park skyscraper concept for a contemporary city. An idea that got tremendous popularity and promotion during the COVID-19 pandemic was the 15-minute city. It is based on developing polynuclear city composed of lived-in, people-friendly neighborhoods—"complete" and well connected—to create "a city of proximities," not only between structures but also people. This idea, apart from the need of the pandemic, also emerged from the desire for promoting people-oriented cities. The emerging interest in the "15-minute city"—with a popular vision of urban living—has already taken many names and shapes around the world with Paris in the lead. The growing popularity of the concept has made it important to test and showcase the concept and its elements in cities of the Global South and those that may be financially constrained before undertaking the extensive urban regeneration exercise there that this kind of planning model demands.

Cities' layout and urban design are also important in climate change mitigation and adaptation. Street and public space layout, building design, and types of construction materials contribute to climate change. They also have role in generating heat and causing heat island effects in cities. Even before the pandemic, forward-looking cities had therefore started utilizing the available potential in the built environment and natural systems to both mitigate and "future proof" their built environment from climate extremes. These included design of compact neighborhoods and climate responsive buildings, as well as green transport infrastructure like investing in pedestrian and cycling tracks and integrating them with green spaces and transit access. These ideas on climate-friendly urban design are in line with the concept of 15-minute city or similar models that became popular during the pandemic. Hence, rather than reinventing the wheel, it will be more beneficial to help cities incorporate or adapt design techniques that would reduce temperatures within the existing city/neighborhood development models like the 15-minute city, techniques such as heat-resistant construction materials, reflective surface coatings, cool and green roofs, solar panels, cool pavements, and increasing vegetative cover. These and other similar measures are necessary to support integrated mitigation and adaptation approach to climate change in future city, neighborhood, and building designs.

Finally, the visions for improving sustainability of cities through selection of right economic, social, and environmental trade-offs or making them smart by adopting appropriate technologies are not isolated ideas. Similarly, making cities resilient to mega risks by appropriate designing of buildings, neighborhoods, or public spaces is not a secluded notion either. All these ideas need to be implemented within one overarching context of sustainable and resilient development. Fortunately, it is available in the framework of Sustainable Development Goals of 2030 Agenda for development, an agenda that is based on inclusive, coordinated, and responsive governance across various jurisdictions—local, subnational, national, and international. Further, it also envisages promotion of institutionalized and participatory approach by involving all stakeholders and citizen groups including vulnerable and marginal communities. Moreover, goal 11 of the agenda, "make cities inclusive, safe, resilient and sustainable," recognizes that cities are the spaces where all SDGs can be integrated to provide holistic solutions to the challenges from climate change and health risks to poverty and exclusion. Hence, there is an urgent need to augment city/local government's capacity and budgets while promoting their transparency and accountability. Regarding preparedness to face risks, there is a need to include and implement multi-hazard resilience aspect in the city-level spatial development plans along with investment in care economy. In terms of overall development, the mega risks particularly the pandemic raised the need for nations and cities to make a judicious choice from among the alternative paths to structural transformation and technological revolution. It will be important for them to take into consideration both their current conditions and future in mind while ensuring that the development is sustained, socially inclusive, technologically smart, and environmentally sustainable.

References

ADB, Asian Development Bank (2019) Effective Approaches to Poverty Reduction, Manila, https://www.adb.org/sites/default/files/publication/540611/approaches-poverty-reduction-cases-adb.pdf

Alkire S, Santos ME (2010) Multidimensional poverty index. Oxford Poverty & Human Development Initiative, Oxford

Allison AL, Ambrose-Dempster E, Bawn M et al (2021) The impact and effectiveness of the general public wearing masks to reduce the spread of pandemics in the UK: a multidisciplinary comparison of single-use masks versus reusable face masks. UCLOE 3. https://doi.org/10.14324/111.444/ucloe.000022

Arya D (2021, June 15) India's vaccine drive: stories from the best and worst districts. BBC News. https://www.bbc.com/news/world-asia-india-57400620

Azunu R, Mensah JK (2019) Local economic development and poverty reduction in developing societies: the experience of the ILO decent work project in Ghana. Local Econ 34(5):405–420. https://doi.org/10.1177/0269094219859234

Baller S (2021) What will a post pandemic economy look like? Here's what chief economists expect. World Economic Forum. https://www.weforum.org/agenda/2021/06/chief-economists-outlook/

Banai R (2020) Pandemic and the planning of resilient cities and regions. Cities 106:102929. https://doi.org/10.1016/j.cities.2020.102929

Barbier E (2020) Greening the post-pandemic recovery in the G20. Environ Resour Econ 76:685–703. https://doi.org/10.1007/s10640-020-00437-w

Barrientos A (2010) Social protection and poverty, social policy and development program paper 24 United Nations Research Institute for Social Development. https://www.unrisd.org/80256B3C005BCCF9/%28httpAuxPages%29/973B0F57CA78D834C12576DB003BE255/$file/Barrientos-pp.pdf. Accessed 10 Feb 2022

Barufi AM, Kourtit K (2015) Agglomeration forces in smart cities. In: Kourtit K, Nijkamp P, Stough R (eds) The rise of the city—spatial dynamics in the urban century. Edward Elgar, Cheltenham, pp 33–54

Baumol WJ, Oates WE (1993) Economics, environmental policy, and the quality of life. Gregg Revivals, Princeton

Bauwens M et al (2020) Impact of coronavirus outbreak on NO_2 pollution assessed using TROPOMI and OMI observations. Geophys Res Lett. https://doi.org/10.1029/2020GL087978. Accessed 31 Jan 2022

Bhardwaj G, Esch T, Lall SV et al (2020, Cities, crowding, and the coronavirus: predicting contagion risk hotspots. World Bank, Washington, DC. https://openknowledge.worldbank.org/handle/10986/33648

Boonyabancha S, Kerr T (2018) Lessons from CODI on co-production. Environ Urban 30(2):444–460. https://doi.org/10.1177/0956247818791239

Borsekova K, Koróny S, Vaňová A, Vitálišová K (2018) Functionality between the size and indicators of smart cities: a research challenge with policy implications. Cities 78:17–26

Budd J, Miller BS, Manning EM et al (2020) Digital technologies in the public-health response to COVID-19. Nat Med 26:1183–1192. https://doi.org/10.1038/s41591-020-1011-4

C40 Cities (2020) How to build back better with a 15-minute city. https://www.c40knowledgehub.org/s/article/How-to-build-back-better-with-a-15-minute-city?language=en_US

C40 Cities (2022) Case Studies. https://www.c40.org/case-studies/

Campbell T (2012) Beyond smart cities. Routledge, London

Chandler DL (2018) Urban heat island effects depend on a city's layout. MIT News. https://news.mit.edu/2018/urban-heat-island-effects-depend-city-layout-0222

Chang V (2020) The post pandemic style. Slate Magazine. https://slate.com/business/2020/04/coronavirus--architecture-1918-flu-cholera-modernism.html

CLC, Center for the Living City (undated) Jane Jacobs and The Center. https://centerforthelivingcity.org/janejacobs#jane-and-the-center

Climate Policy Watcher (2022) The warming effects of the industrial revolution. https://www.climate-policy-watcher.org/global-temperatures/the-warming-effects-of-the-industrial-revolution.html

CODI, Community Organization Development Institute (2020) Citywide housing for all: planned and built by people. https://en.codi.or.th/baan-mankong-housing/baan-mankong-rural/

Connolly K (2020) 'Cleaner and Greener': Covid-19 prompts world's cities to free public space of cars. https://www.theguardian.com/world/2020/may/18/cleaner-and-greener-covid-19-prompts-worlds-cities-to-free-public-space-of-cars. Accessed 6 Dec 2020

CRED, UNDRR (2021) The non-COVID year in disasters. Brussels

Cuttle J (2018) Georges-Eugène Haussmann: the man who created Paris. https://theculturetrip.com/europe/france/paris/articles/georges-eugene-haussmann-the-man-who-created-paris/

Da Silva L, Robins E (2021) A Green reboot for cities: strategies for post-COVID growth. World Bank Blog. https://blogs.worldbank.org/ppps/green-reboot-cities-strategies-post-covid-growth

District Office Berlin (2020) Temporary installation and expansion of bicycle traffic facilities during the pandemic crisis. Available online: https://www.berlin.de/ba-friedrichshain-kreuzberg/aktuelles/pressemitteilungen/2020/pressemitteilung.911780.php. Accessed 6 Dec 2020

Djankov S, Panizza U (2020) Developing economies after COVID-19: an introduction, Vox eBook chapters. In: Djankov S, Panizza U (eds) COVID-19 in developing economies, vol 1, 1st edn. Centre for Economic Policy Research, pp 8–23

Dulal HB, Brodnig G, Onoriose C (2011) Climate change mitigation in the transport sector through urban planning: a review. Habit Int 35(3):494–500. https://doi.org/10.1016/j.habitatint.2011.02.001

Energy Policy Tracker (2021) Energy Policy Tracker - Track funds for energy in recovery packages. https://www.energypolicytracker.org/. Accessed 14 Sept 2021

European Commission (2020) Nature-based solutions for climate mitigation: analysis of EU-funded projects. Research & Innovation, Brussels

Florida R, Rodríguez-Pose A, Storper M (2021) Cities in a post-COVID world. Urban Stud. https://doi.org/10.1177/00420980211018072

Free C, Hecimovic A (2021) Global supply chains after COVID-19: the end of the road for neoliberal globalisation? Account Audit. Account J 34(1):58–84

Gade DS, Aithal PS (2021) Smart cities development during and post COVID-19 pandemic – a predictive analysis. IJMTS 6(1):189–202. https://doi.org/10.5281/zenodo.4903338

Gavas M, Pleeck S (2021) Global trends in 2021, how COVID-19 is transforming international development. Center for Global Development. https://www.cgdev.org/publication/global-trends-2021-how-covid-transforming-international-development

Giacobbe A (2020) How the COVID-19 pandemic will change the built environment. https://www.architecturaldigest.com/story/covid-19-design

Global Recovery Observatory (2021) Key messages. https://wedocs.unep.org/bitstream/handle/20.500.11822/35311/GROKM.pdf

Godfrey N, King R, Stanley-Price F (2021) A green recovery starts with cities. World Resources Institute. https://www.wri.org/insights/green-recovery-starts-cities

Harrouk C (2020) The Gastro Safe Zone: a public space proposal that considers social distancing measures. https://www.archdaily.com/938599/the-gastro-safe-zone-a-public-space-proposal-respecting-social-distancing-measure

Hart H (2020, April 30) Coronavirus may prompt migration out of American cities. Axios. Available at https://www.axios.com/coronavirus-migration-american-cities-survey-aba181ba-a4ce-45b2-931c-6c479889ad37.html. Accessed 15 June 2020

Hassankhani M, Alidadi M, Sharifi A, Azhdari A (2021) Smart city and crisis management: lessons for the COVID-19 pandemic. Int J Environ Res Public Health 18(15):7736. https://doi.org/10.3390/ijerph18157736

Hepburn C, O'Callaghan B, Stern N, Stiglitz J, Zenghelis D (2020) Will COVID-19 fiscal recovery packages accelerate or retard progress on climate change? Oxf Rev Econ Policy 36:3–48, Smith School Working Paper 20-02. https://www.lagone.it/wp-content/uploads/2020/05/STUDIO-STIGLITZ-ART4.pdf

Hill RV, Narayan A (2020) COVID-19, and inequality: a review of the evidence on likely impact and policy option. Prevention Web, UNDRR. https://www.preventionweb.net/publication/covid-19-and-inequality-review-evidence-likely-impact-and-policy-options

Hu W (2020) A surge in biking to avoid crowded trains in N.Y.C. https://www.nytimes.com/2020/03/14/nyregion/coronavirus-nyc-bike-commute.html. Accessed 5 Nov 2020

Hughes CJ (2020, May 8) Coronavirus escape: to the suburbs. New York Times. Available at https://www.nytimes.com/2020/05/08/realestate/coronavirus-escape-city-to-suburbs.html. Accessed 24 June 2020

Ibn Mohammed T, Mustapha KB, Godsell J, Adamu Z, Babatunde K, Akintade DD, Acquaye A, Fujii H, Ndiaye MM, Yamoah FA, Koh SCL (2021) A critical review of the impacts of COVID-19 on the global economy and ecosystems and opportunities for circular economy strategies. Resour Conserv Recycl 164. https://doi.org/10.1016/j.resconrec.2020.105169

Ibold S, Medimorec N, Wagner A, Peruzzo J, Platzer L (2020) The Covid-19 outbreak and implications to sustainable urban mobility—some observations. https://www.transformative-mobility.org/news/the-covid-19-outbreak-and-implications-to-public-transport-some-observations. Accessed 12 Dec 2020

IEA (2021) Sustainable Recovery Tracker. https://www.iea.org/reports/sustainable-recovery-tracker. Accessed 14 Sept 2021

IFC – International Finance Corporation (2021) A Green Reboot for emerging markets. https://www.ifc.org/wps/wcm/connect/26f79a1b-c191-494b-b2d9-c891e138bb37/IFC_GreenReport_FINAL_web_1-14-21.pdf?MOD=AJPERES&CVID=ns1JVaR

IISD, International Institute for Sustainable Development (undated) Sustainable Development. https://www.iisd.org/about-iisd/sustainable-development

ILO (2021) ILO monitor: COVID-19 and the world of work, 7th edn, updated estimates and analysis. https://bit.ly/34KwWuV

Intertraffic (2021) 15 minute city: urban mobility solution to the environment. https://www.intertraffic.com/news/15-minute-city-urban-mobility-solution-to-environment/

IPCC (2022) Climate change, 2022, impacts, adaptation and vulnerability. Working Group II contribution to Sixth Assessment Report (AR6). https://www.ipcc.ch/report/ar6/wg2/

Iradian G (2005) Inequality, poverty, and growth: cross-country evidence. IMF Working Paper WP/05/28

ITC, International Trade Center News (2020) Firms that reduce their environmental footprint are more resilient to crises. https://www.intracen.org/layouts/2coltemplate.aspx?pageid=47244640256&id=47244681925

IUCN, International Union for the Conservation of Nature and Natural Resources, UNEP, United Nations Environment Program, and WWF, World Wildlife Fund (1980) World conservation strategy: living resource conservation for sustainable development. Moreges

Jackson D (2019a) Data cities: how post-internet technology is changing the way we design our world. https://www.foreground.com.au/technology/data-cities-how-post-internet-technology-is-changing-the-way-we-design-our-world/

Jackson D (2019b) Data cities. ISBN: 9781848222748, Lund Humphries

Jacobs J (1961) The death and life of great American cities. ISBN 0-679-60047-7

Kakderi C, Oikonomaki E, Papadaki I (2021) Smart and resilient urban futures for sustainability in the post COVID-19 era: a review of policy responses on urban mobility. Sustainability 13:6486. https://doi.org/10.3390/su13116486

Kang M, Choi Y, Kim J, Lee KO, Lee S, Park IK, Park J, Seo I (2020) COVID-19 impact on city and region: what's next after lockdown? Int J Unity Sci 24(3):297–315

Kaza S (2020, April 9) Waste workers are protecting our communities during COVID-19. https://blogs.

worldbank.org/sustainablecities/waste-workers-are-protecting-our-communities-during-covid-19

Klerkx L, Begemann S (2020) Supporting food systems transformation: the what, why, who, where and how of mission-oriented agricultural innovation systems. Agric Syst 184:102901

Kummitha R (2020) Smart technologies for fighting pandemics: the techno- and human- driven approaches in controlling the virus transmission. Gov Inf Q 37(3):101481. https://doi.org/10.1016/j.giq.2020.101481

Lakner, CN, Yonzan, D, Mahler, G, Aguilar, RAC, Wu H (2021) Updated estimates of the impact of COVID-19 on global poverty: looking back at 2020 and the outlook for 2021. World Bank. https://bit.ly/3peextg

Lang J (1994) Urban design: the American experience. Van Nostrand Reinhold, New York

Lavorel S, Colloff MJ, Locatelli B, Gorddard R, Prober SM, Gabillet M et al (2019) Mustering the power of ecosystems for adaptation to climate change. Environ Sci Pol 92:87–97

Le Quéré C, Jackson RB, Jones MW et al (2020) Temporary reduction in daily global CO_2 emissions during the COVID-19 forced confinement. Nat Clim Chang 10:647–653. https://doi.org/10.1038/s41558-020-0797-x

LeGates RT, Stout F (1998) Early urban planning 1870–1940. Routledge, London

Lennon M (2021) Planning and the post-pandemic city. Plan Theory Pract. https://doi.org/10.1080/14649357.2021.1960733

Lenzen M, Li M, Malik A, Pomponi F, Sun YY, Wiedmann T, Faturay F, Fry J, Gallego B, Geschke A, Gómez-Paredes J, Kanemoto K, Kenway S, Nansai K, Prokopenko M et al (2020) Global socio-economic losses and environmental gains from the coronavirus pandemic. PLoS One 15(7):e0235654. https://doi.org/10.1371/journal.pone.0235654

Lipschutz RD (2012) The sustainability debate: *déjà vu* all over again? In: Dauvergne P (ed) Handbook of global environmental politics, 2nd edn. Edward Elgar, Cheltenham

Liu J (2019) Commuters in this city spend 119 hours a year stuck in traffic. https://www.cnbc.com/2019/09/04 / commuters-in-this-city-spend-119-hours-a-year-stuck-in-traffic.html. Accessed 6 Dec 2020

Livingston E, Desai A, Berkwits M (2020) Sourcing personal protective equipment during the COVID-19 pandemic. JAMA 323:1912–1914

Lubell S (2020) Commentry: past pandemics changed the design of cities: six ways COVID-19 could do the same. Los Angeles Times. https://www.latimes.com/entertainment-arts/story/2020-04-22/coronavirus-pandemics-architecture-urban-design

MacCarthy A (2021) The new normal, the status of outdoor eating across the country. https://www.eater.com/22833407/pandemic-outdoor-dining-america

Mak R, Breakingviews (2020) Wuhan virus will shape China's smart city vision 2020. https://www.nasdaq.com/articles/breakingviews-wuhan-virus-will-shape-chinas-smart-city-vision-2020-01-24/

Martínez L, Short JR (2021) The pandemic city: urban issues in the time of COVID-19. Sustainability 13(6):3295. https://doi.org/10.3390/su13063295

Masuku MM, Jili NN, Selepe BM (2016) The implementation of local economic development initiatives towards poverty alleviation in Big 5 False Bay Local Municipality. Afr J Hosp Tour Leis 5(4):1–11

Mbeba RD (2014) Local economic development and urban poverty alleviation: the case of Buffalo City Metropolitan Municipality. Mediterr J Soc Sci 5(20). https://doi.org/10.5901/mjss.2014.v5n20p347

McPherson G, Simpson J, Peper P, Maco S, Xiao Q (2005) Municipal forest benefits and costs in five US cities. United States Forest Services

Meerow S, Stults M (2016) Comparing conceptualizations of urban climate resilience in theory and practice. Sustainability 8(7):701. https://doi.org/10.3390/su8070701

Meijer A (2016) Smart city governance: a local emergent perspective. In: Gil-Garcia J, Pardo T, Nam T (eds) Smarter as the new urban agenda, Public Administration and Information Technology. Springer, Cham, pp 73–85

Meyer DF (2014) Local economic development (LED), challenges and solutions: the case of the northern free state region, South Africa. Mediterr J Soc Sci 5(16):624. https://doi.org/10.5901/mjss.2014.v5n16p624

Mitchell N (2021) The program: tackling poverty in Thailand. https://borgenproject.org/tag/the-baan-mankong-program/

Moglia M, Frantzeskaki N, Newton P, Pineda-Pinto M, Witheridge J, Cook S, Glackin S (2021) Accelerating a green recovery of cities: lessons from a scoping review and a proposal for mission-oriented recovery towards post-pandemic urban resilience. Dev Built Environ 7. https://doi.org/10.1016/j.dibe.2021.100052

Moraci F, Errigo MF, Fazia C, Campisi T, Castelli F (2020) Cities under pressure: strategies and tools to face climate change and pandemic. Sustainability 12(18):7743

Moreno C, Allam Z, Chabaud D, Gall C, Pratlong F (2021) Introducing the "15-minute city": sustainability, resilience and place identity in future post-pandemic cities. Smart Cities 4:93–111. https://doi.org/10.3390/smartcities4010006

Morris DZ (2020) After coronavirus, bicycles will have a new place in city life. https://fortune.com/2020/06/15/bicycles-coronavirus-cities-lime-citi-bike/

Nagargoje SV, Somani GK, Sutaria MM, Jha N (2016) Smart cities as a solution for reducing urban waste and pollution. https://doi.org/10.4018/978-1-5225-0302-6.ch009

Nathan M (2021, December) The city and the virus. Urban Studies. https://doi.org/10.1177/00420980211058383

O'callaghan B, Murdock E (2021) Are we building back better? Evidence from 2020 and Pathways for Inclusive Green Recovery Spending. https://wedocs.unep.org/bitstream/handle/20.500.11822/35281/ AWBBB.pdf. Accessed 14 Sept 2021

OECD (2020a) Building back better, A sustainable resilient recovery after COVID–19. https://www.oecd. org/coronavirus/policy-responses/building-back-better-a-sustainable-resilient-recovery-after-covid-19-52b869f5/

OECD (2020b) Smart cities and inclusive growth. https://www.oecd.org/cfe/cities/OECD_Policy_Paper_Smart_Cities_and_Inclusive_Growth.pdf

Perry CA (1929) The neighborhood unit, a scheme of arrangement for the family-life community. Monograph one in neighborhood and community planning, regional plan of New York, and its environs. Committee on Regional Plan of New York and Its Environs, New York

Pirages DC (ed) (1977) The sustainable society: implications for limited growth. Praeger, New York

Purvis B, Mao Y, Robinson D (2019) Three pillars of sustainability: in search of conceptual origins. Sustain Sci 14:681–695. https://doi.org/10.1007/s11625-018-0627-5

Raven J (2011) Cooling the public realm: climate-resilient urban design resilient cities. In: Otto-Zimmermann K (ed) Cities and adaptation to climate change: local sustainability, vol 1. Springer, pp 451–463

Raven J, Stone B, Mills G, Towers J, Katzschner L, Leone M, Gaborit P, Georgescu M, Hariri M (2018) Urban planning and design. In: Rosenzweig C, Solecki W, Romero-Lankao P, Mehrotra S, Dhakal S, Ali Ibrahim S (eds) Climate change and cities: second assessment report of the urban climate change research network. Cambridge University Press, New York, pp 139–172

Ravi OP (2021) Energy grid ontology for digital twins is now available. Microsoft Blog. https://techcommunity.microsoft.com/t5/internet-of-things-blog/energy-grid-ontology-for-digital-twins-is-now-available/ba-p/2325134

Redclift M (1994) Reflections on the 'sustainable development' debate. Int J Sustain Dev World Ecol 1(1):3–21. https://doi.org/10.1080/13504509409469856

Redclift M (2005) Sustainable Development (1987–2005): an oxymoron comes of age. Sustain Dev 13(4):212–227

Reid C (2020) Anne Hidalgo reelected as mayor of Paris vowing to remove cars and boost bicycling and walking. https://www.forbes.com/sites/carltonreid/2020/06/28/anne-hidalgo-reelected-as-mayor-of-paris-vowing-to-remove-cars-and-boost-bicycling-and-walking/?sh=ba645d11c852. Accessed 5 Dec 2020

Reimer Y (2020) The 15-minute infrastructure trend that could change public transit as we know it. https://360.here.com/15-minute-cities-infrastructure

Republic of South Africa (1998) White paper on local government. Development of Constitutional Development, Pretoria. Government Printer

Resch E, Bohne RA, Kvamsdal T, Lohne J (2016) Impact of urban density and building height on energy use in cities. Energy Proc 96:800–814

Ríos-Osorio L, Lobato M, Castillo X (2005) Debates on sustainable development: towards a holistic view of reality. Environ Dev Sustain 7:501–518. https://doi.org/10.1007/s10668-004-5539-0

Rogerson CM (1999) Local economic development and urban poverty alleviation: the experience of post-apartheid South Africa. Habitat Int 23(4):511–534. https://doi.org/10.1016/S0197-3975(99)00019-3

Romero-Lankao P, Gnatz DM, Wilhelmi O, Hayden M (2016) Urban sustainability and resilience: from theory to practice. Sustainability 8(12):1224. https://doi.org/10.3390/su8121224

Rubio-Romero JC, del Carmen Pardo-Ferreira M, García JAT, Calero-Castro S (2020) Disposable masks: disinfection and sterilization for reuse, and non-certified manufacturing, in the face of shortages during the COVID-19 pandemic. Safe Sci 129(104830)

Ruggerio CA (2021) Sustainability and sustainable development: a review of principles and definitions. Sci Total Environ 786. https://doi.org/10.1016/j.scitotenv.2021.147481

Rume T, Didar-Ul Islam SM (2020) Environmental effects of COVID-19 pandemic and potential strategies of sustainability. Heliyon 6(9). https://doi.org/10.1016/j.heliyon.2020.e04965

Russell ES (1931) Some theoretical considerations on the "overfishing" problem. Journal de Conseil International pour l'Exploration de la Mer 6(1):1–20

Sanseverino ER, Sanseverino RR, Vaccaro V (eds) (2017) Smart Cities Atlas, Springer, SBN-10:3319473603. https://www.amazon.com/Smart-Cities-Atlas-Intelligent-Communities/dp/3319473603

Sassen S, Kourtit K (2021) A post-Corona perspective for smart cities: 'Should I stay, or should I go?'. Sustainability 13:9988. https://doi.org/10.3390/su13179988

Satterthwaite D (2010) The role of cities in sustainable development. Sustainable Development Insights, https://www.bu.edu/pardee/files/2010/04/UNsdkp004fsingle.pdf

SDI (undated) Slum Dwellers International. https://skoll.org/organization/slum-dwellers-international/

Seddon N, Sengupta S, García-Espinosa M, Hauler I, Herr D, Rizvi A (2019) Nature-based solutions in nationally determined contributions: synthesis and recommendations for enhancing limate ambition and action by 2020. IUCN and University of Oxford, Gland/Oxford

Serhan Y (2020) Vilnius shows how the pandemic is already making cities. Atlantic. https://www.theatlantic.com/international/archive/2020/06/coronavirus-pandemic-urban-suburbs-cities/612760/. Accessed 5 Jan 2021

Sharifi A, Khavarian-Garmsir AR (2020) The COVID-19 pandemic: impacts on cities and major lessons for urban planning, design, and management. Sci Total Environ 749:142391. https://doi.org/10.1016/j.scitotenv.2020.142391

Stewart I, García-Navarro L (2019, March 31) Building for an uncertain future: Miami residents adapt to the changing climate. NPR. https://www.npr.org/2019/03/31/706940085/building-for-an-uncertain-future-miami-residents-adapt-to-the-changing-climate. Accessed 6 June 2020

Stone B, Hess J, Frumkin H (2010) Urban form and extreme heat events: are sprawling cities more vulnerable to climate change? Environ Health Perspect 118:1425–1428

Storper M, Venables AJ (2004) Buzz: face-to-face contact and the urban economy. J Econ Geogr 4(4):351–370

Swenson G, Goga S, Murphy F (2006) Local economic development: a primer, the World Bank. https://documents1.worldbank.org/curated/en/763491468313739403/pdf/337690REVISED0ENGLISH0led1primer.pdf

The climate reality Project (2017) The alliance for climate protection: five sustainable cities making a difference for the planet. https://www.climaterealityproject.org/blog/five-sustainable-cities-making-difference-planet

Tschirret M (2019) Smart cities: the role of smart utilities, SCC. https://smartcitiesconnect.org/smart-cities-the-role-of-smart-utilities/

ULI, Urban Land Institute (2019) Cool Neighborhoods, New York, https://developingresilience.uli.org/case/cool-neighborhoods/

UN ESCAP, United Nations Economic and Social Commission for Asia and the Pacific (2021) Economic and social survey of Asia and the Pacific 2021, ST/ESCAP/2942, Bangkok

UN Habitat (2013) Local economic development in practice. https://unhabitat.org/sites/default/files/download-manager-files/Local%20Economic%20Development%20in%20Practice.pdf

UN Habitat (2020) UN-Habitat guidance on COVID-19 and public space. https://unhabitat.org/sites/default/files/2020/06/un-habitat_guidance_on_covid-19_and_public_space.pdf. Accessed 25 June 2021

UN Habitat (2021) Cities and pandemics, towards a more just green and healthy future, Nairobi

UN REDD (2014) REDD+ and adaptation: identifying complementary responses to climate change. UN REDD programme, Info Brief

UNDESA (United Nations – Department of Economic and Social Affairs) (2020) COVID-19: embracing digital government during the pandemic and beyond, policy brief #61. https://www.un.org/development/desa/dpad/publication/un-desa-policy-brief-61-covid-19-embracing-digital-government-during-the-pandemic-and-beyond/

UNDESA (United Nations – Department of Economic and Social Affairs) (2021a) UN/DESA Policy Time for transformative changes for SDGs: what the data tells us Brief #110, New York. https://www.un.org/development/desa/dpad/publication/un-desa-policy-brief-110-time-for-transformative-changes-for-sdgs-what-the-data-tells-us/

UNDESA (United Nations – Department of Economic and Social Affairs) (2021b) High level political forum on sustainable development. https://sustainabledevelopment.un.org/hlpf

UNDESA (United Nations – Department of Economic and Social Affairs) (2021c) Accelerate action to revamp production and consumption patterns: the circular economy, cooperatives and the social and solidarity economy. https://www.un.org/development/desa/dpad/publication/un-desa-policy-brief-109-accelerate-action-to-revamp-production-and-consumption-patterns-the-circular-economy-cooperatives-and-the-social-and-solidarity-economy/

UNDESA (United Nations – Department of Economic and Social Affairs) (2021d) Sustainable Development Outlook: from anguish to determination. https://www.un.org/development/desa/dpad/wp-content/uploads/sites/45/publication/SDO_2021_Full_Report.pdf

UNDESA (United Nations – Department of Economic and Social Affairs) (2021e) Adapting international development cooperation to reduce risk, enable recovery and build resilience, policy brief # 122. https://www.un.org/development/desa/dpad/publication/un-desa-policy-brief-122-adapting-international-development-cooperation-to-reduce-risk-enable-recovery-and-build-resilience/

UNEP (United Nations Environment Program) (2021a) Are we on track for a green recovery? Not yet. Press Release. https://www.unep.org/news-and-stories/press-release/are-we-track-green-recovery-not-yet

UNEP (United Nations Environment Program) (2021b) Emission gap report 2021. https://www.unep.org/resources/emissions-gap-report-2021

United Nations (2017) Battle for sustainability will be won or lost in cities, deputy secretary general tells high-level general assembly meeting on new urban agenda. UN Habitat. https://www.un.org/press/en/2017/dsgsm1080.doc.htm

United Nations (2018) Progress on the implementation of the new urban agenda, report of the secretary general, A/73/83–E/2018/62, to economic and social council

United Nations (2020) Sustainable cities: why they matter, New York. https://www.un.org/sustainabledevelopment/wp-content/uploads/2019/07/11_Why-It-Matters-2020.pdf

UNODC, United Nations Office on Drugs and Crime (UNODC) (2020) Research brief: the impact of COVID-19 on organized crime. https://www.unodc.org/documents/data-and-analysis/covid/RB_COVID_organized_crime_july13_web.pdf. Accessed 20 Jan 2022

USFS United States Forest Service (2020) Urban Forest. Available at https://www.fs.usda.gov/managing-land/urban-forests

Vargo D, Lin Z, Benwell B, Zheng Y (2020) Digital technology use during COVID-19 pandemic: a rapid review. Hum Behav Emerg Technol 3(1):13–24. https://doi.org/10.1002/hbe2.242

Venter Z, Barton D, Gundersen V, Figari H, Nowell M (2020) Urban nature in a time of crisis: recreational use of green space increases during the COVID-19 outbreak in Oslo, Norway. Environ Res Lett 15(10):104075

Victoria State Government, Australia (2022) Plan Melbourne 2017–2050, 20-minute Neighborhoods. https://www.planning.vic.gov.au/policy-and-strategy/planning-for-melbourne/plan-melbourne/20-minute-neighbourhoods

Vivid Economics (2021) Greenness of stimulus index. https://www.vivideconomics.com/wp-content/uploads/2021/07/Green-Stimulus-Index-6th-Edition_final-report.pdf. Accessed 14 Nov 2021

WCED, The World Commission on Environment and Development (1987) Our common future. Oxford University Press

WHO (2021) Achieving 70% COVID-19 immunization coverage by mid – 2022. https://www.who.int/news/item/23-12-2021-achieving-70-covid-19-immunization-coverage-by-mid-2022

Wickenberg B, McCormick K, Olsson JA (2021) Advancing the implementation of nature-based solutions in cities: a review of frameworks. Environ Sci Policy 125:44–53. https://doi.org/10.1016/j.envsci.2021.08.016

Wilkinson A (2020) Local response in health emergencies: key considerations for addressing the COVID-19 pandemic in informal urban settlements. Environ Urban 32(2):503–522

Wittenberg A (2020) How to design a post-pandemic city, CityLab Design, Bloomberg. https://www.bloomberg.com/news/articles/2020-07-14/10-design-concepts-for-city-living-under-covid-19

WMO (World Meteorological Organization) (2021) State of the climate report, extreme events and major impacts. https://public.wmo.int/en/media/press-release/state-of-climate-2021-extreme-events-and-major-impacts

World Bank (2013) Building resilience to disasters and climate change through social protection. Washington, DC

World Bank (2020a) Urban development. https://www.worldbank.org/en/topic/urbandevelopment/overview

World Bank (2020b) Covid-19 high-frequency monitoring dashboard'. www.worldbank.org/en/data/interactive/2020/11/11/covid-19-high-frequency-monitoring-dashboard

World Economic Forum (2020) Dashboard for a new economy. https://www3.weforum.org/docs/WEF_Dashboard_for_a_New_Economy_2020.pdf. Accessed 8 Dec 2021

WWF – Worldwide Fund for Nature and ILO – International Labour Organization (2020) Nature hire: how nature based solutions can power a green job recovery. https://wwfint.awsassets.panda.org/downloads/nature_hires_report_wwf_ilo.pdf

Yeung P (2021) How 15- Minute Cities will change the way we socialize. https://www.bbc.com/worklife/article/20201214-how-15-minute-cities-will-change-the-way-we-socialise

Zeljic AS (2020) Polycentric cities: the future of sustainable urban growth, Gensler. https://www.gensler.com/blog/polycentric-cities-new-normal-manila-finance-centre

Index

© The Editor(s) (if applicable) and The Author(s), under exclusive license to Springer Nature Switzerland AG 2022
M. A. Khan, *Cities and Mega Risks*, https://doi.org/10.1007/978-3-031-14088-4

Milton Keynes UK
Ingram Content Group UK Ltd.
UKHW050830161023
430687UK00003B/10